D0742299

Risk Management in a Hazardous Environment

A Comparative Study of Two Pastoral Societies

STUDIES IN HUMAN ECOLOGY AND ADAPTATION

AS PASTORALISTS SETTLE:
Social, Health, and Economic Consequences of the Pastoral Sedentarization in Marsabit District, Kenya
Elliot Fratkin and Eric Abella Roth

RISK MANAGEMENT IN A HAZARDOUS ENVIRONMENT:
A Comparative Study of Two Pastoral Societies
Michael Bollig

Risk Management in a Hazardous Environment

A Comparative Study of Two Pastoral Societies

Michael Bollig

Institute of Social and Cultural Anthropology
University of Cologne
Cologne, Germany

Library of Congress Control Number: 2005936164

ISBN-10 : 0-387-27581-9 (Hardbound) e-ISBN 0-387-27582-7
ISBN-13 : 978-0387-27581-9

Printed on acid-free paper.

Printed in the United States of America. (SPI/IBT)

9 8 7 6 5 4 3 2 1

springeronline.com

Studies in Human Ecology and Adaptation

The objective of ***Studies in Human Ecology and Adaptation***, a new series launched by Springer in 2005, is to publish cutting-edge work on the biosocial processes of adaptation and human-environmental dynamics. Human ecology, a broad perspective, views the biological, environmental, demographic, and technological aspects of human existence as interrelated. and spans a large number of disciplines such as archaeology, anthropology, environmental sociology, human geography and demography—to name a few central ones. The series will reflect a growing trend in interdisciplinary study and research. This multidisciplinary approach brings new, often unexpected, perspectives to many topical issues.

Key to making the studies in this series accessible to a broad readership is that they are clearly written and tightly focused. We have quite deliberately taken the journal *Human Ecology* as a model since it focuses on empirically rooted original research addressing a wide interdisciplinary readership. As a consequence, the journal has achieved considerable recognition among academic publications worldwide which deal with social and environmental issues. As with contributions to *Human Ecology*, potential manuscripts in this series are subject to peer review.

Daniel G. Bates
Ludomir R. Lozny
Hunter College, C.U.N.Y.

Preface

This book has come a long way. Data collection for some chapters already began in 1986 with research in the Kenya National Archives in Nairobi. While the manuscript is about to be completed in February 2005, I am currently doing fieldwork in Namibia and a limited amount of data from the latest period of investigations has entered the book. Hence, data collection, analysis and discussions on topics salient to this book span almost two decades. In the course of these twenty years I spent almost six years either in Namibia or in Kenya, mostly staying at the respective field sites. Numerous people and institutions have been of great help in making this possible. First and foremost I would like to thank both the Namibian and the Kenyan government for granting and extending research permits. My sincere thanks go to the Institute of African Studies of the University of Nairobi and the Institute of Geography and Environmental Sciences and the History Department of the University of Namibia as well as the Namibia Economic Planning and Research Unit for hosting me during my studies. The local traditional leaders under whose authority the respective field sites fell, namely Chief Kapkoyo of the Pokot around Nginyang and Chief Kapika of the Ovahimba living in the wider Omuramba/Omuhonga area, accepted me, a stranger, in their community and never objected to my numerous and at times unusual questions. Both communities are organised along gerontocratic principles. Hence, elders – many of them deceased by now -were of great importance not only as repositories of collective memories but also as men in a position to explain the intricacies of local histories and the complex fabric of their societies. Loriko, Wasareng, Kanyakol, Todokin and Lotepamuk among the Pokot were crucial. Among the Himba Katjira, Tjikumbamba and Kozongombe were such lead figures. I hope that this volume may to some extent contribute to conveying their knowledge and their memories to future generations.

In Kenya I would like to thank Yusuf K. Losute and his family for their hospitality. Yusuf's father Wasareng introduced me to many aspects of Pokot life. As an energetic elder he is a major actor in many of my case studies. Yusuf Losute proved to be a competent translator in the early phases of my research as did John R. Konopöchö, Bebe and Simon. Yusuf's brothers Teta, Arekwen and Amos acted as guardians during various periods of fieldwork. Many Pokot friends contributed immensely to the success of the research: Lomirmoi, Lopetö, Akurtepa and Rengeruk, Tochil and the unfortunate

Ngoratepa, Wiapale, Dangapus and Lokomol. In Kenya the early phases of my studies were facilitated by the kind hospitality of the Trittler and Glombitza families. Throughout the past twenty years, Kositei Mission Station was a secure haven. Many aspects of this research were discussed intensively with Father Sean McGovern and Sister Rebecca Janaçeck, both being in charge of developmental programmes in the region. During the major period of fieldwork in Kenya I benefited greatly from discussions with my colleagues Urs Herren and Cory Kratz during brief visits to Nairobi. During the first few weeks of field work Cory once rescued me from the scene of an accident I had been involved in. Her insistence that things like these may happen but do not portend anything for the future perhaps persuaded me to persevere.

In Namibia first of all I would like to thank the families of the late Kandjuhu Rutjindo and his son Mungerenyeu Rutjindo. They welcomed me to their homes and taught me to speak and to behave appropriately and according to Himba norms and values. Like the Losutes in Kenya the Rutjindos provided a secure and comfortable base from which to explore social relations, economic strategies and worldviews. Both households were regarded as successful herding units where I learnt a great deal merely by observation. In Kaokoland, the friendship with Mutaambanda, Maongo, Tako, Motjinduika, Vahenuna, Karambongenda, Kakapa and Kamuhoke were and still are of great relevance. Be it for my greater awareness, be it for lesser emphasis on gender divides among the Himba, I found it easier to enlist women as trusted informants in north-western Namibia than in Kenya. Among the Pokot most of my information on women originated from the immediate social environment I was living in, notably from Chepösait who, adopting me as yet another child, for two years prepared my food while casually voicing her views on Pokot society. Among the Himba the elder women Kozombandi, Mukaakaserari, Kazupotjo and Watundwa were essential during various periods of my work. Anke Kuper shared a camp with me during a part of the study. Discussions with her on methodology and epistemology furthered my research work. Uhangatenwa Kapi furthered the project as an assistant in many aspects of anthropological fieldwork: from survey work to complex explanations of genealogical relations his endurance and practical intelligence were of great help. In Opuwo Father Zaby's guest house provided a quiet and peaceful place to rest. In Windhoek I greatly enjoyed the hospitality of the Baas family. Although Pokot lands and Nairobi are only some 300 km apart while my Namibian field site was about 1000 km away from the capital Windhoek, I always felt that the rural/urban gap was easier to bridge in Namibia: in the cosmopolitan Windhoek environment I benefited greatly from discussions with Wolfgang Werner, Peter Reiner and Antje Reiner-Otto, Werner Hillebrecht and Jekura Kavari.

Obtaining empirical data is one thing, data analysis and interpretation pose an entirely different kind of challenge. Sadly, my long-time teacher Thomas Schweizer died before this book was completed. Epistemology and methodology were greatly influenced by his work, by discussions with him over many years and by the shared conviction that there is a reality out there, which is accessible and interpretable according to analytical principles. Hartmut Lang and Michael Casimir influenced key passages of the book as teachers, colleagues and friends. Michael Schnegg and Julia Pauli, Aparna Rao, Joachim Görlich, Barbara Göbel, Christoph Brumann, Erwin Orywal contributed to the manuscript in various ways: by discussing certain sections, by helping with data analysis or by drawing my attention to aspects I had overlooked. I felt privileged to share not only ideas but also an apartment with the historian Jan Bart Gewald. His intimate knowledge of southern African history was inspiring and helped me to contextualise data along diachronic paths. Polly Wiessner did more than others to transform the manuscript into a book by carefully reading chapter after chapter, correcting minor mistakes as much as pointing out major errors. During discussions with her the manuscript achieved its final form.

I would also like to thank three anonymous reviewers for their comments. Although criticism is sometimes hard to digest, I found the issues they put forward extremely helpful when finishing the manuscript.

As the book is written in English, which is not my native tongue, I had to rely on translators and native speakers for corrections: while Paul Harris worked on the initial manuscript, Eileen Küpper and Luise Hoffmann polished the final draft. Maps were expertly drawn by Monika Feinen. Sonja Gierse-Arsten and Nicole Körkel skilfully assembled the bibliography and the various sections of this book, making sure that from photographs, to tables and to bibliographic entries nothing was omitted. Brigitte Schwinge and Viriginia Suter assisted during earlier periods of tiring data entry. Needless to say, any errors still found in the book are entirely my own responsibility.

This book would not have been completed without the emotional support from many people. During the early phases of research parental help is often necessary and I benefited greatly thereby. During the latter part of the study and throughout the entire phase of writing my wife, Heike Heinemann-Bollig, made many sacrifices to help me persevere. Her assistance ranged from help in the frantic search for "lost" literature or data bits to discussions on the heart of the matter: how to understand an alien culture and how to present it in writing. Last but not least I thank my two children, Antonio and Vivian, for sharing the most recent years of this project and for accompanying me on the long roads to the various the field sites.

Windhoek, February 2005
Michael Bollig

Contents

List of Tables

List of Figures

List of Maps

List of Photographs

Chapter 1

Studying Hazard and Risk in Pastoral Societies

INTRODUCTION

This book centres around the comparison of hazards, risk perception and risk minimising strategies in two African pastoral societies, the Pokot of northern Kenya and the Himba of northern Namibia (see Map 1). Both societies were studied over several years of intensive field research between 1987 and 1999. The central questions guiding the comparative approach are: (1) How are *hazards* generated through environmental variation and degradation, through market failures, violent conflicts and marginalisation? (2) How do these hazards result in *damage* to single households or to individual actors and how does the damage vary within one society? (3) How are *hazards* perceived by the people affected? (4) How do actors of different wealth, status, age and gender try to minimise risks by delimiting the effect of damage during an on-going crisis and what kind of institutionalised measures do they design to insure themselves against hazards, preventing their occurrence or limiting their effects? (5) How is risk minimisation affected by cultural change and in how far is the quest for enhanced security itself a driving force of cultural evolution? Answering these questions in a comparative perspective should lead to generalising hypotheses on the dynamic interrelation of hazards, damage, risk-perception, risk-minimising strategies and buffering institutions in African pastoral societies.[1]

A first case-orientated glance at the difficulties of risk management in a pastoral setting will forcefully indicate the need for a holistic, historically embedded and yet comparative and theoretically-informed treatment of hazards and risk management.

Map 1. Location of Field-sites

1.1. DISCARDED BOREHOLES AND PROTECTED PASTURES: THE WAY TO THE SUBJECT OF THE STUDY

In March 1992 I visited the Pokot (north-western Kenya, Baringo District) once again after having stayed there for two years between 1987 and 1989 and for a short while in 1991. It was hot and dusty to the extreme and there was almost no grazing left. The herds were living on bits of trodden down grass and fallen pods of acacia bushes, the fodder of extremely dry seasons which the Pokot contemptuously call *lal*. In the late 1980s I had been to north-western Kenya during better times; the rains in 1988 had been almost double the normal amount and 1987 and 1989 had experienced good rains too. Good rainfall and the absence of raiding, which had marred the decade before 1984, offered the opportunity of carrying out overdue generation-set rituals (Bollig, 1992a, 1994a).

In stark contrast to this, the situation in 1992 was appalling. The rains had been very bad and to make things worse, interethnic warfare had resumed once again. Pokot herders had given up wide areas bordering the Turkana to the north and were crowding into the southern parts of their grazing lands. Of course, it was not only Turkana raiders, who had brought the Pokot into such a precarious situation! Against the advice of their elders young Pokot men had set off on raids. They were looking for booty to build up their herds and for prestige to establish their fame as brave and daring warriors. Each raid sparked off reprisals from the enemy factions. On top of this, raids were usually followed up by operations of the anti-stock theft unit of the GSU, a paramilitary wing of the Kenyan army. Frequently these operations hit innocent households, neighbourhoods were punished summarily and herds were chosen at random and driven off to compensate the victims. It seemed as if this radically egalitarian society had no institutional capacity to control powerful subgroups who were prepared to accept an increase in vulnerability to the entire community for their own benefits.

This lack of institutional control struck me as important in other instances as well. In March 1992, at the height of the drought, the water-pump at Chesemirion, a small village near Nginyang, had become overcrowded. Next to the people who usually settled there continuously, another twenty or thirty pastoral households had gathered around the pump. They had been forced to move to Chesemirion when the wells and pans in other areas had dried up. Soon an argument arose as to who would have to contribute to the running costs of the motor pump. Whereas the original inhabitants of Chesemirion usually sent somebody to the nearest petrol station once a week to buy diesel for the pump with money each household had contributed, it became virtually impossible to include the newcomers into this arrangement. At the same time it was impossible to exclude the newcomers from using the well. Eventually all households - newcomers and original settlers alike - left the place in the midst of a drought because they could not find a lasting arrangement for the maintenance of the pump.

This short account of the institutional problems of Pokot resource management should not lead to the assumption that all their institutions were generally weak. The Pokot had been extremely successful in sustaining households within a pastoral mode of subsistence despite several catastrophic droughts, violent conflicts and massive demographic growth. To put it in blunt terms, the Pokot were successful when they had to rely on internal solidarity, but when they had to rely on an institutional framework to guarantee a sustainable management of resources and peaceful conflict management they ran into serious problems.

In 1994 I started working with the pastoral Himba of north-western Namibia. At first sight they seemed rather similar to the Pokot. Both economies were livestock based and mobility was an important strategy to guarantee the survival of herds. The economic base of the Himba had been virtually destroyed during a drought in the early 1980's when they lost 90 per cent of their cattle. However, only a decade later the Himba had successfully restocked their herds. Restocking had been at least as efficient as with the Pokot but it ran on other premises. While the Pokot redistributed cattle internally on an egalitarian basis through institutions such as bridewealth and stockfriendships, the Himba redistributed livestock mainly through patron-client networks.

1.2. RESEARCH ON RISK MANAGEMENT IN ANTHROPOLOGY AND THE SOCIAL SCIENCES: AN OVERVIEW

The consideration of hazards, risk perception and risk management are relatively new fields in the social and environmental sciences. For a long time anthropological theory and research preferred to describe social and economic systems as being in a state of equilibrium. More recent advances in cultural ecology (Moran, 1979), evolutionary biology and ecology (Borgerhoff-Mulder, 1991; Boyd & Richerson, 1985), economic theory (Ensminger, 1992) and political economy (Sen, 1981, 1985; Watts, 1983, 1991) have given risk a prominent status in theory building. These new orientations point out that risk minimisation is central to our understanding of individual strategies and social institutions and not just a peripheral and transient moment in a group's history. There are at least three distinct anthropological approaches to the topic: a formalistic, an ethnographic and an interpretative approach.

1.2.1. Actor-Oriented Approaches to the Study of Risk

Anthropologists interested in forager societies have emphasised risk management strategies as a major force shaping hunting and gathering routines and structuring institutions of food sharing and territorial behaviour (Acheson, 1989; Wiessner, 1977, 1982; Winterhalder, 1990; Cashdan, 1983, 1985, 1990a, 1990b; Kaplan, Hill & Hurtado, 1990). These contributions have in common that they apply rational actor-based models to empirical data. They set off with some clearly defined hypotheses with relevance for a larger theoretical framework, quantify the variation of key-resources and then single out specific strategies of risk minimisation. Finally they test why the strategies empirically found are superior to other, alternative ways of handling unreliable food supplies. Kaplan, Hill & Hurtado (1990) have explained the specifics of Aché (Paraguay) hunting strategies and institutionalised sharing by referring to subsistence risks. Instead of relying on widely available food sources such as palms, the Aché go for the high variance game resource, thereby accepting a much higher unreliability in food supply. Institutionalised food sharing levels out the unpredictable supply of meat. However, by institutionalising the distribution of meat, Aché hunters come upon another problem. Good hunters continuously invest more into meat distribution than poor hunters do. They forsake their higher mean returns in favour of group welfare. According to Kaplan, Hill & Hurtado (1990) they are rewarded by more extramarital affairs and a higher number of offspring thus increasing their fitness in a socio-biological sense. Hames (1990) takes up the subject of sharing as a strategy of lowering the variance of food amongst the Amazonian Yanomami. He finds that the higher the variability of a resource the more intensive the sharing, the larger the spatial scope of sharing and the lower the importance of a kin bias in sharing is.

While the theoretical relevance of the issues discussed was clearly elaborated, the paradigm tended to be pinned down to the anthropology of forager populations. Furthermore, research concentrated on specific types of hazards: aperiodic shortages of food or randomly distributed shortfalls in food acquisition. Institutionalised food sharing (cf. Bell, 1995; Kent, 1993; Kaplan, Hill & Hurtado, 1990) and territorial behaviour (Cashdan, 1983; Smith, 1988) could well be explained as institutions of risk minimising foragers. However, these studies did not result in models of risk minimisation with a more general applicability. They did not have lucid answers for understanding more complex issues involving actors in politically unstable, internally stratified and heterogeneous, internationally enmeshed and environmentally degraded environments. Three further points of critique must be addressed to these studies: (1) they offered their data in a historically decontextualised way; (2) emic perspectives on hazards were lacking, (3) actors were mainly presented as constrained only by natural conditions but not by existing norms and values and/or political conditions determining ownership rights, exchange in markets and competition with other groups (cf. for a comprehensive critique of these highly formalised studies see Baksh & Johnson, 1990).

1.2.2. Ethnographic Approaches

The literature on hazards and risk minimisation in Africa's dry belts has increased rapidly since the middle of the 1970s. Pastoralists were frequently seen as culprits and victims of the various dilemmas impacting on them. Their alleged conservatism and propensity to accumulate livestock beyond economic rationality made them obstacles to economic development. For almost two decades anthropological reports on pastoral societies have dealt with economic marginalisation and impoverishment (e.g. Hogg, 1986, 1989; White, 1997; Broch-Due & Anderson, 2000), environmental degradation (Spooner, 1989), the dissolution of communal lands (Galaty, 1994; Ensminger, 1992; Hitchcock, 1990; Stahl, 2000), and land loss (Arhem, 1984; McCabe, 1997), internal stratification (Hogg, 1986; Little, 1987, 1992) and social disintegration (Hogg, 1989).

A number of excellent descriptive studies on hazards and risk management in Africa's dryland areas have been written since the 1970s (Swift, 1977; Colson, 1979; Watts, 1983; White, 1984; Johnson & Anderson, 1988; Downs, Reyna & Kerner, 1991; Hogg, 1986, 1989; Hjort & Salih, 1989; Spittler, 1989a, 1989b; Fratkin, 1991; see Shipton, 1990 for an overview of recent literature). These ethnographically condensed accounts clearly indicated that risk management was a major key to understanding African rural communities. Since the early 1980s several major research projects have been focussed on hazards and risk management: The multidisciplinary South Turkana Ecosystem Project (STEP; Dyson-Hudson, 1983; Dyson-Hudson & McCabe 1985; Ellis & Swift, 1988; Little & Leslie, 1999) showed how migrational routes were carefully planned to limit negative environmental impact and to evade violent conflict (Dyson-Hudson & McCabe, 1985; McCabe, 1994; McCabe, Dyson-Hudson & Wienpahl, 1999), how human reproduction was finely tuned with environmental cycles (Little et al., 1988; Leslie & Dyson-Hudson, 1999) and how nutritional strategies balanced annual shortfalls (Galvin, 1987; Galvin et al., 1994; Galvin & Little, 1999). A major point made in Dyson-Hudson & Meekers (1999) is that the viability of the pastoral system is maintained by excluding the poor from society: poor Turkana tend to migrate out of the Turkana District in large numbers. Broch-Due & Anderson (2000: 3ff) extensively document cases on the social exclusion of the pastoral poor and convincingly show that the maintenance of a pastoral ideal partially rests upon the exclusion of people who do not conform to the ideal.

These contributions are strong on ethnography and provide substantial descriptions of risk management. However, they mainly deal with the economic and ecological side of the problem. We learn little about how people perceive hazards and how they relate them to their belief system. Ritual as a way to cope with unpredictability is not touched upon. Furthermore, no attempt is made to explore generalising hypotheses: the contributions quoted above are first of all excellent descriptions of social and ecological systems and risk management within these environments.

1.2.3. Interpretative Approaches

While Mary Douglas's approach to risk analysis (Douglas, 1985, 1994) has not had much influence on the two paradigms discussed above, it had considerable impact on the study of risk analysis beyond anthropology and is widely accepted as *the* anthropological contribution to the study of risk. Her major contributions 'Risk Acceptability according to the Social Sciences' (1985) and 'Risk and Blame' (1994) embrace the concept of risk perception as a general concept of social analysis. Douglas pinpoints the relevance of the concept in western thought aptly (Douglas, 1994:15): "The idea of risk could have been custom made. Its universalising terminology, its abstractness, its power of condensation, its scientificity, its connection with objective analysis, makes it perfect. Above all, its forensic uses fit the tool to the task of building a culture that supports a modern industrial society." Her approach to the topic is demanding: she develops a theory of risk perception which includes all sorts of risks ranging from nuclear waste pollution to witchcraft in various types of society (Douglas, 1994:22). Her main message is: risk perception is encoded in social institutions.

Although Douglas's approach brings risk perception, as a new sub-field, into the study of risk its analytical focus is vague. The typology Douglas proposes is simplistic and apparently has not influenced empirical studies very much (see Boholm, 1996 for a general critique). In general, studies on risk perception are few. Risk perception as a topic of research has been most frequently analysed by social scientists working in western societies (Beck, 1986; Bechmann, 1993; Douglas & Widlavsky, 1982).

1.3. THEORETICAL SCOPE OF THIS STUDY

The review of recent literature on hazards and risk minimisation suggests some guidelines for this study:

(1) A wide variety of hazards which have effect on well being and wealth will be studied. Hazards not only affect food security but have malign effects on all three forms of capital, economic, social and symbolic.
(2) Hazards and risk management will be contextualised historically. Archival research and the extensive study of oral traditions and biographic accounts allows for a diachronic view on the development of hazards and risk management.
(3) Local perceptions of hazards will be portrayed and their embeddedness in social structures will be traced.
(4) The provision of generalising hypotheses on hazards and risk management in sub-Saharan African pastoral societies is a major aim of this study. The two-community comparison is aimed at generating hypotheses on the change of hazards and risk management in pastoral societies.

1.4. KEY CONCEPTS: HAZARD, RISK, AND UNCERTAINTY

It is important to define key concepts clearly in order to compare phenomena across cultures and over time. Literature on risk has grown exponentially over the last two decades. Technical sciences deal with the issue just as much as the social sciences - and within the social sciences there are again several strands of basically unrelated research on risk. There is little doubt that a general conceptualisation of the term risk is lacking (Bechmann, 1993:240). Nowadays the concept *risk* is used in colloquial as well as in various scientific discourses and we are confronted with various meanings and definitions.

The etymology of the term *risk* is not clear. According to Luhmann (1993:327) the term emerged during the transition from medieval times to the modern era. According to him the word, adopted from Arabic, was used to designate hazards that were connected to an individual's decisions. While most natural hazards were beyond the scope of human influence, risks were potentially manageable or even the immediate result of decisions. In contrast to Luhmann's rather clear-cut history of the concept, representatives of the insurance studies point out the ambivalent character of the term. Helten (1994:1) relates the origins of the term to the Italian *risco* – "cliff " and draws a line to modern applications of the term from the dangers that cliffs presented to ancient seafarers. Bechmann (1993:240) traces the etymology of the term to the Italian *riscare*, "to dare" which connotes "a possibility to cope with a future that is perceived as unpredictable and hazardous". Some basic variations inherent in the concept are discernible in these etymologies: while the equation with a cliff refers to a hazardous object or a dangerous condition, the differentiation between danger and risk brings in cognition, and the line drawn to the verb "to dare" introduces human agency.

However the concept risk may have been defined at the beginning of the modern era, nowadays its conceptual borders are blurred. In colloquial speech the term risk may be applied to (1) hazards such as lightning, fire, storms, earthquakes, (2) objects that may cause hazards such as nuclear power plants, fireworks, and subjects that represent certain categories of risk such as smokers, and (3) hazardous activities such as hang-gliding, expeditions to deserts and urban jungles. Jungermann & Slovic (1993) present six scientific definitions of risk of varying complexity.[2] It was especially the definition of risk as the product of the probability and the extent of specific damage that has dominated insurance studies and sociology for some time. The definition suggests that the phenomenon is easily quantifiable: one looks for the probable frequency of an occurrence and computes it with the expected extent of the damage. One obvious advantage of this definition was its capacity to make risks comparable: after doing the necessary computing, one could compare the risks of nuclear power to the risks of energy supply with coal.

While such clearcut definitions of risk are neccessary in insurance studies in order to allocate premiums differentially, they do not capture the core of the problem in the social sciences. Jungermann & Slovic (1993:171) point out that *risk* is not a directly perceivable phenomenon. They conclude (1993:201):

> "In short, there is not an 'objective risk'. Risk is a multidimensional construct. 'Risk' exists as an intuitive concept, which for most people means more than the 'expected number of future damage'. Its mental presentations are shaped by knowledge on the subject matter, by characteristics of the cognitive and motivational system and finally by social reality with its inherent interests and values."(transl. by author)

In contrast to the sociological approximations of the term, a great deal of anthropological ideas on the concept were tied to the observable. Wiessner (1977:5) defined risk as the "probability of loss or the possibility (or probability) of an unfortunate occurrence....An unfortunate occurrence can be considered to be anything which alone, or in combination with other occurences, can be detrimental to the survival and reproduction of an individual and his family." Halstead and O'Shea (1989:3) equate risk and variability: "In practice, variability may be conceptualised in two different ways: as the actual pattern of variation in food supply, or as the operation of those factors, ranging from climate to micro-organisms to human judgement, that influence the availability of a particular resource. Regardless of the focus, the crucial aspect of the analysis is the timing, frequency and severity of shortages." Cashdan, the editor of one of the most frequently cited anthropological volumes on risk management (Cashdan, 1990b), defines risk in a similar way: According to her terminology risk is the "unpredictable variation in some ecological or economic variable (for example, variation in rainfall, hunting returns, prices etc.) and an outcome is viewed as riskier if it has a greater variance".

To summarise the argument so far: some anthropologists see risk as the unpredictable variation in resources and hence an objective phenomenon which is accessible to quantitative analysis; in stark contrast to this, many sociological accounts see risk as a cognitive phenomenon and rarely treat the objective hazards risks are related to, but confine themselves to studies of the social construction of perception. Understandably so: in Western societies the objective world of hazards is relegated to disciplines such as physics (nuclear power), chemistry (pollution) etc. Sociologists are not deemed to understand the inner workings of a nuclear power plant nor are they thought to be able to analyse how people cope with these problems practically. However, anthropologists researching small-scale societies do not deal only with the cognitive world of the people and the social embeddedness of perceptions. They gather data on their economics and their material world too. They are interested in the frequency of droughts, their consequences for livestock mortality, and the vulnerability of a population. On top of that they gather information on the cognitive frameworks people use to interpret misfortune and they collect data on strategies people adopt to cope with crises. Hence, anthropology is in need of a wider terminology than just sociology, insurance studies and economics. At the same time the holistic anthropological perspective promises interesting insights into the interrelations between hazards, vulnerability, perceptions and risk management.

A further, but strongly related problem arises with the terms *unpredictability* and *uncertainty*. Generally *uncertainty* is defined as a lack of information about the world whereas *unpredictability* is a feature of a hazard itself. Only if future damage is unpredictable, i.e. actors are uncertain about their occurrence, do we speak of risks. The lack of information may relate to the temporal framework (we do not know when we have to cope with specific damage), the spatial framework (we do not know what area will be affected) and the extent of damage (we do not know the relevance of damage). Hence, *unpredictability* is a salient feature of hazards and *uncertainty* is a defining criteria for risk, something with which risk is inextricably linked. It is a matter of perception and not of objective, quantitative measurements. Generally partial and total lack of information are differentiated (Helten, 1994:3; Cashdan, 1990b). If risks are socially and culturally embedded in perceptions of future damage, uncertainty is the perception of unpredictability. Uncertainty is connected to emotions, norms, values and knowledge. The prediction of the future in oracles is one of many means of minimising uncertainty.

In conclusion, it is necessary to tie loose ends together and to differentiate hazards, damage, risk and risk management:

(1) *Hazards* are defined as "naturally occurring or human-induced process(es) or event(s) with the potential to create loss, i.e. a general resource of danger". (Smith, 1996:5)

(2) Environmental and socio-political processes may result in detrimental changes in an individual's and household's assets. While these changes do not result in easily noticeable losses they result in *vulnerability* increasing the chance that future hazards have a disastrous impact.

(3) *Damage* results from hazards and is defined as any negative impact on assets and/or the well-being of individuals and groups. *Damage* is often unevenly spread within one population. The extent of *damage* is not only dependent on the severity of the hazard but also on the vulnerability of the household.

(4) Hazards and the related damage are *unpredictable*. The culturally and socially embedded perception of this *unpredictability* is called *uncertainty*.

(5) *Risk* relates to an unpredictable or hardly predictable event which has consequences that are perceived negatively. Risks are the culturally and socially embedded perceptions of future possible damage. Risks are neither directly observable nor are they directly measurable. They are multidimensional constructions and are linked (through perception) to the living conditions of a people. A formal definition of risk implies:[3]
risk = df

- (a) it relates (i.e. a cognitive process of connecting phenomena) to a specific recognised event X
- (b) X brings about Y which is negatively evaluated (e.g. drought brings about livestock losses)
- (c) X lies in the future
- (d) X is hardly predictable or unpredictable

(5) *Risk minimisation* is always based on the culturally and socially embedded assessments and perceptions of past and future damage. The analysis of prior personal experiences or consensus based models is always a necessary first step for developing risk minimising strategies. Risk minimisation may be based on conscious decisions or may be embedded in custom and refers to (a) attempts at eliminating the occurrence of negatively evaluated events and (b) to strategies to decrease vulnerability and (c) to limiting the impact of damage once it has occurred.

Anthropological research then deals with four phenomena: (a) the causation and effects of hazards (b) the factual distribution of damage in a population, its frequency and extent, (c) mental constructs of hazards on the basis of social and cultural embeddedness of individual actors, (d) the minimisation of risks, i.e. the attempt to minimise losses and to decrease vulnerability.

1.4.1. Hazards and Damage

The causation of hazards and damage to households and individuals in Africa's drylands are discussed controversially in various sciences: while environmental factors have been traditionally emphasised by geographers, demographers see rapid population increase out-pacing the growth of agricultural production as the main cause for a crisis. Social and political scientists point out that there is no simple correlation between drought, demographic growth and hunger and frequently see the commoditisation of production and distribution, the concentration of the means of production with a rural elite, the growing dependency on outside markets (peasantisation) and the marginalisation in a globalised economy as primary causes for a growing vulnerability in Third World rural societies (Glantz, 1987:39 and Watts, 1991 for overviews; see Sen, 1981, 1985 for the most influential writings in this direction). However, a short overview of the literature shows that it is not an issue of either/or but rather one of a growing interdependency of a multitude of factors that increases the vulnerability of many populations living in Africa's dry belts (Shipton, 1990).

The following hazards will be discussed, as causes for disasters, in subsequent chapters: (1) demographic growth out-pacing resources, (2) degradation of resources as a consequence of over-exploitation (frequently termed desertification), (3) changes in access to and command over resources as a consequence of commoditisation, increasing stratification and more exclusive formulation of property rights (usually summarised after Sen 1981 under the term *entitlement decline*), (4) short-term climatic changes (usually droughts), (5) livestock epidemics, (6) violent conflicts interrupting production and exchange. These hazards lead to the loss of property and frequently result in famine.[4] While it is possible to qualify and quantify damage resulting from droughts and epidemics it is harder to estimate damage caused by population growth, degradation and entitlement decline. While e.g. drought causes damage to individual property, the damage caused by rapid population growth outpassing local resources is more abstract and is borne by the community. Demographic growth, environmental degradation and entitlement decline develop their impact over a long time span, whereas drought, violent conflicts and epidemics have sudden, frequently disastrous consequences but are reversible in a relatively short time.

1.4.2. The Perception of Hazards

All cultures have specific ideas about the natural and social problems they are trying to cope with. Usually concepts exist to explain personal misfortune and environmental hazards. There are broad frames for the explanation of negatively evaluated events (e.g. witchcraft, pollution) as well as naturalistic explanations of hazards and growing vulnerability.

Mary Douglas (1994) was adamant in her basic hypothesis that the perception of hazards is socially embedded and that specifics of hazard perception in each culture can be traced back to social institutions. Douglas (1994:5) differentiates societies who prefer moralistic explanations of misfortune and those who attribute misfortune to internal or external enemies. Applying the grid/group analysis[5] she looks at how different "ways of life" define risk perception. On the basis of these concepts Douglas develops a set of "cosmological types", all of which have developed their special relationship towards the environment, to other people and the self and, of course, to risk: for example, while egalitarian people see their natural surroundings as fragile and approach technological innovations cautiously, individualistic societies see abundance and chance in nature. Hierarchical societies tend to emphasise the necessity of rules mediating between society and nature: if these rules are violated the system breaks down, if they are accepted, nature and society will do well. Fatalistic societies see natural processes as unpredictable and ruled by constant change (see Boholm, 1996 for a critique of this approach).

As yet, Douglas has found few followers who could fill her model with empirical data. However, there have been several attempts to show the general trends of risk perception in specific cultures. Göbel (1997) points out that luck (*suerte*) is a key concept of pastoralists in north-west Argentina in describing risk and uncertainty. The term connects environmental and social uncertainty to actor specific management strategies. Economic success and failure are seen as instances of "having luck" or "not having luck". It is thought that not all people possess the same degree of luck with the same things. While some people may be luckier than others in trade, others may have more luck with specific livestock. The Beja herders of north-eastern Sudan interpreted consecutive droughts as a sign of God's wrath. They saw the immorality of urban dwellers as a major instance arousing God's propensity to punish all the living (Hjort & Dahl, 1991:173). Scoones (1996: 151) shows that in a community of Zimbabwean farmers, different concepts of risk were in use at the same time: while one group connected a recent drought to disgruntled ancestors, others perceived unstable political conditions as the root cause of all other hazards. They saw drought and degradation as caused by a corrupt and incapable government. Another group pointed out that risks mainly arose from social conflicts. A fourth party (the church-goers) explained the drought as divine punishment for social misdemeanours. Scoone's example forcefully shows that there is not a culture-specific mode of risk perception per se: various concepts to explain disaster may be in use and may be used by actors according to their personal experience and their personal goals.

In recent years anthropologists have worked on specific fields of risk perception rather than on general conceptual approaches to risk. There is a growing body of literature on indigenous knowledge which closely relates to the perception of environmental risks (see e.g. Brokensha, Warren & Werner, 1989; Warren & Rajasekaran, 1995). Indigenous perceptions of sustainable resource management and of environmental degradation give a good idea of emic accounts of environmental vulnerability. Sollod (1990) conducted a survey on Tuareg perceptions of rainfall variability. Tuareg herders perceived drought as a prolonged process of consecutive years with below normal rainfall. Single years with severely diminished rainfall were not identified as droughts by herders. While actual rainfall data did not imply trends or cycles, Tuareg herders conceptualised droughts as regularly reoccurring phenomena (Sollod, 1990:287f). Ethnoveterinary accounts (McCorkle, 1986; Catley & Mohammed, 1995) report indigenous ideas of how diseases are caused.

Somali herders clearly distinguish between transmittable and non transmittable diseases and have a clear idea on disease aetiology via ticks and flies (Catley & Mohammed, 1995:12).

1.4.3. Risk Minimisation

Halstead & O'Shea (1989:3) define risk minimisation or buffering mechanisms as "practices [that] are designed to lessen the impact of variability by dampening its effects". They group risk minimising strategies into four major practices "mobility, diversification, physical storage and exchange". Colson's typology of risk minimisation (1979:21) points out five strategies as common devices to counter future damage: (1) diversification, (2) storage of food, (3) storage and transmission of information on famine foods, (4) conversion of surplus food into durable valuables which can be stored and reconverted into food during crises and (5) the cultivation of social relationships to allow the tapping of food resources of other regions. Browman (1987:171) in an account on risk management of Andean pastoralists, identifies similar types of risk management: (1) reduction of productive risks (terracing, special pasturing), (2) diversification of productive strategies even within single crops, (3) movement and/or fragmentation of land holdings, (4) social networks and (5) storage technology (in many ways similar typologies are offered by Watts [1988] and Fleuret [1986]). These typologies of risk minimising strategies are fairly close to empirical data and offer little abstraction. Wiessner (1977:6), in a theoretically motivated way, differentiates three ways of reducing risk: (1) prevention (the reduction of hazards), (2) transfer of risks to another party, (3) self-assumption and self-insurance. She sees prevention as attempts at minimising losses and at reducing vulnerability. Transferring risks implies the shifting of probabilities of loss from one party to a politically subordinate party (which has to accept the shift because of power relations) or to a specialised party (which makes profits on taking on risks from others). Self-assumption and self-insurance may include (1) the absorption of losses by previously accumulated food and goods (e.g. grain stores), (2) the sale of assets in order to exchange the gains for food so as to compensate for losses and (3) the distribution of losses over a "large number of independent exposure units so that losses can be more predictable and can be absorbed by the gains of other units" (Wiessner, 1977:8).

Forbes (1989:89) emphasises the different levels of risk minimisation and points out some fundamental differences between first defence mechanisms (or: lower-level hazard response mechanisms) on the one hand and emergency and catastrophe mechanisms on the other hand which he subsumes under the concept higher-level hazard response mechanisms. While lower-level hazard response mechanisms (such as the diversification of economic strategies) are energetically intensive, have a low visibility and are socially acceptable, higher-level hazard response mechanisms (e.g. begging, eating unusual food) have a high visibility, require only low energetic inputs and are frequently socially unacceptable. Strategies applied during an ongoing crisis are frequently extensions of lower level response mechanisms: while mobility is important in pastoral societies in any year, it becomes indispensable in drought years; the sharing of food is important in many social situations, even in normal times, and during a crisis food sharing may increase the reliability of supplies to all members of a group. However, there are differences too: resource protection is essential in thwarting the dangers of a fragile environment, during a drought however, people will not pay much attention to resource protection but rather rely on an efficient harvest of what is left. In a more complex way Shipton (1990:363f) differentiates temporal sequences of responses (1) precautionary strategies: diversifying, rotating crops, planting drought-resistant crops, accumulating herds, storing debts, maintaining friendships in distant groups, (2) earliest or most reversible measures: intensifying production or trade, substituting foods, splitting households into smaller units, (3) immediate or semi-reversible responses: borrowing money, pledging land, stealing, expulsing clients, (4) last or least reversible responses: expulsing elders or dependent kin, selling relatives, permanent out-migration.

Risk minimising strategies have been lauded as the backbone of indigenous economies. Costs of risk minimising strategies were often forgotten altogether. Land fragmentation involves exorbitant (time) costs as a farmer has to move between twenty or more plots (Forbes, 1989). The benefits of fragmentation are that through the distribution of holdings the danger of being hit by a single hazard (e.g. crop pest) is reduced. Poly-cropping has obvious benefits which have been frequently commented upon. However, they also entail costs, as no single crop in a field with many different crops will result in optimal yields. The costs of food sharing systems in forager societies are borne out by successful hunters. Kaplan, Hill & Hurtado (1990) for the Paraguayan Aché, Hames (1990) for the Brazilian Yanomamö and Kent (1993) for a group of San foragers from Botswana show that hunting fortunes do not level out over time. Good hunters contribute consistently more to the common pot than poor hunters do. For them food sharing as such involves costs rather than benefits. In the same vein livestock loaned by wealthy herders to poorer comrades are first of all animals which are no longer of immediate use to the owner. Institutions ensuring the protection of resources bear costs too: free-riders have to be punished and energy has to be spent on screening people who have transgressed the rules. Sometimes such institutions of communal management become so overburdened by transaction costs that they are altered into other less cost-intensive institutions. Only a cost/benefit analysis of specific risk minimising strategies will make changes in management strategies understandable.

1.5. ON CONDUCTING FIELDWORK IN TWO SOCIETIES

Intensive field research was conducted with the Kenyan Pokot and with the Namibian Himba. After a two month period of archival work in Nairobi in

1986, a first two-year long period of field work was conducted between October 1987 and September 1989 amongst the Pokot. Further field stays in 1991, 1992, 1993 and 1996 lasted between four and ten weeks. In Namibia, field work was conducted for a period of 25 months between February 1994 and March 1996. Four further field stays in late 1996, 1997/98, 1999 and 2001 lasted for four to eight weeks each. A two week visit to the Pokot in March/April 2004 and a three month field stay in late 2004 in Namibia resulted in data which has not been fully analysed within the context of this book. However, both fieldwork periods contributed greatly to the long-term perspective of this study. In total a period of 31 months was spent both in Kenya with the Pokot and in Namibia with the Himba.

In both settings I stayed with a wealthy and well-established household. Amongst the Pokot I chose to live with the household of a temporary research assistant, whereas amongst the Himba I decided to establish my camp next to a household I had become acquainted with during an exploratory tour of the area in 1994. In both instances the decision to live with a well-to-do household proved favourable. The dignity and authority of both household heads sheltered me from over-curious neighbours and at the same time supplied me with numerous guests and potential informants. Both men were - although not leading political figures - highly esteemed elders in their respective communities.

The higher spatial mobility of the Pokot also forced me to change my place of living more frequently. During the 31 months of fieldwork amongst the Pokot I lived in at least nine different places. Amongst the Himba, households usually shift between one settlement site in the rainy season and one in the dry season while livestock camps are more mobile. During the 31 months of field work amongst the Himba I only stayed in five places. The higher mobility of the Pokot brought about changes in the neighbourhood we lived in. Right at the beginning of my research in Kenya, I found these shifts rather discouraging. I had just started to feel at home in one neighbourhood, when the shift to a new site forced me to become accustomed to a new set of people. However, after some time this resulted in the situation that I got to know many people beyond the immediate neighbourhood. Amongst the Himba I was socialised within one wider neighbourhood, which consisted of about 40 households. I only left this neighbourhood occasionally for surveys in other communities. During my latter stays in the region in 1997 and 1998 I had the opportunity of starting work with Himba communities across the Kunene river in southern Angola.

Learning the local language was a major task in both instances. Pokot proved more difficult than Otjiherero (the Himba language) in this respect. The linguistic base of Pokot language has not been well described up until now. There were two grammars written by missionaries (Crazzolara, n.d., Hereros et al., 1989) based on language material from West Pokot which constitutes another dialect of the Pokot language. During the first months of my stay amongst the Pokot, I only worked with translators. Only after about three quarters of a year did I became versatile enough to conduct simple dialogues alone. During the latter months of my field work, I worked with an assistant only when transcribing and translating tapes. In Namibia I took more care to invest a lot of time into language training right at the beginning of my field stay. There were several grammars (Ohly, 1990; Booysen, 1982; Überall, 1963) and even a trilingual dictionary (Viljoen & Kamupingene, 1983). After about three months I was able to conduct simple survey interviews by myself and after about a year I had developed enough language capacity to conduct all the interviews by myself.

1.6. COMPARATIVE RESEARCH

Epistemological progress in anthropology depends on the comparison of social phenomena. Comparisons at different levels are needed in order to gain insight into the structural relations between culture and society, the evolution of societies and the relation between individual strategies and social institutions (Schweizer 1998). Only comparative research designs lead to valid explanations beyond the single case. Furthermore comparative research leads to a broader understanding of options and limitations within a specific type of society. The present book presents a two community comparison. Johnson has attempted to delineate the benefits of such an approach to which the present study fully subscribes:

> "... two community comparisons are useful, in that they do produce convincing explanatory analyses....Two community comparisons, to the extent that they involve systems of variables, can be quite plausible, just because the kind of accident that could produce a spurious correlation between two variables is highly unlikely to produce a theoretically predictable correlation between sets of any interrelated variables." (Johnson, 1991:14).

Salzman (1971:104) stated that pastoral studies have been strong on ethnography but weak on generalisation. He aptly warned "what we must not do is to regress to the position that better field work and more and more ethnographic detail will somehow be miraculously transformed into general knowledge, for if theory without data is baseless, data without theory is trivial". However, next to numerous good monographic studies, there have been few comparative studies on pastoralists. These few studies are of a different scope: (1) while some are interested in regional processes, others attempt to find general characteristics of pastoral societies; (2) while some clearly define the variables to be compared, others just aim at a general account of pastoralism with an implicit comparative perspective; (3) while some are based on field work others are based on literature.

In an early attempt Gulliver (1955) compared the pastoral Turkana and the agro-pastoral Jie. Both are neighbouring communities, the Turkana living in the hot and arid plains of north-western Kenya and the Jie on the escarpment just over the border in north-eastern Uganda. Gulliver sees both groups as representatives of two different types of pastoralism: while the Turkana are specialised and highly mobile livestock breeders, the Jie have a mixed economy in which they combine livestock husbandry and rain-fed millet cultivation. How do these differences in economy reflect upon kinship relations and property rights? Gulliver (1955:244) finds that similarities between both societies are due to the close historical relationship of both cultures. In fact, both populations have developed from one earlier population living in northeastern Uganda. Differences are basically due to the different environment. Harsh conditions and a high degree of unpredictability has led to the situation in which the Turkana rely more on a network of widely distributed stock friends, while the Jie rely rather on the solidarity of a localised kinship group. An ambitious effort to compare pastoral and agro-pastoral societies was undertaken by Schneider (1979). He assumes that "where pastoralism occurs, egalitarianism results from the fluidity of this form of wealth and the inability of any person to monopolise its production. Where the rate of production of livestock is lower than 1:1, exchanges between people become characterised to one degree or another by submissiveness due to the monopolisation of material resources, mainly land, by a few chiefs and aristocrats." (Schneider, 1979:10). A carefully designed comparison of four pastoral and agro-pastoral communities in eastern Africa was assembled by Edgerton (1971): four independent field studies were conducted in East African communities that had a pastoral as well as an agro-pastoral section (Pokot, Kamba, Gogo, Sebei). This research frame was designed to trace causes for certain cultural and psychological traits and link them either to the economic specialisation or to the cultural background. Goldschmidt (1971) summarised some of the results of the study in which he gathered ethnographic material on the Sebei. According to the study pastoralists tend (1) to display emotions more openly and are generally freer in their expressions of affection, whether positive or negative, (2) to be more given to direct action in interpersonal relationships and less to deviousness (3) to be more independently-minded in their behaviour, (4) to display more social cohesiveness despite their greater independence of action, (5) to have stronger and more sharply defined social values such as independence, self-control and bravery (Goldschmidt, 1971:132f). Recently scholars have conducted two community comparisons of pastoral societies in several instances. Typically the comparison is based on extensive fieldwork in two societies. Casimir (1991) analyses nutritional strategies amongst the Nurzay Pashtuns of Afghanistan and the Bakkarwal of Jammu and Cashmere. Galvin, Coppock & Leslie (1994) compare diet patterns amongst the Ethiopian Borana and the Kenyan Turkana. Roth (1994) juxtaposes marriage strategies amongst the Kenyan Rendille and the Sudanese Toposa and finds that while the Toposa use polygyny for forming clan alliances, marriage amongst the Rendille is rather an instrument for economic planning. Some years earlier, with a less refined approach, Legesse (1993) compared demographic trends and environmental management amongst the northern Kenyan Gabbra, Borana and Rendille: he found that the Rendille mismanaged their environment grossly, while the Gabbra successfully maintained the viability of the pastures they exploited. Beyond that a large number of recently edited volumes focus on specific problems of pastoralism and attempt some sort of comparison (e.g. Fukui & Turton, 1989 and Fukui & Markakis, 1994 on conflict management; Almagor & Baxter, 1978; Baxter & Hogg, 1990 and Anderson & Broch-Due, 2000 on poverty in pastoral settings; Hodgson, 2000 on gender to give only a few examples). These volumes are usually the result of thematically orientated conferences. However, these contributions do not undertake any strict comparisons with the aim of providing or testing hypotheses. Rather they present cases in order to portray cross-cultural variation of a specific problem or variable. The present study takes another route: two societies are compared under a similar research design and with the focus on a specific, theoretically interesting sub-field.

After giving a rough and comparative outline of Pokot and Himba societies in section 2, the major hazards to the pastoral system and resulting damage to pastoral households are described in a comparative way in section 3. Section 4 presents emic views on hazards in both societies. Sections 5 and 6 compare risk minimising strategies. Whereas section 5 deals with immediate reactions to an ongoing crisis, section 6 deals with precautionary strategies. The final section 7 condenses the results of the comparison and works towards a theory of risk management and social change in African pastoral societies.

ENDNOTES

1. Schweizer (1998) identifies the discovery and testing of hypotheses that are true for many cultures and societies as the basic aim of cross-cultural research. In an introduction to comparative methods in anthropology, he sees the construction of hypotheses as the main goal of comparisons with a limited number of cases, while cross-cultural comparisons with larger samples aim at the testing of hypotheses.
2. They list the following definitions: (a) risk as the probability of damage, (b) risk as the extent of damage, (c) risk as a function (usually the product) of probability and extent of damage, (d) risk as the variance of probability distribution of all possible outcomes of a decision, (e) risk as the semi-variance of the distribution of all negative outcomes with a definite point of reference, (f) risk as a weighed linear combination of the variance and the expected value of a distribution of all possible consequences (Jungermann & Slovic, 1993). (translation by the author)
3. Next to inspiring discussions on the intricacies of defining the concept *risk*, my colleague Hartmut Lang supplied the formal definition of the concept.
4. Shipton (1990:358) defined famine as "severe shortage or inaccessibility of appropriate food (including water), along with related threats to survival, affecting major parts of a population."
5. The term *group* is defined fairly conventionally as a number of people with some sort of common identity and with a definition of its borders (Douglas, 1978:8), while the concept grid is defined as "the cross-hatch of rules to which individuals are subject in the course of their interaction". (Douglas, 1978:8).

Chapter 2

An Outline of Pokot and Himba Societies: Environment, Political Economy and Cultural Beliefs

The nilotic speaking Pokot of northern Kenya (Rift Valley Province, Baringo District) and the Bantu speaking Himba of northern Namibia (Kaokoland, nowadays Kunene Region) are both pastoral nomadic peoples living on the fringe of young African states. To the traveller, the Himba and Pokot may look similar at first sight: both are exotic looking tribal people who adorn themselves with complex coiffures and wear colourful beads, they dwell in picturesque semi-arid environments, wear leather garments and live in traditional huts. However, despite the traditional appearance of both people colonialism has had a grave and lasting impact in both instances. In both regions herders mainly live off their livestock. While milk and meat is produced by the herds, maize is purchased through market sales of livestock or barter exchange. Decisions on production, distribution and consumption are taken at the household level while the management of communal resources (pastures, water) takes place on a neighbourhood level. The social organisation is shaped – although to very different degrees – by patrilineal and matrilineal descent groups and age-based groups. While chieftaincy among the Pokot is something alien and chiefs and their councillors are government personnel, Himba chiefs exert more authority and gain legitimacy both from local traditions and official acknowledgement. The religious system of the Himba is characterised by an ancestral cult whereas among the Pokot neither beliefs in ancestors nor in a divine being feature importantly. The basis of the Pokot religious system is a well formulated morality and a detailed code of honour. Both groups are peripheral minorities within their nations, the Himba numbering some 15,000 to 18,000 people living in north-western Namibia (about 1 per cent of the national population of 1.8 million Namibians)[1] and the pastoral Pokot of northern Baringo District totalling about 60,000 people (some 0.2 per cent of the national population of 28 million Kenyans).

The following sketches of Pokot and Himba societies serf as contextual outlines providing a rough insight into the history, political economy, social organisation and belief system of the two societies.

2.1. THE PASTORAL POKOT

The Pokot are a southern nilotic speaking pastoral group (Rottland, 1982) living in Kenya's semi-arid plains north of the densely populated central highlands. Linguistically they are akin to the Nandi, Kipsigis, Tugen, Marakwet, Sebei, Keiyo, Terik and Okiek who live as agriculturalists, agro-pastoralists and foragers in the west Kenyan highlands. Economically, and in many aspects also culturally, the Pokot are closer to their northern pastoral neighbours, the eastern nilotic speaking Karimojong (cf. Dyson-Hudson, 1966) and Turkana (Gulliver, 1955; Dyson-Hudson-McCabe, 1985; Little & Leslie, 1999). The Pokot are divided into two groups, the agro-pastoral Hill Pokot (Schneider, 1953; Tully, 1985; Dietz, 1987; Bianco, 2000) of Kapenguria and the Cherengani mountains and the pastoral Pokot of the adjoining lowlands of Baringo and West Pokot Districts (Beech, 1911; Bollig, 1992a). This study focuses on the pastoral Pokot of Nginyang Division[2], Baringo District.

2.1.1. The Ecology of the Northern Baringo Plains

Nginyang Division, the home of some 60.000 pastoral Pokot in the late 1990s, has an area of 4,400 km^2. The topography of the area (see Map 2) is characterised by wide savannah plains (at about 500 to 700m NN) – the Loyamoruk Plains in the east and the Kerio Valley plains in the west – which are crossed in a north-south direction by two mountain ranges, the Chepanda Hills and Mount Tiati on the one side and the volcanic Silali, Paka, and Korossi hill (at 1400 to 1700 NN [Reckers, 1992:33]). Towards the east the area is bordered by the Leroghi Plateau (2000NN) and towards the West by the towering Cherangani Range (3500 NN). The close proximity of mountains and plains results in a highly differentiated landscape as regards to precipitation and vegetation.

Lake Baringo is the only perennial water source within Nginyang Division. The Kerio River, at the western border of the Division flows for about nine months of the year, whereas all other rivers in the area are ephemeral. However, subterranean water is stored in sandy river beds and herders rarely have to dig deeper than 1.5 metres to obtain enough water for their herds. Since the middle of the 1980s a non-government organisation has erected numerous earthen dams in the area which somewhat relieved the burden of trekking for water.

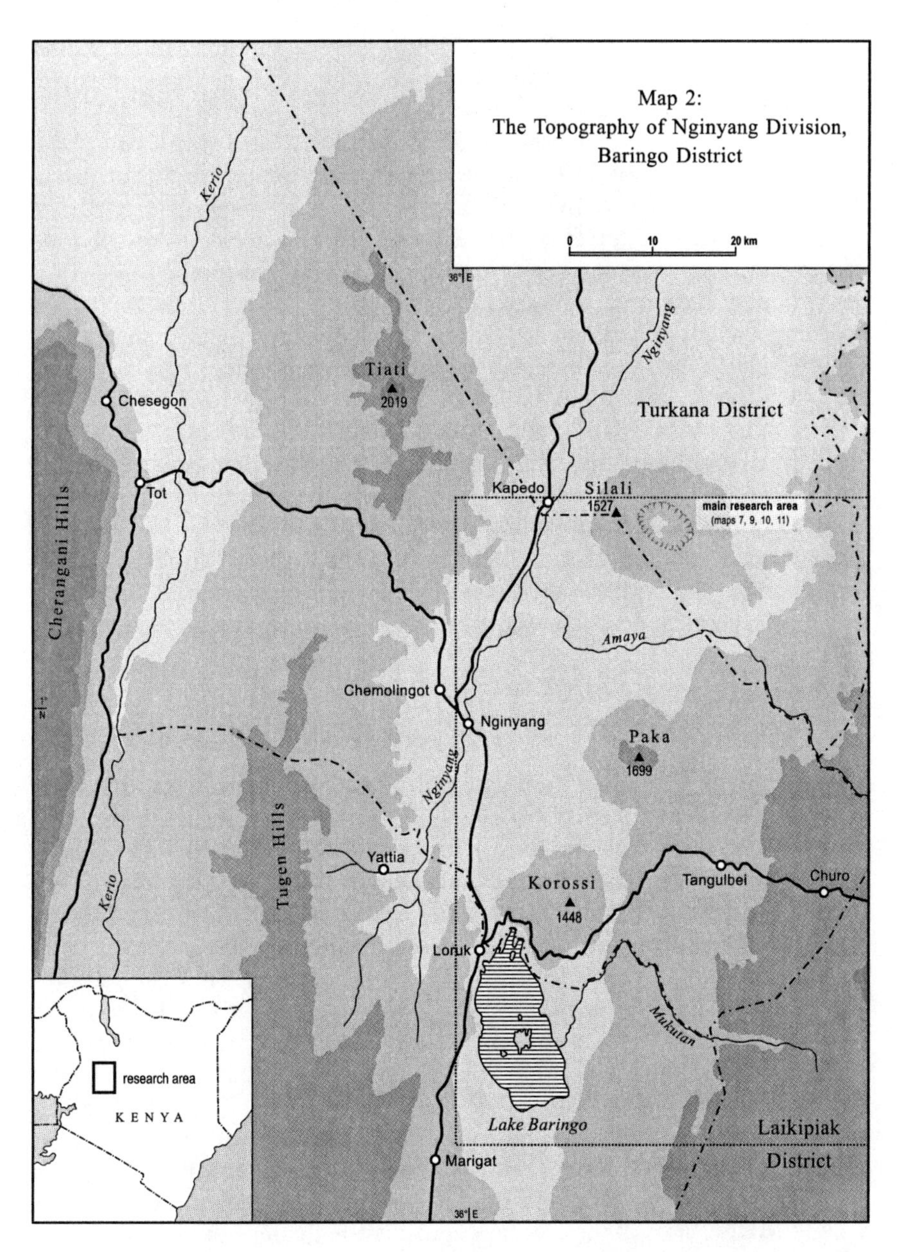

The area is extremely hot with a mean annual temperature of 26 degrees Celsius, but temperatures may rise up to 40 degrees Celsius during the hot season. The mean annual rainfall in Nginyang is 594.5 mm (over 41 measured years) with a variation from a maximum of 1125 mm (1961) to a minimum of 204 mm (1984). The Pokot divide the year into the main rainy season (*pengat*, May to August) and the dry season (*kömöy*, December to February).

In between are two shorter seasons, the unreliable short rainy season (*kitokot*, September to November) and the early rainy season (*sarngatat*, March and April).

Four types of savannah are distinguishable (cf. Barrow & Long, 1981; Pratt & Gwynne, 1977): (1) The largest area of the Division consists of thorn-bush savannah (Acacia refisciens, Acacia mellifera, Acacia tortilis). Perennial grasses are rare and only in the rainy season can a patchy distribution of annual grasses be found here. Hence the thorn-bush savannah is used during the rainy season as a pasture for cattle only, but serves as fodder for goats and camels all year round. (2) Pastures of perennial and annual grasses with only a few trees are only found north of Mount Paka and here, too, bush encroachment has become a serious problem (Reckers, 1992:55). (3) Highland pastures are found around Mounts Paka, Silali, Korossi and Tiati. Here slightly higher rainfall and a certain degree of protection have ensured that stretches of pasture covered with perennial grasses are retained. (4) Gallery forests are found along the major river courses, Kerio, Nginyang and Amaya. These forests are a source of good pasture for goats all year round. All vegetation types show signs of massive degradation which reaches from bush encroachment to invasions of unpalatable grasses and soil erosion.

2.1.2. Pastoral Expansion and Colonial Domination: The Historical Developments of the Pastoral Pokot

There are no unified Pokot traditions which portray migration histories, wars or other major events from a tribal perspective. Pokot history has to be pieced together from many clan and lineage histories (Bollig, 1990b).[3] Fairly regular cycles of generation-sets and age-sets are helpful in the approximate dating of events over the last two hundred years.[4]

The earliest oral accounts refer to events and processes that may tentatively be dated back to the late 18th or early 19th century. Then the plains of what is now Nginyang Division were still populated by Laikipiak and Purko Maasai (Beech, 1911:4; Lamphear, 1994:89). The ancestors of those people, who were to become the pastoral Pokot, were living on the northern edges of the western Kenyan highlands, on the slopes of Mt. Elgon and on the Cherengani range. At that time the pastoral Pokot were not yet detectable as a single unified tribal group. There were the Cheptulel and the Kurut, two mainly agricultural groups, and the Kasauria, an agro-pastoral group living around Mt. Sekerr. All three groups spoke mutually understandable dialects of southern Nilotic. According to traditions their social organisation was based on exogamous patrilineal clans and generation sets.[5] Neighbouring groups such as the Sapiny (Sebei, see Goldschmidt, 1976), the legendary Mtia and Oropom (Bollig, 1992a:52; Lamphear, 1988, 1994) and various Marakwet-speaking sections (Kipkorir & Welbourn, 1973), were culturally and economically fairly similar.

Ethnogenesis and Pastoral Expansion

Since the late 18th or early 19th century livestock husbandry among the Kasauria has expanded rapidly. Cattle became more and more important in a society which had previously relied mainly on small stock husbandry in combination with millet/sorghum based dryland farming. The emergence and expansion of the pastoral Pokot is chronologically (first half of 19th century) and socially (emergence of an age-set organisation, increased prominence of prophets) similar to processes of cultural evolution observed among other nilotic pastoralists. Pastro-centric ideologies made it a prescribed social and economic goal to obtain cattle not only for subsistence but also as objects for exchange. Cattle became the medium to establish and confirm social relations in the form of bridewealth payments, contributions to age-set rituals and the payment of compensation.

The effect of pastoralisation on Pokot society was threefold: (1) clan-based territoriality became dysfunctional within a pastoral economy and access to resources became tribally defined; (2) encompassing tribal institutions became more important than clan-based institutions; and (3) the rapid expansion of pastoralism created the economic base for the integration of non-Kasauria into the emergent pastoral Pokot society.

During the 19th century the Pokot successfully expanded into the lowlands. Obviously one precondition for the successful organisation of large scale raids was the adoption of the Eastern Nilotic age-set system (Bollig, 1992a:55; Peristiany, 1951). The pastoral Pokot expanded by integrating numerous individuals and families from other societies. The traditions of descent groups are similar to some extent: usually there is little information on pre-pastoral ancestors and then some kind of disaster compels an ancestor or a group of ancestors to move and join the pastoral Pokot. Several groups of immigrants can be traced through their family histories. People claiming ancestry with the Sebei and the Kurut fled the Elgon area because of drought and various livestock epidemics. Elders claiming ancestry with the Ngabotok Turkana of the Turkwell valley also report environmental calamities as the reason for moving. Another group of clans emphasises ancestry in the east. They report that, during their westward-bound migration, they settled for some time at Mount Nyiro, a mountain range south-east of Lake Turkana, which is today inhabited by Ariaal pastoralists. Individuals were integrated into existing lineages; bigger groups of people of one origin were adopted as a complete lineage by one other Pokot lineage. All immigrant men became members of Pokot age-sets and generation sets.[6] In a similar way the pastoral Turkana assimilated individuals from groups such as the Oropom, the Siger and the Kor (Lamphear, 1993:90ff). These processes of pastoral ethnogensis are observable in several instances in Eastern Africa. Generally smaller groups of agro-pastoralists, pastro-foragers and "old style pastoralists" (Galaty, 1993:64) gave way to larger tribally organised groups of highly specialised pastoralists in the savannahs of Kenya, Uganda and northern Tanzania during the course of the 18th and 19th centuries.

Within less than eighty years the Maasai had lost their ground in the hills and plains between Kerio Valley and Leroghi Plateau to the emergent pastoral Pokot. Oral traditions describe the battles between Pokot and Maasai warriors as excessively bloody affairs. Pokot elders were convinced that superior weaponry and military strategies as well as unrivalled courage ensured their ancestors' success against an otherwise equal enemy. Successful warriors – defined as those who had killed enemies and had raided cattle – were entitled to honorary *noms de guerre* and were allowed to decorate themselves with red plumes and, most important of all, to adorn their bodies with numerous scarifications on the right or left shoulder indicating the number and sex of their victims. Famous ancestors are depicted as men who were covered with this type of honorary scarification.[7] Pokot traditions relating to the last century are, to a large extent, descriptions of battles and warrior heroes depicting the Pokot as successful aggressors (Bollig, 1992a, 1993, 1995b, 1996a).

Photograph 1. *Suk* War-Dance (*amumur*) (from Beech 1911)

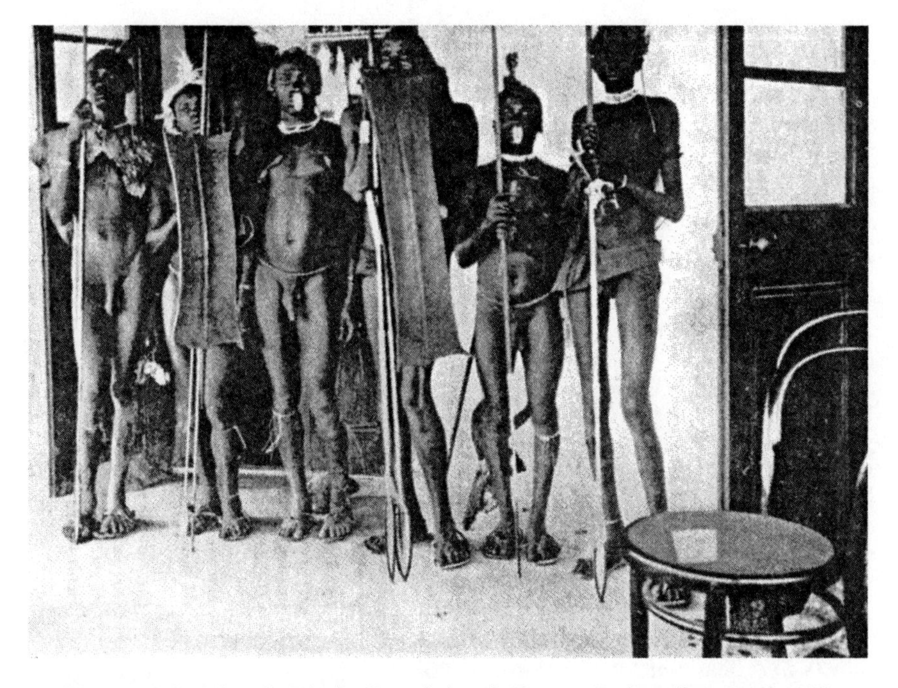

Photograph 2. *Suk* at the District Commissioner's House at Baringo (from Beech 1911)

By the end of the 19th century the Pokot of the Baringo basin were a fully pastoral group. Hunting and gathering were of some importance especially during droughts. Households alternated between rainy season lowland camps and dry season highland camps. Grain, weapons, pottery and jewellery were exchanged for livestock (usually small stock) with neighbouring groups (e.g. Marakwet and Tugen). Social relations – such as bridewealth payments and stock-friendship donations – were increasingly based on the exchange of livestock. Cattle became as much an exchange commodity, enabling herders to form and maintain social relations and to accumulate power, as they were a means of subsistence (cf. Sobania, 1991: 123).

The Pokot became only marginally integrated into the trade system of Swahili and Arab traders. During the late 19th century coastal traders preferred to travel to the Laikipiak plateau (Jacobs, 1979:43) and to eastern Uganda and the Turkana basin (Barber, 1968:91) to look for untapped ivory resources. The two Njemps villages at the southern tip of Lake Baringo became entrepôts for coastal traders. This, however, seems to have had little effect on the pastoral Pokot living further north. The fact that traders tended to avoid the pastoral Pokot was probably due to their reputation of being very "warlike" (Austin, 1903:308). Towards the end of the pre-colonial period, Pokot herds were diminished by Rinderpest (1890/91) and drought and people were at least once (in 1902) subjected to a major smallpox epidemic

(Barber 1968:84). The two decades before the colonial penetration were marked by major disasters. This, however, holds true not only for the Pokot but for most other pastoralists in East Africa. Sobania (1991) writes of severe livestock epidemics and imported human diseases such as smallpox in the 1880s and 1890s affecting north-central Kenya. Waller (1988) describes how the Maasai were diminished by a series of epidemics and droughts. Like the Pokot, many Maasai, Samburu and Borana had to take up hunting and gathering at the end of the last century.

The Colonial Period

The British colonial empire established its administration in the Baringo District in 1902. Early on administrators acknowledged that "... there is little in the district to attract the European settler, and this area may be definitely looked upon as a native reserve" (Hobley, 1906:472). During the first two decades of colonial penetration the northern drylands of Baringo District were only occasionally toured by an officer from the small station at Mukutani. His main interest was to procure tax payments from the Pokot, which they paid up with little hesitation, obviously wishing to establish a good relationship with the administration. A few big men were instituted as chiefs and headmen of the pastoral Pokot, a group which numbered some 6,000 to 7,000 individuals around 1920 (Bollig, 1992:48). In the 1920s and 1930s the northern parts of Baringo District slipped back into a state of slumber. After the administrative headquarters were moved from Mukutani to Kabarnet in the highlands, the administration was even further away. In 1939 a District Commissioner moaned: "The attitude of the Suk[8] towards any change in their present way of living, whether to their advantage or not, is one of complete indifference. They do not wish to cooperate in any policy of rehabilitation ... They wish to be left entirely alone".[9]

However, Pax Britannica brought two major changes to the arid savannahs of Kenya's north-west. Firstly inter-tribal relations became more peaceful after 1920 (Bollig, 1992a:254ff). Secondly, interregional trade became more prominent. Before 1900 the Pokot had engaged in small scale barter trade with neighbouring ethnic groups, but since at least 1909 they had been able to tap into larger trade networks through Somali traders. First these traders brought heifers to the Pokot and exchanged them for sheep which they in turn sold in the nearest town, Nakuru. After 1920 Somali traders brought with them trade goods such as maize, tea, sugar and cloth to exchange for livestock. For a long time exchange was carried on without using money. The traders delivered goods on credit and collected their livestock at the end of the rainy season. Trade was dominated by the Somali until well into the 1970s. Tugen traders have only become prominent in the livestock trade since the 1970s and, by the late 1980s, they had almost taken over from the Somali as livestock traders. However, the majority of shop owners in the area have remained Somali. Most of them do just a little livestock trading but concentrate mainly on the supply of maize, sugar, tea and other consumer goods. The Pokot were only weakly integrated into the colonial economy. While other groups went in search of labour in order to pay taxes, the Pokot barely looked outside their area for work. The colonial government made little effort to change things within the pastoral sector, and those initiatives which were taken were met with passive resistance by the Pokot (Schneider, 1959:149). Apart from occasional vaccination campaigns and forced de-stocking in the 1950s little changed on the ground. Missionaries started work in the 1940s but did not make any lasting impact until the 1970s. A migration out of the area of impoverished herders as observed for the Turkana (Dyson-Hudson & Meekers, 1999) has not been observed for the Pokot.

The political set up was dominated by a few appointed chiefs whose power base was critically judged by the colonial authorities. Colonial tranquillity was interrupted for a short time when a militant movement took hold of the Pokot in 1952 and some warriors attacked British administrative personnel. The Pokot neither took notice of the Mau Mau uprising nor did they participate in the early nationalist movements of their neighbours in the Kalenjin Political Alliance which was founded in the early 1950s. In fact the Pokot frustrated the colonial regime until independence (1963) and an outgoing commissioner moaned: “In spite of great progress in the rest of the District the Suk once more showed little desire for advance in any direction. Every encouragement has been given to them to keep pace with development elsewhere, they spurn it”.[10]

Post Independence

With independence little changed for the Pokot. Major changes only came about at the end of the 1960s when well armed Turkana raided the Pokot. The Pokot soon realised that only if they obtained guns could they defend themselves effectively against Turkana raiders. Throughout the 1970s and early 1980s each raid was followed by a counter-raid. The arming of the Pokot and their pastoral neighbours went on almost uninhibited for about a decade. After massive disarmament operations in the early 1980s, which involved the forcible disarming of individuals and collective punishment of truculent sections (Dietz, 1987:26), an insecure peace prevailed for about five years. In 1990 raiding resumed and was more violent than before. Large parties of well armed men ventured far into enemy territory, raiding cattle and killing people. The years 1998 to 2001 were marred by an escalation of warfare. Pokot now fought with Turkana, Samburu and Marakwet alike. In 2000 the markets in the Kerio Valley, crucial for the exchange of pastoral produce for cereals, honey and iron implements came to a standstill due to excessive violence.

While the economic integration of Nginyang Division into the national economy is still marginal and lags far behind the other divisions of Baringo

District, some changes are notable. Livestock trade has become more important during the 1980s and since 1988 a weekly livestock market in Nginyang has facilitated exchanges. Several development projects have been instigated to try to implement sustainable methods of livestock management. However, lately the activities instigated by these projects have been affected by violent inter-ethnic conflict and army operations. In 1997 German Agro Action, a non-governmental organisation which had heavily invested into infrastructure and livestock improvement for almost two decades withdrew due to a change within their overall programme policy.

2.1.3. The Family Herds: The Household Based Economy of the Pastoral Pokot

Anthropological research on East African pastoralist societies has focussed on adaptation to a semi-arid environment, on household based production and on economic and social change over the last two decades. It has been shown that spatial mobility, flexibility of labour allocation and consumption patterns were major strategies which enabled pastoralists to cope with an unstable environment (e.g. McCabe, Dyson-Hudson & Wienpahl, 1999; Galvin & Little, 1999). Sedentarisation, commoditisation and stratification were described as major trajectories of change (Sperling, 1987; Grandin, 1989; Fratkin, 1991; Herren, 1991; Little, 1992; Ensminger, 1984, 1992).

Household economy

The household is the basic unit of pastoral production and consumption. Pokot households (*kaw*, pl *keston*) consist of several houses (*ko*, pl. *korin*) inhabited by wives or female relatives (mother, sisters) of the household head and their children. About one quarter (25.6 per cent) of all surveyed households (n=442) lived in joined homesteads. There is a fairly close correlation between the age of the household head and the size of the household (χ^2=61.0 p< .00001). Table 1 shows the relationship between age, rate of polygyny and household size.

Table 1. The size of pastoral Pokot households

Generation-set of household head	Av. no. of wives range	Av. no. of children range	Av. total no. of household members
Kaplelach n = 148, (20 to 35 years)	1.7 (1-4)	5.4 (1-21)	8,1
Koronkoro n = 236, (36 to 60 years)	2.5 (1-13)	11.5 (1-42)	14,0
Chumwö n = 53, (over 60 years)	3.6 (1-10)	19.3 (1-40)	22.9
Total	2.6	12.1	15.0

Note: Relatives' children permanently residing in the household are included under "children". Permanent adult guests of the household are not included.

A new household is established when a man enters into his first marriage. This marriage is usually arranged by his father who looks for an appropriate spouse and pays the bridewealth from the household herd. Men usually only marry for the first time in their late twenties, since bridewealth is rather high and every older brother has to get married beforehand. Getting married is one precondition of independence. After his first marriage a man will usually stay on with his father or an elder brother for some time. Only after marrying a second wife has a man acquired enough labour workforce to set up as the household head of an independent homestead. However, he may cut short the route to independence either by convincing his mother and his siblings to move along with him to establish an independent homestead or by joining one of his peers. When a young man leaves his father he takes with him a share of the household herd. These animals are all taken from the herd which has been allotted to his mother during the previous decades. For the next two or three decades the household grows constantly, children are born and, if herd-growth allows, further wives are acquired. A new stage in the domestic cycle is reached when the eldest sons and daughters get married. However, the fact that a senior man still marries young women and has children by them, ensures that the household continues and frequently even grows. After the death of the household head, the household dissolves as surviving wives join their sons with the remaining livestock.

Pastoral Production

The household herd is fundamental to survival in the savannah of Kenya's semi-arid north-west. Pokot households hardly engage in any agriculture. Milk and, to a lesser extent, blood and meat make up an important part of the Pokot diet. All other food (mainly maize, tea, sugar) is procured via livestock sales. Livestock herds guarantee that social obligations can be met: The household head pays bridewealth, engages in livestock exchanges with stockfriends and acts as a host for age-set related rituals. While the household head formally owns all the animals in his herd, the usufruct rights of his wives in the animals are fairly strong. Virtually all animals are in the care of one of the women of the household and the household head cannot easily rearrange user rights between his wives. These animals are the core of their sons' future herds.

Pokot herds ideally consist of camels, cattle, goats, sheep and a few donkeys. In all types of herds, female animals make up the larger part of the herd (52 per cent in cattle, 62 per cent in camels, 60 per cent in goats and 60 per cent in sheep)[11], whereas males (young males, castrated males, breeding stock) form a smaller but equally important part as they are the media for market exchange (48 per cent in cattle, 38 per cent in camels, 40 per cent in goats, and 40 per cent in sheep).

Pokot cattle are of the Zebu breed. Taking data from the South Turkana Ecosystem Project the Zebu has a lactation period of 8 to 9 months and yields some 1.3 litres of milk as a daily average (Dahl & Hjort, 1976:146). My own data on milk yields was obtained over a period of about 12 months in 1988/1989 (in a year of very good rain). Average figures indicate that, under favourable conditions, milk yields are much higher and can reach about 2 litres per day. The lactation period of camels lasts 12 to 15 months and Pokot camels yield some 2 to 3 litres of milk per day. Goats have two kidding peaks per year and supply milk – although sometimes in minimal quantities – almost all year round. The milk is either drunk soured or is turned to butter. Clarified butter is stored to be added to food in the dry season. Soured goats' and camels' milk is also stored in giant calabashes as *chepösöyö* and eaten in a cheese-like form during the early dry season. Occasionally cattle, camels and goats are bled.

Wealth Distribution

By contrast with other pastoral communities in East Africa, a major differentiation between rich cattle owners cum traders, pastoral commoners and stockless paupers (see accounts of Dahl, 1979:210ff; Ensminger, 1990:670; Hogg, 1989; Broch-Due & Anderson, 2000) was not found amongst the Pokot in the 1980s and 1990s. There were still few Pokot traders of any consequence and very few men and hardly any women had migrated to central Kenya for work. Young Pokot men with a school education found it hard to get jobs within Nginyang Division. They either opted for permanent transfer to the bigger towns or they went for one of the few jobs in local NGOs.

The cattle herds of wealthy people comprised about 80 to 120 animals, while the herds of poor pastoralists consisted of only 10 or 15 heads of cattle. The biggest small stock herds surveyed numbered somewhere between 260 to 320 animals while the smallest herd had just 39 animals. Large camel herds consisted of 30 to 40 animals. About 60 per cent of all households did not possess any camels. Even the rich herders did not own other assets such as houses in towns, stores or cars. Wealth hierarchies are unstable amongst the Pokot. In 1987, 1991, 1992, 1993 and 1996 wealth rankings were conducted in the community: several men who were among the richer pastoralists in 1987 were regarded as medium rich or even poor herders in 1996 due to heavy stock losses during drought, disease or raids, whereas several herders who had been assigned the status `poor' in 1987 were regarded as medium herd owners in 1996. In contrast to these observations in 2004 a small group of wealthy Pokot traders and entrepreneurs had established itself. They still maintained a stake in livestock husbandry but generated their wealth mainly through trade and transport businesses.

Spatial Mobility

Herds are moved to ensure an ample supply of pasture and water. At the end of the rainy season young men move the major, non-lactating part of the cattle herd to dry season pastures while lactating cows, camels and small stock stay with the main household. While cattle camps are highly mobile, the households themselves move less frequently. At the height of the dry season further units of the herd are split off to be herded separately. A household may then break up into three or four units, each one migrating independently from the others. The reasons for leaving a place are not only ecological; inter-tribal violence and martial counteractions by the army are just as important a reason for moving as the scarcity of fodder or lack of water.

Map 3 shows patterns of spatial mobility in a normal year when nomadic moves are largely determined by fodder and water scarcity. From some central settlement areas people move out to more hilly areas such as Mounts Paka, Tiati, Silali and Korossi and even towards the northern fringes of Pokot territory (see Map 3).

Patterns of Consumption

A detailed analysis of progeny histories (n= 988 camels, 443 cattle, 321 goats, 203 sheep)[12] gives an insight into patterns of consumption. Of all the animals listed, about 50 per cent in all herds were still within the household herd at the time of inquiry, while roughly 20 per cent had died prematurely due to drought or disease.[13] About 30 per cent of all the animals were slaughtered, exchanged or sold.

The popular *cattle complex* stereotype[14] maintains that African pastoralists slaughter rarely and, when they do so, only for ceremonial purposes. They are said to be even more reluctant to sell cattle. By contrast, the data on the Pokot shows that, even if slaughtering is frequently only carried out in a ceremonial context, these rituals are plentiful. About 4.3 per cent of cattle, 10.0 per cent of camels, 9.4 per cent of goats and 19.2 per cent of sheep surveyed in progeny histories were slaughtered. Animals were slaughtered for neighbourhood festivities, healing rituals, marriage ceremonies and, perhaps most frequently, for age-set and generation-set related rituals. Small stock were frequently slaughtered just to obtain food, no ritual justification being necessary in this context. Many animals (26.0 per cent of camels, 17.3 per cent of cattle, 11.3 per cent of goats and 19.5 per cent of sheep) change hands over their lifetime. Livestock is mainly exchanged for three reasons: bridewealth, stock-friendship and direct exchange. An average bridewealth comprises about 12 cattle, 2 or 3 camels (if available) and about 30 goats and sheep. Friendship ties are sealed with presents of livestock. Additionally animals are exchanged directly: an ox is exchanged for a heifer, a camel for goats, or goats for a heifer etc.

Animals were sold at various livestock markets and occasionally to itinerant traders (13.0 per cent of cattle, 20 per cent of camels, 13.1 per cent of goats, 13.4 per cent of sheep). Most animals are sold in order to procure food (maize, sugar and tea). Pokot herders prefer to sell one ox in the early dry season. Then male goats are sold in order to take them through the rest of the

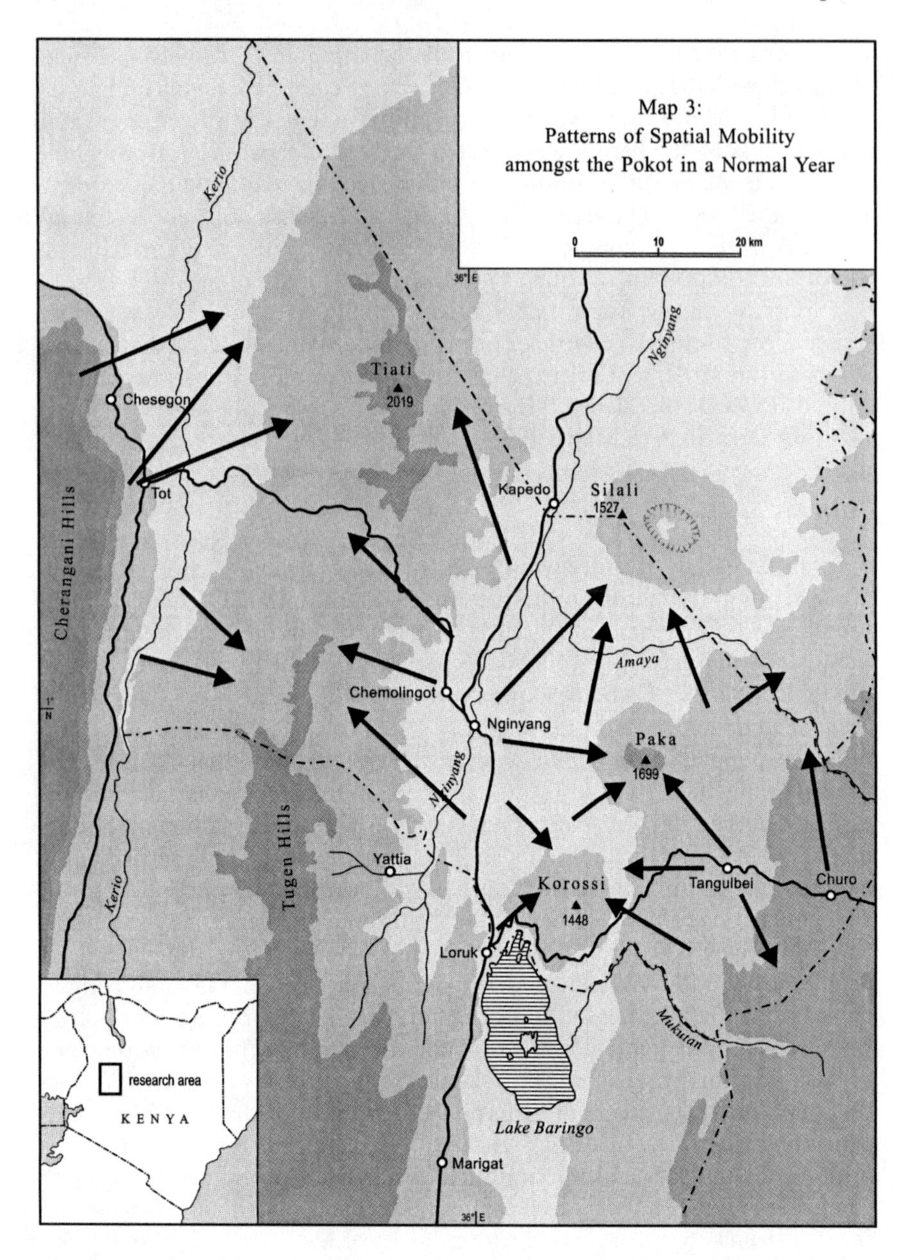

dry season. However, many households are forced to sell premature or even female stock due to the lack of males. During the rainy season food consists mainly of milk, blood and meat and only a little maize is needed, while during the dry season food may consist entirely of maize together with some sugared tea. Further money from livestock sales is spent on beads, second hand clothing, contributions to Harambee funds[15], school money, internally imposed fines and on veterinary and human medicine. Recently a lot of money was spent on the acquisition of guns and ammunition. Selling cattle became easier after a weekly livestock market was established in Nginyang in 1988. It was mainly the herders themselves or, to a rising degree, Pokot intermediary traders who offered livestock at Nginyang's Monday market to Tugen and a few rich Pokot traders who in turn took the livestock south using hired personnel.

2.1.4. Descent and Age: Social Organisation amongst the Pastoral Pokot

The social organisation of the pastoral Pokot is based on patrilineal descent groups, age- and generations sets and individual networks. Every person is a member of a patrilineal descent group by birth and both men and women remain members of one descent group for life. There are twenty two Pokot clans which in turn consist of three to seven lineages. Clans are exogamous and are united under a specific symbol, such as rain (the *Ngusur*), frog (the *Kiptinkö*), or fire sticks (the *Koimö*). Cattle brands and a set of symbols and songs (*enwait*) are clan specific. Beyond that there is little formal organisation at clan or lineage level. Patrilineal descent groups do not own any common property. There is no formal authority vested in lineage elders nor is there any formal decision-making at descent group level. Although numerous non-Pokot lineages have been integrated into the descent system, there is no ranking of lineages or clans, i.e. a lineage claiming Sapiny descent or Borana descent will have as much say in communal decision-making as any other lineage. Due to the fragmentation of descent groups lineages are numerically rather small and are spatially distributed. The Pokot do not put much emphasis on descent ideology. Communal rituals and celebrations are connected rather to the dynamics of the age-set cum generation-set organisation than to issues concerning the descent system.

Through separate initiation rituals each man becomes a member of an age-set (*asapantin*) and of a generation-set (*pïn*). Membership of one set involves egalitarian, almost brotherly relationships with one's peers on the one hand and submission under an age-based hierarchy on the other hand. Initiation into generation-sets only takes place every twenty-five to thirty-five years[16] when numerous young men and boys are circumcised together and from then on constitute one generation-set (Bollig, 1990c). At each stage three generations are alive. The system needs a junior or warrior generation, a generation of seniors and a generation of elders. At the time of my survey, about 40 per cent of all households were headed by juniors, 53 per cent by seniors and only 7 per cent by elders. Once most members of the generation of ritual elders have died, a new set has to be initiated thus the former warriors become seniors and the former seniors become elders (Bollig, 1990c, 1992a:85, 1994b; Sutton, 1990).

By contrast, membership of an age-set is attained at a fixed age. Boys of seventeen to twenty years of age have to undergo the *sapana*-ritual (Bollig, 1990c). In this celebration they have to slaughter an ox in a ritually prescribed way and are "washed" with the stomach contents of the ox which they have slaughtered. Later they are formally endowed with the signs of manhood: they are given the men's colourful head-dress. Each initiation set, i.e. those boys who were initiated in one particular year, is given a name which is reminiscent of a special event of that particular year or initiation period. After some ten to twenty years, between four and seven sets are united into one set and the set with the most members will usually give its name to the united set.

Age-sets and generation-sets constitute a complex gerontocratic system. Men achieve status not as members of descent groups but as members of age groups. During each and every festivity, even if it is minor, the gerontocratic hierarchy is depicted in the seating within the half circle (*kirket*) where all men are seated according to seniority. Age-sets compete for prestigious symbols. Junior age-sets have to fight for certain feathers and colours and senior age-sets have to ask elders for the grant of ritual powers symbolised in adornment (Photograph 3). Each set has to undergo a prescribed set of rituals in order to climb up the gerontocratic hierarchy.

Extended personal networks are features of the social organisation of many pastoralists (Moran, 1979:231; Dyson-Hudson & Dyson-Hudson, 1980:26; Schlee, 1985; Bollig, 1998a). Relations are strengthened by exchanges of livestock: friendship (*kongot*) always involves the transfer of animals between men (Schneider, 1953:240), paternal relatives are given a

Photograph 3. Pokot Man wearing *atöröntin* and *alim* Feathers

share of incoming bridewealth payments and affinal relatives and friends participate in these distributions too. While the gerontocratic age-based system is founded on the principle of respect (*tekotön*), solidarity (*tilyontön*) is the basic ideological principle of friendship.

2.1.5. Councils, Ritual Experts and Chiefs: Political Organisation amongst the Pastoral Pokot

In anthropology, East African pastoralists have been renowned for their egalitarian political organisation which has been connected to the specific character of mobile livestock husbandry and to its ideological and psychological foundations (Schneider, 1959:144, 147; 1979:10; Edgerton, 1971). Amongst the Pokot, neighbourhood councils (*kokwö*) are important for decision making at a local level. Nominated chiefs and councillors represent the interface between the local community and the national administration.

A neighbourhood council consists of all the initiated men in a local community (Conant, 1965). Seniors and elders dominate the decision-making process within the council but every adult man is entitled to contribute his own opinion. Since households are mobile, one might wonder who is entitled to membership of a certain local council. The Pokot have a very easy answer to this: every initiated man living in the locality has the right to speak out on issues dealing with the management of resources. Decisions within the council are usually consensus-based and are sanctioned by the blessing of senior ritual experts (*kapolokyon*). Prophets (*werkoy*) are essential during times of war (Peristiany, 1975). Through visions they foresee possible dangers ahead, such as pending raids by the enemy or promising targets for their own attacks. Prophets are also asked to give advice on the appropriate time to begin initiation rituals. A prophet's authority, however, is strictly confined to these fields.

Locations as administrative units have their own chief, and each sub-location has its sub-chief. Both chiefs and sub-chiefs are nominated by the central government in Nairobi. The job is advertised formally and any man who can read and write and has some knowledge of Swahili and English can apply for the job. Most of the present Pokot chiefs have been in formal employment before becoming chiefs.[17] None of them had been a herder for most of their adult life. Needless to say, under these conditions chiefs and sub-chiefs do not pass on their powers to their sons. Communities complain frequently that their chiefs have no contact with them at all and are basically looking to enrich themselves. Chiefs are legally entitled to use the administrative police to enforce their orders and it is on account of this that many Pokot are very critical of the institution.

Besides the chiefs, the local Member of Parliament plus five officially appointed councillors who assist him are influential at the interface of local and national politics. Chiefs, councillors and members of parliament communicate with local people through *barazas*. These are large formal meetings with a prescribed order of seating where the government representatives sit in the centre, frequently on chairs, and their citizens (the so-called *wananchi*) squat in front of them in a large semi-circle. Although all participants are Pokot the language of the *barazas* is Swahili and therefore usually no real discussions take place in these meetings. *Barazas* are rather meant to provide a platform for political agents and are orchestrated to inform the local community of the administration's plans (cf. Haugerud, 1995:102ff).

A local elite formed by traders, entrepreneurs and senior government employees is, if at all, only incipient. Due to a shortage of jobs many secondary school leavers are forced to look for employment outside the local setting. Senior jobs in development organisations are few in Nginyang Division. In the 1990s there were not even ten Pokot employed in permanent and influential positions in this field. Trading has definitely risen in importance in the 1990s. While in the late 1980s there were very few Pokot traders of some standing, in the 1990s many young Pokot men have tried at least some minor trading in veterinary drugs, beads or livestock. A few of them have become wealthy and, in order to diversify their assets, have established shops in the local centres.

2.1.6. Solidarity and Respect: The Belief System of the Pastoral Pokot

Pokot beliefs in supernatural beings and forces are neither canonised nor formulated by ritual specialists (Schneider, 1959:157). However, there is consensus on a limited set of ideas.[18] The Pokot believe in the existence of a supreme being, *Töröröt* (see also Schneider, 1953:104). Stories about *Töröröt* are variable and sometimes even contradictory. While some people maintain that he created the world, others think that he has always been there, just as the world has always been there. All informants relate God to climatic events but, while some maintain that rain is just another emanation from God, others are of the opinion that God makes the rain. However, lack of rainfall, i.e. drought, was habitually connected to God's displeasure with sinful human conduct.

After death human beings become spirits (*oy*). There is no clear differentiation between ancestral spirits, nature spirits and demons – all are termed *oy*. In everyday life the *oy,* as ancestral spirits, are of some importance. Soon after their death they roam about in the bush near to human settlements. They like to be given little presents, such as tobacco, a drop of milk or pieces of green grass, which are put on the grave. If they are forgotten too quickly, they may cause disease in humans and livestock although diseases and other worldly problems are more often attributed to witchcraft, the evil eye or simply the envy of others than to malevolent ancestral spirits. A small gift of tobacco may be enough to appease them. Belief in ancestral spirits does not feature significantly in the Pokot belief system, and despite the fact that the Pokot believe that presents should occasionally be taken to the graves, most graves are neglected and forgotten.

If we find Pokot religious beliefs vague and sometimes inconclusive, their system of norms and values is all the more pronounced. Even if the concept *moral economy* is derived from western sociological thought, Pokot herders would have a very clear idea what is meant by the term. Ideas about morality form the backbone of the belief system. Leading concepts here are solidarity (*tilyontön*), respect (*tekotön*) and bravery (*sirumoi*). Solidarity implies a set of behavioural rules. A person acting on the principles of *tilyontön* bases his deeds on deep-seated feelings of peace and harmony with his social surrounding (*kalya*). Respect (*tekotön*) dominates the relationship between husband and wife, between in-laws and between members of different age-sets. On the emotional side, respect always involves a certain amount of anxiety (*nyokoryö*) and uneasiness. Respect is most emphatically demanded during age-set and generation-set rituals when the elders demand services and presents from the juniors. The younger ones have to "work like women for them", fetching water and collecting firewood and being obedient throughout the ritual.

Allusions to the overall importance of warrior ideals for the moral system of East African pastoralists are frequent in ethnographic literature (Fukui & Turton, 1979:4; Dyson-Hudson & Dyson-Hudson, 1980:44; Mazrui, 1977b:71; Bollig, 1993, 1995b). The ideals of warriorhood are also an important feature in the moral system of the Pokot. The ideal Pokot warrior is a *nyakan*. The *nyakan's* outstanding character traits are bravery (*sirumoi*), determination (*körömnyö*), and solidarity (*tilyontön*), and his ability to handle weapons, to sing and dance rousingly at contests as well as his superior intelligence.[19] A man who cares about his good reputation will eat separately from women and children. He will live cleanly (*tïlïl*), which means that he only eats "clean food" (e.g. no meat from donkeys) and tries to avoid any contact with menstruating women. At the same time he is a very sociable character, who loves the company of his peers, goes to nightly dances where he shows himself to be a perfect dancer. Envy (*ngatkong*) and hatred (*osonöt*) are unknown to him. However, ideal warriorhood does not only entail gallantry. It implies a specific conduct in conflict behaviour as well. Many *nyakan* are men who have killed enemies and have undergone lengthy rituals to become *kolïn*, men who have killed and have later become initiated as healers for pregnant women. Scarifications on their shoulders, red feathers and their own specific praise songs make their heroic deeds both visible and audible to the public.

2.2. THE HIMBA OF KAOKOLAND

The Himba of north-western Namibia (Kunene Region) and south-western Angola (Kunene Province, Namibe Province) belong to a group of south-western Bantu-speaking agro-pastoralist and pastoralist peoples.[20] Their languages are basically dialects of *otjiHerero* which is mainly spoken in

Namibia's north-west, central Namibia and western Botswana (Möhlig, 1981:83). While the Himba were regarded as a subgroup of the Herero and are addressed as such in colonial documents, their independence as an indigenous group has recently been emphasised. There has been some discussion on Himba ethnicity (see Miescher & Henrichsen, 2000a, b): a Himba identity separate from a Herero identity emerged in the late 19th and early 20th century only as a consequence of early colonial changes in northern Namibia and southern Angola.

Kaokoland was inhabited by some 26,176 people according to the census published in 1991 (Government of Namibia 1991) of those about 15.000 to 18.000 were Himba (Talavera et al., 2000:32ff). The northern and western parts of Kaokoland are predominantly settled by Himba pastoralists while the southern parts are used mainly by semi-nomadic Herero households. Opuwo, the administrative centre, is inhabited by some 4,000 people of multi-ethnic origin. In 1992 Kaokoland ceased to exist as an administrative unit. The former Kaokoland became part of the Kunene Region and within this administrative unit is addressed as Kunene North or as Epupa and Opuwo constituencies. However, the term Kaokoland is still in wide use and especially the locals hardly use the new administrative terminology. Therefore the term Kaokoland is maintained here.

2.2.1. The Ecology of Northern Kaokoland

Kaokoland is a sparsely settled stretch of semi-arid savannah, tree and bush savannah in the east and shrub and grass savannah in the more arid western parts (for an overview see Becker & Jürgens, 2002a; Craven, 2002). Perennial and ephemeral rivers, mountains and different soil types create a multitude of microhabitats (see Map 4). The Kunene, which rises in the well watered highlands of central Angola, is the only perennial river in the area. Ephemeral rivers drain towards the Kunene basin in the north and towards the Atlantic Ocean in the central, western and southern parts of Kaokoland. Topographically the western pro-Namib plains, the interior highlands and the eastern sandveld are distinct units. Within the context of this study the northern parts of the interior highlands are of central importance. The landscape is characterised by broad plains divided by high ridges and mountain ranges (Malan & Owen-Smith, 1974:138). While temperatures may reach freezing point in winter, in summer the heat soars and temperatures around 40 degrees are not uncommon.[21] Rainfall occurs between October and April. The highest precipitation is recorded in February and March. Average rainfall in central Kaokoland is around 300mm while places further west towards the Namib Desert record annual averages of 100 to 150 mm and lower (for an overview on the ecology of Kaokoland see studies in Bollig, Brunotte & Becker, 2002).

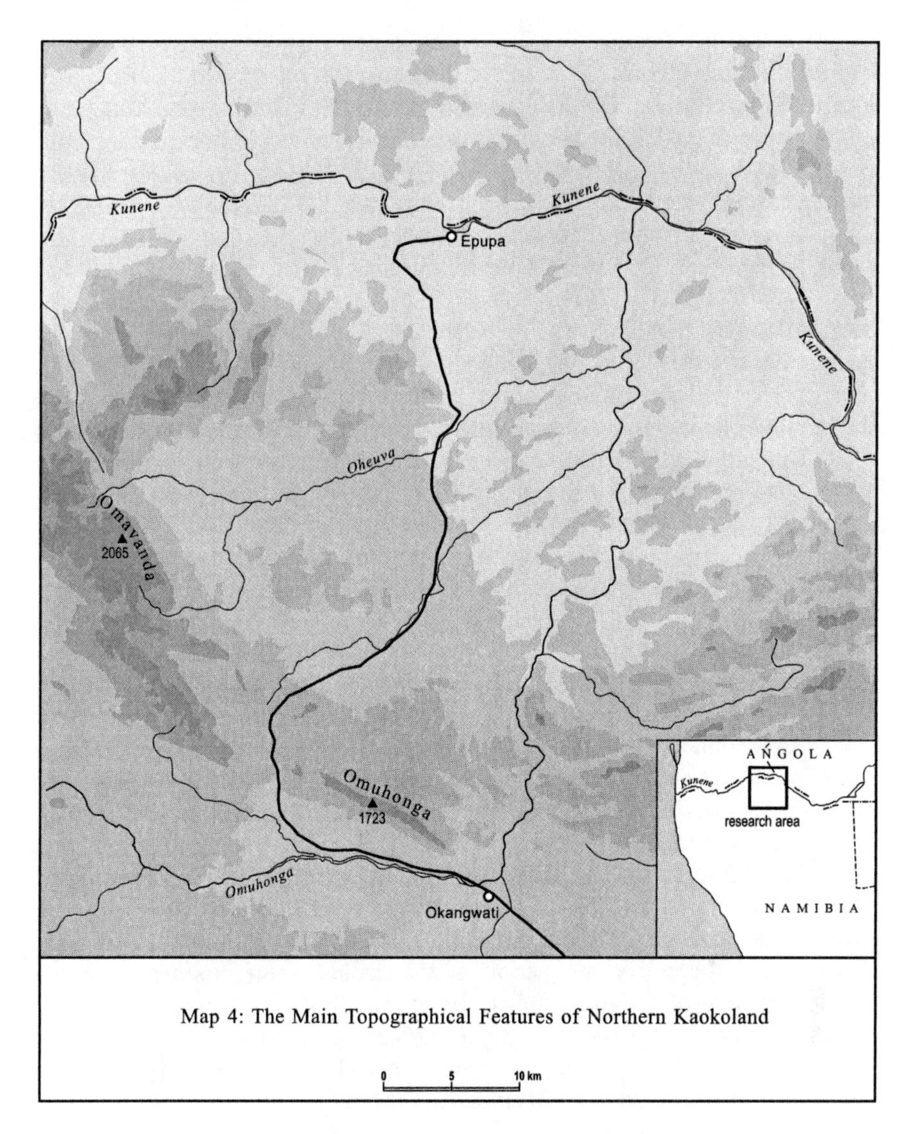

Map 4: The Main Topographical Features of Northern Kaokoland

2.2.2. From Early Integration into the World system to Colonial Encapsulation: The Historical Development of the Pastoral Himba

The history of north-western Namibia can be reconstructed on the basis of oral accounts (Bollig, 1997c), archival sources (Bollig, 1998a) and comparative evidence from the historiography of neighbouring regions (Siiskonen, 1990; Notkola & Siiskonen, 2000; Hayes, 1992; McKittrick, 1995; Williams, 1991; Miller, 1988; Clarence Smith, 1976, 1978; Gewald, 1996, 2000; Werner, 1998).

Himba elders frequently trace their origins to a small mountain, known as Okarundu Kambeti, near Ruacana at the north-eastern edge of Kaokoland.[22] From this place the ancestors migrated westwards following the Kunene. It might very well be that this migration was connected to the stock raiding politics of the Ovambo kingdoms (Williams, 1991:132-141) and an increase in slave trading in south-central Angola in the 18th century (Miller, 1988:222; Williams, 1991:141). Powerful chiefdoms were established in the Huila highlands of southern Angola at that time. Based on early travelogues, Estermann (1981:15) says that in 1787 the chief of Huila "ruled" the peoples of the lower Kunene basin. In the 1760s the highlands had the first direct contacts with the expansion of mercantile capitalism and were drawn into the Atlantic slave trade (Clarence Smith, 1978:165/166) bringing about a significant increase in warfare and social disruption. The mountainous and arid lands of the Kaokoland may have offered more protection than the open and fertile plains around Ruacana at that stage. Dating, based solely on genealogies, has proven to be weak; however, backed by the chronology of the expanding Portuguese colonial empire it is suggested that these traditions refer to migrations that took place about 300 to 200 years ago. The early migrants lived as pastoralists but relied heavily on hunting and gathering as well. Obviously their settlements were concentrated along the major rivers. Estermann (1969:77, 1979:8 based on Nogueira,1880) translates the ethnonym *Himba* with "people settling on the banks of a river".

Oral accounts become more detailed when talking about the raids of the so-called Ovahuahua (e.g. Katjira Muniombara in Bollig, 1997c:221f). These "people of the shields" were of Ovambo origin.[23] Their attacks on Kaokolands' inhabitants may have been connected with slave raiding and the general upsurge of violence in southern Angola between the 18th and the 19th centuries (Miller, 1988: 222/223) or may have been the result of the political centralisation of Ovambo communities. However, oral accounts do not mention that people were taken as slaves by the Ovahuahua nor do they talk about the conquest of land. According to the elders the "people of the shields" came in search of cattle which they drove off in large numbers. By the middle of the 19th century the pastoral society of north-western Namibia was based on patrilineal and matrilineal descent groups. There was no overarching social institution (as provided by chieftaincy or an age-set system) embracing the entire group.

Integration into the World System: Violent Conflicts and Trading Networks

Local herds were just about to recover from the depredations of Ovahuahua raids, when the herders were attacked again, this time by Oorlam commandos from central Namibia. The earliest raids can be dated fairly reliably to the beginning of the 1850s, lasting until the early 1890s.[24] Well-armed Oorlam raiders preyed upon the livestock of Kaokoland's herders. The booty was mainly sold towards the Cape or to Walfish Bay-based livestock traders (Lau, 1987:41). When the Swartboois and Topnaar, two prominent Oorlam groups, settled at Franzfontein around 1882 (Moritz, 1987:30) and some years later at Sesfontein, raiding became even more intense. By the 1880s, at the latest, most herders had fled from the Kaokoland to southern Angola. There the refugees came into close contact with the Portuguese colonial economy which had rapidly started developing after 1830. After the independence of Brazil in 1833, numerous rich Portuguese settlers left their plantations to establish new enterprises in southern Angola. Since the 1870s commercial hunters were based in Mossamedes and operated mainly in the Kunene basin to procure ivory, ostrich feathers and other tropical commodities for the world market (Bollig, 1998c, e).[25] The capitalisation and expansion of the farming sector was enhanced by a group of Dorsland Trekkers who had been settled in the Huila highlands since the 1880s. Their businesses and farms in the highlands, the large-scale sugar cane plantations in the lower Coroca valley and commercial hunting throughout the Kunene basin were major attractions for people looking for work. Oral accounts leave little doubt that Herero/Himba men eagerly looked for opportunities to earn money to invest into commodities such as guns, blankets and clothes and to restock their herds. However, after 1890 the single most important employer for the refugees was the Portuguese colonial army. Year after year so-called punitive expeditions were sent out against "rebellious natives" who had done little more than not to except the yoke of colonial exploitation. Himba and Herero participated as mercenaries in these raids. Especially after the Rinderpest of 1897 they saw income and livestock generated from commercial soldiering as a way to curb losses. João de Almeida's publication (1935) reports on the authors experiences as Governor of Mossamedes district between 1908 and 1910. Next to photographs of tribal costumes and coiffures he reports on and gives photographs of Himba mercenaries, heavily armed young men and a *corpo de irregulares* with their leader Vita Tom (Oorlog) (Almeida, 1935: 151/152).

The liberal revolution in Portugal in 1910 greatly changed the conditions for the former refugees in south-western Angola. After the liberal revolution had changed the political set-up, it was decided to extract labour and capital more systematically from the native population (Clarence-Smith, 1978:168). The military administration was changed to a civil administration. There was no need for mercenaries anymore and well-armed natives were a thorn in the side of the administration. Soon former mercenary leaders had criminal charges running against them.

Kaokoland under South African Rule

When the South African army first entered Kaokoland in 1917, many pastoralists had crossed the Kunene river once again and settled in the northern parts of Kaokoland. The South African government immediately disarmed the pastoralists and decided to put them under three chiefs, Vita Tom, Muhonakatiti and Kakurukouye. All through the 1920s the colonial government implemented boundaries along the Kunene river and on the northern edge of the Police Zone between Kaokoland and Ovamboland, and created a major buffer zone which was cleared of inhabitants in order to close off Kaokoland (for political reasons for these enforced moves see Bollig, 1998e). Mobility was additionally curtailed by interior boundaries between chiefdoms.

Since the middle of the 1920s the regulation of trade became even stricter. All trade involving products from livestock husbandry (live animals, skins, meat) crossing national boundaries (to Ovamboland and to the commercial ranching area) and international boundaries (to Portuguese Angola) was categorically prohibited. The Kaokoland became isolated with very little contact with the outside world. Visitors had to undergo a lengthy process of applying for permits. Missionaries were not allowed to operate in the area. The Odendaal plan, which proposed a homeland status for Kaokoland in 1964, described the area as very isolated and traditional with its economy based completely on livestock husbandry. It was not in the interests of the compilers of this influential report to acknowledge that Kaokoland's isolation was rather the product of colonial policy than the result of its geographical position.

Plans for the creation of a homeland in Namibia's north-west were never fully realised. Due to the lack of financial resources and the beginning of the civil war, plans to make Kaokoland an independent homeland were neglected. Since the 1970s the military had been stationed at several points in Kaokoland to ward off PLAN (Peoples Liberation Army of Namibia) fighters who allegedly tried to infiltrate the area. While the war took place inside Angola, the atmosphere in Kaokoland was tense and the army forcefully resettled pastoralists near their stations in order to cut off guerrilla support.

Since independence in 1990 several major development projects have been envisioned for the area. A large scale hydro-electric dam, which is in the planning, is opposed by local herders. Several mining companies are prospecting the area for future exploitation. In the late 1990s the IFAD sponsored Northern Regions Livestock Development Project tried to implement livestock development programmes including range management, livestock marketing and the extension of veterinary services. However, the project did not have a major impact on the ground. The rapidly growing importance of livestock marketing has probably been the major economic change since 1990. Talavera et al (2000:254ff) differentiate formal and informal livestock markets in Kunene North. The National Meat Corporation of Namibia (Meatco), a parastatal set up after independence, was the main formal buyer of cattle until 2001 when it withdrew from Kaokoland. Due to quarantine regulations private traders are still handicapped, as livestock has to be stationed in a quarantine camp for three months before it can be transported south of the Veterinary Fence. Cattle bought at formal auctions by Meatco are slaughtered at the Meatco abattoir in Oshakati. With 49 per cent of all animals slaughtered in Oshakati, Kunene North (the former Kaokoland) is the major supplier of meat to the abattoir (Talavera et al., 2000:271). In

Okangwati (a village of about 500 inhabitants with school, police, health station and shops) the auction place nearest to the community I worked with, some 161, 838, 587 and 602 cattle were sold in the years 1995 to 1998 (Talavera et al., 2000:271-273). These are comparatively low off-take figures given an estimated number of 20,000 cattle in the auction's catchment area. At the same time an informal meat market developed in Okangwati (as it did in other places in Kaokoland). A survey of livestock sales in Okangwati area reports that some 180 cattle were sold to traders between July 1998 and June 1999 while some 32 head of cattle were slaughtered in Okangwati. In contrast to this the 238 goats sold alive in Okangwati are matched by some 203 goats slaughtered in Okangwati's informal butcheries (Talavera et al., 2000: 264).

2.2.3. Household Economy and Pastoral Production

The economy of the pastoral and agro-pastoral people of north-west Namibia and south-west Angola has not been described in any detailed way up to now. Descriptions of the economy in older accounts (Vedder, 1928; Abel, 1954; Estermann, 1981; Baumann, 1975; Malan, 1972, 1977) are very general and lack any quantitative data on production, exchange and consumption.[26]

Household Economy

The basic unit of production and consumption is the household (sing. *onganda*, pl. *ozonganda*). Most Himba households surveyed (n=94) were rather small, the average household having some 8.5 members (range 2-32). Most households were extended. If the average Himba household consisted of some 8.5 people, members of the nuclear family made up for just 5.0 persons while 3.5 persons were relatives of some sort. The household herd usually consists of an amalgam of animals owned and borrowed by various members of the household (see Table 2).[27]

Table 2. The size of pastoral Himba households

Age of household head	No. of wives and range	No. of children and range	Total
under 35 years (n=4)	1.5 (1-2)	2.0 (1-4)	8.3 (5-11)
35 to 60 years (n=40)	1.7 (1-4)	3.2 (1-10)	8.3 (3-16)
over 60 years (n=22)	1.3 (1-2)	2.2 (1-10)	8.9 (2-32)
Total	1.5 (1-4)	2.5 (1-10)	8.5 (2-32)

Note: Relatives' children permanently residing in the household are included under "children". Permanent adult guests of the household are not included. A survey conducted in 2004 included more sedentary households in the Omuhonga basin. They showed a much higher average number of household members.

Young men start their career as herders, usually as shepherds of their fathers' or their maternal uncles. Whilst in charge of the paternal cattle camp, they will add some of their own cattle to their father's herd which they have either borrowed from relatives or were given as presents. These animals are the nucleus of their future household herd. Men marry at a late stage in their lives; most of them are about thirty by the age of their first marriage. Usually the first marriage is arranged and the preferred partner is a cross cousin of about three to ten years. This precludes a young man from setting up his own household at an early stage. He will not be able to depend on his child-wife as a source of labour for at least another decade, as she will go on living with her parents for another ten to fifteen years. Between the age of 35 and 50 men usually succeed in slowly building up their herd with animals they have borrowed or were presented with. For a long time their households are dependent on the paternal household, a fact which is underlined in ritual prescriptions. Only the paternal household has an ancestral fire and the junior household has to consult the paternal household for all matters concerning the ancestors. Herd growth and independence may be speeded up considerably if a herd is inherited from a deceased brother or mother's brother. A household stays together until its owner dies. Then the entire estate, including livestock, women and children, is inherited by one heir. The splitting of inheritance between various heirs is avoided.

Pastoral Production

The Sanga cattle breed herded by Himba pastoralists is well adapted to the arid conditions of north-western Namibia. The age at first calving is four rather than three years (mean: 50.31 months, median 48 months).[28] Intercalving intervals are quite long (mean: 20.37 months; median: 15 months). During dry years cows do not get pregnant easily and annual calving rates differ quite a lot according to fodder availability. It is not rare that cows do not calf for three or four years in a row.

Livestock produce milk and meat and are used as exchange items to barter for maize, cheap alcohol, cloth and other goods. Himba pastoralists milk cows and goats, whereas sheep are not milked. The production of milk is determined by the herder's decision as to how much milk he leaves to the calf or the kid and how much he is prepared to consign to human consumption. The data presented in Figure 1 therefore, do not represent the total milk yield but rather the amount which goes for human consumption. Most cattle calf from November to January. However, milk production only peaks some weeks after the first heavy rains when more grasses become available (usually January to April). The lactation peak is quite short and with the onset of dry conditions in May lactation decreases considerably. However, cattle are sometimes milked for more than twelve months, even if they produce minimal amounts of milk in the end. Goats have three kidding peaks per year (November, March, June) so that milk production varies over the year.

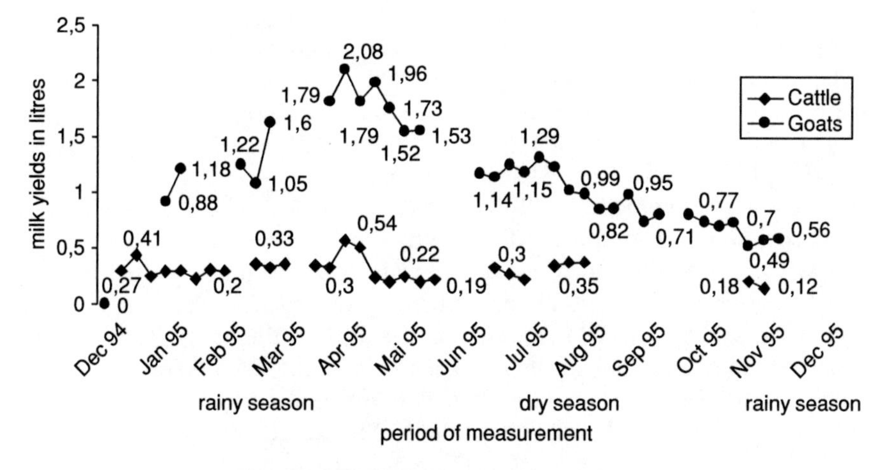

Figure 1. Milk yields in cattle and goats herds

Milk is consumed as soured milk or as buttermilk. Butter is boiled to produce clear butterfat. This product is stored in large quantities in leather sacks or in metal containers and serves as a store of proteins and fat for the meagre months of the year. A lot of it is used for cosmetic purposes as well. Goats' milk is usually drunk in a soured state and is not buttered.

Herd structures vary according to the overall wealth of the household (here the percentage of oxen is much higher) and according to the herd type. Table 3 presents figures on the structures of three different herd types. In herd type A the entire cattle herd is together, in herd type B – the household herd – the herd has been reduced as a considerable number of male and young stock have been taken to cattle camps. Herd type C represents herds in cattle camps. The figures show that the percentage of females is highest in pure household herds, whereas the percentage of oxen is highest in cattle camps. Counts in a few small stock herds suggest that in goats' herds perhaps 60 per cent to 65 per cent of all animals are females, while in sheep herds the ratio between males and females is more or less equal. Sheep are raised not for their milk but for meat.

Table 3. Structure of cattle herds (in per cent)

	Female Stock			Male stock			
	in milk	not in milk	heifer	tollie	ox	bull	calf
Type A	18.8	7.4	18.1	12.0	20.3	1.2	18.9
Type B	21.9	13.4	19.9	12.0	9.2	1.8	21.9
Type C	8.9	6.3	19.3	16.7	41.6	1.9	8.9
Average	16.5	9.0	19.1	13.6	23.7	1.6	16.6

Note: 22 cattle herds were counted

The size of herds varies considerably. The cattle herd of one household may number anything between 3 and 500 cattle and there were a few herders who were said to possess more than 1000 cattle distributed over several camps. In a survey of cattle of 36 households (comprising altogether some 4,634 cattle), rich households (n=7, 19.4%) herded some 35.6% of all cattle, medium households (n = 13, 36.1%) herded 43.2%, while poor households (n = 16, 44.4%) herded 21.3%. These figures do not yet give a clear idea of the distribution of wealth. In those herds kept by poor and medium households there are still numerous animals which are formally owned by rich herders. Rich herd owners distribute their wealth over several widely spread cattle camps. Small stock herds vary in size as do cattle herds. However, differences are not as dramatic as in cattle herds. While rich herders possess up to 400 head of small stock (approximately two thirds being goats, one third sheep) poor households will have around 100 head of small stock.

Land tenure and spatial mobility

Settlements are concentrated along seasonal river courses or at permanent wells and boreholes. Himba locate their households preferably in places where there is good access to water, pasture and arable land all at the same time. The spatial organisation of the pastoral community shows four types of utilisation zones: (I) Population centres with more than three households in the vicinity. Opportunities for gardening are found in these places and reliable water resources are nearby. (II) Villages with one to three households. These places frequently have good, but limited gardening opportunities and long lasting water resources. A specific hallmark of zones I and II is that they have graveyards in close proximity to the village which marks them as permanently occupied places. (III) A major part of the land is used by mobile cattle posts only. There are usually no gardening opportunities in these areas and water resources are limited. (IV) Grazing reserves (like the Baynes Mts.) are used during periods of intense stress when other pastures are depleted.

Usually an "owner of the land" (*omuni wehi*) can be named for places in zones I and II. He should be approached with any decision in connection to the management of the land. Land beyond the zone of settled households, (zones III and IV) is managed jointly by the men of the community. Specific settlement and gardening sites are connected with specific dry season grazing areas so that most households make use of the same dry season pastures every year. Himba pastoralism depends on independent movements of livestock camps (*ozohambo*) and households (*ozonganda*). After a few weeks of heavy rain (usually January to March) the entire household herd gathers at the main homestead in zone I or II. In an average year they stay together for three to four months while the major gardening work is done. However, a cattle camp for oxen, tollies, heifers and non-lactating stock will be established long before grazing resources become depleted around the homestead in May or June. If a household has enough herders, oxen are herded separately from young stock (tollies, heifers), as both types of cattle have different grazing requirements. Later, in July or August, male goats and sheep are separated from the household and either a separate small stock camp is established or the small stock herd joins the cattle camp. During the drier parts of the year small numbers of livestock are constantly shifted from the main household to the various livestock camps and vice versa. At the height of the dry season, between September and December, a number of households shift all their remaining cattle to their cattle camp to ensure that the cows have enough fodder when calving starts towards the end of the year. During these latter parts of the year, the household may also change its place. These moves of the main household rarely cover large distances (see Map 5).

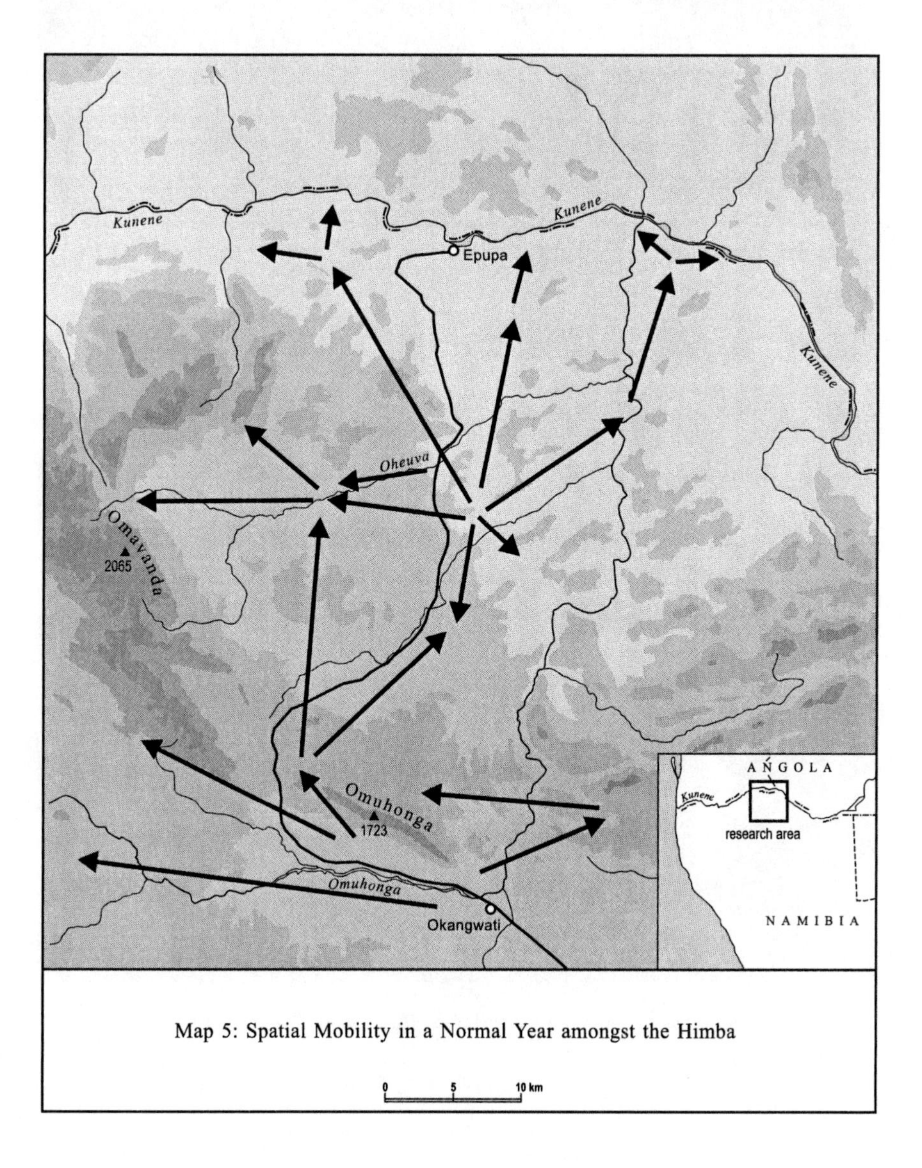

Map 5: Spatial Mobility in a Normal Year amongst the Himba

Photograph 4. A Pokot Homestead (*kaw*)

Photograph 5. A Pokot Girl grinding Hematite in front of her Hut

Photograph 6. A Young Himba Man milking a Cow

Photograph 7. A Himba Woman grinding Maize

Patterns of Consumption

Livestock is slaughtered, exchanged and sold on various occasions. The statistical analysis of a survey of about 4,600 cattle from 36 households resulted in a quantitative documentation of livestock transfers. Looking at the off-take only (32.4 per cent of all recorded animals): 35.3 per cent had died, 7.9 per cent were slaughtered, 30.4 per cent were loaned and given as presents and 16.4 per cent were sold. Previously it was said that the Himba slaughtered cattle mainly for ritual purposes (e.g. Paskin 1990). However, the main reason given for slaughtering was old age of the animal (43.3 per cent of all animals slaughtered, n=164). Ritual activities ranked only second. 37.2 per cent (of all cattle slaughtered) were killed for funerals, initiation rituals, marriage and healing rituals. While huge oxen are slaughtered at major festivities the major meat producers for everyday life were small stock. An astounding 30.4 per cent of all cattle leaving the herd are loaned or given as presents to relatives. This is more than four times the number of animals slaughtered and almost double the number of animals sold. These figures indicate that the transfer of cattle within social relations is the major form of investing livestock. Out of the 30.4 per cent only 3.4 per cent were livestock presents. About 16.4 per cent of all cattle were bartered or sold for goods (maize, liquor, medicine etc.) and money. While bartering still accounted for about 70 to 80 per cent of this type of off-take, the remaining animals were sold at auctions. However, goats and to a lesser extent sheep are the major objects for bartering for maize and other commodities.

All the Himba households surveyed over a 22 month period bartered substantial amounts of maize: the average ran at 27 sacks per household (1.2 per month). Major differences in the amounts of maize traded were due to wealth differences and differences in household size. Small stock sales (in contrast to slaughter) are highly seasonal: during the rainy season hardly any maize is bought (the Himba produce substantial amounts of maize themselves in good years) while at the peak of the dry season sacks of maize are bought in plenty. The major mode of exchange with superior markets is the bartering of livestock with itinerant traders. These traders (most of them of Ovambo origin) drive into the area with a wagon load of maize and cheap alcohol. Occasionally they will also bring other goods such as blankets, cloth, tobacco, simple tools and human and veterinary medicine. All products are bartered according to fixed customary rates. Most traders are well known in the area they trade and also hand out goods on credit.

Agricultural Activities

About 75 per cent of all households practice agriculture. The gardens are situated on the alluvial soils of riverbanks or in depositional areas (sedimentary basins). A first phase of planting takes place in November after the first rains. Planting activities follow every major rainfall and the field is only fully planted by February, at the height of the rainy season. The main crops are maize, millet and sorghum; additionally beans are planted in many gardens (van der Behrens 2004).

Migratory Work

Migratory work has never featured importantly in the Himba economy. SWANLA started its recruiting efforts in 1951 only to stop them again three years later. Until the 1970s almost no Himba men went to work on the farms or in the mines of central Namibia. The low labour migration rates make the Himba an exception within the Namibian context. Neighbouring Ovambo and Damara were engaged in labour migration at a very early stage. Around 1910 thousands of Ovambo men were already looking for work in central Namibia (Clarence-Smith & Moorsom, 1975; Notkola & Siiskonen, 2000) and contributed to the dynamics of stratification and the emergence of a labour class. Explanations as to why Himba opted against migrant labour are manifold. Many households were rich and needed all their labour to manage their herds. Then a few early migrants clearly had bad experiences with their employers. Their wages were cut arbitrarily and they had practically no chance of opposing their employer legally. After 1970 and even more so after the disastrous drought of 1981 quite a number of young Himba men sought employment in the South African army. This form of employment came to an end in 1989 when the South African army left the area. Nowadays there is a great shortage of formal employment in the region. The government acts as the main employer. Several smaller mining operations employ 10 to 50 people temporarily but also recruit staff from other regions.

2.2.4. Double-Descent and Patron-Client Networks: Social Organisation amongst the Pastoral Himba

The double descent system of the Herero and the Himba has had wide coverage in anthropological literature (Viehe, 1902; Irle, 1906; Vedder, 1928; Gibson, 1956; Malan, 1972, 1973; Crandall, 1991, 1992). The system rests on matrilineal (sing. *eanda*) and patrilineal descent groups (sing. *oruzo*).[29]

The Matriclan

The matrilineal clan is a named, non-totemic and non-residential group.[30] Due to the rule of patri-local residence after marriage members of matriclans are constantly dispersed (Malan, 1973:84; Crandall, 1991). The high rates of divorce and subsequent shifts of divorced women further contribute to the dispersal of matrilineal relatives. Individuals obtain membership to matriclans by birth and there are very few cases of adoption. Matriclans overlap into other ethnic groups, i.e. a man from the Omukwenatja *eanda* will find support not only with other Ovakwenatja within Himba society but also with

Ovakwenatja clan members in neighbouring Ngambwe, Zemba and Hakaona societies. There is no form of corporate decision making on the clan level and there are no clan leaders.

Clans are subdivided into lineages. Members of these lineages trace their descent to a common ancestress who may be rated some three to five generations ago. Matrilineal lineages are unnamed but one may refer to a unit by naming it "the house of the ancestress X within the matriclan Y". These matrilineages control the bulk of the livestock. All livestock is inherited matrilaterally, i.e. the herd of a deceased man will be inherited by his brother (of one mother) and in absence of a brother by his sister's son. Whereas the Herero and the western Himba inherit a smaller part of the herd, the so-called ancestral cattle (*ozondumehupa* and *ozomwaha*) patrilaterally and the rest of the herd matrilaterally, the northern Himba inherit the entire herd matrilaterally. Matrilineal lineages have an acknowledged leader, a senior male member. Without him no major rituals concerning members of the segment may be conducted. Although he does not infringe upon day to day decision making he has an important say in the timing of rituals and together with his peers in the allocation of livestock in inheritance transfers.

Patrilineal Groupings

Patrilineages are residential units in the sense that members of one patriline live predominantly in one region. Membership in a patriclan is obtained by birth. Women change their patriclan membership at marriage and become members of the patriclans of their husbands. The Himba patriclan is a named, non-totemic group which traces its ancestry to a remote and unknown person. Members of patriclans share a set of taboos, favourite colours for cattle, and specific ear clippings for cattle and goats.[31] Like matriclans, patriclans are unranked although some patrilineages may be regarded as dominant in specific areas: the Kapika lineage of the Ohorongo clan is politically dominant on the Namibian and the Angolan side of the Kunene basin and the Tjambiru lineage of the Oherero clan is dominant in western Kaokoland. To be dominant in this context means that the chief of the area comes from their lineage and many of their senior male members are regarded as *oveni vehi*, as owners of places. The oldest member of a lineage is in charge of the ancestral fire (*okuruwo*) which symbolises the presence of ancestors. At this fire he conducts healing sessions (*okuhuhura*) for clan members who feel sick, bewitched or just unlucky.

2.2.5. Chiefs and Councillors: Political Organisation amongst the Pastoral Himba

Early on this century Portuguese, German and South African colonial administrations established a few Himba and Herero bigmen as chiefs within their colonial set-up. The colonial chiefs and/or their headmen heard dispute cases and received their orders from the colonial headquarters. They were entitled to judge most offences and to administer corporal punishment. The Himba are politically organised into chieftaincies. The traditional set up of a few major, powerful chiefs – who were products of the early South African administration of the 1920s – was gradually changed when the administration named more chiefs during the 1960's thereby restricting the power of each single chief. Above these chiefs there is no generally accepted overarching tribal authority[32] although there was (and still is) the idea that all chiefs are under the Herero Paramount Chief. Although specific boundaries of chiefdoms may sometimes be vague, places and with them people, can generally be connected to a chief and a neighbourhood. Since independence, Nambia has become a multi-party state. Kaokoland's chiefs soon became involved in party-politics and expressed internal friction through alignment to different political parties.

How does political office and wealth run together? Wealth rankings of 49 households within a part of one chieftainship showed that the politically dominant households were firmly established in the group of very rich households. They did not only exert political influence, but were also able to substantiate their quest for power with cattle herds well above 400 heads of cattle. Up until today differences in wealth are mainly based on differences in the number of livestock a herder owns. Himba big men and chiefs have not invested significantly outside the livestock sector.

2.2.6. Death, Commemoration and Ancestor Worship: The Himba belief system

Himba religious beliefs centre around the worship of ancestors. Harmonious relations with ancestors (*ovakuru*) ensure good luck and good health to all the living of a descent group. Symbolically ancestral beliefs are condensed in rituals of commemoration at gravesides and at the ancestral fire. Physically the *okuruwo* is a fireplace framed by a half-circle of stones between the main hut and the main entrance of the stock enclosure. In commemoration rituals the symbolic relation between holy fire and ancestral graves is enforced (Bollig, 1997b). Graves are, next to and in connection with the ancestral fire, the most important places for carrying out ancestral rituals. A grave is established as a ritual site during the funeral. If a senior man dies, numerous oxen have to be slaughtered to provide skulls for the decoration of the grave.[33] After the activities directly connected to the burial have ceased, the grave "is closed" until the first commemoration ritual (*okuyambera*) takes place about one year later. The year between the burial and first commemoration is a period of mourning. The first commemoration ritual marks an important turning point in the history of the household. Not only is the mourning gear changed back to a normal gear once again, at this festivity the inheritance is also finally decided upon and the heir is formally introduced to the ancestral fire as the rightful successor.

Beliefs in a divine being are less articulated. Himba rarely evoke their God in prayers or in ritual practice. God (*Mukuru*, *Karunga* or *Ndjambi-Karunga*)[34] does not interfere much with affairs on earth and people spend little time on theorising if he created the world or not (Estermann, 1981:144). Vedder (1928:164) reports on the Herero, that the Supreme Being is not worshipped and that missionaries only became aware of this concept after they had worked for more than thirty years amongst the Herero.

In recent decades spirits have become the focus of rituals. For perhaps two or three decades spirits of deceased people mainly from neighbouring groups to the north (frequently of Kuvale and Ngambwe origin) have found their way south into regions settled by Himba. It is mainly younger people who become possessed. These spirits are neither evil nor good but may cause health problems to the person possessed if they are not made "to speak out on exactly what they want". Possessed people cannot be treated at the ancestral fire by the clan elders but need the treatment of a ritual specialist (*otjimbanda*). In lengthy sessions the *otjimbanda* tries to bring the sick person into a state of trance and then makes the spirit speak out. In contrast to these more or less harmless spirits there are the dangerous *ovirara* spirits. The *ovirara* originate from people who died through witchcraft and are now seeking revenge not only on the witch but also on other human beings. There is the belief that witches can direct *ovirara* against their enemies and thereby cause them harm. The *ovirara* cause fatal diseases and need a very specialised treatment. Interestingly it is mainly healers from the two despised minority groups, Thwa and Koroka, who are thought to be the most powerful in curing these evil spirits.

Male values of the Himba are orientated in two directions: on the one hand the ideal man is a wealthy and politically influential *omuhona*, a big man. He does not shy away from bullying others and play out his advantages as a wealthy patron. His qualities are underlined by rhetoric brilliance and a keen interest in the politics of power on the local, regional and even national level. He has to involve himself in community affairs. In many localities of Kaokoland one finds two or three rich herders quarrelling amongst themselves for dominance within the community. On the other hand there is the male ideal of a wealthy, introverted elder who lives in complete harmony with the living and the dead. Female values are mainly orientated towards the domestic sphere. Young women hope to become the first wife within an influential household. Here they will get the chance to build up their personal herd gradually and then give livestock loans to younger male relatives and their own sons. Old women are central in addressing the ancestors at the holy fire and are quite influential in domestic politics.

2.3. COMPARING HIMBA AND POKOT SOCIETIES

The following section is meant to supply an initial comparative view on both societies. While in general the level of comparison is the society, in the paragraphs on environment and history it is the wider regional ecological and economic setting.

Environmental Variables

In both cases a semi-arid climatic regime sets constraints on economic activities. However, while the rainfall in Baringo (with an average of about 600 mm) merges into a moister tropical climate, the precipitation in Kaokoland (with an average of about 300mm in the centre and averages as low as 100mm in the western parts) is akin to more desert like conditions. Astonishingly the rates of climatic variability (CV) for both systems are similar and are rated at about 30 per cent. While this CV rate conforms to the average of about 300mm measured in Kaokoland, it is abnormally high for a rainfall regime with an average of about 600mm. This indicates that in absolute terms, probably because of specific orographic features, rainfall is more unpredictable in northern Baringo than in Kaokoland.

Both habitats have been exploited by pastoralists for a long time and natural savannahs have been transformed into pastures. Annual grass species dominate over perennial species in the herb layer and a marked loss in biodiversity is observable in both instances. While the vegetation cover of the northern Baringo District has changed tremendously over the last thirty years, the environment of Kaokoland has remained more stable. Nowadays the dominating formation in northern Baringo District are Acacia dominated thornbush communities with few annual grasses and herbs while in the past the very same areas were characterised by perennial grasses with few trees and bushes. Although there have been remarkable environmental changes in Kaokoland (loss of diversity, replacement of perennials by annuals), these processes have generally not been as dramatic as in Pokot land. Devastating degradation is only found in a few places where there is a population concentration.

Historical Development of Economy and Social Organisation

The pre-colonial history of both groups could not be more different. According to oral traditions the ancestors of the Himba migrated peacefully in small groups into the sparsely populated Kaokoland. In stark contrast to this, the early history of the pastoral Pokot was marred by endemic warfare. Justifiably, one may claim that an expansionist war against the Maasai "made the pastoral Pokot".[35] The Pokot raiding economy (next to livestock, female children and women were also stolen) and the easiness with which outsiders were integrated into Pokot society made for rapid community growth.

While there is virtually no tradition that the Himba (or their ancestors) conquered land violently or organised large scale military activities against their neighbours, they repeatedly became the victims of raiders. Oorlam commandos attacked them for about three decades and finally forced the remaining population to flee over the Kunene river into southern Angola. There the refugees came into close contact with the expanding Portuguese trading system. Although most of them kept a stake in pastoralism, they worked on plantations, scouted for commercial hunters and, from the 1890s onwards, acted as mercenaries for the Portuguese colonial forces.

The Pokot, on the other hand, had very limited contact with the expanding world system in the 19th century. Major trading routes passed their area and apart from the (very) occasional commercial hunter to whom they sold some ivory, there was little commodity trade going on in the Baringo Basin. When the British finally took hold of Baringo, the Pokot had very little experience with traders and foreigners. Men still fought their battles with spears and bows, and guns were only of very limited importance, while at the same time the Himba held an astonishing number of modern guns which they made use of as mercenaries. While the Himba were fully engaged in the early colonial mercantile and predatory economy, the Pokot did not seek contact to long distance trade networks.

The British administration saw its first task as establishing peace in Kenya's northern lowlands. Until the 1950s the colonial power remained elusive amongst the Pokot and they thwarted all attempts at modernising their economy (grazing schemes, import of breeding animals, agricultural schemes etc.) and culture (missionaries, schools etc.). The colonial set up however, enabled traders to engage in barter exchange with local pastoralists and after c. 1900 the livestock-grain trade gained in importance. In contrast the South African administration exerted a much more authoritative and drastic governance over the Himba. After disarming a well-armed group of erstwhile mercenaries, a large part of the population was segregated into three distinct reserves. Borders were drawn around and within Kaokoland and these newly sanctioned inventions of western administration were strictly controlled. All types of exchange involving barter and commercial transactions were prohibited. Kaokoland remained isolated until the 1970s. Then the area became a battlefield within the Namibian War of Independence.

The Pokot were plunged into another period of open hostilities with their pastoral neighbours in the late 1960s. By the 1990s most pastoral groups in northern Kenya were very well armed. Large scale raiding has become the order of the day and influences the economic development of the area heavily. Since Namibia only became independent from South Africa in 1990, Himba history within the independent state is rather short. With independence, the commercial interests of private enterprises and of the mining and energy sector were allowed to develop strategies for the advancement of so-called backward areas.

Pastoral Production

Comparing the pastoral economy of both societies necessitates an analysis at different levels. Household economics and herd dynamics must be compared at the micro-level first.

Household Economy: In both societies the household is the basic unit of production and consumption. Spatial mobility of the household and of affiliated camps is tuned to the seasonally changing availability of fodder and water. In both societies labour is gender specific: while women are responsible for preparing food, collecting water and firewood, men control the mobility of herds and water the livestock. As in many other pastoral societies children are integrated into the work process at an early age. In both societies men over the age of 30 still participate in herding and watering livestock. However, their major task becomes the organisation of labour within the household and the management of communal resources. Himba women invest a considerable amount of time into gardening. In contrast to the Himba, Pokot rarely invested any time in agriculture until the middle of the 1990s. Pokot women spend a lot of time on transporting maize and other supplies from far-away shops to their homesteads.

The size of households in both societies differs profoundly: the average Pokot household has some 15.0 members while the average Himba household has only 8.6 members. The age of the household head is the major determinant of household size with the Pokot. The average household size rises from an average of 8.1 persons with household heads under 35 years of age to 22.9 persons with household heads over the age of 60. Himba households differ little in size according to age. Here the size of households is rather related to wealth and political power than to the age of the household head. Differences in household size are related to variations in the rate of polygamy. While Pokot men marry 2.6 (range 1-13) women on average, Himba men marry only 1.5 women (range 1-4). Divorce rates are high with the Himba and contribute to the constant rearrangement of households. In contrast to this divorce is rare among the Pokot.

The household cycle in both communities differs profoundly. While young Pokot men gain independence quite early on (between 25 and 30), Himba men obtain their independence much later. When a young Pokot man marries for the first or second time, usually at the age of 25 to 30, he frequently leaves the paternal household together with his mother and his siblings and takes the major part of his and his siblings' inheritance with him. While Himba men formally stay under paternal authority as long as their fathers are alive, they aim at gradually extending their personal share in their father's herd by borrowing animals from relatives.

Compared to Himba households the distribution of wealth amongst Pokot households differs only slightly. There is rarely a household with more than 100 head of cattle and more than 300 head of small stock. Poor herders own at least ten to thirty animals. In contrast rich Himba households own more than 400 head of cattle which are spread over several livestock camps while poor households usually do not own any cattle at all but subsist solely on small stock. In wealthy Himba households a major part of the physical work is done by workers who are usually non-Himba and come from southern Angola. They build huts and fences and dig water holes but generally do not participate in herding. Political power with the Himba always goes hand in hand with old age and wealth. These wealthy and influential men hold a considerable percentage of the regional herd and are major sources for livestock loans. Rich Pokot elders wield a certain amount of power within the gerontocratic system but beyond that they have to rely on consensus based decisions to influence resource allocation and the management of communal resources. Wealth is not instrumental in communal decision making. Even poor herders may be acknowledged for their bravery, their rhetoric skills and their capacity to handle cases.

The ecology of herds: Pokot herds are made up of cattle, goats and sheep and about a third of all households own camels too. The number of small stock is usually about three or four times higher than the number of cattle. In Himba herds, which are generally larger throughout, the number of cattle frequently equals the number of small stock. Herd structures are fairly dissimilar: while in the Pokot as well as in the Himba herds cows and heifers predominate numerically, the percentage of oxen in Himba herds is much higher: whereas Himba herds have 19 per cent oxen on average, Pokot herds have only 11 per cent.

Rates of productivity in both herd types differ according to the differences between Sanga and Zebu breeds (see Table 4). Completed fertility rates were found to be higher in Zebu cattle: the Pokot Zebu attained a completed fertility rate of 5.35 while the Himba Sanga only attained 3.49. The inter-calving intervals are shorter for Zebu than for Sanga: a 16 month inter-calving interval as compared to 20 months. On average Sanga cows only get their first calf after four years while the Pokot Zebu calves for the first time after three years. While most calves (88 per cent) in Kaokoland's cattle herds are born optimally at the beginning of the rainy reason, the calving peak in Baringo's Zebu herds is moderate: 52% of all cattle are born at the beginning of the rainy season. Himba herds are slow in growth and are furthermore susceptible to high rates of calf mortality.

The average milk production of the Sanga cow is also lower than the production of the Zebu: while the Zebu produces some 1.3 litres on average (Galvin, 1987), the Sanga produces less than 1 litre per day during a normal year. This data may underline the assumption that the Zebu is better adapted to arid conditions than the Sanga breed. However, the Sanga is extremely hardy under drought conditions. It may live on bush browse for several months and they are extremely enduring when walking long distances.

Table 4. Major variables of reproduction in cattle herds

	Completed fertility	Calving intervall (in months)	Age at 1st birth (in months)	Per cent born at calving peak
Pokot/Zebu	5.35	16	36	52
Himba/Sanga	3.49	20	50	88

Patterns of Consumption: In both societies the main capital is livestock. This may be put to use to supply the family with food (milk, meat), to engage in market transfers and to invest in social relationships. The major product obtained from regional markets is grain in both economies. The Himba slaughter more frequently and sell cattle more often. Internal livestock transactions are of enormous importance in both societies. While among the Pokot 17.3 per cent of the off-take was exchanged with other households, among the Himba a staggering 30 per cent was mainly loaned to relatives. An important difference between livestock exchanges in both societies is that livestock presents amongst the Himba are rare and the majority of cattle exchanged are transferred on a loan basis, while amongst the Pokot transfers indicate changes of ownership and use rights.

Trade and Diversification of the Pastoral Economy: Livestock trading in Kaokoland today shows many signs of the colonial past. There are almost no Himba livestock traders and most Himba rely on itinerant traders who barter commodities, food and cheap liquor for livestock. As the transaction costs of these traders are high, the overall exchange rates are generally quite bad for livestock producers. In northern Baringo District, barter trade was the major mode of market exchange until the 1950s. Since then, livestock transactions have been mainly monetary. Pokot are used to selling their livestock for money and buying the goods they need in shops. Although prices are connected to seasonal ups and downs, Pokot are regularly assured of selling their livestock at favourable rates.

The overall assumption that there has been little diversification in the pastoral sector amongst the Himba is justified as long as one looks at diversification beyond agricultural activities. The Himba make few handicrafts themselves. Even in earlier times, clay pots were bought from the Kuvale and the Thwa. Medical treatment is frequently supplied by Hakaona, Zemba or Koroka people. Neither Himba men nor women apply for ten-cent jobs such as char-coal burning or liquor brewing. However, gardening constitutes an important additional asset to Himba households. Diversification of the pastoral economy has become important for Pokot households during the last decade. While they take very little interest in gardening, they have gone for small-scale trading during the last ten years. Women brewed beer and produced maize and bean meals near water points to supply hungry and thirsty herders. A lot of the money earned goes into food. However, other consumer goods, such as plates, knives, second-hand clothes and luxury food items (rice, cool drinks) are also bought with this money. Labour migration in both societies is negligible.

Kinship, age and locality

In both societies descent groups, age groups and social networks are of crucial importance. Himba patrilineages and matrilineages are corporate units which structure economic, social and ritual life. Neither clans nor lineages are ranked. Most livestock transfers are channelled through the matrilineage. Members of matrilineages are dispersed throughout the area and constitute a widespread support network for individuals. In spite of their dispersal, members of one matrilineage accept the leadership of one senior male member. He has to be present at all major rituals and guides the inheritance of livestock. Patrilineal descent is frequently the basis for residential units and guarantees access to land. Patrilineages, too, have acknowledged leaders. These are the guardians of the ancestral fire and they conduct the major rituals at funerals and ancestral commemoration festivities. Preferential cross-cousin marriages contribute to the corporateness of descent groups.

Pokot patrilineages – there are about two hundred of them – are fragmented and small. Besides the adoption of common symbols and rules of exogamy, membership in a patriline has few implications for the individual. There are no leaders within these patrilineages as there is no common property. Livestock transfers are channelled through ego-centred networks. While marriages with the Himba follow a predictable cross-cousin pattern, Pokot men have to find their future spouses in descent groups from which none of their close relatives have already married. While descent groups are fragmented, age-sets are tribally encompassing groups amongst the Pokot. All Pokot men are members of age-sets and generation sets. These age-based groups form a gerontocratic system. Although the Himba have a rudimentary age-grade system, these sets do not feature importantly in the social organisation of the group. There is no social institution which comprises the entire tribal group.

The Quest for Power: Pokot Egalitarianism and Himba Patronage

The Pokot political system is characterised by an egalitarian, gerontocratic ideology. Age-set rituals are not only stages to reconfirm the gerontocratic hierarchy, but they are also important arenas for emphasising egalitarian conditions. The Himba, on the other hand, despite an egalitarian bias, acknowledge that some form of leadership is necessary to run community affairs effectively.

Chiefs amongst the Pokot are government personnel and have major difficulties in carrying out the orders they get from the government. The inability of Pokot chiefs to work within a system of indirect rule was noted early on by colonial officials who had selected them from a range of prominent Pokot big men. The South African administration basically did the same in the early 1920s when they chose three chiefs from a number of big men, all of whom had had prior experience as local warlords. However, unlike the British officers, the South African administration strongly supported their nominated chiefs. In 1923 they granted each chief his own reserve. All chiefs were paid modest salaries by the government. In the late 1930s they were integrated into a tribal council. The number of chiefs was expanded in the 1960s and again in the 1970s, so that by the 1980s 36 Himba and Herero chiefs were firmly established. While in Kaokoland the administration succeeded in establishing administrative chiefdoms, in northern Kenya they failed to do so and chieftaincy remained an alien, imposed institution there.

In both societies new elites have not yet developed fully. Only very recently have some rich Pokot men gone into trading. For the first time herders invested excess money in villages to build shops and small restaurants. However, compared to pastoral elites amongst the Borana of Isiolo (Hjort, 1979), the Maasai of Kajiado and Narok (Galaty, 1994) or the Orma of the Tana District (Ensminger, 1992), the Pokot elite is still unstructured and suffers from a lack of capital and experience in dealing with national institutions. Establishing good connections to national power groups seems to be much easier for the Kaokolanders than for the incipient Pokot elite. Traditional Herero and Himba leaders are well represented in Herero political circles operating on a national level. All of them have a clear idea of party politics and some are active in one of Namibia's major opposition parties. Under the guidance of non-governmental organisations Himba chiefs have visited Europe and herders of the Epupa area have established the Epupa Development Committee which controls the money paid to the community by tour-operators.

Norms, Values and Ideas on Humanity and the Supernatural

In both belief systems ideas of a supreme being are not of major importance for everyday religious life. The Pokot *Töröröt* and the Himba *Ndjambi Mukuru* or *Karunga* are evoked only under exceptional circumstances. Occasionally God may punish humans for sinful deeds – but for the Himba such conjectures are aimed rather at offended ancestor spirits and with the Pokot such assumptions tend to be vague, and misfortune is connected rather to witchcraft and personal conflicts than to an incensed divine being.

While for the Pokot ancestor beliefs are almost as vague as ideas about a supreme being, ancestral worship is the essence of Himba religion. Good relations with the ancestors ensure the well being of the living. Rituals which connect the living with the dead and ensure the ancestors veneration and thankfulness of the living are frequent. The major rituals of the Himba circle around ancestral graves and the ancestral fire within the household.

If Pokot ideas on God and ancestors are ambiguous and vague, their system of norms and values establishes a well formulated morality. Leading concepts within the normative system are solidarity, respect and bravery. Harmony should ideally constitute the emotional basis of social life within the Pokot community. An old man once put this idea into a metaphor and said "Pokot lick each other like young calves". Harmony – *kalya* – is a leading cultural theme that is emphasised in daily life as much as in ritual. Respect and discipline frame the relationship between people of different ages, between the sexes and between in-laws. The Himba code of honour and solidarity is less pronounced. Generosity is a major value, ideally guiding the transfer of goods, and a miser is openly ridiculed.

Both religious systems are, as yet, little influenced by Christianity. Although missionaries have been active amongst the Pokot for some decades very few members of the pastoral society have actually been converted. Pokot Christians are usually pupils and young adults in government and development organisations. Among the Himba, missionaries have only become active since 1994 and have not yet had any noticeable influence.

ENDNOTES

1. About 10,000 to 15,000 Himba are living in southern Angola. Here they live interspersed with other south-western Bantu speaking groups such as the Hakaona, Zemba and Thwa.
2. Nowadays the area is divided into three divisions: Nginyang, Kolloa and Tangulbei.
3. East Africanist historians usually have to rely on clan histories, in order to piece together tribal or regional histories. Histories of the Kikuyu (Muriuki, 1974) or the Giriama (Spear, 1978) and ethnographic descriptions of Rendille, Somali, Boran interactions (Schlee, 1989) rely on the comparative analysis of numerous descent-group related histories. Osogo (1970) describes the "clan history approach" as a basic principle for historical research in East Africa's stateless societies.
4. Vansina (1985:180) discusses the relevance of age-set lists for the reconstruction of chronologies on a general level. Jacobs (1968) uses age-set lists for dating events and processes in Maasai history and Spencer uses similar lists for the writing of Samburu and Njemps history (Spencer, 1965, 1973, 1998).
5. This information is confirmed by Peristiany's (1954) report on the clan based social organisation of the Cheptulel Pokot.
6. The assimilation of non-Pokot even went so far that, for example, non-circumcised Turkana men and women were circumcised according to Pokot tradition when integrated into Pokot society.
7. Neighbouring groups such as the Karimojong, Turkana, Mursi and Dassanetch have the same custom of honorary scarifications (e.g. Almagor, 1979; Turton, 1979; Fukui 1979).
8. *Suk* is an earlier name for Pokot. The Pokot interpret it as a derogatory term and attribute it to haughty Maasai ridiculing the Pokot. In early colonial papers the term *Suk* is frequently used.
9. Kenya National Archives (KNA) DC BAR 1/3 Annual Report 1939 "Native Attitude towards Rehabilitation".
10. KNA DC BAR 1/5: Annual Report for 1959.
11. In total 12 camel herds, 12 cattle herds, 16 goat herds and 16 sheep herds were recorded.
12. Progeny histories for all four types of livestock were obtained. Each herder was first asked to name as many mother-animals in the herd as he liked. Then the informant was asked to list the offspring of each animal and state whether they were still in the herd or not. If they had left the herd, the fate of each animal was elicited: did it die of disease or drought, or was it slaughtered or exchanged?
13. This 20 per cent include a smaller number of animals which were stolen. The survey took place some 6 years after major inter-ethnic hostilities had ended and about a year before they resumed so that the number of animals reported as stolen in this survey was rather small. Today this number would be much higher.
14. The "cattle complex" stereotype has frequently been criticised. While Herskovitts (1926) was only saying that East African herders belonged to a complex of societies which stood out for the overall importance which they attributed to cattle in social transactions and

rituals, later epigones used the concept as if the people themselves had a complex, i.e. a somewhat abnormal close relation to livestock. This “complex” was thought to be the cause of herders’ reluctance to sell cattle or to slaughter them for non-ritual reasons.

15. Harambees (literally translated as “let us work together”) are a more or less indirect form of taxation. Originally the late President Kenyatta founded Harambees as a means to generate money from citizens without taxing them directly. In recent years the pressure to contribute has risen: the President would ask the Member of Parliament for each area how much his region could contribute for a specific goal. Each parliamentarian would then promise a certain amount of money and see to it that his constituents contributed enough. Fund raisers (who are, in many regions, local policemen) are then sent to all households to gather the contributions.
16. In this respect other Kalenjin groups are different. Among the Hill Pokot and Marakwet, circumcision-based initiations into the generation-set system take place approximately every seven years (see Peristiany, 1951; Kipkorir & Welbourn, 1973).
17. The current chiefs in Nginyang division have previously been a prison warden, pastor, teacher, employee of the veterinary department, employee of the ministry of agriculture, commercial fisher-man at Lake Baringo.
18. Van der Jagt’s (1989) study of Turkana religion is one of the few detailed accounts on the religion of East African pastoralists. In his detailed account he stresses ambiguity and vagueness in terms and concepts. Concepts like *akuj* (God) and *ekipe* (spirit) have many different layers of meaning. Van der Jagt sees the essence of Turkana beliefs embodied in communal rituals.
19. Edgerton (1971:117) emphasises the high status Pokot attribute to physical beauty. However, if physical beauty was the only thing a man had to offer he would be scorned and ridiculed. His personal pride should be accompanied by a dignified, generous and respectful behaviour (*anisa nyo pö ötop*).
20. Estermann (1981, see also Malan, 1974) lists several Otjiherero speaking groups living in the area: the Himba, the Ndimba, the Tjavikwa, the Hakaona, the Zemba, the Thwa and the Kuvale.
21. For Opuwo the following figures are given: Hot season/December: Mean Max. 34.1 Mean min. 14.9 Mean 24.5; Cold Season/June: 26.6, Mean Min 6.1 Mean 16.4, Mean Annual 21.5, Absolute max. 40.4, Absolute min. -4.0 (Malan & Owen-Smith, 1974).
22. For a critical summary of accounts on Herero origins see Bollig & Gewald 2000:8ff.
23. Duparquet gives *Onkouankoua* as a smaller Oshiwambo speaking group near Ombalantu. It could well be that these are the *Ovahuahua* of Himba oral traditions. Himba informants tried to pronounce Ovahuahua with the nasalisation typical for the Ombalantu dialect. Duparquet F. “Notes sur les tribus de l’Ovampo”. Archives Générales. Congregation du Saint-Esperit. Chevilly Larue.
24. Anderson (1863:271) reported that the Herero of the Kaokoland had been wiped out by Oorlam raiders in 1863, as does another early traveller (anonymous 1878:309) who described Kaokoland as deserted.
25. After the elephant populations of central Namibia and the Kalahari had been erased by the 1860s and the Herero – Nama war had made regular trade impossible, major trading companies decided to resettle from Walfish Bay to Mossamedes (Siiskonen, 1990:136ff).
26. A spate of new research is currently being conducted in Kaokoland. An overview on tendencies in recent research is offered in the contributions of Warnlof, Wolputte, Crandall, Ohta and Miescher in Bollig & Gewald (2000).
27. In a household survey conducted in 2004 the populous and more sedentary households were ennumerated comprehensively. If these households are taken into account the average household size would be much higher.
28. This data is based on the statistical analysis of a survey of about 4600 cattle from 36 households
29. Although there are named age-groups (*otjiwondo*), they do not feature importantly in Himba social organisation.

30. Malan (1973:84) distinguishes seven major matriclans and for some of them distinguishes sublines: (1) Omukweyuva: (i) Omukweyuva Woyamuzi or Woktenda (ii) Omukweyuva Woyahawari (iii) Omukweyuva Woyapera (iv) Omukweyuva Woyamutati ; (2) Omukwendjandje, (3) Omukwendata (i) Omukwendata wondjuwo onene, or Omukwendata wozongombe or Omukwaruvara (ii) Omukwendata wondjuwo okatiti, or Omukwatjitupa (4) Omukwenambura, (5) Omukwandongo or Omukwauti, (6) Omukwatjivi, (7) Omukwenatja.
31. The following example (cf. Malan 1973:94) lists the major taboos and symbols of one patri-clan, *Ombongora*: Nowadays people of the Ombongora clan live mainly in the area west and south of Etanga. Their favourite colour in cattle is *ombongora*. They do not keep black dogs. Horses of *ekondo* colour are forbidden. Their main families of ancestral cattle are *Mbamba*, *Tjivandeka* and *Handura*. They may eat all types of meat but do not eat the tissue covering the heart (*oruamba*). Furthermore, members of this lineage do not eat blood.
32. Recently, however, there was a significant shift towards more centralisation. During the debate on the disputed Epupa scheme, two main parties established themselves. While one group rallied under the label Kaoko Royal House and appointed a chief as King of Okaoko, the other side called itself Toms Royal House. The leaders of both parties were addressed as *ombara*, as kings.
33. Estermann (1981:59) refers to a case where about 70 oxen were slaughtered for their skulls at a burial.
34. In ethnographic literature on the Herero several translations are given. Vedder (1928:165) and Estermann (1981:144ff) hold that only *Ndjambi Karunga* is the right term for the divine being and that *Mukuru* denotes a venerated ancestor. Amongst the Himba I found little differentiation between both terms. While the Himba from Angola used the term *Njambi Karunga* or *Karunga*, Kaokolanders preferred the term *Mukuru*.
35. In the volumes of Fukui & Turton (1979) and Fukui & Markakis (1994) several case studies are presented that describe the role of violent conflict on the emergence of distinct ethnic groups. The most intriguing examples are the Mursi and the Bodi described by David Turton and Katsuyoshi Fukui in both volumes. For both groups one could claim that war against their neighbours caused them to arise as distinct ethnic units.

Chapter 3

Hazards and Damages

Pokot and Himba pastoralists face environmental as well as political hazards. These hazards are clearly distinguishable along the lines of duration, spatial scope and reversibility of impact. Environmental degradation, for example, is slow working and has a gradual impact but, at the same time, is highly irreversible. A drought on the contrary, is a sudden and dramatic event but rarely has long lasting effects. Long - term processes and short term events are interlinked via negative feed backs: droughts accelerate processes of degradation, violent conflicts almost inevitably result in entitlement loss. The following chapter will first outline hazards that lead to an increased vulnerability of households. Then short term hazardous events are discussed. The chapters are concluded with a comparative view on the extent and importance of specific hazards in both societies.

3.1. DEMOGRAPHIC GROWTH AND A NARROWING RESOURCE BASE

While there are many studies on population growth comparing the growth rates of states, there is very little information on growth rates in pastoral populations and on demographic growth in relation to specific production systems (Roth, 1986:63, 1994:133). A first set of studies suggests a variety of demographic patterns in pastoral societies (Roth, 1986:133; Little 1989:235; Leslie & Dyson-Hudson, 1999) and a variable impact on local resources. McCabe (1997) shows how the population of the Ngorongoro Maasai has risen constantly over the last two decades: at the same time the ratio of livestock per person has decreased massively, making the reliance on pastoralism an ever more scanty affair. However, not all pastoral populations grow at the same rate. Spencer (1973) suggested that the pastoral Samburu attained higher growth rates than the neighbouring pastoral Rendille. He connected differential growth rates to different growth rates in camel and cattle herds:

while the camel herding Rendille adjusted their population growth to their slow growing camel herds, the Samburu were able to grow quicker due to the rapid growth of their cattle herds. He suggested that pastoral populations like the Rendille control their fertility in order to adapt the growth of the human population to growth rates of livestock herds.[1] Legesse (1993:264), comparing pastoral Borana and Gabbra of northern Kenya, described that infanticide, post-partem taboos and culturally prescribed delayed marriages of women had a "sizeable impact on population and subsequently set a limit on the rate of population growth". Hypotheses claiming that population growth in pastoral societies is adjusted to herd dynamics are highly interesting within this context, but evidence on the issue is conflicting. Little (1989:221) states this hypothesis as proven in the Turkana case: "Human population structure and dynamics are linked closely with those of the livestock. Numbers and growth rates of both livestock and people must be balanced for the system of subsistence to operate successfully." Still I would maintain that there is little evidence that population dynamics and herd dynamics are actually tuned. In general it is assumed that sedentary populations grow quicker than pastoral groups. However, Brainard (1981) found that nomadic Turkana had higher birth rates than Turkana settled on agricultural projects. Pennington & Harpending (1993) show that demographic trends amongst the pastoral Herero of Botswana/Namibia have changed tremendously over the last decades due to historical conditions. While growth was low early this century, after the German onslaught on the Herero, it gained momentum in the 1980s with improved health facilities.

If we know little about demographic processes in pastoral populations, we know even less about the link between demographic growth and growth and/or intensification in pastoral economies. Ethnographers, such as Spencer and Legesse believe that pastoral populations control their fertility in order to adjust it to slow growing economies. However, pastoralists could also improve and extend their resource base (expansion, intensification) or intensify the exploitation of resources in order to allow for population increase. Boserup (1965) showed that population growth leads to agricultural intensification. Do such processes also take place in pastoral societies in response to a growing population? Based on archival sources and data on female fertility, trends in population figures among the Pokot and the Himba are followed up and the changing ratios of people to land and, probably more important, to livestock are traced.

3.1.1 Rapid Growth: Demographic Development of the Pokot

While at the turn of the century the Pokot shared the Baringo plains with Turkana pastoralists[2], they became the sole inhabitants of northern Baringo in the first decade of this century. The British created an ethnically homogenous reserve[3] expulsing Turkana households to the Turkana District and controlling the influx of traders and would-be settlers. Hence census figures on

Baringo East or later the Nginyang Division of Baringo District relate to pastoral Pokot only. Up until today, there are no other people, beyond a handful of Somali traders and a few labourers from other destinations, living in this area.

Growth Rates based on Census Figures

The annual reviews for Baringo give population figures for almost all the years between 1914 and 1947. These figures were usually based on counts for tax payments of the local population. Figures vary a lot with maximum numbers in 1925 of 7,776 people and minimum numbers in 1934 of 4,572 people. Obviously population figures changed due to different rates of taxation in neighbouring districts and drought-related migrations.[4] The rather sudden decline of the population from 1933 to 1935 (from 7,307 to 4,572) is explained by the heavy emigration of herders to West Pokot District and Turkana District where taxation was considerably lower. However, figures for the period 1914 to 1947 always remained within the margin of 4,572 to 7,776 people.

For the period after independence (1963), census-based population figures are only given for 1979 (26,881). The Baringo District development plan of 1989 gives projections for 1988 (31,741 people) and 1993 (37,238 people) (Government of Kenya, Ministry of Development and Planning, Baringo District Development Plan 1989:24). A census conducted in 1989 obviously did not lead to accountable results. Even governmental offices kept on referring to projections based on the 1979 census. A consultancy report (Ahuya & Odongo 1991), however, cites from the unpublished 1989 records: Ahuya & Odongo gives 37,000 people. While this may indicate that population figures for the Pokot were somewhat higher than estimated in the projection for 1988, they are nearly identical with the projection for 1993. A further consultancy referring to the lack of exact census data claims "the population of the study area is variously given as 37,500 to 40,000" (Saltlick 1991). Unofficial estimates in 1996 gave a population figure of 45,000 people. The Baringo District Development Plan for 1997-2001 gives 49,451 people for 1999 and 52,122 people for 2000 (McGovern 1999, 2000). It is not clear on what kind of data the figures for 1999 and 2000 are based. The national census of 1999 which was then published in 2001 gives a population of 63,659 (Government of Kenya, National Census 2001). While details of the population development may be somewhat unclear, archival sources, recent censuses and projections leave little doubt - the pastoral population of Nginyang Division has been growing rapidly throughout the century! (see Figure 2)

Intrinsic Growth Rates: Fertility and Rapid Population Growth

Figures on intrinsic growth rates were obtained through interviews with individual women. In 38 households - these were the households I had worked

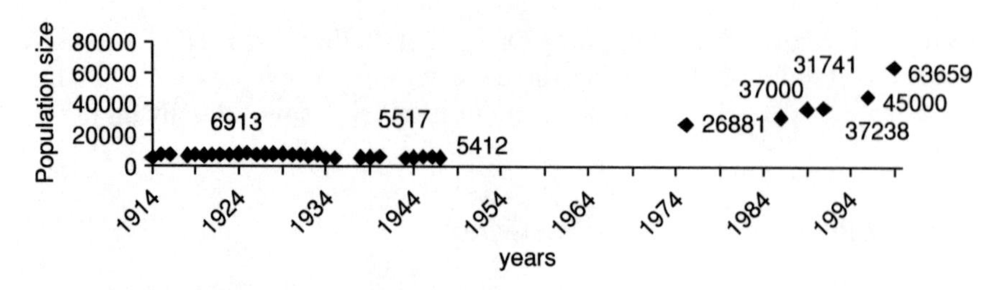

Figure 2. Population figures of the Pokot, 1914 to 1999

with over eight years - 106 women were interviewed in 1993 and 1996 on births and the fate of their children. 43 women were between the age of 15 to 30, 38 between 30 and 45 and 25 older than 45. While it is culturally not appropriate to talk or even to mention dead people, the long duration of contact with the respective households and a compact data frame on household membership (household censuses on varying numbers of households had been taken in 1987, 1988, 1991, 1992, 1996) helped to fill gaps and make it easier to talk about deceased people. Other difficulties were presented by the fact that hardly anybody knew his or her exact age. As women from distant communities are preferred marriage partners, the age of a woman could rarely be determined by discussing the life history of her mother. For computing the intrinsic fertility rate, age at marriage was assumed as taking place at seventeen years of age. This assumption is warranted on the basis of ethnographic observations.

A first glance at the data shows that Pokot women bear many children. On average the completed fertility rate for Pokot women was 4.7 children. The mean interval between births was 3.3 years. Young Pokot women are circumcised shortly after their first monthly period and usually marry soon after. The aim of both spouses is to conceive the first child as soon as possible. Premarital pregnancies are not rare and are hardly commented on negatively. Hence, the mother's age at her first birth is usually between 16 and 20. The computation of female fertility on the basis of all the samples resulted in a growth rate of 2.4 per cent per annum.[5] This figure is not very high compared to the overall Kenyan level which was at 3.8 per cent per annum throughout the 1980s and at about 3.4 per cent in the 1990s. However, given a static or even dwindling resource base, this growth rate must be regarded as very high.

Population Growth and Resource Base

The census figures indicate that the population density rose from about 1.2 persons/km^2 in 1914 to 8.5 in 1993 and probably to 14.5 in 1999. Compared to other East African pastoral groups, this population density is overwhelmingly high: the highly arid districts of Turkana and Marsabit had densities of 1.7 and 0.4 respectively in the 1970s, and even the semi-arid south Ethiopian highlands inhabited by the agro-pastoral Borana had a density of only 7 persons/km^2 in the 1980s. However, other publications on Kenyan and Tanzanian pastoralists report similar trends in population development. The Maasai of Ngorongoro (McCabe 1997) and Kajiado (Grandin 1987) and the Mukogodo Maasai (Herren 1991) have to cope with the same problem: rapid population growth accompanied by a steadily degrading resource base and a stagnant or even dwindling local herd.

Livestock numbers are given for the years 1916 to 1933. Livestock was counted in connection with tax payments. All through the 1920s and the 1930s a hut tax (based on population counts) as well as a poll tax (based on livestock counts) was taken. While some of the annually acquired numbers on livestock are rather unreliable (e.g. from 1930 to 1931 the cattle numbers rose from 114,226 to an unbelievable height of 305,054), the major portion of the data is consistent. Livestock numbers change considerably between 1916 and 1933 and range between 60,180 (1920) and 114,226 (1930).[6] The average cattle number for these years is 87,144, excluding those years with obvious miscounts. For small stock the average is 173,000 animals and for camels 1,183[7]. These figures indicate soaring livestock densities of 19.8 cattle per km^2 and 25.0 TLU per km^2 even before the 1940s in areas where a density of 5 to 10 cattle is recommended nowadays. This is enormously high and tantamount to massive overstocking. After 1950, figures on the regional herd are few: there are rough estimates for 1950 (100,000 cattle, 275,000 small stock) and for 1991 (50,000 cattle, 100,000 goats, 20,000 sheep, 3,500 camels, 7,000 donkeys) and - disputed in its accuracy - based on a cattle-plague vaccination of 1988 (73,323 cattle, no data on small stock) (Saltlick 1991:22/23). Although livestock figures have decreased, the area must still be regarded as highly stocked. The ratio of TLU/ km^2 is still at 15.4 as compared to 5 in the neighbouring Turkana District.

For the computation of person/livestock ratios the average population number for the years from 1914 to 1947 is taken, which stands at 6,380 people. This number is related to the average livestock population for these years. The ratio of person/cattle is 1:14 and of person/TLU 1:17 for this period. These figures emphasise that the Pokot were extremely rich livestock owners. Early colonial accounts contain many indications of the Pokot pastoralists' tremendous wealth. A report connected with the tax collection of the year 1911, for example, describes the "lowland sections" (i.e. the pastoralists) as "very wealthy".[8] The figures for recent years leave little doubt that these riches have dwindled considerably - almost catastrophically. Based on the projection for 1993 of 37,238 people, the person/cattle ratio has gone down to 1:1.3 and the person/TLU ratio to 1: 1.9. Given population figures of about 63,000 for the year 1999 we may safely assume that the ratio of people to cattle has dropped far below 1:1. The data suggests a dramatic decline in livestock wealth amongst the pastoral Pokot. In 2000 the Pokot were hit by a major drought which caused a high mortality in livestock herds and made the people to livestock ratios even more unfavourable.

Other Kenyan pastoralists have tried to cope with a decreasing wealth in cattle by investing mainly in small stock (Grandin, 1987; Ensminger, 1992; Herren 1991). These are able to use bushy vegetation and are highly marketable: amongst the Pokot the same trend is observable though to a lesser degree. While the ratio cattle/small stock was at 1:2 in the 1920s and 1930s, it had gone up to 1:2.8 in 1950 but fell again to 1:2.4 in recent years. Also the total number of small stock has become less over the last few decades: while 175,000 small stock were estimated to be the average early this century nowadays only 120,000 head of small stock are thought to live in northern Baringo. The significant increase in camels in Pokot herds during the last few decades - while there were only about 1000 camels in the 1930s, nowadays about 4000 camels are herded - is mirrored in increasing numbers of camels amongst other Kenyan herders (Spencer 1998:265). However, the increase in camels obviously does not compensate for the decline of cattle and small stock herds.

The Pokot rarely comment on this catastrophic development beyond stereotypic utterances that the good old days are gone. Ostensibly they talk about the riches of their ancestors. In the past many rich Pokot households used hired herdsmen to manage their herds. In the 1990s many households just herded 20 or 30 cattle and nobody in the sample owned more than 120 head of cattle and in 2004, for the first time, stockless households were recorded. It goes without saying that nobody needed to hire a herdsman for these small herds. The subsistence economy depends more and more on the exchange of male cattle and small stock for grains. Most Pokot households have to sell livestock on a monthly basis in order to acquire sufficient grains. During the dry season many households subsist entirely on store bought maize.

3.1.2. Slow Growth: Demographic Development in Kaokoland

Population statistics for Kaokoland are fairly complete over several decades. Although the population of Kaokoland was not taxed, the area was frequently toured by officials who have been counting the number of people since the early 1920s. In 1923 reserves proclaimed by the South African administration were allotted to three leaders: Muhona Katiti, Vita Tom and Kasupi. Accordingly census figures up until the 1950s relate to these reserves.[9] Until the 1950s census figures differentiated - though with a slightly changing categorisation-between the three main Herero speaking groups of Kaokoland (Herero, Himba, Tjimba). After the 1960s, the population of the region was subsumed under the term *Kaokolanders*, referring to the homeland status of the region. Figures from livestock counts can be rated as fairly reliable too. Since the early 1930s vaccination campaigns brought officials into direct contact with herds. Since the 1950s, vaccinations against anthrax and CBPP have taken place on an annual basis. After some reluctance from the local farmers, herders apparently took part in the vaccination campaigns eagerly.

Growth Rates based on Census Figures

Before 1927 all figures on the population of Kaokoland were based on assumptions and more or less well informed guess work. While Hartmann (1897, 1902/03) described Kaokoland as almost uninhabited, Vedder, some 15 years later, estimated the population to be somewhat less than 5,000 people (Vedder 1914:28). According to oral traditions major migrations to Kaokoland took place during the first two decades of the 20th century when refugees from southern Angola crossed the Kunene again to resettle their former homelands in north-western Namibia. Manning, in his report on his first tour to Kaokoland in 1917, likewise estimated population figures to be around 5,000.[10]

A first population count was conducted in 1927. In total 3,182 people were counted (see also Stals & Otto-Reiner, 1990:70/71). People were registered along ethnic lines and according to affiliation to one of the three chiefs. Of 3,182 people counted in 1927, 26.1 per cent were residing in Oorlog's reserve, 13.4 per cent in Muhonakatiti's reserve, 11.9 per cent in Kahewa Nawa's reserve and 48.7 per cent outside the reserve. The ratio between men and women was fairly equal which suggests that labour migration was negligible in those days. While only the Himba and Tjimba were counted for Muhonakatiti's and Kahewa Nawa's reserve, Oorlog's reserve was more heterogeneous with Herero and Oorlam living together with Himba and Tjimba. The Herero made for about 16.4 per cent of the total population, Himba for the majority of 62.8 per cent, Tjimba for 18.6 per cent and Oorlam for 2.5 per cent.[11] In connection with the 1927 survey, the number of livestock was rated to be 6,000 to 9,000 cattle and about 14,000 small stock. These figures apparently underrated both human population and livestock figures grossly. The police station Tjimuhaka, from where the survey was conducted, had only been established one year earlier and the patrols had more or less just taken up their duties. Only two years later some 1,200 people more were counted and the number of livestock was more than double the number assumed in 1927. A second census in 1929, in connection with the forced relocation of southern Kaokoland Herero to the central parts of Kaokoland and a dreadful drought, for which famine relief was planned, probably came nearer to the truth and noted some 4,309 people.[12] The data shows that the inhabitants of the southern Kaokoland were wealthy pastoralists. The ratio of people to cattle in southern Kaokoland stood at 1:6.5 while it was at 1:3.9 in northern Kaokoland. If TLU are used for figuring out population/livestock ratios, the difference was even more pronounced: 1:8.9 in southern Kaokoland compared to 1:4.5 in northern Kaokoland.[13]

Another two years later, in 1931, a further 300 people were counted reaching a total of 4,669 inhabitants. The colonial records kept the 1931 figure in their files for about five years. Another comprehensive enumeration was conducted only in 1942 registering 5,173 people. For some reason, people in all three reserves were suddenly counted as Himba or Tjimba and no longer as Herero (even in the late Oorlog's reserve). The category Herero was simply dropped for the three reserves but maintained for southern Kaokoland. There, people were mainly enumerated under the category Herero. The vagueness of such colonial surveys is conveyed by looking at the number of children in relation to adults in various population groups. This rate varies tremendously and variations remain unexplained.[14] Only gender categories were fairly unambiguous.

For a long time after that Kaokoland's population was not counted again. The next census took place 18 years later: in 1960 some 9,234 people were noted. These figures were used by the Odendaal report four years later, in 1964/65. Another 21 years elapsed before the next comprehensive count. 1981 figures were at 15,570 people and again 10 years later, in 1991, at 26,176. The last three counts do not differentiate ethnic affiliation. In recent times the total population of Kaokoland has been estimated to be about 30,000 people. A recent report by Talavera et al. (2000:36) gives detailed population estimates for Himba settlements only and sets the total number of Himba at 16,070. However, for accounting population/livestock ratios the overall population figure for Kaokoland is taken, as livestock counts also give figures for the entire region. It is mainly the township of Opuwo which is growing rapidly (at the moment at about 7 per cent per year). Immigrants originate mainly from neighbouring (former) Ovamboland and from war torn southern Angola.

Figures on the Intrinsic Fertility of Himba

In a survey, 78 Himba women were questioned on the number of their off-spring. All women lived in households which were known to the interviewer for about two years. Hence, there was enough trust to imbue the history of fertility within an abridged life history. Women were asked about their own date of birth, their first marriage, about divorces and remarriages and then about births. All the events can usually be related to specific years via the Himba system of naming years (Gibson, 1977). In the case where a child died early on, I inquired as to what age the child died at but did not try to discuss the possible causes of death. Women from the age of 18 to the age of 80 were included in the survey: in total information on the fertility of 140 women is available, 72 (51.4 per cent) women were between the age of 15 and 30, 30 between 31 and 45 (21.4 per cent) and 38 (27 per cent) above the age of 45.

Although quite a number of women marry early on in childhood, Himba women have their first child rather late: only two women younger than 20 had children already. The completed fertility rate was 3.2 children.[15] The average interval between births was 3.4 years. For women who were born earlier than 1951, i.e. those women whose fertile phase was already over, an average of 2.6 children (completed fertility rate) was computed. Figures on population growth amongst the Herero in Botswana point to a similarly low fertility rate (Pennington & Harpending,1993:103/130). According to

Pennington & Harpending the low fertility rates are related to the prevalence of sexually transmitted diseases (STDs) and not to specific cultural practices. Younger women reached an average of 4.9 children - this signals that the impact of STDs on fertility was becoming less due to the availability of antibiotics. However, a high number of women (about 26 per cent) did not have any children at all, probably due to chronic infection with gonorrhoea. Because of the high infertility rates the growth rate of the population was lower that 1 per cent per annum.[16]

Population Growth and Resource Base

The figures taken from the census imply that the population density rose from some 0.07 person/km^2 in 1929 to 0.53 person/ km^2 in 1991 (or 0.61 taking the estimate for 1997 of 30,000 people into account). Figure 3 shows that the population of Kaokoland has apparently increased at a more rapid pace since the 1980s.

What factors contributed to the rather steady increase in population despite low intrinsic growth rates? Two factors are rather obvious: Due to war in southern Angola numerous people from south-western Angola fled to Kaokoland. At the same time regulations for migrations between various regions of Namibia were dropped. It was mainly Ovambo speaking peoples from the more densely settled areas of north-central Namibia who took the chance to look for employment in Opuwo. During the 1980s Opuwo grew from a small administrative centre of some 300 inhabitants into a town of about 4,000 people. These recent immigrants are mainly confined to the town and only a few of them own any major numbers of livestock.

How did population growth correspond to the increase of the regional herd? The census data for 1929 and parallel livestock counts indicate a rate of people to cattle of 1:7 in southern Kaokoland and of 1:4 in northern Kaokoland. Population and livestock figures for the year 1942 are given at 5,174 people and 36,200 cattle and 56,600 small stock, which results in ratios of 1:7 (person/cattle) and 1:8 (person/TLU) respectively. It has to be taken

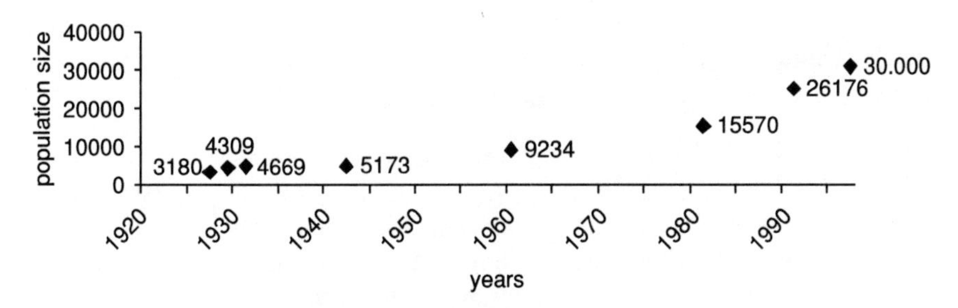

Figure 3. Population growth in Kaokoland, 1927-1999

into account that these figures relate to a year which was preceded by a dreadful drought with losses of about 20 - 40 per cent. This suggests higher overall ratios for the period before the drought. Activities of the South African administration to combat CBPP through massive vaccination campaigns and to drill boreholes and blast wells on pastures that were inaccessible before due to water shortage contributed to the increase in the regional herd. In 1960 next to 9,234 people some 65,500 head of cattle and 160,000 small stock were counted. Again these figures refer to a year which was preceded by two consecutive bad years. The livestock figures for 1958 were given as 122,495 cattle and 200,000 small stock resulting in a ratio of 1:13 (person/cattle) or 1:16 (person/TLU), falling then, due to drought related mortality, to 1:7 and 1:9 respectively (the population figure for 1960 is taken as a baseline). Within a period of about 20 years the bovine and small stock populations grew almost fourfold and much quicker than the human population. This apparently had effects on the environment. Several reports of the late 1950s comment on serious overgrazing in the central parts of Kaokoland (see next chapter).

Demographic figures and information on the regional herd are again available for 1981, once again a year that followed a dramatic drought with soaring mortality rates in cattle. Meanwhile the population of Kaokoland had grown to 15,570 people. Taking livestock figures from the pre-drought year 1980 as a baseline (about 110,000) this results in a ratio of 1:4 and 1:5 respectively. The first demographic survey for Kaokoland after independence counted 26,176 people in 1991. For the same year a number of 90,702 cattle is given. Cattle numbers rose quickly throughout the 1990s and reached soaring numbers towards the end of the decade: for 1998 some 173,473 cattle, 65,640 sheep and 40,484 goats were accounted for. These figures suggest a ratio of 1:5.8 (person/cattle) and 1:8.4 (person/TLU) respectively. If the 4,000 town-dwellers who are mainly stockless are subtracted from the computation the ratio is 1:6.7 and 1:9.7.

Over the century livestock numbers dwindled drastically due to drought and rose again rapidly during good years. Nowadays the pastoralists of Kaokoland maintain a similar ratio to livestock as they did in the 1920s. The livestock population of the area grew gradually and in correspondence with the growing population.

3.1.3. Comparative Discussion of the Pokot/Himba Demographic Trends in Relation to the Resource Base

The demographic development of the Himba and Pokot is very different. While the Pokot are a rapidly growing population the Himba are a slow growing one. The growth rate for the Pokot established on the basis of various historical censuses matches that of the growth rate deduced from figures on intrinsic fertility, i.e. intrinsic growth explains the total growth rate. In contrast to this, population development in Kaokoland and amongst the Himba has been different. While the Himba population only grew gradually, the total population of Kaokoland grew more rapidly, due to the immigration of people from neighbouring war torn areas.

Figures on the intrinsic growth of the Himba community suggest an annual growth rate of below 1 per cent. In contrast the 2.4 per cent growth rate of the Pokot is rather high for a pastoral population. The completed fertility rate for Pokot women was at 4.7 while it was 3.2 for Himba women (2.6 for older Himba women and 4.9 for younger Himba women). Only the interbirth intervals were fairly close to each other: 3.4 among the Himba and 3.3 among the Pokot.[17]

This contrast cannot be explained by cultural preferences; the Himba like children as much as the Pokot do: at this stage the high prevalence of STDs amongst Himba must be seen as a major cause for the difference. STDs were not treated in Kaokoland until the first small hospital was opened in Orumana in the 1960s. Obviously this hospital did not care intensively for STDs and had a limited spatial scope. In stark contrast to this neighbouring Ovamboland, more populous and from the beginning of colonialism regarded as the most important labour reservoir, had already received medical institutions in the 1920s. Since the middle of the 1920s regular anti-gonorrhoea activities took place with about 1500 to 2500 people being treated per year.[18] However, there are some cultural features which may add to this explanation. In Pokot society everything seems to be set on rapid growth. Infertility is seen as the greatest misfortune a person can meet. Infertile women and men become social outsiders. There are numerous rituals that are meant to ensure the fertility of humans and animals. Women and men want to have as many children as possible. The high rate of polygamy is partially explained by the wish of wealthy household heads to increase their number of children. Several institutions encourage the growth of the population. In former times children, especially girls, were stolen on raids. These children were easily integrated into the Pokot clan system and soon became marriageable spouses. Throughout the second half of the 19th century Pokot constantly absorbed major numbers of non-Pokot and integrated them swiftly into their society. Throughout the 20th century Pokot men married Turkana, Tugen, Marakwet and Samburu women; but it was very rare that a Pokot woman married an outsider. In contrast the Himba do not seem to put so much emphasis on population growth. Infertile people are not scorned and may participate in any ritual just as other people do. Infertility as such is not a reason for divorce. Himba society never expanded through incorporation of outsiders. Either one is born a Himba, i.e. one belongs to a Himba matriline and patriline or stays an outsider - which in the Himba case does not exclude one from using pastures alongside Himba herders.

Different growth rates of the population condition different degrees of pressure on basic resources. The ratio of people to livestock gradually declined amongst the Pokot. While the ratio was above 1:10 during the first part of this century, it declined to a low of 1:1 and even below that. These figures give a strong indication of the vulnerability of the Pokot pastoral system. Not only did the total number of livestock in the area become greatly reduced, but there is ample evidence for a decrease of wealth differences as well. Very rich herders with herds of more than 200 cattle are extremely rare in Pokotland and in the sample of herders that I worked with there was nobody in this category. Listening to oral testimonies and screening archival files, there is little doubt that such wealthy herders existed during the first half of the 20th century. In contrast demographic growth among the Himba has been slow and has not exerted such an intense pressure on the resource base. While livestock ratios were roughly at 1:5 earlier this century, this ratio went up and reached values above 1:10 around the middle of the century. Due to droughts, which caused excessive mortality in livestock, ratios decreased repeatedly and recovered swiftly thereafter. Towards the end of this century rates again reached a value of 1:6 to 1:7.

A further point is worth mentioning: although both groups underwent severe disasters with soaring livestock mortality, population growth was not greatly influenced by drought related human mortality. The Himba lost a large percentage of their herds in 1914/15, in 1928/32, in 1951, in 1958-61, in 1975 and 1981. While in 1915 about 50,000 people died in neighbouring Ovamboland (Hayes 1992:178ff) few people succumbed to famine amongst the Himba. Pokot pastoralism underwent similar tribulations: massive losses were noted in 1917, in 1927, 1940 and 1984. However, herders unanimously state that few people (if any) died due to outright starvation during these disasters. Livestock herds seem to buffer the human population exceptionally well against major disasters.

3.2. ENVIRONMENTAL DEGRADATION

Environmental degradation[19] has been analysed as one of the main factors leading to increased vulnerability of pastoral systems. In many volumes, range ecologists, anthropologists and development workers have discussed the detrimental effect of livestock on natural vegetation (Lamprey, 1983; Coppock, 1994; Pflaumbaum, 1994; Mainguet, 1994). The negative effects of overstocking have dominated the debate on the development of arid and semi-arid zones for a long time (for case studies see Mortimore, 1998:55ff; Gartrell, 1988; Krzywinski, Vetaas & Manger, 1996:62ff; Van Dijk, 1995:74). Recently, the ecologically-oriented volume "Range Ecology at Disequilibrium" (Behnke, Scoones & Kerven, 1993) cautioned the public against the rash conclusion that pastoralists per se damage the environment: they claimed that savannahs with rates of climatic variability (CV) higher than 30 per cent (this corresponds roughly to 300 mm of rainfall) were not driven by stocking rates but by the erratic character of rainfalls; overgrazing, i.e. degradation of pastures due to overstocking, generally occurred only in savannahs with CV rates lower than 30 per cent (or annual average precipitation higher than 300mm). Anthropologists hinted at indigenous systems of pasture protection and showed that communal pasture management and sustainable modes of exploitation were compatible (Galaty and Johnson, 1990; Fratkin 1997). They saw the failure of such systems grounded rather in the demise of communal management institutions than in high stocking rates. The two cases studied here developed different degrees of vulnerability over the last decades.

3.2.1. Degradation in the Nginyang Division

Already colonial accounts have emphasised that the pastures of the northern Baringo District were degraded due to overstocking. However, throughout the colonial period the state of range-lands in northern Baringo was never recorded in detail. The annual reports on Baringo contain some details on degradation in the Tugen hills and in the Njemps flats (Anderson 1984), but near to nothing on the Pokot plains. This lack of administrative concern is explained by the fact that all grazing schemes (registered as "land betterment schemes") were located in the southern part of the district: in 1959 all 15 projects were located in the Tugen area in South Baringo.[20] In stark contrast to administrative neglect of the topic, oral accounts treat the topic of degradation quite widely.

Oral Accounts on Degradation

Pokot discuss environmental processes in a detailed way. Like every disease (Nyamwaya, 1987), degradation is dealt with on two levels: it has a physical face (specific plants are lost and/or substituted by others of lesser value) and a supernatural cause. In several interviews with older men of the community I tried to elicit processes of environmental change. The following is a summary of an interview with Todokin, an elder of about 75 years in 1993:[21]

> When Todokin was a herdsboy, the mountainous parts of Ribkwo were still covered by forest: *cheptuya* (Euclea divinorum, E. racemosa), *kerelwö* (Croton dichogamus), *siyoyowö* (Ficus dekdekana), *rotï* (Kigelia africana?), *tuwit* (Diospyros scabra), *chowoway* (unspecified), *koloswö* (Terminalia brownii) were the dominant trees, bushes and grasses. The plains below were covered by grasses such as *seretion* (Cynodon nlemfuensis), *puressongolyon* (Aristida mutabilis), *churukechir* (Eragrostis superba) and *adwariyan* (unidentified). From the hill of Parpelo one could view the grass-plains down below up as far as the Kerio river.
>
> The vegetation started to change in the following way: one day an old man realised that his wife had committed adultery with a much younger man. The old man said to himself that it was just the abundance of milk which caused these young men to commit adultery with the wives of their elders. And in the end it was the abundance of grass that caused the milk to be in plenty. The young man was caught and beaten terribly. He had to hand over an ox to his seniors. However, the old man was still furious, he was angry with the juniors, he was angry with the abundance of milk and he was angry with the plentiful grasses. So he cursed the

plains and the grasses. Within one year the plains started to change into a dense bush and after three years the bush had become almost impenetrable. Lions, hyenas and leopards entered the bush and devoured livestock. Elephants and buffaloes came too. The grass plains vanished, all became one big dense bush.

All places changed for the worse: Moruase hill was covered by a splendid grass called *chaya*. It was replaced by a bitter grass called *chemnganya* (Cymbopogon caesius). *Chemnganya* covered the entire mountain. Due to burning of the sacred mountain Paka - burning had been prohibited hitherto - that mountain deteriorated too. Due to burning the soil became bad. While *chaya* (Sehima nervosum) and *churukechir* (Eragrostis superba) dominated Paka's pastures in the past, nowadays *chemnganya* has taken over there too. *Chelowowis* (Aristida adscensionis) and *puressongolyon* (Aristida mutabilis) are still there but they have become very few in numbers. Only *seretion* (Cynodon nlemfuensis) is still abundant and of course, *chemnganya*, is becoming ever more. *Chemnganya* even grows when there is little rain.

All bushes are increasing, especially those that are not good for animals. *Pelel* (Acacia nubica) has increased everywhere - however, there is something good about *pelel*, around its roots some grasses grow and the ants (*togh*) which eat the grass do not like it because they are repelled by the scent of the bush. *Pelel* is good fodder for camels and goats. The tree that is eating the grass is *panyarït* (Acacia reficiens), *panyarït* bushes become dense and eat the grass. It is increasing rapidly and is eating the grass everywhere. There is no livestock species that eats *panyarït*, it is completely useless. Consequently its growth is not checked and the seeds are carried away easily by the wind.

At the same time water resources dwindled rapidly. Naudo always had some pools. They are not seen anymore. The dry season only lasted two months - *mu* (January) and *terter* (February). Rains started in p*oykokwö* (March) already. *Kweghe* (December) still had good grass and was counted as a month of *kitokot* (late rainy season) Nowadays the entire *kitokot* is becoming dry season, *kweghe* (December), *kokelyon* (November), *tapach* (September) are now dry season and even p*oykokwö* (March) has become dry season. (Todokin, September 1993)

Todokin's account on degradation represents the commonly accepted idea on the causes of degradation. However, there are alternative presentations. These describe the physical changes of the landscape in a similar way but give different explanations for them. Tochil, who is much younger than Todokin, related the rather sudden decline of vegetation to the disappearance of the fabulous rain people.

In the year of Lokïsïsyö, which had a bad drought, Paka still had grasses which are no longer there. During these days the *Yilat* (rain spirits) were still living at Paka. These *Yilat* behaved like proper human beings. If you were near them you could hear the noises of cattle and goats and their bells, you could hear ringing. You could climb Mount Paka and see them high in the upper regions of that mountain. One could not observe them

> for long. Soon after you had spotted them your eyes would become dim for some seconds and when you opened your eyes again after some rubbing the *Yilat* were gone. The *Yilat* left Paka and with them the power of the steams at Paka decreased. This is why the grazing at Paka has become bad. During these good times the main grasses were *koserinyon* (probably Triumfetta rhomboidea), *chaya* (Sehima nervosum), *nyeswö* (Loudetia flavida), *kericheyan* (unidentified), *pekonyon* (Chloris virgata or Heteropogon contortus). There were only these four grasses - there was no *seret* (Cynodon nlemfuensis) and no *churukechir* (Eragrostis superba). The grasses which have taken over since the *Yilat* left are *chemnganya* (Cymbopogon caesius), *seretion* (Cynodon nlemfuensis), *chelowowis* (Aristida adscensionis), *puressongolyon* (Aristida mutabilis). Another grass which has taken over Mount Korossi is called *chepkomötyan* (probably Aerva lanata). It now starts to invade Paka too. In earlier days all these rocks were covered by grasses, only some few stones were to be seen. Now the entire mountain seems to be covered by stones. (perhaps an indication of soil erosion! add of author) In earlier times there were very few trees. Now everything has changed. And these changes started during the drought of Cheruru. In the times the *Yilat* still inhabited the upper slopes of Mount Paka, thunderstorms lasted for several hours. Now these thunderstorms are short-lived events of a few hours.
>
> On the plains between Paka, Lake Baringo and Nginyang there were hardly any trees. The plains were covered by grasses such as *amirkwayan* (Cenchrus ciliaris or Setaria pumila), *churukechir* (Eragrostis superba) and *amosongwa* (unidentified). There was *seretion* (Cynodon nlemfuensis) up to Chemolingot. From Silali to Chemolingot to Loyamuruk all was one rolling grass plain. The pools in Nginyang river lasted all year round. Even Mount Korossi was covered only in grass. There were no trees. The very first *panyarït* (Acacia refisciens) and *pelelay* (Acacia nubica) bushes were coming up at Lokenoi and down at Nginyang, then covered the areas towards Cheptaran and Chemöril. Only a few years ago these bushes started to invade Loyamoruk and places on Paka like Adomeyon. (Tochil, September 1993)

Oral accounts leave little doubt that the vegetation cover of the Nginyang Division has changed profoundly over the last few decades. Elders enthusiastically reported on the prior state of the Pokot savannah: the plains were covered with grasses and only a few trees marked the landscape. According to their presentations grasses were of high quality. Dominant grasses on the plains were Cenchrus ciliaris, Eragrostis superba, Sehima nervosum, Sporobulus cordofanus, and Aristida mutabilis. On hilly pastures grasses such as Euclea divinorum and Croton dichogamus were prominent. Hilly pastures showed a variety of bushes and trees. When Pokot talk about these pastures, one may gain the idea that pastures in the 1930s and 1940s were in a prime state. This is not the case: the landscape the Pokot refer to as prime grazing was already a pasture in a degraded state (communication A. Linstädter). A type of grass, such as the highly valued Eragrostis cilianensis, is an indicator of overgrazing and can be regarded as a ruderal weed

(Bollig & Schulte, 1999). The same holds true for Aristida mutabilis. Other grasses however, such as Sehima nervosum and Cenchrus ciliaris are, in fact, high quality grasses. Hence we may conclude that the landscape which Pokot traditions refer to already showed signs of overexploitation but was still a diversified and very productive grassland.

More visible and detrimental changes have taken place since the 1940s and 1950s after at least 40 years of massive overstocking. The tribal reserve of 4,400 km^2 had been grazed by about 90,000 cattle since at least 1910 (and probably longer than that). Oral traditions suggest that by the 1950s a change was noticeable everywhere. High quality grasses were replaced by hardly palatable grasses such as Cymbopogon caesius which soon started to cover the best pastures in the plains and on the hills. Other grasses such as Aristida adscensionis, Sporobulus festivus or the palatable herb Tribulus terrestris increased as well, but it is mainly Cymbopogon caesius which, in the eyes of the Pokot, symbolises the destruction of their prime grazing lands. Cymbopogon caesius is an unpalatable perennial grass which is hardly used because of its strong turpentine taste. After a fire or heavy overgrazing, it tends to recover faster than other grasses and hence invades the areas of other grasses. While it still has a low grazing value when it is young, in its mature state it is highly unpalatable. Apparently only the hardy Cynodon nlemfuensis could compete against Cympopogon caesius and still dominates some of the better pastures.

Next to a change from palatable to unpalatable grasses, bush encroachment led to a further degradation of the environment. While Acacia nubica and Acacia refisciens used to be extremely rare - many people even said that they were not seen at all - now these two bushes dominate much of the plains pastures. Especially the encroachment by Acacia refisciens has been disastrous: the bush does not only destroy grazing but in contrast to other Acacia bushes is of little use for goats and camels. Other bushes that have contributed to bush encroachment are Acacia nilotica (*köpkö*), *arekeyan* (unidentified) and Acacia mellifera (*talamogh*). With bush encroachment, trypansomiasis became more frequent. The fly Glossina transitans, which acts as a vector for the disease, finds much better breeding grounds in bushy terrain than in treeless plains (see Waller, 1990:90).

Formalised Accounts on Degradation

After collecting plant names and accounts on degradation over a period of several years, more formalised methods were employed to obtain a deeper insight into Pokot conceptualisations of environmental change. I concentrated the approach on herbaceous species and presented six informants independently with a list of 26 grasses and herbs. The names of the plants were written on small cards and reading the cards slowly to the informants, I asked them first to rank the grasses according to their quality ("What grass do cattle like most?") and then to rank the extent to which these grasses had disappeared, increased or stayed at the same level over the last few decades. After I had tried a free scaling during the first test interview, I then decided to give informants four categories of degradation to sort the plants into. These were (1) plant has disappeared and is hardly seen anymore, (2) plant has disappeared to a large extent, (3) plant population has stayed stable, (4) plant population has increased. Information on both questions was transformed into a multi-dimensional scaling (MDS) matrix (see Figure 4).

The MDS shows a relative dense cluster of plants in the upper right quadrant. These are all grasses which have a high grazing value and have disappeared to a large extent over the last few decades. Of 14 plants which were judged to have a high grazing value; 11 (or 79 per cent) were sorted into this category. The highest ranking grasses such as Sehima nervosum, Eragrostis ciliarensis, Tetrapogon cenchriformis, Cenchrus ciliaris and the unidentified *kericheyan* are almost absent from pastures in Pokot land these days. Tribulus terrestris, a palatable invader herb and the robust Cynodon nlemfuensis are the only fodder plants with a high grazing value that have not decreased. In the lower left quadrant Cymbopogon caesius stands out as a very efficient invader grass. It has increased massively and, as a perennial, has superseded annuals. Aristida adscensionis and the unidentified *tuimöt* are further grasses that have increased or stayed stable. Both have a low grazing value. The ranking shows, in a more general perspective, that grasses and herbs are on the retreat: of 26 herbs and grasses, 19 (73 per cent) were judged to have decreased over the last few decades. This observation directly matches the knowledge of western scientists: grass savannahs are transformed into bush and dwarf-bush savannahs of a more desert like type (Belsky, 1990). Figure 4 also suggests the substitution of high quality fodder grasses by grasses of lower grazing quality.

Recent State of Pastures in the Nginyang Division

In a recent report the consultancy Saltlick (1991:6) described the degraded vegetation of the savannah around Nginyang:

> "Most of the vegetation in the area can be classed as Desert Grass-Bush verging on Desert Scrub in the drier more barren areas. ...Perennial grasses have virtually been grazed out in all except the most favoured areas and even after rain annual grass flushes are limited, due to the fact that all annual grass seed that has been germinated, is eaten before it in turn is allowed to seed. ...A large portion of the project area is predominated by Acacia spp., although other shrub and bush types are present. ...This general state of the range is almost certainly due to overgrazing and the encroachment of bush which chokes out the grasses that survive the overgrazing. ...The dry season grazing areas also appear to be overgrazed at the time of the visit. At least one would expect at the beginning of the dry season for these hill areas to be carrying a good crop of standing hay. This was not the case. There is an over riding need for more care in the utilisation of the grazing resources in order to maintain good pastures."

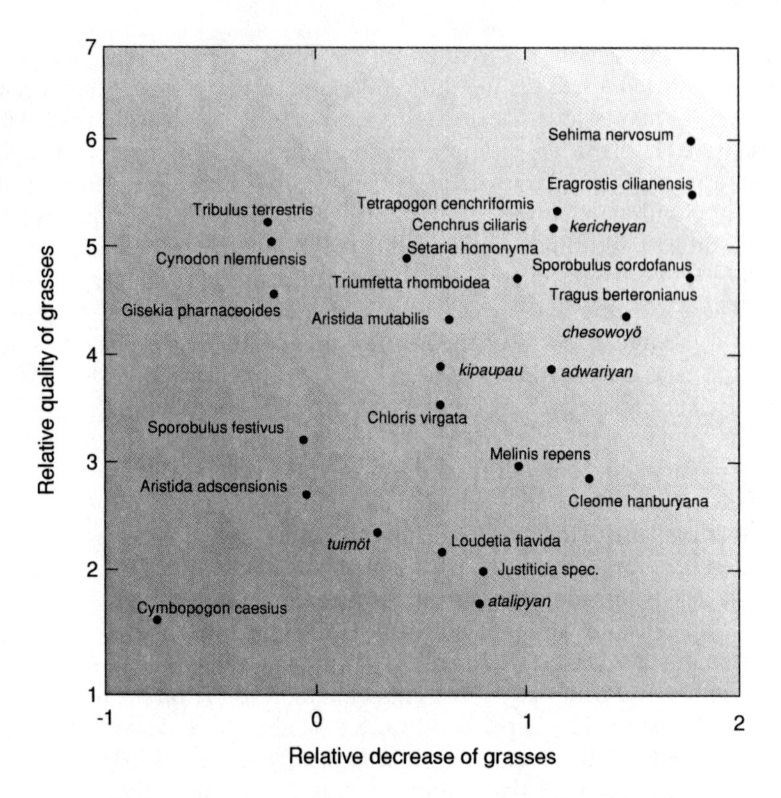

Figure 4. Grazing qualities and plant successions: an emic view

According to a range-land survey conducted by Jätzhold & Schmidt (1983) six types of savannah can be distinguished in the Nginyang Division with a total carrying capacity of 39,680 TLU. In the early 1990s it was estimated that about 50,000 cattle, 120,000 small stock 3,500 camels and about 7,000 donkeys use the area (Saltlick 1991:24).This means that about 72,000 TLU are using the area at this stage. Even if the 1,000 sq km lacking in the classification were all in the high potential area UM4, Pokot herds would be very near the carrying capacity of the area. However, it is more likely that the unclassified land is in the categories LM5 and L which means that Pokot land is still overstocked although the total regional herd has declined from about 130,000 TLU (in 1950) to about 72,000 TLU (in 1990). High stocking rates will drive on the process of environmental degradation. Even during the last ten years massive degradation has been observed: pastures which in 1988 were still grassland had been heavily encroached by Acacia nubica and Acacia refisciens in 1996.[22]

The picture of environmental change depicted for the habitat the Pokot live in is gloomy. Open grasslands have turned into thornbush thickets and the carrying capacity of the area has been drastically reduced. One consequence has been a decline in the livestock population. However, livestock figures have tended to stay higher than the steadily decreasing carrying capacity. The process of degradation will carry on and the last good grazing resources in hilly areas and along the Pokot/Turkana boundary will give way to thornbush.

3.2.2 Degradation in Northern Kaokoland

Data on degradation in northern Kaokoland stems from different sources. Just as in the Pokot case, numerous interviews were conducted on vegetational change and degradation. More formal approaches to describe degradation remained inconclusive and resulted in rankings of fodder, while the elucidation of indigenous ideas on environmental decline remained problematic for the reasons discussed below. The Cologne based ACACIA project is currently analysing the impact of heavy stocking rates on the ecosystem through on-ground botanical research and remote sensing approaches (Schulte, 2002, Casimir & Bollig 2002).

Degrees of Degradation in Different Grazing Zones

Pastures in northern Kaokoland can be broadly differentiated into five categories which show various grazing potential and are marked by different degrees of vulnerability. Major pasture categories fall into five types (1) pastures on sediments along riverine basins, (2) pastures on flat or slightly sloped stony undergrounds, (3) hilly pastures, (4) pastures on Baynes Mountains and (5) riverine gallery forest along the Kunene and its main tributaries. While pastures of type (2) and type (3) make up the major part of the land (probably about 80%), type (1) pastures are of great importance as they are more intensively used than all other types. It is this type of pasture which shows severe signs of degradation.

(1) **Riverine Basins:** Sedimentary basins along river courses have been favourite places for fairly sedentary settlements for long (Bollig & Vogelsang, 2002). Gardens are usually located along the banks of perennial rivers to make use of the fertile alluvial soils there. The population concentrates in these settlements during the rainy season when many people are busy with gardening. During these months a major part of the livestock population is herded around these settlements too. The concentration of people and livestock has a triple impact on the environment: (1) clearing of bushes, trees and herb/grass layer for gardening, (2) massive extraction of wood for hut and fence construction and (3) permanent grazing of cattle and small stock herds. The following case story of the Omuhonga basin describes particular forms of degradation and looks for causes. It represents a worst case scenario in the Kaokoland context. The following two cases show different scenarios for human exploitation of basins:

In early reports, the Omuhonga basin, which is the largest sedimentary basin of northern Kaokoland, has been described as a Garden of Eden. Kuntz described the basin enthusiastically as the "Tjimba paradise".[23] Oral accounts corroborate this emphatic description of Omuhonga's riches. Many informants described how the basin, until about twenty years ago, was overgrown with high quality grasses. Nowadays the vegetation is depleted catastrophically. There are hardly any grasses to be found in the basin, the tree/bush cover has been seriously depleted and soil erosion has taken away the better sediments. Free-standing roots of mopane trees bear witness to the tremendous scale of soil erosion. The process and degree of vegetational degradation and soil erosion has been described in Sander, Bollig & Schulte (1998) in some detail. Here only the more important aspects will be pinpointed:

(1) Heavy soil erosion is observable in most places in the basin. Almost all roots show signs of removal of 20 centimetres (on average) of topsoil. Soil removal of 20 to 50 centimetres is frequent and in some places the soil removal is up to 100 centimetres. Soil erosion in the sandy substratum of Omuhonga is characterised by sheet flooding. Hence, erosion is not restricted to the immediate vicinity of watercourses which are flooded after rains but to a much larger terrain. A horizon of cemented calcite covered with stones remains. Large parts of the seed bank have been irrevocably destroyed by soil erosion.

(2) Almost all Colophospermum mopane trees (more than 90 per cent) show marks of heavy woodcutting for hut construction and fencing. When mopane trees are cut, at least 30 per cent of the tree crown is lost; the mean loss was estimated to be about 58 per cent. In Terminalia prunioides nearly all adult individuals have been cut at least once in their life. While mopane trees are solely used for construction, Terminalias may be cut to provide fodder for young goats too. In Omuhonga, a pronounced differentiation of re-growth abilities after cutting or intense browsing was observed. Terminalia prunioides and other species suffer the most from constant wood cutting while Colophospermum mopane seedlings and juveniles are quite hardy. This has led to a high proportion of mopane in the vegetation of the Omuhonga basin.

(3) The grass cover in the basin is usually less than 10 per cent and frequently the remaining soils and gravels are completely devoid of any grasses. The species composition on tiny alluvial spots at the edge of erosion sheets still indicates the diversity of grasses which grew here only a few years ago. It is just the hardy Aristida spec and the short-lived Enneapogon desvauxii which manage to grow under these conditions over wider areas. With the loss of the sedimentary cover the basis for extensive grass growth has been destroyed. The carrying capacity of the Omuhonga valley as a cattle pasture nowadays is minimal to non-existent.

The reasons for the rather sudden decline of degradation lie in enforced changes of settlement patterns in the late 1970s and the parallel decline of precipitation. Up until then about 15 to 25 households had settled in the basin.[24] Then the South African army forced households living north of Okangwati towards the Kunene river to resettle in the Omuhonga basin near the army barracks of Okangwati, allegedly to protect them against infiltrating guerrillas. The number of households went up to about 70 or 80 households. The sudden increase of households was paralleled by a rather dramatic decline in precipitation - all the years between 1977 and 1983 were recorded as drought years. The number of households never decreased to the pre-1979 level and up until today the number of households is about 70. Several non-Himba households, namely Hakaona and Zemba, settled on the river to garden there. Their household herds are fairly sedentary and add to the problem of overgrazing.

Aerial photographs of 1975 and 1996 show the degradation of the Omuhonga settlement area clearly: along the river the bright colours indicate the absence of vegetation. Given that the LandSAT scene was taken at the end of the rainy season it is beyond any doubt that severe degradation marks the entire valley[25] (see Photographs 8 and 9).

Himba households need a tremendous amount of wood for the construction of their homestead and gardens. Just for the sturdy inner enclosures numerous trees have to be cut down. The measurement of fences on the basis of aerial photographs and ground observations suggest that about 173 trees are used to build one inner enclosure, 69 for an outer enclosure and 115 for a garden fence, suggesting that per newly built homestead on average 360 trees have to be felled.

The Omuhonga basin only makes up a minor fraction of the area. Other basins are much less degraded. The Ombuku valley for example - the

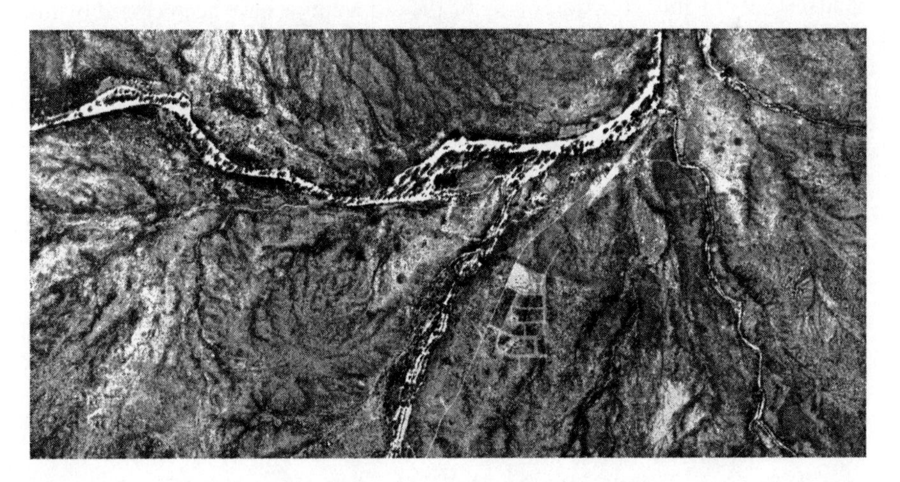

Photograph 8. Subsets of Orthorectified Aerial Photos of the Omuhonga Region (from 1975) processed by Torsten Welle

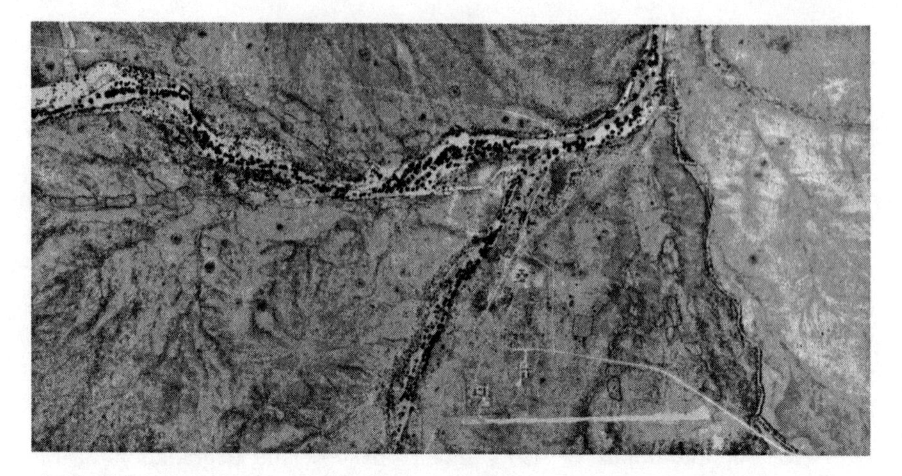

Photograph 9. Subsets of Orthorectified Aerial Photos of the Omuhonga Region (from 1996) processed by Torsten Welle

extension of the Omuhonga valley towards the Kunene - had been settled densely up until the 1940s. The Ombuku valley is much narrower than the Omuhonga valley and sedimentary deposits are restricted to a one kilometre radius of the river. Many Himba households took advantage of the close proximity to splendid pastures, good gardening places and abundant non-domesticated food and fodder resources in the earlier part of the 20th century. However, the basin was never populated to such a large degree as the Omuhonga basin has been since the 1970s. The maximum number of households would have been about 20. Since the 1950s many households have left the valley and nowadays there are usually less than five households using the basin. Many trees today show old cuttings, but the range of destruction is restricted to the immediate vicinity of the basin and many of the old wounds have been closed by the recent growth of bushes. The grass cover shows some signs of degradation: perennial grasses are rare in the immediate vicinity of the river.

(2) **Pastures on Stony Undergrounds:** There are many pastures in northern Kaokoland that are situated on flat but stony grounds. Degradation in this zone - if not situated in the immediate vicinity of homesteads - is restricted to a loss in biodiversity in the grass and herb cover as well as in the tree and bush cover. Stipagrostis spp. is dominant in the better parts of this area while pastures on calcite basements have frequently only a very low potential and are overgrown with Enneapogon desvauxii or Aristida adscenionis. The overall potential for biomass production is low and over the last decade herders marked an increase of lowly valued Aristida grasses at the cost of other higher valued grasses.

(3) **Pastures on Hill Slopes:** Many parts of northern Kaokoland are hilly and even mountainous. On many slopes, plant communities of

Commiphora trees and Stipagrostis grasses are characteristic. There has been a distinct loss in biodiversity, especially in the grass and herb layer and in more intensively used areas perennial grasses have been replaced by annuals.

(4) **Pastures on Baynes Mountains:** The Baynes Mountains are an escarpment plateau which, from a level of about 700 metres rises steeply to about 1,500 to 2,000 metres. Due to their remoteness the ecology of the Baynes Mountains has not been studied in any detail. The details given here are impressions of three tours to this mountainous area. The pastures of the Baynes Mountains look splendid: they are high-yielding and present a lot of diversity in the tree as well as in the grass and herb layer. Most grasses are perennials and of high fodder quality. The pastures of the Baynes Mountains are rarely used by large herds of cattle. There are very few signs of environmental change. Biodiversity and a cover of perennial grasses have been maintained.

(5) **The Kunene Valley:** The Kunene valley has been settled by pastoralists for a long time (Bollig & Vogelsang, 2002).[26] The pastures along the river are degraded to some extent. However, the sturdy *onenge* (Cyperus longus spp. tenuviflorus) and *otjiuu* (Phragmites mauritianus) grasses that mainly grow in the riverine inundation zone are still a reliable pasture. The banks of the river are lined with important fodder trees such as Faidherbia albida. Pastoralists use it mainly at the height of the dry season when cattle herds are driven into the riverine forest and are left there for weeks to graze freely.

Although humans have constantly taken wood from the riverine forest, there are as yet no stretches where the wood and bush layer has been completely destroyed. The constant availability of water guarantees a good rate of regrowth. However, palm trees, which have always been a major food resource for the Himba, have become severely depleted during the last decade in some stretches of the river. Poor, frequently even stockless, people settled along the river to live off its abundant resources. While living on palm nuts, they frequently also tap palms for the slightly alcoholic palm wine. In order to do so they have to chop off the crown of the palm, which eventually means the death of the tree. In more densely settled stretches more than 30 per cent of the palm trees have been destroyed.

Estimations on Carrying Capacity

Up to now there have been no estimations on the carrying capacity of the study area and figures presented here are preliminary. I estimate that per tropical livestock unit (TLU) about 15 ha of land are needed. In the middle of the 1990s there were about 12,000 cattle and about 20,000 smallstock grazing in the study area, in total about 15,000 TLU (Casimir & Bollig, 2002). These animals needed at least c. 225,000 ha of pasture (2250 km^2). The total area under consideration is 340,600 ha. If 25 per cent of the total is reduced for areas that are not grazed (i.e. the Zebra Mountains, steep slopes and highly degraded lands such as the Omuhonga basin), there are still 255,500 ha of pasture land. This would indicate that livestock numbers were nearing the carrying capacity threshold during the middle of the 1990s. Since then livestock numbers in Kaokoland have increased rapidly. Currently an estimated number of 20.000 to 25.000 TLU graze the same area, a number well above the carrying capacity for the area accounted for in Casimir & Bollig (2002). These estimations are rough and need further elaboration. We can conclude that the area has not been heavily stocked after the disaster of 1981 for many subsequent years. The carrying capacity was reached about 15 years after the disaster. Livestock numbers then rapidly transcended carrying capacity. At the moment cattle numbers in Kaokoland exceed 170,000 cattle and small stock numbers are well above 450,000. The area has never experienced such an intense grazing pressure in its documented history and environmental consequences are felt in several places.

Oral Accounts on Degradation

Oral accounts on degradation are few.[27] Most informants stated that there has been less rain in recent years than previously. This impression matches with the data from rainfall stations: during the last twenty years rainfall has dropped by about 20 per cent below the long term average (Bollig & Schulte, 1999; Sander & Becker, 2002). People attribute changes in vegetation to this lack of rains. In general informants were of the opinion that vegetational change is solely dependent on the rains. As soon as rains fall again abundantly, grasses will grow. There are only a few areas that the Himba consider as massively degraded. Omuhonga is one of these areas: here the more valuable grasses have disappeared and the bush and the tree cover have been destroyed to a large extent. However, the vegetational change around settlements is seen as an inevitable consequence of livestock husbandry.

Six informants were asked to rank 29 grasses and 15 trees and bushes according to their fodder value. In a second step they were asked to rank the same plants according to degradation. While the Himba found very little problem in ranking plants according to their value for livestock economy, all the questions on ranking them according to degradation remained inconclusive. Informants maintained that the growth of grasses was entirely dependent on the abundance of rainfall in any one year. High ranking grasses, such as Schmidtia kalahariensis, Enneapogon desvauxii (Microchloa caffra) and various Eragrostis species, are still prevalent in most grazing habitats of northern Kaokoland. A major replacement of high quality grasses by lower quality grasses had apparently not taken place on a broad scale in the mid 1990s.[28] This does not preclude that locally, as described in the Omuhonga case, the grass cover has been virtually destroyed. Bushes that rank high such as Rhigozum brevispinosum, Catophractes alexandri and Terminalia prunioides are still abundant on many pastures. Only in settlement areas have their numbers decreased due to browsing.

There is little doubt that grazing over several centuries has shaped the vegetation of northern Kaokoland. A loss of biodiversity and a thorough change from perennial to annual grasses can be diagnosed when comparing rarely grazed stretches of land with more regularly grazed ranges. However, most ranges are still productive pastures. While the change from perennial to annual grasses and the loss of biodiversity must be attributed to grazing, changes in the present vegetation are driven rather by rainfall variability than by stocking densities. It is mainly the basins in which next to grazing pressure, massive wood-cutting and the complete clearing of undergrowth present further stresses.

3.2.3. A Comparison of Degradation in the Two Pastoral Areas

The differences between both areas are very obvious. While the region the Pokot live in has been degraded massively during the last few decades and has changed its character from an open grassland into a dense thornbush, northern Kaokoland has not changed to such an extent.

The carrying capacity of the northern Baringo savannahs has decreased due to a nearly complete change from grassland savannah to thornbush savannah. Unfortunately, early reports do not describe the state of the savannah north of Lake Baringo in detail and we have to rely on oral evidence to reconstruct the prior state of ranges. The grasses and trees mentioned by the Pokot suggest a savannah type as it is found in the moister highland savannahs: grasslands consisting mainly of perennial grasses with only a few trees. However, even the grass species that older Pokot informants mention for this type of savannah (which they describe as prime grazing) indicate heavy grazing. Nowadays, the vegetation of Nginyang Division is rather similar to the savannah which is found some 200 km north in the arid stretches of the Turkana region. It is suggested that the changes in vegetation have been brought about by constant overstocking, as there is no evidence for a climatic change.

In contrast to that the character of Kaokoland's savannah has not changed by such an extent over the last few decades. Kaokoland comes near to what Behnke, Scoones and Kerven (1993) describe as a disequilibrium system: after initial changes in the vegetation due to grazing (loss of biodiversity, change from perennial to annual grasses and even an impoverishment in the layer of annual grasses), environmental changes are induced rather by rainfall stochasticity than by stocking densities. It is only the continuously settled sedimentary basins that show massive degradation due to overexploitation. On other pastures vegetational change is dominated by the variability of rains.

Obviously the massive changes noted in Baringo were easier to observe and their consequences more important to the Pokot than the more subtle but less consequential changes in Kaokoland. Hence, it is not surprising that while Pokot traditions comment on environmental change at length, Himba traditions (correctly) hint at an overall decrease of rains. While in interviews in the mid 1990s Himba respondents pointed out to environmental variability in 2004 several herders connected the increase of the lowly valued *ohoke* (Aristida spec.) at the cost of other higher valued grasses to high stocking rates.

3.3. REGIONAL MARGINALISATION, EMERGENT INTERNAL STRATIFICATIONS AND THE LOSS OF ENTITLEMENTS

Sen (1981, 1985) saw famines as being caused by the loss of access to resources and decreased capacities to exchange products for food. This process he labelled "entitlement decline" which he defined in the following way: "Entitlement is determined in two ways. It consists first of an original bundle of ownership (the endowment) which accrues to the status of the individual, and second, of alternative bundles that are acquired through trade and production (exchange entitlement mapping). Famines occur through entitlement failures related either to endowment decline (e.g. alienation of land or loss of grazing rights) or to exchange entitlement decline (e.g. loss of employment, failure of money wages to keep up with food prices...)." (Sen, 1985:208f). Herders have to exchange milk, hides and livestock for cereal food and commodities and therefore their vulnerability to detrimental impacts on regional exchange systems is high. On the one hand commoditisation and increased stratification are named as the main causes for the loss of entitlements in African pastoral societies (Hogg, 1989; White, 1991; Hitchcock, 1990). On the other hand commoditisation obviously establishes new opportunities: if livestock prices are good, food can be acquired easily and major sums can be spent on other commodities. Increased stratification usually goes hand in hand with the decline of entitlements of the poorer parts of the population. In many regions of Africa rich herders have started to fence off major parts of the land (Galaty, 1994; Hitchcock, 1990; Stahl, 2000). Former communal lands have decreased rapidly due to the expansion of private and highly commercialised farms (Grandin, 1989; Krings, 1991). In Kenya the number of rich absentee-herd owners has risen steadily over the last two decades (Little, 1987:205f; Hogg, 1989). Rich livestock owners, themselves living in towns and engaging in trade or in jobs as high ranking civil servants, manage their herds via hired shepherds (Hogg, 1986, 1989; Little, 1987; Herren, 1991; Galaty, 1994). In Botswana and Namibia rich herd-owners have fenced off high quality rangelands and have invested excess capital into boreholes and dips in order to improve their assets. Poorer herders have been forced to make use of pastures of lesser quality (Hitchcock, 1990; Stahl, 2000; Werner, 2000). Farmers from densely populated agricultural areas desperately seek resources in pastoral lands. Nowadays almost 30 per cent of the population of Kajiado District, the largest Maasai District in Kenya, is constituted by agricultural Kikuyu. Other pastoral areas of Kenya such as West Pokot, Tana River and Narok have experienced similar inflows of farmers (Little, 1987:196). Governments seeking for, as yet, under-exploited resources target rural areas for major mechanised agricultural schemes that rob the local population of land (cf. Manger, 1988:169). The Tanzanian Maasai and Barabaig have lost large parts of their ranges to wheat growing farms during the last decade (Ndagala, 1991, 1994; Lane 1994). The Sudanese Hadendowa lost important grazing in the fertile Gash valley to food and cotton growing agricultural schemes (Salih, 1994). Other pastoralists lost access to prime pastures with the establishment of game parks: the Ethiopian Afar, for example, were barred from the area proclaimed as the Awash National Park (Gebre-Mariam, 1994). Maasai were denied access to the Serengeti plains in Tanzania, and to the Mara and Amboseli Parks in Kenya. Their activities were severely curtailed by regulations in the Tanzanian Ngorongoro conservation area (Galaty, 1994; Arhem, 1984; McCabe, 1997).[29]

The relations between herders and the state had already been strained in colonial times. In many parts of Africa pastoralists were barred from commercial livestock markets and in this way lost important exchange entitlements (Kerven, 1992). In order to protect the interests of white settlers, black pastoralists were prevented from bringing their cattle to well-established livestock markets (Waller & Sobania, 1994). The encapsulation and at the same time marginalisation of livestock based economies relegated pastoralists to the fringes of the emerging national economies.

In the following chapter the colonial encapsulation of two pastoral economies is described. While encapsulation was total and efficient for the Himba, for a period of about 70 years, it was a much shorter (only about 40 years) and contradictory amongst the Pokot.

3.3.1. Capricious relations: Colonial Encapsulation and Trade in Pokot Land

While the Maasai and Samburu (Waller & Sobania, 1994; Sobania, 1991) were actively engaged in the pre-colonial coast-bound trade, the Pokot apparently were reluctant to do so. On the other hand regional exchange with agriculturalists was of importance as the Pokot neither cultivated nor practised any form of craftsmanship. From about 1910 onwards, itinerant Somali traders roamed the area. Generally trade was seen to be beneficial for the area by the colonial administration but regulations were frequently changed. The change from British administration to an independent Kenyan administration altered very little in substance. The arid northern parts of Baringo remained marginal to Kenya's economy and the livestock market of Baringo was beset with imponderabilities which range from climatic variation to price instability and violent conflict.

INVOLVEMENT IN PRE-COLONIAL LONG DISTANCE TRADE AND EARLY COLONIAL TRADE

From the 1840s onwards Swahili and Arab caravaniers roamed the vicinity of Lake Baringo in search of ivory (Anderson, 1981:8). Contact with the pastoral Pokot seems to have been marginal at least until the 1880s. According to early travellers in the region the Pokot were regarded as hostile and belligerent (von Höhnel, 1892:765). While trade had a profound impact on the neighbouring Njemps community on the southern tip of Lake Baringo, it apparently had little influence on the Pokot. Only when Pokot herds were depleted by CBPP in the 1880s and then again by cattle plague in 1891, did the Pokot apparently look for ways to rebuild their herds via long distance trade. One informant reported that Pokot bartered ivory for cattle at a rate of 2 to 5 cattle for one tusk. Itinerant Somali traders replaced Swahili traders at the beginning of the 20th century. The Somali no longer sought ivory but small stock and cattle which were sold in the emerging towns of the Kenyan highlands. The British took control of the area in the first decade of the 20th century. While the area was integrated into the Uganda Protectorate in 1902, it soon (1906) became the Baringo District of the Kenya Colony. From early on the administration tried to assert some control over the trade in livestock, hides and skins - though with contradictory measures. While trade was regarded as a parasitic and sometimes even seditious activity (Kerven, 1992:17), its gradual expansion was seen as a necessary precondition for the development of the entire colony. Trade and taxation were seen as means of forcing indigenous groups to integrate themselves more fully into the labour economy of the colony.

The Pokot were taxed early on and after 1909/10 hut and poll taxes were regularly collected.[30] They were praised as regular and reliable tax payers. Their attitude towards taxation apparently stood in contrast to their participation in other government programmes. Pokot did not leave to seek waged labour and it was almost impossible for the administration to hire Pokot labourers for their stations (Tully, 1985:93).[31] Unlike other regions of Kenya, a native colonial elite did not develop amongst the pastoral Pokot. The institution of chiefs and councillors did not become the basis for an incipient internal stratification. While Pokot involvement in administrative initiatives remained tentative, trade was developing rapidly. Since early this century itinerant Somali traders have roamed the area. In a trade report in 1909 an officer described the situation: "large number of sheep & goats have been principally exported by Somalis, from Nov. 1909 - Aug. 1910 as many as 55,000 sheep have been exported by Somalis alone; Suk sell their sheep for cattle at very favourable prices."[32] Obviously the Pokot responded very well to the new opportunities offered by the Somali trade. They needed to provide themselves with money in order to pay their taxes and to buy a few commodities. It was mainly small stock that the Pokot sold to traders. This may be due to reluctance on the Pokot side to sell cattle but must also be connected to the immense difficulties of selling cattle to the highlands. A major argument in those days was that African cattle could infect farmer cattle with all sorts of diseases. Hence quarantine stations were opened up, in which all cattle, which were to be traded to the highlands, had to stay - sometimes for months. Entire districts were closed to livestock trading for years. Especially the 1920s were marked by a very restrictive handling of livestock trade from native areas. The administration tried to inhibit fixed trading settlements in the area until the 1930s in order to maintain control over traders.[33] During the first three decades of the 20th century Baringo became a periphery to the White Highlands of central Kenya and the arid northern Baringo plains became marginal within Baringo district.

While Somali traders dominated the trade in hides, small stock and cattle, Indians carried out the retail trade (cf. Tully 1985:131). Traders were looking for cheap livestock in the pastoral areas of northern Kenya. While the costs for buying livestock were low, they had to bear high transaction costs as the roads were few and bad and the major markets far away. The traders only became firmly established in the area by the 1940s. Most of the trade was conducted on a barter basis. Frequently traders would hand out goods on credit and collect the livestock in the rainy season when the animals were fat. Already during the 1933 drought, the trade network was sufficiently established to assist the Pokot to cope with famine conditions.[34] Although the administration changed the set-up for trade regulations frequently, at times favouring Somali traders at other times disfavouring them (see also Kerven, 1992: 23 on similar conditions amongst the neighbouring Samburu), the integration of Pokot herders into the regional trade network gradually grew. However, in the network they remained solely producers and did not engage in trade themselves. Since at least the 1950s it was fairly standard that Pokot families bought large amounts of grains, sugar, tea and cloth from traders. Monetary exchange gradually substituted barter trade.

THE DESTOCKING DEBATE

Pokot herders and Somali traders did not only suffer from restrictions in trade. In the 1950s enforced destocking had a detrimental impact on livestock trade resulting in a massive loss of exchange entitlements on the herders' side. The Worthington Plan of 1946 identified degradation as the major problem of African pastoral and agricultural activities in Kenya (Tully, 1985:128). Overstocking was connected to the irrational attitude African farmers had towards livestock.[35] Since 1946 the ALDEV programme (African Land Development Programme) instituted grazing schemes and organised auctions. The government instituted parastatals as the sole buyers. This, of course, was another blow for the small-scale livestock trade of local Somali merchants. Colonial environmental policy in northern Baringo entailed forced destocking rather than activities directed to the degraded ranges.

Quotas for destocking were fixed for each pastoral area and chiefs were ordered to put pressure on people to sell their cattle under-priced. Destocking became the major government activity in the northern Baringo District during the 1950s. While the 1948 report still ponders on the negligence of local herders, the 1949 report praises them for selling more stock then ever before. Of some 10,078 animals sold in the District, the Pokot sold 3,111 cattle (about 3% of the total number of cattle enumerated for the Pokot at that time). In order to encourage the herders willingness to destock further, taxes were constantly increased throughout the 1950s: from 11/-Shilling in 1947 to 19/-Shillings in 1956. The profits of this taxation were only rarely reinvested into the area. Tully (1985:149) showed that about 50 per cent of the taxes originating from the West Pokot reserve were reinvested in the highly productive commercial zone. In spite of all these initiatives, the sale of cattle from Baringo never stabilised. Destocking activities were as prone to changing government attitudes as trading regulations had been in the 1920s and 1930s. After enforced destocking resulted in massive sales in 1949, the numbers of cattle sold dropped tremendously in subsequent years. By the end of the 1950s officials decided to institute compulsory destocking. Once again this resulted in major "sales" - about 14,000 cattle were sold in one year only. However, the next year's sales decreased once again and in 1961 the whole programme was disbanded during a serious drought.

INVOLVEMENT IN THE REGIONAL TRADE SINCE INDEPENDENCE

When Kenya became independent in 1963, the more unpopular measures of the administration were immediately withdrawn. The grazing scheme programme was cancelled summarily and compulsory destocking stopped. Taxation was abolished. However, trading patterns changed only marginally up until the 1980s. The Somali took over the retail trade from the few Indians. In the 1960s and 1970s the Somali controlled the trade in livestock and commodities in Nginyang Division under similar conditions as during the colonial period. Herders came to Nginyang and sold their animals to local traders and later bought whatever they needed with the cash they had obtained.

Major changes occurred during the middle of the 1980s. Due to massive demographic growth and decreasing livestock numbers, the Pokot grew ever more dependent on exchange with the market. Trade in livestock rapidly developed. In 1988 a weekly livestock market opened in Nginyang and only a short time later a second one opened in Tangulbei. Herders frequently voiced their doubts as to the honesty of the livestock traders. However, what herders rarely took into account was that the livestock they sold had to be driven about 100 kilometres up country to the Nakuru abattoir. The livestock traders bore the risk of losing livestock on the way. Furthermore they had to link their prices with the variable pricing for livestock on a national level.

Livestock prices reached disparaging lows during periods of drought. The Pokot took many of these sudden changes in prices as the ill will and sheer profit seeking of non-Pokot traders. Only within the last ten years have a number of young Pokot men gone into the business of livestock trading. They buy livestock in the rural areas and sell them at the weekly market in Nginyang.[36]

PRICING

Two processes have been discernible since the beginning of this century. On the one hand the colonial administration was concerned with guaranteeing white settlers privileged access to the market. This entailed that the marketing of livestock from pastoral areas was strictly controlled. On the other hand the administration was interested in opening up pastoral areas for trade. Taxation was meant to induce local farmers and pastoralists to participate in the colonial economy. Pax Britannica and a growing infrastructure paved the way for traders who reached out from the centres to acquire livestock in the peripheral pastoral areas. Did this process of marginalisation entail a loss of exchange entitlements? Did the Pokot get less for their livestock in the 1990s than, for example, fifty years before? Figure 5 showing livestock/maize exchange ratios suggests that this is not the case. Figure 5 combines data presented by Ensminger (1992:70ff) and data procured from interviews in Baringo and the analysis of archival material.

Ensminger (1992:70ff) offers data on livestock prices and prices for maize and various commodities, since the 1920s. Her data indicates that in general there was no decline in cattle prices in relation to maize prices. In 1933 a herder could acquire 286 kilos of maize for an ox, in 1944 it was 273 kilos and in 1954 it was 294 kilos. In the late 1950s prices for oxen increased and up to 400 kilos of maize could be obtained for one ox. The ratio was at 341 kilos maize per ox in 1980 and 329 kilos in 1987. The trend of the

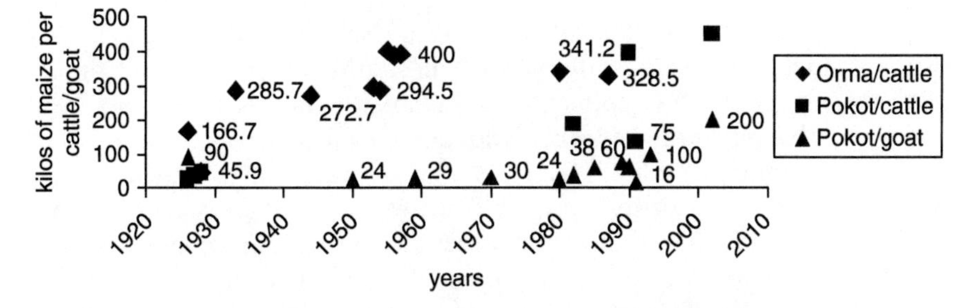

Figure 5. Livestock/maize exchange ratios in Kenya, 1926-1995

Note: The numbers presented in Figure 5 indicate the exchange ratio of livestock for maize, e.g. the number "166.7" indicates that for an average priced ox, one could obtain 166.7 kilos of maize in 1927, or the number "45.9" that for an average priced goat one could obtain 45.9 kilos of maize in 1928.

cattle/maize exchange ratio is clearly moving upwards. The Orma gained from being progressively included into the market. The data for cattle pricing in northern Baringo is not as detailed. However, the trend is similar: in general the ratio of maize for livestock increases favourably for pastoral producers (see also Dietz, 1987:127ff). The upward trend is clearly shown in the ratios for goats to maize. While in 1954 only 24 kilos of maize were obtained per goat, in 1960 it was 29 kilos, in 1971 it was 30 kilos, in the early 1980s 60 kilos, in the middle 1980s 75 kilos and in the middle 1990s up to 100 kilos for one goat. In 2002 a good goat could fetch up to 200 kilos of maize (pers. com. Sean McGovern, Kositei Mission). The relative price of goats has increased by a factor of almost ten over the course of the last 70 years. It is this positive trend in livestock pricing which has induced many herders in East Africa to expand their small stock herds.

There are two drawbacks to this general image of increasingly better pricing for livestock. During droughts the prices for livestock fall rapidly. The development of livestock pricing during the 1990/92 drought is a case in point: while before the drought almost 400 kilos of maize were obtained for an ox, only one year later a herder received just little over 100 kilos of maize per ox.

The analysis indicates that in spite of marginalisation in trade networks, Pokot herders profited from the inclusion in the national livestock market and the transfer from a barter market to a cash-based livestock market. As demand was steadily on the increase, prices increased too. Although the colonial government was concerned with a regulation of the market, it guaranteed that profits of increasing livestock prices were handed down to the producers. Pax Britannica, trade regulations and an emergent infrastructure of roads and livestock markets gradually lowered transaction costs of long distance livestock trade.

3.3.2. Unmaking a Market: The Repression of Trade in Kaokoland[37]

The history of the Himba has been characterised by several major turning points. In the second half of the last century they were systematically raided by Swartboois and Topnaar from central Namibia. Most people fled Kaokoland and looked for safer grounds in south-western Angola. Contact with the emergent settler colony there was intense. Between 1900 and 1930 many Himba returned to Kaokoland due to an increasingly oppressive Portuguese administration. There the wealthy pastoralists of Kaokoland were encapsulated in a tribal reserve. Trade with the outside was prohibited and the internal exchange of commodities discouraged. Since independence Kaokoland has been in the press time and time again: the government is planning a major hydroelectric power scheme in the area and mining prospectors are combing the area for mineral riches. Many colonial trade restrictions are still in place as the northern areas are still episodically infested with CBPP.

THE EXTENT OF TRADE BETWEEN THE 1860S AND 1915

From the second half of the 19th century onwards, two separate trading spheres were established in the lower Kunene region: on the one hand, Himba pastoralists in northern Kaokoland and the bordering areas of southern Angola were engaged in exchanges with the Portuguese colonial economy, with the Dorsland Trekker settlement at Humpata and various native communities to the north of the Kunene river (such as the Ngambwe, Thwa and Kuvale); on the other hand pastro-foragers who had remained in Kaokoland traded intensively with the western Ovambo kingdoms (see Map 6).

Trade Relations with the Portuguese: Traders had started to explore the Kunene basin since the 1860s (Estermann, 1969:68; 1981:15f; Wilmsen, 1989:96; Siiskonen, 1990:146). At the same time the two coastal towns, Mossamedes and Porto Alexandre (nowadays Namibe and Tombua), boomed as Portuguese settlers established plantations in the hinterland of Mossamedes along the Koroka river, and the fishing industry gained ground in Porto Alexandre (Esser, 1887). The flourishing trade attracted commercial hunters who roamed the lower Kunene region between the 1870s and the 1890s (Siiskonen, 1990:131).[38] Ivory achieved maximum prices on the world market in the second half of the 19th century (Clarence Smith, 1979:170). Employment with commercial hunters (both Boer and Portuguese) became a profitable activity.[39] An industrious community of Dorsland Trekkers settled in the Huila highlands during the 1880s.[40] Trading, commercial hunting and farming became the main economic activities.

The impoverished refugees who had fled Kaokoland due to frequent Nama raids were interested in exchange with the Portuguese and the Boers. Both Portuguese and Boers were in need of cheap labour as they needed scouts for hunting expeditions, workers on plantations and for the transport of goods from Mossamedes to the interior. Since 1896 native mercenaries supported the operations of the Portuguese colonial army against native groups. Oral accounts emphasise that Himba men mainly traded for cattle. Katjira Muniombara recollects traditions on encounters with the Portuguese:

> "It depended upon the whites who summoned them, who asked for young men from the homesteads to work for them, to build houses, to look after the horses and to work in the fields. There were some fields belonging to the Portuguese which were called *ozongarandja* (i.e. plantations)...This is why the young men were sought out and they were given hoes in order to plant. This is the type of work they did. There was not one kind of job, there were different jobs from several employers ... When they were paid money, they paid it back to buy the settlers' cattle, like these of the farms, they bought cattle in the farms and brought the cattle to their fathers and mothers here in Otjiku. The work was like that". (Katjira Muniombara in Bollig, 1997: 234)

In order to cope with losses due to raiding and drought they went into waged labour and ivory trade. Several accounts report that it was mainly young men

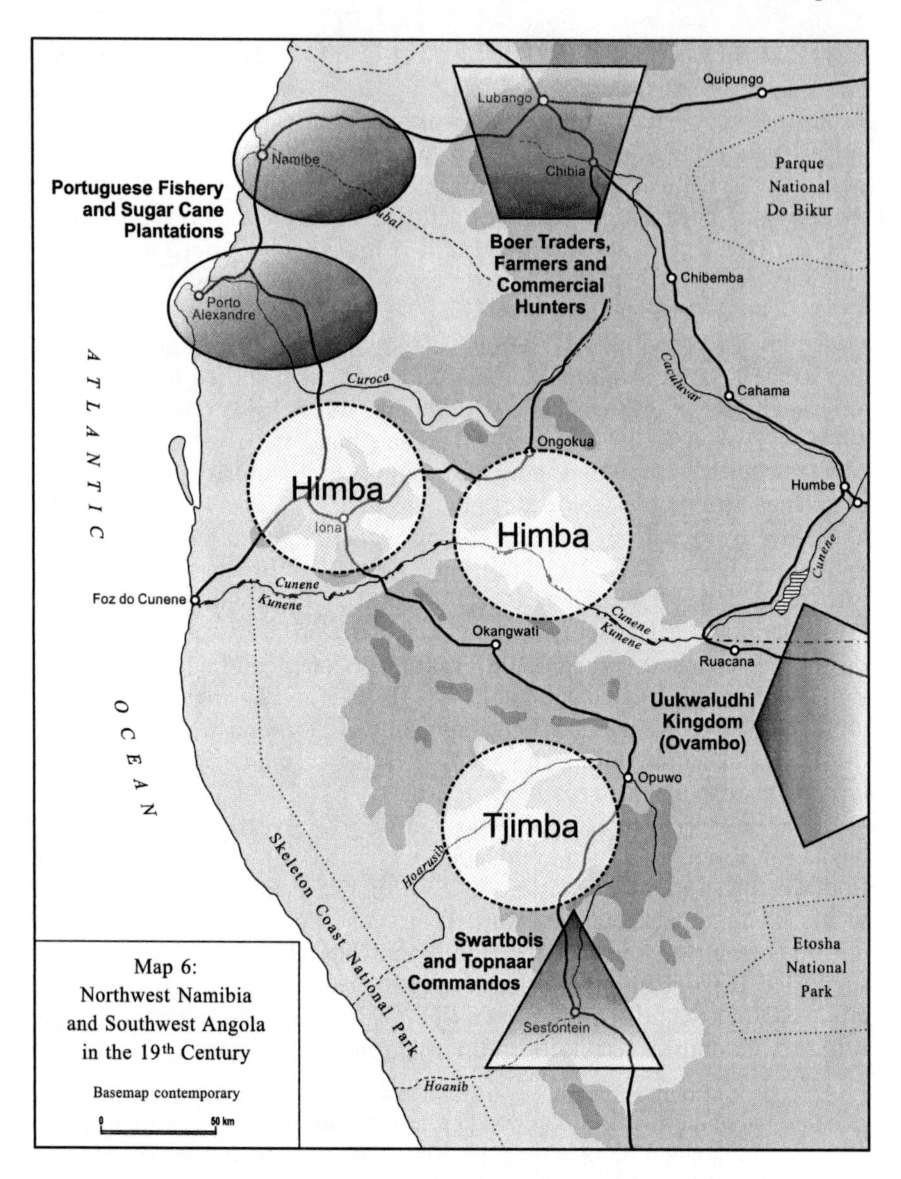

Map 6:
Northwest Namibia
and Southwest Angola
in the 19th Century

engaging in this trade. Trade then opened an opportunity to get possession of cattle at an early stage. Obviously it was mainly during times of stress that Himba pastoralists traded for maize with their agricultural neighbours. They obtained commodities and grains from exchange either for exotic commodities (such as ivory), for livestock or by offering themselves as servants on hunting trips and workers on plantations. Young men tried to build up their herds by investing salaries into the sale of livestock or exchanged oxen for heifers with Portuguese livestock traders.

The cattle plague epidemic of 1897 altered these exchange relations profoundly. Up to 90 per cent of the cattle of south-western Angolan pastoralists died (Clarence-Smith, 1978:170; Siiskonen, 1990:164; Hayes, 1992:100). Most herders were only left with small stock. The understaffed Portuguese army presented new options for work. The tribal groups in south central Angola had not been brought under the effective control of the Portuguese government by then and polities such as the Mbandja and Kwanyama kingdoms were wielding considerable power in the region. Traders had to enter into agreements with local rulers in order to obtain permission to trade in their areas or just to pass through them. From 1895 to 1906 almost every year saw at least one major military expedition against Nkumbi, Ngambwe and various Ovambo groups. Part of the loot was divided by the mercenary leaders among their men.[41] Within this early colonial setting, which was characterised by a predatory expansion of colonial interests, new forms of stratification developed among the Himba and Herero. War lords wielded significant power. They commanded armed groups of men which they hired to the Portuguese army. Oral traditions leave little doubt that mercenary leaders enriched themselves massively. The raiding of neighbouring communities at the command of the Portuguese military officials became a profitable business. The leaders of these mercenary bands formed commando-like groups with a strong leadership and a council of leading men.[42] When these groups returned to Namibia they were the owners of huge herds of cattle. Back in Kaokoland the emergent elite sought to transfer political power into economic dominance by seeking close contact to the South African government.

Trade with the Ovambo Kingdoms: A group of pastro-foragers, usually termed *Tjimba* in reports, had remained in Kaokoland in spite of Nama raids.[43] Tjimba actively traded with the Ovambo of Uukwaludhi. There they obtained iron implements for their weapons (knives, spears), their tools (axes) and their ornaments (iron beads, arm and foot rings) and seeds. They bartered these goods mainly for ivory, ostrich egg shells and herbs. Additionally some Tjimba around Kaoko Otavi were given cattle to herd by rich Ovambo. Oral traditions unanimously state that the Tjimba in those days only traded with the king of Uukwaludhi who had succeeded in obtaining and maintaining a royal monopoly on all external trade in luxury goods (see Siiskonen, 1990:191f). In contrast to the Himba living further to the north, the Tjimba only obtained a few outdated muzzle loaders from their Ovambo patrons who in turn had traded these guns from the Portuguese some decades earlier. While the Himba, who mainly traded with Portuguese traders at that time, had to offer about 10 oxen for one gun and one ox for 100 bullets (Vedder, 1914:22), Tjimba pastro-foragers obviously had little to barter with beyond their ivory to obtain guns. Exchange rates were fixed, and one old informant claimed that two heifers and two oxen were given in exchange per two tusks (Eengombe Kapeke in Bollig, 1997:202). The Tjimba of Kaoko Otavi also obtained guns on a loan basis from their Ovambo patrons. It was mainly the Tjimba who sent trading caravans to Uukwaludhi's king, while the Ovambo rarely ventured into the vast Kaokoland. These unequal trade relations implicated political dependence. The inhabitants of Kaoko Otavi for example regarded themselves as subjects of the king of Uukwaludhi. Kuntz, a German prospector, reported from a visit to Kaoko Otavi in 1909 that the Tjimba at Kaoko Otavi recognised the Ovambo king Omuaru (Mwala) as their leader "by whose command they collect ostrich eggs and shoot elephants with large ... muzzle loading rifles which he lends them".[44]

In the period between 1860 and 1920, barter obviously featured very importantly both for the Himba and the Tjimba. Both communities curbed losses within the livestock sector due to droughts, epidemics and raids by investing more energy into trade. The southern Angolan Himba were able to recover from the shock of the cattle plague epidemic fairly quickly, because they used the venues of an expanding market and opportunities for commercial soldiering. The Himba were advantaged in this respect. They had direct contact with Portuguese traders and owned enough livestock to invest several oxen into one transaction, i.e. they could afford to buy modern guns at the exorbitant price of ten oxen while Tjimba were dependent on obtaining old fashioned guns on a loan basis. Furthermore, the proximity to Portuguese and Boer employers in the Coroca Valley and the Huila highlands offered them chances of direct employment. The option to act as mercenaries for the Portuguese army opened up yet another opportunity of accumulating livestock quickly.

Trade opened new venues for restocking but it would be inadequate to glorify these forms of exchange. The pastoralists were at the far end of global trade networks. Major gains were made by merchants operating internationally and perhaps local traders. The Himba were used to exploit the area efficiently for the world market. By the 1890s elephants had been exterminated from the Kunene Basin too. Vedder remarked that the inhabitants of Kaokoland were exploited by their Ovambo overlords, by Portuguese traders and the Nama at Sesfontein. He emphasises the extractive nature of trade: "The inhabitants of the Kaokoveld are exploited by all sides, so that any development (literally up-lifting) is truly impossible. They are stripped by the Portuguese, cheated by the Ovambo and in Sesfontein they do not get what they deserve as payment". (Vedder, 1914).[45] However, there is no doubt that the diversification of the pastoral economy had become essential by the turn of the century. The South African colonial rule turned back the clock and forced the pastoralists of Kaokoland back into subsistence pastoralism.

EXCHANGE AND TRADE UNDER THE SOUTH AFRICAN ADMINISTRATION, 1917 - 1940

The South African administration set out to control the border trade with Portuguese traders immediately after it had taken over from the German administration. In order to guarantee colonial military supremacy it confiscated 155 modern firearms on two expeditions in 1917 and 1919.[46] In order to control the vast area efficiently and with little financial input they installed tribal chiefs. Hunting became a crime ("poaching"). Natives were urged not to hunt anymore, while white officers went on hunting for their supplies and probably for their pleasure unabatedly. Restrictions on trade and hunting became more effective when in 1926 a police station was established at Tjimuhaka right on the Kunene river.

Restrictions on trade with Portuguese traders: Trade with Portuguese traders had already been prohibited along the Kunene boundary in 1917. The official reasoning behind keeping strict control over trade was the fight against contagious livestock diseases. Other reasons were mentioned as well: the administration still feared that modern guns were being sold to Himba and Herero by the Portuguese. Additionally, Portuguese traders were found to barter cheap liquor for livestock along the boundary which was regarded as detrimental to "tribal health". The police station at Tjimuhaka was to control, not only the movement of livestock and people but to eradicate cross-border trade as well. In the past Kaokoland's pastoralists had exchanged cattle for grain with Portuguese traders and southern Angolan agriculturalists in times of drought. South African prohibitions of trade across the border made this route to cereals obsolete.

In late 1926 the prohibition of the grain/livestock trade had to be partially lifted to allow some emergency maize to be traded into the area. Under the supervision of the Post Commander of Tjimuhaka some grain was sold for cash or in exchange for heifers.[47] In the 1929/30 drought local Portuguese traders were allowed by both the Portuguese and South African authorities to carry out some trade at Tjimuhaka under the auspices of the Post Commander. Originally the Portuguese authorities had hesitated in authorising the introduction of cattle from South-west Africa to Angola, and no trade in other commodities was allowed, meaning that local herders either had to pay cash or could not obtain grain at all.[48] However, one month later the traders were able to obtain a permit to import 500 heifers to Angola. Oxen, the main asset for exchange for the local pastoralists, were not allowed to be imported into Angola. The Post Commander of Tjimuhaka was dismayed by the restrictive handling of trade by Portuguese authorities.[49] The stern handling of trans-border trade by both the South African and the Portuguese administration meant a significant loss of exchange entitlements for local pastoralists. Oxen, a major trade good in the past, were practically devalued as objects of exchange. Trade regulations forced herders to supply heifers. This they had to do at very unfavourable prices which were forced upon them in times of stress.

In 1932 the administration once again felt unable to provide relief maize and asked Portuguese authorities to allow local traders to come down to the river to barter maize for heifers. However, Portuguese authorities flatly denied, arguing that the maize was urgently needed within southern Angola.[50] The Native Commissioner of Ovamboland, Hahn, was advised to buy maize directly in Huila district. Livestock had become completely defunct as an asset for exchange. Hahn then applied for funds from his headquarters. Probably due to the drought, the Administrator for Southwest Africa paid a hurried visit to Kaokoland and decided that a bit of relief maize would be paid for. He was determined to make political use of the situation:

> "The Administrator blames people not to have done enough; e.g. the people at Otjitundwa and Ombathu did not plant. ... The natives of the Kaokoveld contribute nothing whatever to the labour supply of the territory at any time. They lead a particularly lazy and carefree life and they must therefore as far as it lies in their power to do so look after themselves".[51]

He then decided to deliver a ridiculous 26 sacks of maize and 6 sacks of seed maize for Kaokoland's total population. Rations of this maize were only to be given in return for labour such as road construction.[52] The encapsulation of Kaokoland was felt most viciously during times of drought.

TRADE TO THE SOUTH

Trade with the commercial ranching area was hampered massively as well. The sale of livestock, livestock-based products, and for some time all other products (including tobacco and tools) had been prohibited since the early 1920s. Although this was in accordance with the general aims of the government and the farmers association throughout the thirties, several traders tried to find alternative ways to conduct trade with Kaokolanders. Especially after the Karakul boom of the 1930s local sheep were highly valued for cross-breeding. Farmers, traders and speculators were eager to buy cheap livestock in Kaokoland and sell it at much higher rates in central Namibia to farmers or abattoirs.[53] In the late 1930s the government tried several times to initiate some controlled livestock trade from Kaokoland to the Police Zone. Herders were allowed to import sheep through a quarantine station to the Police Zone and then had one month to dispose of them among farmers. After that time unsold stock had to be taken back to Kaokoland.[54] Due to complaints from the farmers who feared infection of their herds (the quarantine station was near Kamanjab), this practice had to be stopped in 1938. Most of the pastoralists from Kaokoland were excluded from any form of trade.[55] A complaint from 1949 makes the disastrous situation very clear.

> "We have difficulty. We cry. We are imprisoned. We do not know why we are locked up. We are in gaol. We have no place to live. This is our difficulty which we report to our Master, Nakale. Here our living is our cattle, sheep, goats, tobacco, buchu. Our donkey wagons do not fetch anything from Kamanjab. We cannot get meat from the south or even mealiemeal. Our sleeping skins cannot be sent out. We have to throw them away on the border. We enter the Police Zone with hunger. We have no money ... Ovamboland is closed for us. We lived on (in) Ovamboland for a long time. We want to take our cattle there, also our sheep and goats. Here in the Kaokoveld we live only on our livestock. The borders are closed. The borders press us heavily. We cannot live. We are in a kraal".[56]

In 1949 there was not a single store in Kaokoland and people had to travel several hundred kilometres to Kamanjab, Outjo or Ondangwa to obtain goods. Ovambo petty traders sometimes came to Kaokoland with donkeys to barter tobacco from local herders in exchange for beads and other commodities.[57] Trade was further hampered by laws prohibiting barter between natives and white or black traders. The legal act was advertised as an initiative to protect local producers against unfavourable trading. The new law obviously had its value in forcing even more people into waged work in mines and on farms. Producers without any contact to the labour economy were excluded from obtaining commodities.[58]

Reacting to many local complaints, the Administration proposed that SWANLA (South West African Native Labour Association), the officially acknowledged labour recruitment organisation, should start its activities in Kaokoland.

Labour Recruitment: Official labour recruitment only started in 1950, when SWANLA's recruiting officer went to Opuwo for the first time.[59] After discussions with the local elite he left Kaokoland under the impression that the local people would flock into SWANLA's labour office. However little happened in the first year. Instead of recruiting 500 men per year plus some 300 men from Angola, only about 100 labourers were recruited.[60] Recruiting officers were confronted with numerous reluctant and frightful statements. One young man voiced his anxiety towards the official: "We are afraid to go to the farms in order to work there, we may get lost and never come back again."[61] Some reasons given for the obvious reluctance of the pastoralists to seek work were that returning labourers commented unfavourably on labour conditions in the Police Zone. Frequently farmers reduced wages considerably if herdsmen lost cattle, which resulted in some labourers returning without any income. Men were ordered to do menial jobs, which they regarded as below their dignity. Many potential labourers thought that contracts of a duration of twelve or eighteen months were too long; they claimed that they would accept six month contracts maximum. Finally SWANLA closed its Opuwo office. People from Kaokoland who wished to leave for work had to contract via Ombalantu, some 150 kilometres away. From the middle of the 1950s onwards there was a small but steady outlet of about 100 workers per year from the Kaokoland. Furthermore an unknown number contracted work through farmers. Nevertheless, the Kaokoland remained marginal as a source of labour well into the 1970s.

As SWANLA's activities in the area remained minimal, the government decided in the early 1950s to give out a few trading licences to black traders from the Police Zone and Ovamboland.[62] These traders had to guarantee that all transactions were strictly in cash or in exchange for tobacco. Bartering for livestock was not allowed. They had to prove that they owned a vehicle and had a lump sum of 150 Pounds at their disposal to start the business. Until the middle of the 1950s four licences were given out, one to an Ovambo trader (whose permission was withdrawn only a month later) and three to Herero traders from the Police Zone. Nobody from Kaokoland applied for a licence at that stage.

COLONIAL ELITE, EMERGENT STRATIFICATION AND THE DECLINE OF ENDOWMENTS

From the beginning the South African administration had based its government of the vast northern areas on the local elite. While a stratified society was firmly established in Ovamboland and kings were chosen as colonial agents, in Kaokoland leaders had to be established first of all. All three chiefs appointed in 1920 had a career as war lords behind them and had little claim to traditional forms of leadership. The South African administration sustained claims for power by these chiefs substantially in its own interests. In the 1920s there were numerous complaints by local herders against these leaders. Frequently they were accused of stealing livestock from herders or of mingling into inheritance affairs to their own advantage.

However, there is little doubt that stratification in Kaokoland remained incipient. None of the chiefs succeeded in creating a stable power base. Complaints of colonial officers on the powerlessness of local leaders were frequent. One obvious reason for the fact that the elite did not become more permanently structured was that an investment of pastoral assets into other domains remained impossible. Even the most powerful colonial chief Vita Tom with all his power and good relations to the administration could not establish himself as a trader. Colonial records give some evidence that he tried to engage in trade with white livestock traders from the Police Zone. However, his efforts were subdued by the interests of the colonial administration to close off Kaokoland.

EMERGENT COMMERCIAL TRADE: KAOKOLAND SINCE 1954

Since the middle of the 1950s the colonial administration somewhat changed its policy on encapsulation. In South Africa the National Party, which has ruled since 1948, implemented its Apartheid policy. For South Africa the Thomlinson Plan had shown the way to how reserves should be transformed into semi-independent homelands. In 1954 Kaokoland was placed under the Ministry of the Bantu Administration. While exchange with the commercial ranching area was still strictly controlled, restrictions on trade with Angolan traders and Ovambo were alleviated. Legally prior prohibitions were kept but they were no longer enforced with the same rigour. During a drought at the end of the 1950s an estimated 10,000 cattle were brought to Angola and exchanged for grain.[63] Several shops were opened in Kaokoland in the late 1960s and early 1970s: in 1975 seven shops and six itinerant traders were registered. However, it was mainly the parastatal Bantoe Beleggingskorporasie (BBK) which handled livestock sales. The near monopoly of the BBK and an apparently low price policy discouraged producers.[64] The figures offered in a report on the development of Kaokoland speak for themselves. In the early seventies 189 (1971), 2,469 (1974) and 1,739 (1975) cattle were sold to the BKK. These are some 0.1 per cent, 2.0 per cent and 1.4 per cent of the total livestock population counted for the respective years. Presumably many more head of livestock were sold and bartered to Angola along illegal or semi-legal channels.

A change in livestock trade came about towards the end of the 1970s when three army camps were stationed in Kaokoland. Especially the camps at Opuwo and Okangwati needed livestock for the provision of soldiers. Many traders stepped up their businesses during these days. They bartered maize, cheap alcohol and clothes for livestock with Himba and then sold the livestock to the army camps. This trade came to an abrupt standstill in 1989, when the South African army retreated and the Namibian army did not take over the army barracks. Traders had to re-orient their sales activities towards Ovamboland. The abattoir at Oshakati became the biggest buyer of livestock from Kaokoland.

The restrictions on trade during colonial times - a past which has only recently gone - are echoed in recent trade patterns. Almost all itinerant traders who operate in Kaokoland are of non-Himba origin. Many are Oshivambo-speaking or stem from ethnic groups with an Angolan background. Most of the exchange is conducted as barter trade. The trading partners have frequently developed attitudes of mutual trust and credit is given in both directions, i.e. the traders hand out maize without receiving any direct payment or the pastoralists hand out livestock without immediately getting the sacks of maize due to them. However, a lot of barter runs on a tit-for-tat-level: the trader delivers certain goods and takes the goat due to him immediately. The following case study of one trader shows to what extent local livestock producers are disadvantaged in barter trade.

> Henry, a forty-four year old trader from Ombalantu takes his goods far into the bush, brings goats back to Okanguati and finally to Oshakati with his old car. The goods are bought in Oshakati or Ondongwa. To Otjiwarongo he sometimes has to go to buy cloth and beads. On one tour he can bring back 40-50 cases of Castello, Clubman or Tassenberg (cheap alcoholic beverages) and 25 to 30 bags of maize meal. Furthermore he sells blankets, cloth, beads, cooking oil and sugar in the bush. All goods are preferably exchanged for goats. The fat male goats are directly sold to Oshakati, thin or small ones he fattens in his own herd. The livestock debts he collects personally and sometimes he sends a "boy" to far away places.
>
> The exchange rates are the following: one box of Clubman (a cheap mint punch) goes for a fat goat (1 box - > 60-70 kg goat). If no goat of this type is available, two smaller goats will be accepted. For one box of Castello (a cheap sparkling wine) a 50-60 kg goat has to be given. In the

shop Clubman sells for 106 N$ the box and Castello for 65 N$ the box. One bag of maize of 50 kg goes for a 50-60 kg goat, 60 kg bags are sold for the same amount, only for the 80kg bags he asks for a considerably fatter goat. The returns are very good. For a 70kg goat he got 180 N$ to 200 N$ at the abattoir in Oshakati. All animals are accepted at the abattoir, and are paid immediately in cash there.

Livestock sales at auctions, which are organised by the parastatal MeatCo, are rare. Auctions at Okangwati take place twice a year and at Opuwo five to eight times a year. However, few herders from the Epupa region will walk as far as Opuwo in order to sell livestock. There is nothing which comes near a regular market place for cattle. All cattle are put on a scale and then sorted into one of the three quality classes. Prices are paid according to rates fixed on a national level. There is no room for bargaining. Pastoral producers frequently bemoaned the low prices MeatCo paid realising that the prices for livestock were somewhat better in other areas of northern and eastern Namibia. This negative perception of MeatCo's activities has resulted in a boycott of MeatCo auctions by all pastoral producers in Kunene North since early 2001.

Stratification and Endowment Decline since the 1980s

Since the early 1980s trade in livestock has become prominent in Kaokoland. A local trader elite rapidly evolved in Opuwo. Some traders succeeded in amassing considerable riches. They own fine houses in Opuwo, usually possess a commercial farm in the commercial ranching area, invest into shops, vehicles and other forms of trade. Few of them seem to hold considerable herds of livestock in Kaokoland. They mainly use their Kaokoland base for trading with local livestock producers. This elite is dominated by Oshiwambo speakers both from Angola and Namibia, some other Angolans and a few Herero. Himba are hardly represented in this elite up to now.

To analyse differences in wealth within Himba communities is more intricate. Wealth is still mainly based on livestock. Only a few Himba own further assets of any importance. Even chiefs and councillors still live in traditional mud houses and to the outside show little effort in documenting their wealth in more sophisticated patterns of consumption. Recently cars have become a target for investment for wealthy herders. However, those few second hand cars which have been bought up to now are rarely used for trade and are mainly non-productive capital.

Internal stratification is most clearly shown in different patterns of consumption. Table 5 shows what 24 Himba households invested and what they obtained for their sales. The data was taken on a monthly basis over a period of 22 months (April 94 to January 96). The table only gives a rough idea of exchanges and relates the total volume of exchanges over this period with the wealth of the respective household.

Table 5. Wealth, livestock sold and goods bought

No.	Wealth	Sold Goat	Sheep	Cattle	TLU	Bought Maize	Alcohol	Blanket	Cloth	Veterinary	Other
1Mu	3	15	13	3	6.5	26	0	5	0	1	1
2Mu	1	27	1	2	5.5	13	8	0	0	2	2
3Mb	3	9	2	4	5.4	22	1	1	1	1	1
4Va	2	17	0	2	4.1	14	0	1	1	2	1
5Ka	4	20	29	18	24.1	22	20	3	0	0	5
6Ka	3	9	8	10	12.1	42	16	3	1	1	4
7Hi	4	13	1	4	5.8	28	10	0	0	5	2
8Tw	3	23	1	6	9.0	32	2	2	6	1	6
9Mi	4	7	0	1	1.9	4	0	1	0	2	3
10Tj	1	11	8	0	2.4	7	0	1	3	0	3
11K	2	25	3	6	9.5	25	15	2	0	2	4
12K	3	40	2	4	9.3	40	3	0	2	0	4
13U	1	47	17	2	10.0	23	18	1	2	0	3
14W	2	27	2	4	7.6	17	8	3	1	1	1
15H	2	7	9	0	2.1	14	2	0	0	0	0
16M	2	13	5	3	5.3	19	4	0	0	3	2
17K	1	5	3	2	3.0	8	0	2	0	0	0
18K	3	24	3	4	7.4	21	0	2	2	0	1
19M	3	21	2	9	11.9	46	10	2	1	0	1
20K	2	35	4	2	6.9	29	6	3	0	0	0
21M	4	59	45	19	32.0	49	13	3	10	2	4
22K	1	29	0	3	6.6	15	7	0	0	3	0
23N	3	12	0	5	6.5	49	4	0	0	0	0
24V	3	6	2	13	14.0	79	12	3	0	0	1

Notes: wealth status: 1 = poor, 2 = medium, 3 = wealthy 4 = very rich; TLU = 1 cow = 1 TLU; 1 goat / 1 sheep = 0.125 TLU

Generally speaking wealthy herders sell more livestock than poorer herders (Pearson's square 0.48). However, there are exceptions. The very rich herder No. 9 sells less than all other herders simply because his salary as an auxiliary game warden is paid out in maize and sugar and he rarely sells additional animals in order to obtain food or other commodities. For the same reason the rich herder No. 7 sold few animals. He obtains a salary from his job and spends some of his income on buying food. But why do poor households like No. 2, No. 13, and No. 22 sell so many goats, although their households are rather small? A look at what these households get for these goats shows that they had to sell many goats prematurely, i.e. almost always when they needed maize, they had to pick one or two year old goats and frequently needed two goats to buy one sack. Rich households will usually offer one adult goat to obtain one sack per animal. How can the enormous expenses of the rich household No. 5 be explained? No. 5 is one of the richest households in the area. Its head is a councillor and a politically very active figure. He has hosted many meetings of his followers. To host people at such occasions adds tremendously to the prestige of a leader. Furthermore, household No. 5 has good contacts to traders who buy larger numbers of livestock for money. Once he sold 25 sheep and at another time 9 oxen for money. In other households (No. 6 and No. 24) a number of cattle had to be sold due to rituals which necessitate the acquisition of maize and alcohol for numerous guests.

Although a generalising interpretation of Table 9 is problematic, we may conclude: (1) All households need large amounts of maize. Big households need substantially more maize: while household No. 10 got along with 7 sacks of maize, No. 21 and No. 23 needed 49 sacks and No. 24 even 79 sacks. (2) Festivities are major instances of expenditure and it is mainly the politically dominant figures which heavily invest into these occasions. However, ceremonies which are deemed necessary such as funerals and ancestral commemorations require some investment from all households. (3) The range of commodities bartered is rather small. Himba do not invest heavily in luxury items. The investment in a car in household No. 21 is exceptional but indicates that the rich may invest more in luxury items in the future. (4) Himba obtain comparatively little for their livestock if we put barter rates in relation to monetary exchange rates. An adult male goat is rated as being equivalent to a 60 kilo sack of maize. If prices paid at the abattoir were paid to the Himba, they should receive three sacks per adult goat.

While off-take and sales patterns indicate some degree of differentiation according to wealth categories there is no indication that the rich get richer at the account of poor herders. Endowments have not declined over the decades. Rich people loan substantial numbers of cattle to poorer relatives. Trade in livestock, which apparently is the major way of transforming wealth in livestock into other forms of wealth, has only been gaining momentum since the early 1990s.

3.3.3. A Comparative Perspective on Marginalisation, Stratification, and the Loss of Entitlements

Both groups, Himba and Pokot, were marginalised during the colonial period, though to a different extent and with different consequences. While the Pokot increasingly gained from livestock sales and the growth of a national livestock market, the Himba were excluded from livestock markets by the colonial government and were reduced to subsistence pastoralism. While Pokot trade mainly suffered from frequent and unpredictable changes in the British colonial policy towards local trade, the Himba experienced more than 60 years of nearly total encapsulation and heavy state controlled trade.

Trade in the arid parts of north-west Kenya grew gradually at the beginning of this century. Pax Britannica allowed for peaceful co-operation and guaranteed a certain degree of stability to markets. Itinerant Somali merchants brought various types of consumer goods and staple foods and traded them in for small stock and hides. The British administration sometimes encouraged this form of trade and sometimes discouraged it. Enforced destocking in the 1950s hampered local traders considerably. Carol Kerven (1992) has pointed out for the neighbouring Samburu District that the capricious decisions of the British had a malign effect on livestock trade in the area. The frequent changes of regulations led to a definite increase in transaction costs. The Pokot are conspicuously absent from the scene as traders until the 1980s. Only in the 1990s have more Pokot men tried to make a career as livestock traders. Although many of the Pokot claim that prices are constantly low due to the fact that trade is dominated by outsiders, over the course of the century livestock prices have not declined but have increased. This holds especially true for the pricing of small stock: prices for goats have increased throughout the century due to a growing urban demand.

The history of trade in Kaokoland in the 20th century is very different from that of north-western Kenya. The South African administration brought an abrupt and sudden end to a flourishing regional trade network in the 1920s. All trade over the boundaries became prohibited: Owambo traders and Portuguese merchants were prohibited from entering Kaokoland. All transactions with the commercial ranching zone to the south were inhibited. Trade patterns which emerged in the 1980s were mainly based on barter transactions. Traders dictate prices which are far below the market price. Due to their lack of experience with trade, the Himba are frequently cheated. Many times I observed that they were sold outdated veterinary drugs and medicine at grossly exaggerated prices.

While the decline of exchange entitlements has been drastic among the Himba, and to a lesser extent amongst the Pokot, local changes in the access to resources have been rather limited in both societies. Among the highly egalitarian Pokot, no local elite was able to establish itself. Neither colonial nor post-colonial chiefs were able to transform wealth into more permanent forms of political dominance. Although the Himba are clearly more stratified and differences in wealth are much more profound, local elites have not succeeded in dominating the access to resources.

3.4. SHORT TERM CLIMATIC VARIABILITY - DROUGHT AND ITS EFFECTS ON LIVESTOCK HERDS

Temporal and spatial variation of precipitation frequently leads to drought in the world's arid and semi-arid areas. These variations massively affect livestock husbandry and agriculture. Meteorological droughts, hydrological droughts and agricultural droughts are differentiated (Mainguet, 1994:24; Glantz, 1993:57f). Meteorological drought occurs as a "result of any relatively unexpected shortfall of precipitation" (Smith, 1996:28). Smith defines drought as a "function of the length of period of abnormal moisture deficiency, as well as the magnitude of this deficiency" (Smith, 1996:289). Whereas Smith (1996:288) rejects a quantifiable general definition of drought, Glantz (1987:45) defines drought as precipitation below 25 per cent of the average.[65] Long lasting dry conditions may lead to a decrease of groundwater levels and will result in a hydrological drought. Agricultural droughts occur when soil moisture is insufficient to maintain average crop growth and yields or average biomass production on pastures (Glantz 1993:57).[66] It is not the lack of water but the lack of fodder that matters for pastoralists. In droughts livestock rarely die of thirst but rather of a lack of fodder. In years with bad rainfall, fodder production is heavily reduced. There are records which point to a 90 per cent decrease of the total grass yield in years with a precipitation which is 25 per cent below the normal for the Sahel (de Ridder & Breman, 1993).

The lack of biomass production results in increased livestock mortality rates and rapidly decreased milk yields. For Africa's arid and semi-arid lands we have several accounts on livestock mortality due to drought: van Dijk (1997) reports losses of 62 per cent for cattle and 55 per cent for small ruminants during Sahelian droughts, whereas Hjort & Dahl (1991:132) describe livestock losses of 95 per cent amongst the Sudanese Beja during the 1984 drought. Legesse (1989:269) reports 60-70 per cent losses of livestock amongst the Kenyan Borana in the 1973 drought. Coppock (1994:169/70), reporting on the 1984 drought in southern Ethiopia, gives average losses of 54 per cent in cattle herds. Scoones (1996:205) recorded livestock mortality during three droughts in Zimbabwe: in 1983/84 losses were at 39.2 per cent, in 1987/88 at 11.8 per cent, in 1991/92 at 21.9 per cent and in 1992/93 at 19.8 per cent. Causes for death during these droughts are given as outright starvation (47.7 per cent), slaughter before starvation (30.3 per cent), loss (5.7 per cent), accidents (7.2 per cent) and disease (4.5 per cent).[67] Sandford (1983:75) points out that rich Maasai families are usually less affected than poor ones by droughts: while rich families lost 42 per cent of cattle during a drought, poor families lost 52 per cent. The same holds true in Ensminger's Orma case study (1984:65): rich Orma herders lost 59 per cent of all cattle while poor Orma families lost 74 per cent. High stock losses have lasting effects. While a farmer will be able to reap a full harvest from his fields once it rains, it takes years until herds have fully recovered. White (1997) points out that Fulani herders in Niger have never fully recovered from the major droughts of the 1970s (see also Fleuret, 1986:226 for the agro-pastoral Kenyan Taita).

In order to describe the dynamics of droughts I will first depict rainfall variability in both regimes and then briefly discuss the variable effects of decreasing precipitation on fodder production. The consequences of declining fodder production for livestock husbandry (increased mortality, decreased reproduction, declining milk yields) are discussed, taking droughts of the 1980s and 1990s as examples. The final paragraph is devoted to describing the differential damage in the households sampled.[68]

3.4.1. Rainfall Variability

Lack of rainfall and/or badly spaced rains have a variable impact on grasses, herbs, bushes and trees. While annual grasses only start growing after an initial heavy downpour and are subsequently dependent on well-spaced rains, bushes and trees (e.g. Acacias and Commiphoras) start flowering after the very first light rains and are frequently able to cope with extended dry periods thereafter. Hence the effect of rainfall irregularities on biomass production is varied and droughts may look very different from the herder's point of view. This is portrayed vividly in oral accounts - each drought has its idiosyncratic characteristics.

Rainfall Variability in North-western Kenya

Data on rainfall for the Nginyang Division is available for several stations. For Nginyang there are records for the last 28 years. The average rainfall for these years is 577.9mm with the rate of climatic variability (CV) at 28.2 per cent. This CV ratio is very high, usually CV ratios of around 30 per cent are found in regions with an average precipitation of 300mm. The rainfall at Kositei, which is slightly higher and is situated at the bottom of a mountain ridge, is at 742.5mm, with a still astoundingly high CV of 22.1 per cent (18 years). If we take Glantz's definition of a drought as a guideline (25 per cent), droughts occurred in 8 out of 28 years (28.6 per cent) in Nginyang. Out of these, 3 years (10.6 per cent) were severe droughts with a precipitation of more than 40 per cent below the average. The worst recorded droughts occurred in 1955 (–61.6 per cent), 1980 (–42 per cent) and 1984 (–64.8 per cent). The 1990/91 drought, which will be considered later on in detail as a case study for assessing mortality rates, was booked with a low of 33.6 per cent below the average.

Figure 6 shows deviations from the annual average at Nginyang over 28 years. The pattern indicates that there is no clear long term trend. Bad years are regularly followed by good years.[69]

Rainfall Variability in northern Namibia

Although we do not have data from rainfall stations that are very close to each other, there is sufficient long term data from several stations located in north-west Namibia. The annual average for Opuwo is 299 mm (30 years), for Kamanjab 303.3mm (54 years), for Otjovasandu 306.4 mm (30 years), for Sesfontein 95.4 mm (24 years) and finally for Ruacana 299.3 mm (8 years). The overall rainfall regime is mainly characterised by an east-west gradient with precipitation rapidly declining towards the west. The rates of climatic variability (CV) are fairly similar as well: 28 per cent for Opuwo, 29 per cent for Kamanjab, 29 per cent for Otjovasandu. Sesfontein, which has a considerably lower average rainfall, has a CV rate of 58 per cent (see Figure 7).

Unfortunately the rainfall charts for Opuwo are incomplete for the last few decades. When all the other stations recorded several consecutive drought years, especially for the late seventies and early eighties, no data for Opuwo was available. Hence, when considering droughts, data on Kamanjab, Sesfontein, Otjovasandu, Ombalantu and Tsandi will be analysed separately. Kamanjab experienced 14 droughts (25.9 per cent), Sesfontein 11 (45 per cent), Otjovasandu 8 (26.7 per cent) and Tsandi 20 (37 per cent). Severe droughts with more than 40 per cent below the average rainfall were measured

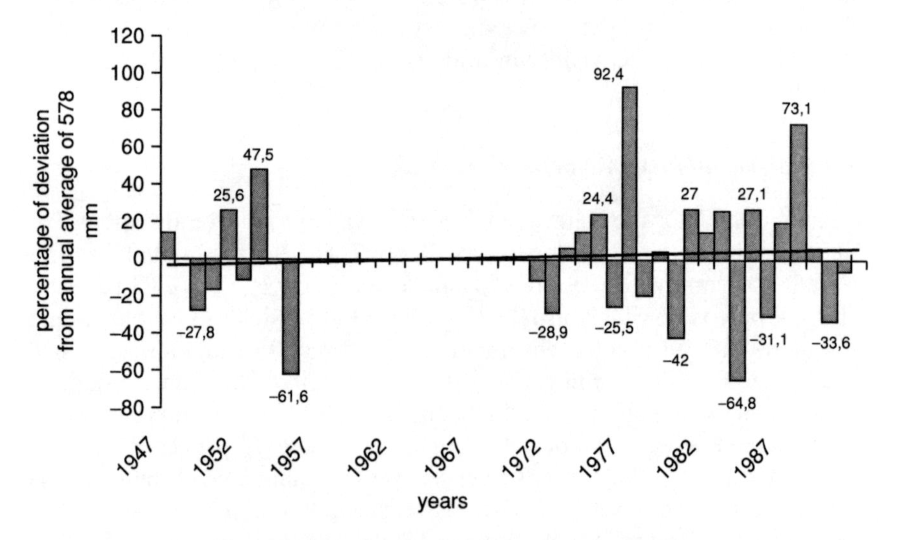

Figure 6. Climatic variation in Nginyang, 1947-1991

Source: Metereological Service Kenya, Nginyang Station

Note: There was no data available for the 1960s. Source: Archival Data (KNA BAR) and Records from Nginyang Station.

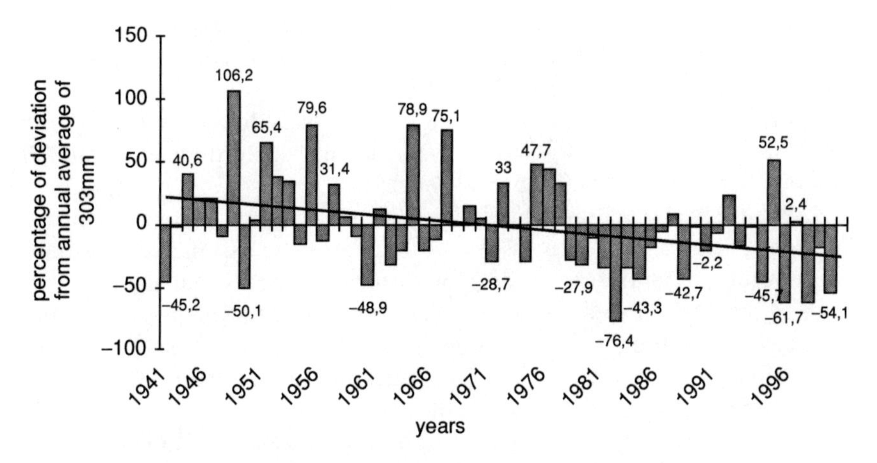

Figure 7. Climatic variation in Kamanjab, 1941-2000
Source: Metereological Service Namibia, Kamanjab Station

6 times (11.1 per cent) in Kamanjab, 5 times (20.8 per cent) in Sesfontein, 2 times (6 per cent) in Otjovasandu, and 5 times (9.2 per cent) in Tsandi. The worst drought recorded was the drought of 1980/81: while there is no data for Opuwo for this crucial year, Kamanjab recorded a low of 76.4 per cent, Sesfontein of 100 per cent (no rainfall at all) and Otjovasandu of 47.1 per cent. The lack of rainfall in 1981 was aggravated by the fact that this disastrous year was preceded by four years of below average rainfall and was followed by another four years of below average precipitation.

A look at Figure 7 shows that between 1941 and the middle of the 1970s drought years were usually followed by good years. Since the rainy season of 1976/77, nine consecutive years have been booked with below average rainfall. If the 1985/86 rainy season, which was measured with 7.7 per cent above average is excluded then there have been thirteen years with below average precipitation in a row. The 18 years between 1976 and 1994 had an average of 21.7 per cent below mean rainfall altogether. The same holds true for Sesfontein, Otjovasandu, Tsandi and Ombalantu rainfall statistics. Records for all north-west Namibian stations indicate the same negative trend from the middle of the 1970s towards the present.

Comparison of Climatic Variability in north-western Namibia and north-western Kenya

The total averages of both areas are quite different. While Nginyang nets an average of about 600mm, Opuwo is noted with only about 300mm. The rates of climatic variability for both regions are fairly similar. For Nginyang and Opuwo they are at about 28 per cent. This implies that although rainfall is higher in north-western Kenya, the rate of variability is much higher in absolute terms too. Despite a much lower annual average, the rainfall regime in north-western Namibia is more stable than it is in north-western Kenya. Even a place like Sesfontein, situated in the arid pro-Namib semi-desert, which only gets some 95mm of rainfall per year has a lower CV rate (58 per cent) than Lodwar in north-west Kenya which has about double the rainfall (180mm) and a CV rate of above 60 per cent.

The frequency of droughts is fairly similar in both regions. Nginyang recorded 28 per cent of all measured years as droughts, Kamanjab did so for 26 per cent and Otjovasandu for 27 per cent; that means that slightly more than a fourth of all years tended to be droughts in both regions. The frequency of bad droughts is fairly similar as well: 11 per cent of all years for Nginyang, 11 per cent for Kamanjab, indicating that roughly a tenth of all years turned out to have very severe droughts.

The severity of droughts seems to be more pronounced in Kaokoland. The devastating drought of 1981 was booked with 75 per cent below average rainfall and with no rainfall at all at stations further west such as Sesfontein. In contrast to this the bad drought of 1984 in north-west Kenya was booked in Nginyang with 64 per cent below average. While for the Pokot data no clear trend was discernible[70] - bad years were usually followed by good years, the trend was clearly negative in the data for north-west Namibia. The last two decades were booked with massively reduced rainfall. Out of 14 drought years in Kamanjab, 8 occurred between 1977 and 1993.

3.4.2. The Effects of Droughts on Fodder Production

The effects of rainfall variability on biomass production are tremendous. Usually the decrease in biomass production lies much higher than the decrease in rains. De Leeuw et al. (1993) show that a CV ratio of 37 per cent (at 144mm) translated into an 86 per cent variability in biomass production in the northern Gourma region and a CV ratio of 28 per cent (at 204mm) into a variability of 64 per cent in biomass production in the southern Gourma region of Mali. LeHouérou (1989:108) summarises these discrepancies: “the variability of primary production is 20 to 80 per cent larger than the variability of rainfall on an annual basis”.

Data in northern Namibia[71] was gathered on five selected plots, presenting different compositions of grassland communities. The rainfall in 1993/94 was at 302.5 mm (1.2 per cent above average), 1994/1995 rainy season was at 452.5 mm (51.3 per cent above average) and the 1995/96 rainy season at 227.2 mm (24.0 per cent below average, and 49 per cent below the 1994/95 figure). A word of warning has to be added to these figures: according to the amount and spacing of rainfall in different years, different grasses grew on the selected plots. While for example in a bad year a plot would only have Aristida spec., in the next good year the plot was mainly overgrown by Stipagrostis hirtigluma, a grass of much higher grazing quality and producing more biomass, and only few Aristida spec. could be found (see Table 6).

Table 6. Changes in biomass production on selected plots in northern Kaokoland

	1994	1995	1996	1997	1994/95	1995/96	1995/97	1996/97	1994/97
Div. from av.rainfall	+ 1.2	+51.3	– 24	+4.7					
Plot 1	86.2	147.5	20.0	32.9	+71.1	–86.4	–77.7	+64.5	–61.8
Plot 4	59.9	65.4	39.4	44.0	+ 9.2	–39.8	–32.7	+11.7	–26.5
Plot 5	236.5	331.7	30.0	90.1	+40.3	–91.0	–72.8	+200.3	–61.9
Plot 11	–	91.3	16.0	55.3	–	– 82.5	–39.4	+245.6	–
Plot 12	–	90.2	17.9	37.3	–	– 80.2	–58.6	+108.4	–
Average	127.5	145.2	24.7	51.9	+ 40.2	–76.0	–56.2	+126.1	–50.1

The comparison of figures on biomass production for the 1994/95 rainy season (51% above average rainfall) and 1995/96 rainy season (24% below average rainfall) shows an average decrease in palatable grasses of 76 per cent. Even if the normal rainy season 1993/1994 (1 per cent above average) is taken as a baseline, the decrease amounted to 69.8 per cent. The decrease is most pronounced in patches where high-yielding annual grasses are found: in plot 5, which to a high percentage consists of the high yielding Schmidtia kalahariensis, the decrease between the two years is at 91 per cent. In patches where lighter grasses predominate, the decrease is somewhat less. However, the fact that high-yielding grasses suffer most from the lack of rainfall indicates that the overall loss of biomass is much higher than 76 per cent and will tend rather towards a value of 90 per cent. This is tremendous given the fact that the 1995/96 rainfall figure of 227.2 mm or 24 per cent below average would not even qualify to call this year a proper drought. We may well envisage how much biomass production is reduced during a year with a 40 per cent or even 75 per cent (as in 1981) decrease in precipitation. During these maximum drought years, no significant growth of annual grasses takes place at all.

3.4.3. The Effects of Reduced Fodder Production on Livestock Mortality

Livestock mortality increases rapidly due to fodder scarcity during drought years. The figures available on the dynamics of Pokot and Himba herds reflect massive changes in stocking density over the last few decades. These changes are connected to severe droughts.

Livestock Mortality in the Regional Herd/ Kaokoland

Livestock counts in Kaokoland since the 1930s give some indication on the effects of drought on the regional herd. From 1958 to 1960 the total herd reduced from 122,425 to 82,495 (–32.6 per cent) in 1959 and to 65,500 (–46.5 per cent) animals in 1960. The collapse of the regional cattle herd was even more marked in the centennial disaster of 1980-1982. The number of animals dropped from 110,580 in 1980 to 60,276 (–45.5 per cent) in 1981 and finally collapsed to 16,000 (–85.5 per cent) in 1982 (see Figure 8).

The regional small stock herd was apparently less vulnerable to drought and did not fluctuate by such an extent. The number of small stock fell from some 200,000 animals to 160,000 (–20 per cent) animals in 1960 and from 163,478 in 1980 to 112,000 in 1982 (–31.5 per cent).

Minor droughts have their effects on livestock mortality as well: the rainy season of 1992/93, which was recorded with a low of 29.3 per cent at Otjovasandu (again no data was available for Opuwo for this year) shows a rapid increase in livestock mortality. Drought-related deaths increased, paralleled by an increase in disease-related deaths. Due to the weakened state of cattle, predator-related deaths show a modest increase too (see Figure 9).

The 1992 mortality were mainly related to grossly increased calf mortality. Adult cattle do not die in great numbers during a modest drought like that of 1992/93.

Livestock Mortality in the Regional Pokot Herd

There are few reliable figures on the dynamics of the regional herd of the Nginyang Division. However there is little doubt that Pokot herds suffered tremendously during the drought year of 1984/85. Homewood & Lewis (1987) record net losses of 50 per cent for the Baringo basin. Livestock losses were higher in the densely populated Njemps flats (60 per cent) than in the Nginyang region, where losses of 25-40 per cent were recorded. In sheep herds, losses were even higher and reached some 50 to 68 per cent.

Comparison of Livestock Mortality Rates

In the Kaokoland sample of 85 drought-related deaths, for which the age of the deceased animal could be established, figures indicate that calves suffer

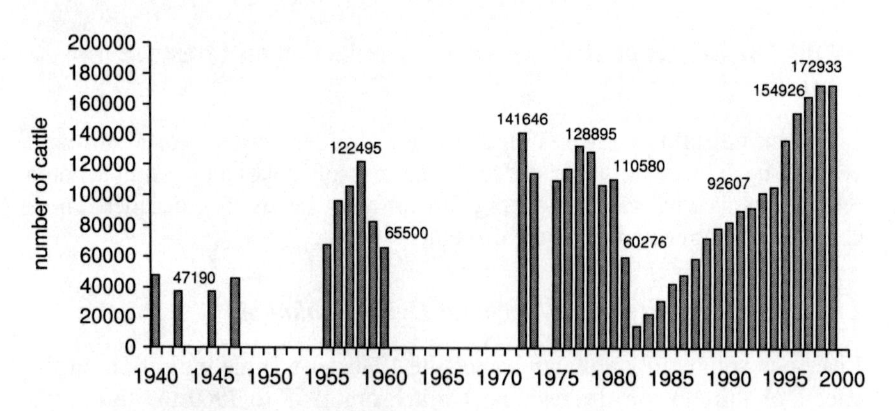

Figure 8. The dynamics of the regional herd of Kaokoland, 1940-2000
Source: Directorate of Veterinary Services; Page, 1976; van Warmelo, 1951

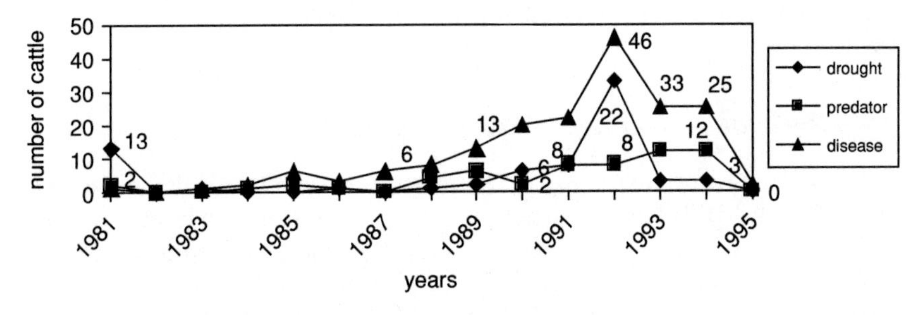

Figure 9. The 1992/93 drought as reflected in mortality rates

overwhelmingly in droughts. Out of the 85 animals, which mainly died in the 1992/93 drought, 63 (74.1 per cent) were calves below the age of 1 year and 13 (15.3 per cent) were weaned calves between 1 and 2 years of age. Only 7 (8.2 per cent) animals were above the age of 2 years.[72] For the Baringo Basin, Homewood & Lewis (1987:628) report losses of 79 per cent for calves between 1 and 2 years, while in adult cattle the death rate was only 45 per cent. The figures for both samples are similar to those given in the literature. During a drought calf mortality goes up to 80 per cent or even 90 per cent. Oxen obviously survive the longest. Their fat reserves carry them through prolonged times of starvation. They are the capital which can be cashed in for post-drought restocking.

In absolute numbers livestock losses due to drought are more disastrous in Kaokoland than they are amongst the Pokot. The Pokot never experienced losses of up to 90 per cent as the pastoralists of Kaokoland did in 1981. The losses both communities experienced due to drought seem to be fairly comparable at first sight according to this data. While in Pokot herds 31 per cent of all animals died due to drought, in Himba herds 32 per cent died due to the same cause. However, Pokot progeny histories were recorded in 1988, four years after the 1984 disaster. Hence, in the progeny histories gathered for the Pokot, most drought-related deaths fall into the year 1984. In contrast to this, Himba progeny histories reflected the very high drought-related mortality of 1981 only vaguely. Data was gathered 14 years after this catastrophic event. This suggests that the Himba experience higher and more constant losses due to drought. Drought regularly takes a toll on animals while amongst the Pokot losses due to drought are concentrated in a few bad years.

3.4.4. The Distribution of Losses amongst Households

In order to understand how damages are distributed amongst households it is important to take a look at the distribution of damage during a single drought. For both communities household specific data on losses was gathered. For the Pokot this data relates to losses experienced during the 1990/92 drought. For the Himba, information on household specific losses was extracted from progeny histories.

The Distribution of Losses in Pokot Households

The following report on the 1990/92 disaster introduces briefly the sequence of events which led to severe losses in livestock in most households: The rainy season of 1990 was 33.6 per cent below normal. The sparse rains were badly spaced so that almost no germination of grasses took place or grasses withered early on. The following rainy season (1991) was again substantially below the average. The effects of drought coincided with and were exacerbated by epidemics in all livestock species: Contagious Caprine Pleuropneumonia (CCPP) in goats, East Coast Fever in cattle and an epidemic in camels. Herd counts and interviews with herders suggested that some 50-70 per cent of cattle and up to 30 per cent of goats died in some households. In the Loyamoruk section it is presumed that 50-75 per cent of all camels perished due to disease, while in other areas camels did fairly well. In the dry season of 1991 most people were without milk and most calves and goat kids died. While this scenario describes the general trend and consequences of the drought the following paragraphs will show that households were differently affected.

Losses in cattle: Pokot cattle herds suffered heavy losses all over the Nginyang Division. Informants reported that one prominent sign of the 1990/91 drought was the complete lack of grass germination in the plains and reduced grass growth on the mountains. As grazers cattle and sheep were hit the hardest. During the early stages of the drought (August to December 1990), it was more the lack of fodder than the lack of water which caused problems. Only from January onwards did the lack of water become a serious problem and long treks to the few waterholes and dams weakened animals further. The situation was aggravated when some Pokot herds entered the Samburu and Laikipiak Districts and were affected by East Coast Fever there. Many cattle herds were still far away in the highlands in April 1991 when the data was collected and I was only able to obtain exact data for seven households. The average decrease in cattle herds was about 43 per cent. However, there were remarkable differences in losses in cattle between the seven households interviewed. Whereas one household's herd decreased by 77 per cent from the 1989 level, another had only one animal less than in the previous count; mainly those who took the risk of trekking their herds to the Leroghi plateau lost the most, as the herds were infected with East Coast Fever there. Whereas East Coast Fever killed cattle regardless of age and sex, drought killed more selectively: first calves, weaners and old cattle perished. Furthermore, the reproduction of the herd was affected massively: herd-boys reported in early 1991 that only very few cows were pregnant at the time due to the drought.[73]

Losses in camels: While camels were generally doing well in spite of the drought an epidemic decimated camel herds in most parts of the Loyamoruk Location - previously the stronghold of Pokot camel husbandry - with losses as high as 65 per cent. In 1989 I counted 184 camels in eleven survey households in the Paka region. The same eleven households only owned 61 camels in April 1991. Interviews with other camel owners in Loyamoruk left little doubt that they had suffered comparable losses in their herds. Camels were in a sad state and in early 1992 only about 20 camels in the eleven households were left. Losses were high in all households. The wide range of losses, from 80 per cent to 24 per cent, is rather an artefact of small numbers. It was the smallest herds which lost the least. As in other epidemics losses were high throughout and differed profoundly from non-affected households in other regions whose camel herds increased during the very same period.

Losses in small stock: Although there were also reports on considerable losses in goat herds, the decrease was apparently not as bad as compared to cattle and camels. In fact, Pokot herders identified the 1991 drought as a cattle drought. Some households had even succeeded in increasing their goat herds since 1989. Goat herds had reached a peak by the middle of 1990. Only from October 1990 onwards did famished animals die. In early 1991 the weakened animals were affected by Contagious Caprine Pleuropneumonia (CCPP) (see Figure 10).

While households 1 and 3 were extremely lucky, with the herds not showing any signs of CCPP, all other herds had losses. The losses of household 5 were exceptionally high compared to the other survey households. In all the herds, goat kids were virtually absent - the entire last litter had perished. As a consequence goat's milk was exceptionally rare. Sheep were as severely affected as cattle.

Summarising the results of the 1990/91 disaster, it is important to note that, however devastating the disaster was, households suffered very differently from it. While all households lost cattle, some lost the major part of their herd while others only lost a few animals. The reasons for these considerable differences in losses are not easy to explain. Do rich households loose more in

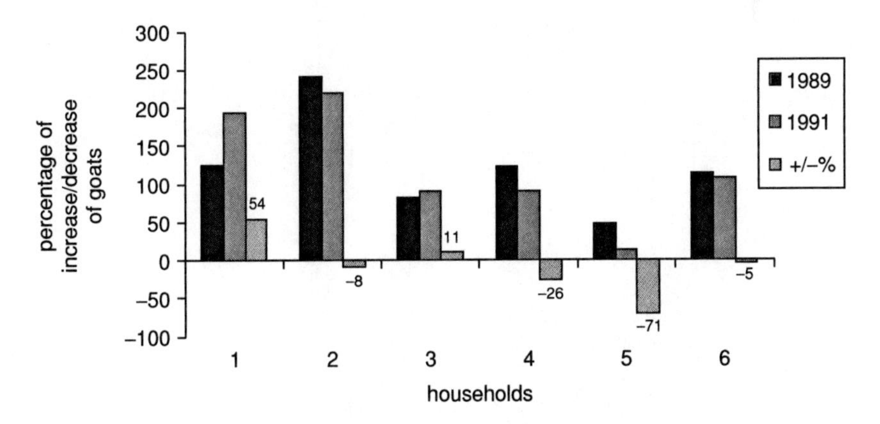

Figure 10. Losses to goat herds 1989/91

these droughts or do poor people complain about higher losses? The number of sampled households was too small to find quantitative evidence for a relation of household wealth and livestock losses. My impression was that losses varied randomly. A relation between losses and labour input was not discernible. Informants, too, claimed that rich people could become poor rapidly and poor people could become completely stockless - the randomness of losses is mirrored in emic perceptions of the volatility of livestock wealth. Wealth rankings which were conducted for the community in 1987, 1989, 1991, 1993, and 1996 showed this instability of livestock property.

Herd Losses in Northern Kaokoland between 1982 and 1995

Himba data on differential losses in households was taken from progeny histories and not from herd counts and only gathered for cattle herds. For most households losses were recorded for a period of about 15 years. Hence, accounts on the variation in household-specific mortality rates are more systematic. Data was recorded for 35 cattle-owning households (see Figure 11). On average these households had experienced losses (due to drought, disease, predators and accidents) of about 14.5 per cent (of a total of 4,496 animals recorded 651 animals had died prematurely) over the entire period. Losses varied between 4 per cent and 27.7 per cent between households.[74]

Most households had experienced losses which were fairly close to the average. Out of 34 households 19 (55.9 per cent) were within a range of +/–5 per cent of the mean losses, 7 (20.6 per cent) households had higher losses and 6 households (17.6 per cent) had smaller losses. In order to find out why some households experienced considerably smaller losses than others, we have to take a closer look at the households. Frequently it has been claimed that losses were related to the wealth of households (see Figure 12).

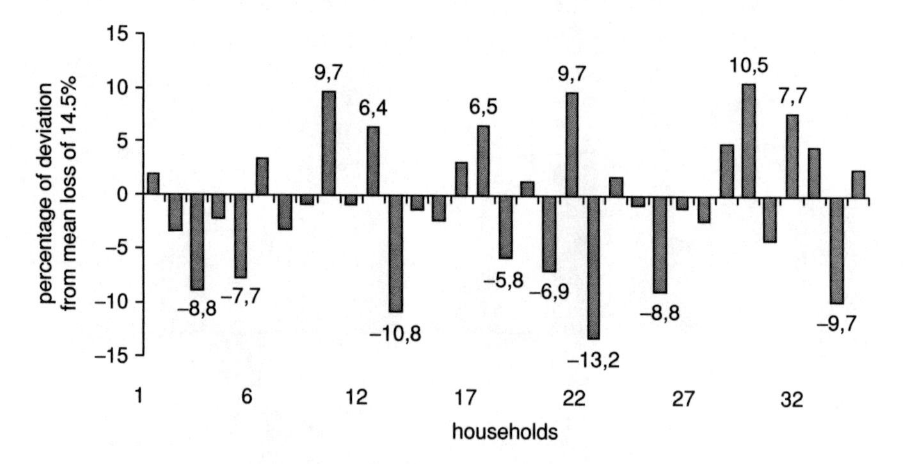

Figure 11. Losses of cattle in 34 households

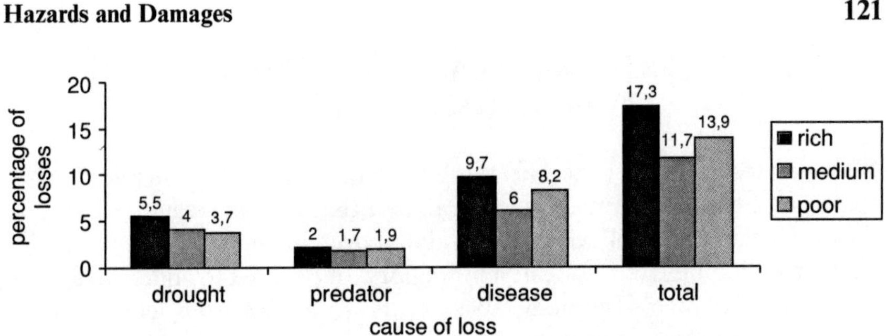

Figure 12. Losses in 34 households according to wealth and cause of loss

Figure 12 shows that rich households have somewhat higher losses than medium and poor households. Rich households lost 5.5 per cent of their livestock to drought, while medium households lost 4 per cent and poor households only 3.7 per cent. Due to predators, rich households lost 2 per cent while medium households lost 1.7 per cent and poor households 1.8 per cent. Due to disease and accidents, rich households lost 9.7 per cent, medium households 6 per cent and poor households 8.2 per cent. Overall losses amongst rich households were 17.3 per cent, amongst medium households 11.7 per cent and amongst poor households 13.9 per cent. The Pearson's square for the relation between household wealth and losses is at 0.35 with a significance of .00038. This is a high value underscoring the relation between wealth and livestock losses suggesting that wealthy herders have higher losses than poor herders. However, while the link between wealth and losses is systematic, differences are not great and certainly do not suffice to transform wealthy herders into paupers within a short period of time.

A Comparative View on the Distribution of Damages

In both economies herders experience high losses of livestock during droughts. Losses of up to 50 per cent are not rare and the Himba suffered from livestock losses of about 90 per cent during the drought of 1981. The distribution of losses in both economies has a strong random affect. However, whereas randomness overrules other influences on livestock losses in Pokot herds, losses are more patterned among the Himba. Among the Pokot, wealth hierarchies are constantly probed. Due to the randomness of losses and due to the fact that wealth differences are not great anyway wealth hierarchies are unstable. Among the Himba wealth hierarchies are more stable despite the fact that wealthy herders have somewhat higher losses than poor and young herders.

3.5. LIVESTOCK DISEASES AND THEIR EFFECT ON LIVESTOCK MORTALITY

While pastoralists can cope fairly well with the impact of endemic diseases, epidemic diseases cause massive losses all over the semi-arid regions of the continent. In many parts of Africa with rainfall over 500 mm Trypanosomiasis is a major inhibiting factor for cattle husbandry. In contrast to endemic diseases, epidemics are highly unpredictable events with disastrous consequences: Rinderpest, Anthrax, East Coast Fever and Contagious Pleuropneumonia may erase herds within a very short time span.

Livestock epidemics of a disastrous scope were prominent in Africa in the second half of the 19th century, when indigenous herds were confronted with exogenous strains of diseases. In the 1890s Rinderpest caused massive losses all over eastern and southern Africa. The epidemic was imported in 1889 on board of an Italian steamer bringing supplies to troops from India to Eritrea. Within two years the epidemic had affected most parts of eastern Africa. The Maasai probably lost more than 90 per cent of their livestock, as did other East African pastoralists (Waller, 1988; Sobania, 1991). In Ethiopia agricultural production came to a standstill due to the lack of draft oxen (Pankhurst & Johnson, 1988). All over eastern Africa the epidemic resulted in famine and high mortality figures in human populations. In 1896 the disease crossed the Zambesi river and spread through southern Africa causing the death of 2.5 million head of cattle in South Africa alone (Schneider, 1994:149). Contagious Bovine Pleuropneumonia has a similar history: the disease was introduced to southern Africa by a Friesland bull imported from Holland and landed in Mossel Bay in South Africa in 1854. Over 100,000 cattle died within the next two years. The disease caused a high mortality in cattle herds in Namibia and southern Angola and is still one of the major threats to livestock husbandry in the northern communal areas of Namibia (Schneider, 1994: 127).

3.5.1. Livestock Mortality due to Diseases amongst the Pokot

Cattle, camels, goats and sheep are affected by numerous endemic and epidemic diseases amongst the Pokot. Some epidemics such as Rinderpest, East Coast Fever, CBPP or Anthrax swept through Pokot herds and caused high rates of mortality during the course of the last century. The first outbreaks of CBPP in northern Kenya probably occurred around 1886/87. Oral traditions do not mention this epidemic, probably because its effects were overshadowed by the more dramatic impact of Rinderpest that ravaged Pokot herds in 1891. However, evidence from von Höhnel's travelogue (1897:768) suggests that the Pokot he met in 1888 were encountering a livestock epidemic as they were eager to trade drugs for livestock with him. The Rinderpest panzoetic of 1891 is widely mentioned in Pokot oral traditions. Obviously Rinderpest (*kiplok*) co-occurred with or was followed by human diseases (notably smallpox) and drought. This period of disaster, which probably lasted several years, is called *asis* - the sun. Traditions point out that the string of interrelated disasters caused a temporary change in the entire society. People were forced into a life as foragers surviving on roots and berries. Bridewealth payments were no longer paid. Although Rinderpest reoccurred in West and East Pokot several times (1920/23, 1940/41, 1947/48, 1960, 1981) it did not cause sweeping damage again over such a wide area. An outbreak in 1928 in the Korossi location caused mortality rates of some 70 per cent (the epidemic was paralleled by a serious drought and an invasion of locusts) but high losses were restricted to one location.[75] Losses were kept at bay during the following decades, probably due to a strengthened immune system and improved veterinary care. East Coast Fever (ECF) is prevalent in the hills of West Pokot and the mountainous areas adjoining East Pokot to the Leroghi Plateau up to the present day. Nowadays, ECF is the single most important epidemic in the area. Anthrax and Rinderpest infections have become less relevant in recent decades.

There is some evidence that disease patterns have changed during the past few decades due to profound environmental changes. According to informants the increase of dense bush led to an upsurge of Trypanosomiasis (see also Conant, 1982:119 for a scientific account of this process). The various subspecies of *Glossina*, the vector of trypanosomes, need bushland to multiply and are usually less frequent in open grasslands. The informant cited in the following passage claims that worms and flies transmitting diseases also increased. Diseases such as Contagious Caprine Pleuropneumonia (*lowkoi*) were recently brought in with herds raided from the Turkana.

> During earlier days *lokurucha* (Trypanosomiasis) was not widely spread. It just occurred in a few bush spots at the foot of the Leroghi highlands, places like Kanakirwa (beyond Amaya), Lochorangasuwia, Chepïrkö had *lokurucha* and the gallery forest along the Kerio Valley had *lokurucha*. But in between on the plains and on the hills *lokurucha* was *not* usually found. And even in Lochorangasuwia *lokurucha* only affected animals during the rainy season. During the dry season it was free of the disease. Also worms increased. These days *kalalyan* (generic term for several types of flies) are all over the place. *Lowkoi* (CCPP) is a new disease. *Kipsosoy* (unspecified, probably a non-contagious form of Caprine Pleuropneumonia) was there but it just killed one or two goats and not more. These days *lowkoi* wipes out almost an entire herd within a few days. Ticks were also fewer in the past. First they started to increase in Serunï and now they are increasing everywhere. (Tochil, October 1993).

A regular veterinary service never worked in the Baringo plains and even worse, what did actually work in the early 1980s, has been dropped due to governmental budgetary constraints in the 1990s. The Pokot try to help themselves by buying broad spectrum antibiotics from itinerant traders. This may bring other problems in the future as farmers usually under-dose the antibiotic and therefore create immunities. Roughly 10 - 15 per cent of all animals of each species (camels 9.9 per cent, cattle 10.0 per cent, goats 14.6 per cent, sheep 13.6 per cent) die prematurely due to some disease. Losses among small stock are higher than among large stock, goats being noted as the species with the highest mortality rates. Table 7 quantifies the reasons Pokot herders gave for the premature death of an animal, hence the reasons for death reflect Pokot disease categories rather than a diagnosis based on western veterinary medicine.

Camels: Tick-related diseases score the highest mortality rates in the sample (22.5 per cent). Diarrhoea, especially in camel calves, and *cholera* in

Table 7. Causes of premature death in Pokot herds

Scientific term	Vernacular term	Camels	Cattle	Goats	Sheep
Trypanosomiasis	*lokurucha*	8	10	–	–
Diarrhoea	*kiyitagh, 'cholera'*	14	1	17	3
Mange	*simpiriyon*	–	1	–	–
Worms	*tiompö mu*	8	20	11	4
Unspec., probably Camel Pocks	*lomitinaa*	7	–	–	–
Ticks	*tilis*	29	–	–	–
East Coast Fever	*lï ïsl Cheptikon*	10	27	–	–
CBPP and Pneumonia	*psosoi*	4	–	7	1
CCBP	*lowkoi*	–	–	5	–
Foot and Mouth Disease	*ngoriyon*	11	14	–	–
Bloat	*lessana*	2	5	2	–
Bloat	*musarer*	1	–	–	–
Evil Eye	*wutöt*	2	16	–	–
Abnormality	*solwö*	1	–	–	–
Orf	*ngïrïmen*	3	–	2	3
Coughing	*rolyon*	4	–	1	–
Unspec.	*cherempes*	–	–	4	–
Enterotoxeamia	*pngerat*	–	–	36	–
Unspec.	*pkoghogh*	–	–	7	–
Unspec.	*kipsiki*	–	–	1	1
Foot-rot	*chepkelyon*	–	–	1	–
Plant poisoning	*lotïl*	–	–	–	44
Too much milk		9	1	–	–
Swollen knees		2	1	–	–
Unknown disease		14	9	5	3
Sub-total		129	105	89	59
Non disease related cause of premature death					
Drought		31	61	52	24
Injury		32	8	–	–
Snake bite		1	1	–	–
Cold shock from rain		1	–	–	–
Predator		52	25	47	20
Other		9	4	–	1
Sub-total		126	99	99	45
Total		255	204	188	104

Note: John Young identified the livestock diseases according to western terminology (Young 1989).

adult camels (10.9 per cent), *nogoriyon* (in cattle the equivalent of Foot and Mouth) (8.5 per cent), *lipis* (in cattle East Coast Fever) (7.8 per cent) and *lochurucha* (Trypanosomiasis) (6.2 per cent) are of importance. Similar results were recorded in another sample (Young 1988): tick-related diseases had the highest mortality rates followed by diarrhoea and *cholera. Ngoriyon* and *cheptikon,* Foot and Mouth and East Coast Fever, are given as further deadly diseases for camels. Both diseases are cattle diseases and the terms probably denote diseases in camels with similar symptoms. Skin diseases add to health problems in camels and weaken the animals.

These figures do not include the extremely high mortality rates of camels due to a recent epidemic called *chepirpirmöt,* as data was gathered before the outbreak of this disease. After an exceptionally wet year (1988) camels in Loyamoruk were infected by this lethal disease (caused by a rare strain of Trypanosomiasis, probably Trypanosomiasis evansii). Within a period of 24 months about 75 per cent of all the camels in the study area had died. Strangely, in other regions of East Pokot camels remained almost unaffected.

Cattle: Among cattle East Coast Fever is rated with the highest mortality rate (25.7 per cent) followed by worm-related diseases (19.0 per cent) and Foot and Mouth Disease (13.3 per cent). East Coast Fever is prevalent, especially in the higher altitudes, where the ticks transmitting the disease survive well. Pokot herders visit these higher lying pastures frequently during a drought where emaciated animals easily succumb to ECF. Trypanosomiasis and worm-associated diseases cause high mortality in calves. Anthrax is rare in the study area but several outbreaks have occurred over the last few decades. In Young's report ECF (Young 1988), Foot and Mouth Disease and Trypanosomiasis are rated as the most dangerous cattle diseases in Pokot. The fact that Pokot attributed 15.2 per cent of all deaths to the Evil Eye shows that western categories of disease causation and emic veterinary theories do not always fit neatly.

Goats: In goats the single most important killer disease is *pngerat* (probably Enterotoxaemia) (36.7 per cent), followed by diarrhoea in kids (17.3 per cent) and worms (11.2 per cent).[76] Young's report emphasises the high mortality rates of Enterotoxaemia and worm-related diseases too. In goat herds Contagious Caprine Pleuropneumonia (CCPP) occasionally flares up and causes high mortality. In areas where goats are not dipped regularly, mange is a severe problem. Worms also cause high mortality in goat kids.

Sheep: In sheep, plant poisoning (by a plant called *lotil,* probably Ruellia patula) was by far (73 per cent) the most important reason given for premature death due to disease. Just as for goats, worms cause high mortality in lambs. Mange is another serious health problem for sheep.

3.5.2. Livestock Mortality due to Diseases amongst the Himba

Livestock diseases do not seem to play such an important role in Himba livestock husbandry. Epidemic as well as endemic diseases are less prominent than in northern Kenya. Losses in small stock due to disease tend to be lower too. Historically livestock epidemics have caused severe losses over a wide region only in 1897/98 when about 90 per cent of all the cattle died as a result of Rinderpest. While early vaccinations helped to save 60-80 per cent of the cattle population in the Windhoek and Otjimbingwe Districts (both occupied by white settlers), only 5-8 per cent of the non-vaccinated Herero cattle survived (Schneider 1994:151). Before the end of 1897 the epidemic had caused massive losses in the neighbouring areas of south-western Angola. Due to constant veterinary checks Rinderpest has not occurred in northern Namibia since 1900.

The first cases of CBPP in Namibia were noted in 1860. The year later became known as *otjipunga* in the Herero calendar (Schneider 1994:128), the year of the lung, alluding to high cattle losses. CBPP spread to the north and by 1875 the large herds of the Ovambo had been severely reduced. Outbreaks of CBPP occurred all over Namibia during the last decades of the 19th century. However, information is lacking as to whether the disease affected the cattle herds of the Himba living in south-western Angola. Oral traditions do not report the spread of the disease at this early stage. Effective control and massive vaccination campaigns, however, resulted in the eradication of CBPP in the commercial sector by 1904. However, in Ovamboland and the adjoining areas of southern Angola the disease has become endemic. In Kaokoland the first outbreak of CBPP was officially noted in 1925. From then on there were severe local outbreaks until about 1940. Vaccination campaigns and severe restrictions on mobility helped to eradicate the disease. It only reoccurred in 1975 but was kept at bay in the 1980s. In 1993 and 1994 there were isolated outbreaks in the north (all Schneider, 1994:127ff). Losses due to CBPP from the 1920s to the 1940s were severe in certain localities but never affected the entire Kaokoland. Since the late 1930s, regular vaccination programmes have helped to considerably reduce livestock mortality due to diseases. Nowadays, Contagious Bovine Pleuropneumonia is occasionally reintroduced from southern Angola. Usually the affected herds are quickly isolated which limits the losses to a few households.

Case study: CBPP outbreak in Omuramba 1994

On 1st April 1995 *epunga* (CBPP) was noted for the first time in Vahikura's household. He had inherited cattle in Angola and had recently brought a small herd to his homestead in Namibia. In the beginning it was just one cow which was coughing and mucus was oozing from its nose. The veterinary department reacted swiftly and only a day later a veterinarian was there who shot the infected cow and cut out the lungs to send them for testing to Windhoek. However, the Himba did not need the results from the Windhoek laboratories. They were sure that the animal had been infected by CBPP. They immediately started to restore the communal livestock enclosures in anticipation of a major vaccination campaign. A major vaccination campaign, however, took place only about six weeks later. During the following weeks Himba

> herders tried to isolate the affected herd. The first idea was to drive the infected herd to an old quarantine camp at the river and isolate it there. The first section of the herd was already brought there when the households living nearby protested and were able to exert enough pressure on the unlucky herd-owner to force him to return to his homestead. Now some of his immediate neighbours decided to leave the area, others started to build long fences to prevent their cattle from mixing with the infected herd.
>
> Eventually the vaccination campaign started in the second and third week of May. For many cattle in Vahikura's herd, these vaccinations came too late. By the middle of July some 15 cattle had died in Vahikura's herd. Towards the end of July a first death case due to CBPP was reported from another household. This news caused extreme irritation as it became clear that all attempts at isolating the herd had been in vain. Only a few days later coughing cattle were reported from yet another herd. Some herders were accused of hiding away infected cattle in order to prevent measures of isolation by their neighbours. Suddenly CBPP spread rapidly and within a few more days about eight households were reporting sick animals. However, losses in these households remained low. They only lost one or two animals. Apparently the vaccination campaign had reached them early enough to prevent major losses.

In a survey on the recent distribution of livestock mortality due to specific livestock diseases it turned out that epidemic diseases are of little importance nowadays. It is mainly endemic diseases and high rates of calf mortality that make up the losses in herds. The most important killer disease in Himba cattle herds covered in the survey was Botulism (*ombindu*) (23.4 per cent). Botulism results from the toxin caused by Clostridium botulinum which is reproduced in the decomposing carcasses of affected cattle (Schneider, 1994:38ff). The main factors leading to the high prevalence of Botulism are phosphorus-deficient soils, resulting in pica (a strong appetite) in cattle which have a craving for bones. Minor quantities of bones infected with C. botulinum result in death. Calf mortality is astonishingly high: diarrhoea in calves (16.9 per cent), late abortions (11.3 per cent) and deaths immediately after birth (13 per cent) make up a major percentage of all losses. The causes for high mortality rates in calves are unknown (see Table 8).

Small stock herds in Kaokoland on the whole seem to be very healthy. There are very few epidemic diseases of any importance. The major killer disease in goats and sheep is *omutima* (probably Enterotoxaemia) and mange (*ongana*).

3.5.3. Comparison of Livestock Mortality due to Diseases

On the whole Pokot herds are more greatly affected by diseases than Himba herds. The aridity of the climate in Kaokoland prevents ticks and flies from becoming a major problem in north-western Namibia. Furthermore, veterinary services in northern Namibia compare fairly well with Kenya, where

Table 8. Major diseases in Himba cattle herds

Scientific term / translation of vernacular term	Vernacular term	No. of deaths	Percentage
Botulism	*ombindu*	56	24.5
CBPP	*epunga*	4	1.8
Anthrax	*eteva*	5	2.2
Diarrhoea	*ohano*	40	17.5
Ticks	*ongupa*	3	1.3
unspec.	*okatikura*	1	0.4
Too much milk		10	4.4
Eye disease	*omutjise womeho*	11	4.8
Lumpy Skin Disease	*ozomburu*	1	0.4
'Intestines'	*okaura*	5	2.2
'Heart'	*omutima*	2	0.9
'Bladder'	*onango*	1	0.4
Unspec.	*okandjumba*	2	0.9
'Part of stomach'	*ombumba*	3	1.3
'Liver'	*ehuri*	4	1.8
Retained afterbirth	*omusepa*	2	0.9
'Urine'	*omatuta*	1	0.4
Death immediately after birth		31	13.5
Late abortion		27	11.8
Cold		1	0.4
Unknown		19	8.3
Total		229	100.1

Note: The total number of recorded cattle was 4634

veterinary care in communal areas is on the retreat due to budgetary constraints. A look at species-specific patterns highlights the differences.

Rinderpest had a profound impact on both economies in the 1890s. Mortality rates in Himba as well as in Pokot cattle herds were at about 90 per cent. In both systems the livestock-based economy broke down for several years. However, while Rinderpest did not reoccur again in Himba herds, Pokot herds were affected by the disease several times during the course of this century. Pokot cattle herds are frequently infected with other epidemic diseases: ECF, Anthrax, Black Leg and Foot and Mouth Disease decimate herds drastically. Endemic diseases such as Trypanosomiasis are prevalent too and take a heavy toll on herds. In stark contrast to this Himba cattle are rarely affected by epidemic diseases. In recent years CBPP infections have flared up occasionally but remained restricted to a few herds due to rapid veterinary measures and isolation of the affected animals from healthy ones.

Camels offered new opportunities to the Pokot when first adopted at the beginning of this century. However, camels offer problems too. In Pokot land they are kept in a climatic border zone: usually camels thrive in rainfall regimes lower than 600mm. The high level of Trypansomiasis infections, which reached epidemic dimensions between 1989 and 1992, caused losses of up to 75 per cent in some regions.

Among the Pokot goats and sheep show high mortality figures. Goats are affected by epidemic (e.g. CCPP, mange) as well as endemic diseases (Enterotoxaemia). Sheep frequently succumb to plant poisoning. Among the Himba small stock tend to be quite healthy. Mange and probably Enterotoxaemia are the only epidemic diseases of any importance.

The distribution of losses was found to be random for both societies. Generally wealthy people do not seem to have an advantaged access to veterinary drugs. Among the Pokot nowadays drugs are widely spread and are sold by the cubic millimetre from a syringe, so that even poorer households are able to afford the necessary quantities of the drugs. Under-dosing of antibiotics is frequent and may cause severe problems in the future by creating unwanted resistancies. In Kaokoland veterinary services are mainly free of charge. Even if the political elite are occasionally able to manipulate the course of vaccination campaigns to their advantage, these benefits seem to be quite low.

3.6. VIOLENT CONFLICT

In many parts of Africa violent conflicts have caused major famines during the last few decades. With violent conflicts in Somalia, the Sudan, Chad, Niger, Mali, Uganda and northern Kenya, countries with pastoral nomadic populations have been hit hard by violence. The massive armament of wide parts of the population has led to an inextricable net of violence, intermeshing intra-community disputes, intertribal conflicts and civil wars (Fukui & Turton, 1979; Fukui & Markakis, 1994). Herder communities such as the Mursi, Nyangatom and Dassanetch of southern Ethiopia were armed with guns in the early 1970s leading to an explosive increase in violent conflicts with high mortality (Turton, 1979; Tournay, 1979, Almagor 1979). In the early 1990s, in a second wave of militarisation, these groups gained access to automatic weapons (Turton, 1994; Fukui, 1994) unleashing another round of destructive warfare. Violent conflicts have been a threat to pastoral viability for both of the populations under study here. Warfare has been endemic in the pastoral areas of northern Kenya and neighbouring countries.[77] In contrast to this rather violent character of interethnic relations in northern Kenya, major conflicts have shaped history only twice in north-western Namibia and south-western Angola: between the 1850s and roughly 1900 and again between the 1970s and 1990. Both times violent conflicts were carried into the region by outsiders. There are no traditions in which the Himba as a people undertook raids themselves on neighbouring populations.

3.6.1. The Gains and Spoils of Violence: Interethnic Violence in North-western Kenya

Pokot history has been marred by violent warfare. Seen over a longer period one may say that interethnic violence has only been interrupted by about

40 years of colonial peace. The introduction of guns around the turn of the century, the increase in the availability of guns since the 1950s in connection with the civil wars in Sudan and later Uganda and another rapid increase in armament in connection with civil wars and disintegration in Somalia, Ethiopia and the Sudan in the early 1990s, has led to escalations of local conflicts.

The results of violent conflict are increasingly felt in northern Kenya. In the early 1990s the Turkana were unable to exploit about 50 per cent of their pastures because of warfare. The same holds true for Pokot and other pastoral groups in Kenya's north. In recent years raiding has become partially commoditised (Gray et al, 2003; Fleisher, 1998; Hendrickson et al., 1998; Hutchinson, 2000).

The Pre-colonial and Early Colonial Period

Until about 1880 the Pokot had mainly fought with the Laikipiak and Purko Maasai who had settled in the Baringo basin (Jacobs, 1979:43f).[78] By about 1890 the Pokot had obviously succeeded in ousting the Purko from the valley.[79] What did pre-colonial warfare look like? In all pastoral groups a broadly similar pattern of raiding had developed: the Pokot contrast sharply between livestock theft (*chorisyö*), which is a clandestine act carried out by a few men and usually non-violent, and raid (*setat, luk*), which is open and involves many, sometimes several hundred men and is exceedingly violent. According to their traditions the Pokot were the net winners in this game. They succeeded in amassing their herds and in expanding their territory. This did not preclude vicious raids by Laikipiak, Purko and Turkana, which resulted in the loss of livestock and human life.

By around 1910 the fortunes of warfare changed. The Turkana had been armed by Ethiopian border-men who were active in trading exotic commodities such as ivory and ostrich plumes and expanded the influence of the Ethiopian empire. Interethnic conflicts between 1910 and 1920 foreshadowed a new pattern of interethnic conflict: access to firearms through trade and the impact of external power groups (Ethiopians, British colonial forces) became essential variables in determining the fortunes of warfare. With the help of the Ethiopians and under the charismatic leadership of two prophets the Turkana organised devastating raids preying on Dodos, Karimojong and Pokot livestock. The following case study gives an abbreviated account of this new pattern of violent interaction:

> In 1908 a Rinderpest epidemic had diminished Turkana herds (Lamphear, 1976b:232, 1976a). A year later in 1909 they attacked the pastoral Pokot at the Turkwell river. In 1910 the Kings African Rifles struck back and raided the Turkana capturing about 10,000 cattle and killing 30 Turkana.[80] Pokot, Njemps and Tugen levies took advantage of the situation and attacked the Turkana on their own account. Since 1911 conflicts between the Turkana and their pastoral neighbours escalated. In 1912 the Turkana undertook several major raids. At the same time some Turkana openly rebelled

> against the colonial administration, declined to pay taxes and refused to accept administrative staff as authorities. British officers estimated that the Turkana had a total fighting force of about 25,000 men and about 1,000 men with guns.[81] When the British had to withdraw troops because of the First World War,[82] the Turkana, who were obtaining ever more arms through trade with Ethiopia, forcefully attacked the Samburu, Pokot and Ugandan Karimojong. The British reprisals in 1915 were drastic: 407 Turkana were killed, 86 wounded, 63 captured, about 20,000 cattle, 8,500 camels, 127,000 goats and 6,600 donkeys were "confiscated".[83]
>
> However, the Turkana still struck back: in February 1917 they raided some 13,000 cattle from the Karimojong. In May they attacked the Pokot of Mösol and the Baringo basin. Their loot was immense: in the attack on the Mösol Pokot they took 5,000 cattle and in the attack on the Baringo Pokot 13,000 cattle, 35,000 small stock and 300 donkeys.[84] The total number of Pokot cattle was given as 79,600 and that of small stock as 135,200 for the year 1914/15 (ibid.) This means that in a day the Pokot lost about 16 per cent of their cattle and 26 per cent of their small-stock. British punitive expeditions in 1918 recovered some of the cattle. About 370 Pokot and 200 Tugen and Njemps warriors participated in these attacks.[85] The northern Turkana lost most of their cattle during these officially sponsored raids. 498 Turkana were shot dead (see also Lamphear 1976b:241).[86]

The drastic measures of the British in the 1920s and 1930s guaranteed a period of peace. Although livestock rustling went on to some degree in West Pokot and adjoining parts of Uganda the following decades are still remembered as years of peaceful interactions. The maintenance of peace fully relied on the strong state. District officers always regarded the maintenance of intertribal peace as their special duty. Each annual report contained a lengthy section on intertribal relations reporting minor skirmishes, misunderstandings and governmental activities along the tribal borders, which then usually coincided with district boundaries.

Interethnic Violence since the 1960s

Already in 1953 there had been heavy fighting between the Pokot and Karimojong. In 1958 these conflicts reached a climax when in one year more than 700 people were killed. In 1962 Pokot engaged police patrols in combats along the Kenyan/Ugandan border. The administration installed "blood money committees" to end the vicious circle of raid and counter raid. In 1964 (now under an independent Ugandan government) police troops were stationed along the border and a curfew was imposed. By the middle of the 1960s violent conflicts had extended to the Pokot/Turkana boundary. The Turkana repeatedly raided the Pokot living near the tribal boundaries and in 1967 there was a first spate of attacks against the Pokot of the Baringo plains. Several Pokot were killed. These raids continued until the early 1970s. At that time the Turkana were armed with guns while the Pokot were still fighting with spears. Attempted counter-attacks by the Pokot were thwarted off successfully by the Turkana. In view of their inferior weaponry and the apparent governmental reluctance to engage in punitive measures the Pokot felt forced to buy guns.[87] From the middle of the 1970s the Pokot counter-raided. The Baringo District Annual Report of the year 1982 gives a vivid picture of the degree of violence (see Table 9). Heavy fighting kept on until 1984 when a major and at times ruthless disarmament campaign was conducted. The army made use of a disastrous drought. Units of the newly founded Anti Stock Theft Unit were stationed at waterholes. Herders were only allowed to water their cattle after they had handed in their weapons. Dietz (1987:191) describes the destitution brought about by army operations in the pastoral areas of West Pokot. While many weapons were handed in, more than just a few remained with their owners. The disarmament campaign of 1984 and probably major losses due to drought in the same year calmed down the situation for a couple of years. Since 1991, however, a new series of interethnic raiding has started.

The demise of the Barre and Mengistu regimes in Somalia and Ethiopia caused a dramatic increase in the availability of modern automatic firearms. While in the 1970s, guns had been bartered for up to 30 oxen, now an automatic gun was exchanged for a maximum of ten oxen and frequently even only for two or three. Well-armed Pokot warriors not only raided the Turkana but also undertook raids on the Samburu - a thing which had never been heard of before and of which elders were utterly critical. Many Pokot households migrated to areas where they hoped to be out of the reach of Turkana

Table 9. Interethnic violence -the year 1982 (Baringo District)

Number	Raiding in the Baringo District
1	31/3/82 Pokot raided Kapedo and made away with 119 head of cattle.
2	21/5/82 Pokot cattle rustlers again raided the same manyatta and made away with 856 goats, 100 goats, 100 cows and 200 sheep; they were intercepted and 32 cattle, 114 goats and 98 sheep were recovered.
3	27/5/82 Turkana took revenge and raided Nakoko. They made away with 40 cows.
4	3/7/82 Pokot attacked the Kapeddo Turkana and made away with 988 cows.
5	17/8/82 Turkana attacked Pokot at Ngoron and made away with 100 cattle. They were all later recovered.
6	24/8/82 Pokot attacked Kapedo Turkana and made away with 2,000 cows, 600 goats, 900 donkeys and 800 camels
7	26/8/82 Pokot attack Kapedo again and made away with 9 donkeys and 200 goats and sheep. All were later recovered.
8	25/9/82 Turkana attacked Nakoko and made away with 23 cattle and 360 goats.
9	27/9/82 Pokot raided Kapedo and swept away all the animals. This was the worst raid as there was a severe exchange of fire.
10	25/11/82 Turkana raided Nakoko and made away with 770 sheep and goats, 770 cattle, 100 camels and 89 donkeys. The cattle were later recovered

Source: Baringo District Annual Report for 1982

raiders. During a visit to the area in the middle of 1996, just after a good rainy season, raiding and counter-raiding had become the major factor determining the course of pastoral activities. Towards the end of the 1990s it escalated once again. McGovern's report on the fate of a development project concludes: "Increasing social unrest has marked the year of 1999. The fighting among the Pokot and Marakwet escalated and acts of violence have been frequent and often brutal." (McGovern, 1999:3). In the late 1990s intertribal violence degraded into acts of ethnic cleansing and outright banditry (Bollig & Österle forthcoming).

Damage Resulting from Violent Conflict Management

Losses due to low intensity warfare were huge: large areas of pasture land were no longer used and other areas degraded even more rapidly due to massive overstocking. Market exchanges broke down in several instances. Livestock was lost to raiders or "had to be" spent on guns. Worst of all, many people died due to the escalation of violent conflicts. The following paragraphs describe the damage in some detail.

Diminished Access to Grazing: In order to evade violent conflicts herders moved away from border areas. The Pokot call these outlying areas that are constantly endangered by raiders *serim*. These areas are evacuated as soon as rumours on pending raids are circulated. Occasionally prophets may give a warning to all the people living at the *serim* and urge them to leave the area summarily. A no-man's land of at least some 100 kilometres length and some 50 kilometres width was no longer used by the herds. If a modest stocking rate of one head of cattle per 20 hectares of land is taken, 5,000 km^2 would offer space for roughly 25,000 head of cattle. This is almost half of the entire Pokot cattle herd. Throughout the 1990s these pastures were not grazed.[88] These are the high opportunity costs of interethnic raiding! And these costs are still higher if (a) increased degradation in overstocked areas is taken into account and (b) the negative impact of under-grazing on the development of the vegetation is rated:

(a) Degradation due to overstocking has been rampant on the northern Baringo Plains for a long time. However, desertification is extreme in the southernmost locations of Pokot land and around the few centres. Since the early 1970s Pokot herds have been concentrated along the Loruk - Maralal road, near the police station at Loruk, around the centres Nginyang, Chemolingot and Tangulbei. The proximity of police stations promised some protection against raiders. However, this extreme imbalance led to a situation in which probably more than two thirds of all livestock were kept on a fraction of the land.

(b) One may be tempted to hypothesise then that pastures that are not used for some time recover and perennial grasses re-establish themselves after the stress of grazing by large herds of herbivores is diminished. However, the opposite seems to happen: Conant (1982:117ff) shows on the basis of satellite imagery, aerial photography and ground surveys that grassland communities were replaced by Acacia thornbush on a large scale in areas that were no longer grazed in the north-western Baringo plains. In the absence of herding, Acacia spp. doubled its distribution and grass retreated from 38 per cent in 1973 to 13 per cent in 1978. The loss of grassland was tremendous and the Pokot, who began to use parts of the area again after raiding had stopped, avoided stretches of bushland altogether. They decided that there was too little grazing for cattle left and that with bush encroachment the Tsetse flies, the vectors of Trypanosomiasis, had become prominent in the area.

Losses due to Livestock Theft: It is hard to quantify the losses from raids. Each raid of one party provokes the counter attack of another party. Hence, the raiding economy offers gains and losses at the same time. Furthermore army units are in the game as a third party. On the one hand they confiscate livestock to compensate Turkana who have been raided by the Pokot and on the other hand compensate Pokot who have been raided by the Turkana. However, actors from both parties claim that a major part of the confiscated animals remains with the army units.

The overview on stock losses and recovered stock (i.e. livestock confiscated by the army and given back) for the years 1976 and 1977 proves that net losses were high. In 1976 the Pokot lost some 700 cattle and in 1977 almost 5,000 (about 10 per cent of the total herd). Goats that were stolen in both years were hardly compensated for: more than 7,000 goats were lost within one year, about 5 per cent of the total herd. Subsequent annual reports did not contain further statistics on livestock losses. According to the Pokot losses due to raiding increased as well as gains from their own raids on the Turkana.[89] However, households which lost livestock did not necessarily gain from raids and vice versa. During raids it is especially young men who gain livestock and it's the older, well-established households who loose most.

Losses due to the Acquisition of Guns: When asked who gained the most from raids, Pokot, Turkana or weapon dealers, young Pokot men unanimously admitted that the highest gains were probably made by the gun dealers. Throughout the 1970s and the 1980s the Pokot had to pay very high prices for guns. The better weapons cost thirty head of cattle and above and even an old fashioned gun was still sold at 10 heads of oxen and more. Table 10 gives an overview of cattle spent on guns in five households (household no 3 had to acquire a second gun after the first one was lost during a raid). The figures indicate that the Pokot were forced to pay heavily for guns. These expenses forced men to raid in order to compensate for the extraordinarily high losses due to the acquisition of guns.

Some young men recounted how they had to postpone marriages because they lacked the adequate number of cattle for paying bridewealth after buying a gun. Obviously they were interested in regaining some of their losses. Did they succeed in doing so? When the loot is distributed it is mainly those men who participated in the raid with a gun who get the larger share of the loot. Men who take part without guns only get a few animals. It is said

Table 10. Livestock paid for guns in the 1970s and early 1980s

Household	Price paid Oxen	Cows/heifers	Smallstock	Camels	TLU
1	15	36	0	1	52.2
2	20	5	0	0	25.0
3a	24	1	0	0	25.0
3b	2	10	30	1	17.0
4	25	12	0	1	38.2
5	18	22	0	2	42.4
6	9	11	0	0	20.0

that men owning a gun contributed disproportionately to the success of the raid. However, these guns cost tremendous numbers of livestock throughout the 1970s and the 1980s. Hence, raiding begins with profound and rather definite losses due to the expensive purchase of a gun and ammunition. Table 11 links these expenditures to gains made in raids and shows that all the men but one made net losses. They had not been able to raid sufficient numbers of cattle to compensate for the purchase of the guns over a span of five years and more. Only the three participants who did not invest in a gun made a marginal profit. Nevertheless, many herders felt forced to buy guns when they saw their enemies arming themselves.

The data on losses and gains derived from raiding has made it clear that violent interethnic conflict plunges the Pokot - especially young Pokot men - into a dilemma. Even when the prices for guns dropped significantly in the 1990s they were still a sizeable investment. On the one hand young men may win numerous heads of livestock in a very short period of time if luck is on their side and they may gain prestige as fierce and brave warriors. This may enable them to establish independent households early on and

Table 11. Gains and losses due to raiding and purchases of guns

Household	Spent in TLU	Received in TLU Total	Per raid	Total Gain/Loss
1	52.2	13.0	1.9	−22.2
2	25.0	30.0	5.0	+5.0
3	42.0	28.3	3.5	−13.7
4	38.2	27.0	4.5	−11.2
5	42.4	24.2	3.5	−18.2
6	20.0	15.4	2.2	−4.6
7	0	1.8	0.6	+1.8
8	0	2.2	1.1	+2.2
9	0	0.4	0,2	+0.4

Note: Figures vary slightly from a table published in (Bollig 1990a). Small stock units were counted in a different way and raided donkeys were also included here.

guarantee them prominent treatment by age-mates and seniors. They are favoured spouses and many men who have accumulated prestige as warriors have in fact married a second and third wife early on.[90] On the other hand losses are more definite: grazing is lost as each raid inevitably provokes a counter-raid, livestock has to be paid to weapon dealers to obtain guns and more continuously to obtain ammunition, and last but not least, livestock is lost to raiders and to army operations. However, those few people who gain from raiding are not necessarily those who loose. Losses of grazing land are borne by the entire community, losses of livestock due to the purchase of guns are generally born by seniors and losses due to raiders hit households unpredictably.

3.6.2. Caught in the Middle: Raiders, Administrators and the Military

In the introductory chapter, I delineated Himba traditions on migrations, settlement history and pre-colonial contacts with other communities. By and large exchanges with other communities have been peaceful. Violent conflict erupted when commando groups started to raid the Herero of Kaokoland in the 1850s (Lau 1987). Local herders eventually fled. Only towards the end of the 1970s did Kaokoland become a battleground again. Once again it was external forces which drew the local population into a war.

The Kwena Raids

The raids of Swartboois and Topnaar groups, collectively addressed as Kwena in oral traditions, form a prominent part in Himba oral history (Katjira Muniombara, Kozongombe Tjingee & Tjikumbamba Tjambiru in Bollig, 1997c). Battles, destruction and escape are portrayed at length. Traditions do not allow for a detailed reconstruction of what happened: battles are not connected to specific years, not even specific decades. While places and actors are mentioned, single skirmishes are not put into a larger perspective.

Although there are no figures on mortality rates, we may assume that many people lost their lives in these raids. Cattle were abducted in large numbers. The raiders were mainly after cattle which they could sell profitably at Walfish Bay or in central Namibia. Their absolute superiority in weaponry (guns) and logistics (horses, well-established trade contacts) made it easy for them to drive off major herds of cattle. In the late 1880s the Swartboois and Topnaar from Franzfontein and Sesfontein undertook several raids into southern Angola harassing pastoral refugees there and causing even more destruction (Estermann, 1981: 17/18). However, in Southern Angola they met with well-armed adversaries. In 1890 a combined Boer and native militia forced the commandos to retreat. In Namibia the German colonial administration extended its control to Franzfontein and Sesfontein in the 1890s and put an end to Swartboois and Topnaar raids.

The Violent Conflicts of the 1970s and 1980s

Only in the 1970s were the Himba once again in the reach of violent conflicts. They were caught in the middle between fighters of the Peoples Liberation Army of Namibia (PLAN) and soldiers of the South African colonial army.[91] Again it is hard to reconstruct a detailed history on what happened in Kaokoland. As far as I can see no major battles actually took place in Kaokoland besides a battle in 1989 in the Zebra Mountains which had numerous casualties on the side of PLAN but did not involve the local population. In contrast to this PLAN bases in south-western Angola were repeatedly bombed and at least one major battle took place near Iona on the western edge of Himba land. PLAN fighters tried to infiltrate Kaokoland and attack military targets several times. The worst effects on the pastoral economy, however, did not derive from violent combats but from land mines and forced resettlement.

While only very few people were killed and not much livestock was stolen, the pastoral economy was seriously affected by enforced changes of settlement patterns. The South African army and its paramilitary wing Koevoet forced the Himba to resettle near villages and army barracks. During the invasions of the South African army into southern Angola in 1976 and 1981 the local infrastructure was largely destroyed. It has not been re-established up until now due to continuous civil war and the general demise of the Angolan state.

3.6.3. Violence as a Hazard to Pastoral Viability: the Pokot and Himba Compared

A comparison does not have to go far in order to find that both cases are very different. For the pastoral Pokot raids have been a constant problem for the last 25 years or so. Besides a period of about 40 years of colonial peace (roughly 1920 to 1960) interethnic hostility has always been of importance throughout Pokot history. Pokot herders have suffered regularly from depredations caused by interethnic hostilities. The hazard of interethnic violence, however, has an ambiguous touch in Pokot views: warfare and raiding are as much a matter of chance as they are a hazard. All men today see the acquisition of fame and the rapid increase of herds as a major chance in raiding. However, the Pokot are cautious enough to analyse *tapan*, the danger of being raided, as the major threat against their livelihood.

The Himba have a very different outlook on violence. Stories about Nama raids in the second half of the 19th century are only remembered by some old men. Recent recollections of warfare depict the Himba as a population caught in the middle between warring South African government troops and SWAPO freedom fighters. The Himba joined Portuguese colonial forces at the turn of the century, and have recently joined the South African Defence Force and its paramilitary wing Koevoet in Namibia. The Himba joined foreign armies after devastating epidemics and droughts in order to reconstitute depleted herds: young men became mercenaries after the Rinderpest epidemic of 1897 and opted for employment in the South African Defence Force after ninety per cent losses in the livestock sector in 1981. Men joining the army never became heroes and especially those who joined the South African army were regarded with some suspicion.

The Pokot have had to evacuate important pastures over several years because of the fear of Turkana raids. Up to a third of the Pokot land is not used during periods of intense raiding. Those areas where pastoralists search for safety and congregate are overstocked. Himba did not loose access to pastures because of the fear of violent conflicts with neighbouring people but rather because of the pressure of colonial forces. The same holds true for losses of livestock: Pokot households frequently lost livestock to raiders. A few men also obtained livestock through raiding. The general picture however is one of losses to raiders and government units. The Himba occasionally lost livestock in colonial police operations, but never in recent history to violent raiders.

ENDNOTES

1. Roth (1994:139) questions this assumption on the basis of detailed demographic data on the Rendille. Taking into account emic ideas on population growth, he quotes Rendille informants denying any relation between *sepaade* and demographic trends.
2. Archer, one of the first officials to tour the area, reported that Pokot and Turkana were living together in "Nagoretti" (probably Nakorete) and that Turkana were in possession of the springs at Nginyang. Both places, Nakorete and Nginyang, are today in the midst of Pokot territory and almost one hundred kilometres away from the nearest Turkana settlements (KNA DC BAR 3/5 Archer, Report on a Journey in the Suk Country, July 1904).
3. Broch-Due (2000) gives a comprehensive account on how British colonialism formed and transformed ethnic identities and boundaries.
4. KNA DC BAR 3/4 1936: Annual Report 1936 Revised Census Baringo District for 1936: 1934 figures showed a decrease of 629 men and 896 women. This population shift was a result of heavy stock losses due to drought. Many people moved to West Pokot "and still more settled in out-of-the-way areas near the boundary and thus avoided having their names put on a Census Register".
5. The figure compares well with growth rates estimated for the neighbouring Samburu (Spencer, 1998:214).
6. Excluded are figures for 1923, 1924, and 1931 which are obviously much too high and are probably mismatches.
7. For camels only figures for 1928-1933 are given. The number of camels in those days was around 1,000 for the entire area. Camel husbandry amongst the Pokot apparently only gained momentum after the Pokot were paid with camels after a punitive expedition against the Turkana in which they participated in 1917 as auxiliaries of the British army (Bollig, 1992b).
8. KNA DC BAR 3/4
9. Possibly data are less reliable for the western most parts of Kaokoland. Several documents state that the number of "wild Tjimba", was estimated at about 500 individuals.
10. Namibia National Archives (NAN) SWAA 2516 A552.

11. Oorlam is a designation of pastoral cum raiding groups of mixed Nama and coloured descent who controlled large parts of central Namibia in the 19th century (Lau, 1987). The term was used until the 1920s.
12. NA NAO 28 Extract from Monthly Report for months of May and June 1929, Post Commander SWA Police to Native Commissioner Hahn: Kaokoland census 1929.
13. These differences may be partly due to more accuracy in herd counts in the south - these were the people that had to be relocated - and partly due to the fact that in the north stockless people, designated as Tjimba, made for some part of the population.
14. The sex ratio for the grand totals of the 1942 census is at 1:1.3. There is no difference in these ratios between Himba, Tjimba and Herero. It is not clear as to what factors this imbalance may be attributed.
15. Based on other data Talavera et al. (2000:32) also gives the figure of 3.2 children as completed fertility rate.
16. The figure accounted for on the basis of fertility accounts indicates a growth rate as low as 0.08 per cent per annum. This figure is exceptionally low and deserves some scepticism.
17. There is few comparative data on these aspects from ethnodemography: Lang (1997:10ff) reports an inter-birth interval for the !Kung of 4 years and for the Sudanese Mahria of 2 to 3 years. With the American Hutterer, a highly fertile population, the interval is down to 2.1 years. The !Kung reach a completed fertility of 4.3 children,
18. At this stage it is not possible to contextualise demographic growth historically in a more detailed and complex way. However, demographic growth is far from being "natural phenomenon" and is closely tied to the political economy of a region.
19. A definition of degradation by Abel and Blaikie stresses the effects of irreversibility (Abel & Blaikie 1989:113): "Rangeland degradation is an effectively permanent decline in the rate at which land yields livestock products under a given system of management. 'Effectively' means that natural processes will not rehabilitate the land within a timescale relevant to humans and that capital or labour invested in rehabilitation are not justified."
20. Kenya National Archives (KNA) District Commissioner (DC) Baringo (BAR) 3/4 Annual Report 1959
21. Many interviews among the Pokot were not taped. I rather tried to scribble down respondents' answers during the interview.
22. In an article on pasture degradation in West Pokot District Conant (1982:114) calls bush encroachment justifiably "green desertification". For further literature on this little understood problem which obviously results from overgrazing as much as from undergrazing see Belsky (1990).
23. In a report to the mining company he was working for, Kuntz said "Under high acacias in thick green grass grazed hundreds of cattle and thousands of sheep; evidently I had got to the hitherto legendary Ovatschimba paradise ... For several hours I rode up the river, finding village on village and water everywhere in the riverbed." 12/10/1910 Kuntz's Report to the Kaoko Land und Minengesellschaft, 1910 (NA Accessions 327).
24. Oral evidence as well as aereal photographs of 1975 gave evidence into this direction.
25. The analysis of aerial photographs from 1975 for the Omuhonga basin shows 23 homesteads for which the size of the inner enclosure, circumference of inner and outer enclosure, and additional fences could be measured. Additionally the size of some 10 gardens and the length of their fences could also be determined. The average size of homesteads was 3,907 m^2 (range: 1,216m^2 to 12,988m^2) and of inner enclosures 110,5m^2 (range 52m^2 to 268m^2). Inner enclosures had on average a length of 36 metres (range 23 to 58 metres) and outer enclosures an average of 228.5 metres length (range: 127 to 469 metres). For 15 homesteads the circumference of severe denudation around the homestead could be determined: it was at 22,807 m^2, which indicates that per household about six times more ground is severely exploited (i.e. severe cutting of all trees, complete denudation of bush and grass cover) than the homestead structures actually cover. In 1975 23 households were counted in the basin. Some 85.8 hectares were covered by household structures and 514.8 hectares were exploited in an extreme way. The number of households rose sharply from

1979 onwards. Around 1985 it had reached a number of about 80. Then some 288 hectares were covered by structures and 1,728 hectares were exploited in an extreme way (Sander, Bollig & Schulte, 1998).

26. Brunotte & Sander (2000) give radiocarbon dating for charcoal remains from the Kunene basin. They assume that the burning of riverine vegetation since c. 1200 indicates increasingly dense settlements along the river
27. Malan & Owen-Smith (1974:167) report old people as saying that tall perennial grass used to grow in many places. I obtained very few traditions hinting in this direction. The general idea was that the cover (of annuals) changed according to annual rains and that these had become less over the years.
28. However, in interviews conducted in 2004 herders indicated a decline in good annual fodder grasses and an increase of the lowly valued *ohoke* (Aristida spec.) for many pastures.
29. For a general discussion on the conflict-laden relationship between states and African pastoralists see Azarya (1996)
30. Obviously there had been some taxing before. Archer reports in 1904 that Pokot had not paid taxes regularly. What kind of taxes these were is not stated clearly (KNA DC BAR 3/5 Report on a Journey in the Suk Country July 1904).
31. KNA DC BAR 3/5 Labour Report 1910: "Labour was unobtainable locally as the natives of the district were refusing to undertake any work other than that of herding stock. Labour was recruited as follows: Swahili and Munyamwezi as porters, Kavirondo as shopkeepers, Kikuyu as Shamba workers, Njamusi and Kamasia as herders."
32. KNA DC BAR 3/5. Trade Report 1909 by A. Bruce.
33. Evidence on this aspect is contradictory: While in 1924 an officer deplored that the Local Native Council of Baringo II (which mainly consisted of the Pokot area) had no revenues because there were no trading centres in the area (KNA DC BAR 3/4 Local Councils, 1924) a further report on a meeting with the Local Native Council Baringo II in 1931 stipulates that traders were only permitted to remain seven days in one camp. Although being in favour of trade in general, the government prohibited the establishment of trading posts for several decades. PC/RVP/6A/1 11B/2 Local Native Council, Report on a Meeting, Nginyang 26.1.1931).
34. During the 1933 drought Pokot and Njemps were able to feed themselves largely by exchanging hides for maize (KNA DC BAR 2/11/4).
35. Anderson (1984) analysed the changing environmental policies of the colonial administration in the Lake Baringo region. He proves that administrators' perceptions of environmental decline were more dependent on government campaigns than on local empirical observations.
36. During the short field trip in March 2004 it became obvious that, especially during the last four years, livestock trading had expanded considerably with secondary trading sites springing up in remote areas. This process is not analysed here.
37. The title of this paragraph alludes to Jean Ensminger's book "Making a Market: The Institutional Transformation of an African Society" on the emergence of market relations amongst the Kenyan Orma pastoralists (Ensminger 1992). The history of Kaokoland points in the opposite direction; while trade flourished for some decades prior to the advent of colonialism, it was categorically inhibited by the South African colonial administration later on.
38. Many had decided to shift their premises from Walfish Bay and Omaruru to Mossamedes as the ivory resources of central Namibia were extinguished and the Herero-Nama wars had made trade risky (Siiskonen, 1990:146-148).
39. A side story of the booming ivory trade in south west Angola, which cannot be covered here, is the history of the Thwa, nowadays a group of despised craftsmen. They claim that they were professional elephant hunters during that period and that, only when the elephants were depleted, did they turn to smithing and pottery. During the earlier days they called themselves "Thwe", which they translate as "the courageous ones" (interview with Tjihandura 05/07/1995 at Ohikara).

40. This group of Dorsland Boers originally stemmed from the Transvaal. They had made their way to southern Angola moving from the Transvaal along the western fringe of the Kalahari through Kaokoland where, in 1879, they stayed for at least one season at Kaoko Otavi (Immelman, 1978: 131/132; William Chapman, n.d.).
41. Notably the two mercenary leaders Vita Tom and Muhona Katiti controlled large followings of armed men. Through them the Portuguese engaged the services of local militias.
42. Lau (1987:41 ff) has described the political organisation of Nama/Oorlam commando groups in great detail. Like the groups in southern Angola, these commandos relied heavily on livestock raiding.
43. Knowledge on these groups is derived from oral traditions and few reports of German travellers (see Ngeendepi Muharukwa and Eengombe Kapeke in Bollig, 1997; Hartmann, 1897, 1902/03; Kuntz, 1912, and Vedder, 1914).
44. Kuntz 22/10/1910 von Kaoko Otavi, NA Accessions 327.
45. The original text says: "Die Bewohner des Kaokofeldes werden von allen Seiten ausgenützt, so dass ein Emporkommen eigentlich ausgeschlossen ist. Von den Portugiesen werden sie ausgezogen, von den Ovambo betrogen, und in Sesfontein erhalten sie nicht, was sie billigerweise für ihre Zahlung erhalten müssten".
46. NA ADM 156 Major Manning: Kaokoveld Tour Diary Notes, November 1917, and NA ADM 156 Second Tour Kaokoveld by Major Manning: Disarmament Native Tribes, August 1919.
47. SWAA Kaokoveld A 552/1 Monthly reports 1926-1938, here Post Commander Tshimhaka to Native Commissioner Hahn, 11/1926. In the colonial records the place name *Tshimhaka* is used. However, when not directly referring to a colonial document I will use the correct Herero spelling *Tjimuhaka*.
48. PTJ Native Commissioner Ondangwa to Police Commander Tshimhaka 21/10/1930
49. PTJ Commander Police Post Thsimhaka to Native Commissioner Ondangwa 3/11/1930.
50. SWAA Kaokoveld A 552/1 Monthly reports 1926-1938, NC Ondangwa to Sec for SWA 14/9/1932.
51. NAO 28 Administrator Windhoek to Native Commissioner Ondangwa 19/11/1932.
52. NAO 28 Native Commissioner in Charge Outjo to the Commandant of SWA Police Windhoek 6/2/1933.
53. The Kamanjab-based trader Gärtner invented a coupon system: his agent received livestock in Kaokoland and gave out coupons for them. With these coupons people travelled to Kamanjab and bought things in Gärtner's shop in Kamanjab. Occasionally Gärtner imported major herds of sheep apparently observing governmental quarantine regulations. Gärtner's business seems to have been extremely successful for a short period of time. At the time he was intercepted, his agent was herding about 930 sheep at Otjondeka. Gärtner's efforts were obviously not the exception. Another trader, Mr. Borchers was allowed by the Administrator to introduce 350 sheep in 1935 and another 450 in 1936 from Kaokoland to the Police Zone (NAO 30 Secretary for SWA, Courtney Clarke, to Senior Veterinary Surgeon WDK 2/12/1935; NAO 30 Administrator WDK to Veterinary Surgeon WDK 29/1/1936). Chief Oorlog obviously was one of Borcher's main suppliers - an indication that the local elite were able to circumvent trade inhibitions to some extent.
54. NAO 30 Magistrate Outjo to Station Commander SWA Police Kamanjab 7/5/1938.
55. NAO 30 Native Commissioner Ovamboland to Commander Police Post Tshimhaka. n.d. probably March or April 1938.
56. NA SWAA 2513 A552/1 Inspection Report: Kaokoveld Native Reserve, 10th October 1949, Native Commissioner Ovamboland to Chief Native Commissioner Windhoek.
57. NAO 29 Officer-in-charge, Native Affairs, Ohopoho to Chief Native Commissioner Windhoek 22/12/1945.
58. NAO 62 In a letter Native Commissioner Eedes comments critically on the consequences of this new law but asserts that it will help to control the flourishing barter trade of mission stations in the area. (NAO 62 Native Commissioner Ondangwa Eedes to Chief Native Commissioner WDK 30/6/1948): The law says: "Any person who gives or tenders

any credit note, token, goods, wares, merchandise, livestock or instrument, other than cash or a negotiable instrument, in payment of or in exchange for livestock or produce offered to him for sale by any native, shall be guilty of an offence, and shall be liable, upon conviction, to a fine not exceeding 25 Pounds or, in default of payment to imprisonment for a period not exceeding three months." (Government Gazette 23.5.1947: Draft Proclamation: Prohibition of Barter Trade in Native Areas)

59. SWAA Kaokoveld A 552/1 Monthly reports 1938-1952, here Post Commander Opuwo to Native Commissioner Hahn, 9/1950.
60. SWAA 2515 A 552/18 Native Labour, here Report of a visit by a SWANLA Offical to the Kaokoveld 6/1/1952.
61. The original quote in the document is in Afrikaans: "Ons is bang op die plase te gaan werk, uns verloor altyd en kom nie weer terug nie. Die Goewernement bring ons nie weer terug nie." (SWAA Kaokoveld A 552/1 Monthly reports 1938-1952, here NC Opuwo to Chief NC in Windhoek 26/11/1951 Notes from a Meeting in Opuwo).
62. NAO 64 17/4 Trading in Kaokoveld, Officer in Cahrge Native Affairs Ohopoho to Nat. Com. Ondangua 1.12.1951 Re.: Application for Hawkers Licence: Simon Kaatendekua. NAO 64 17/4 Trading in Kaokoveld, Officer in Cahrge Native Affairs Ohopoho to Nat. Com. Ondangua 25/2/1952 Re.: Application for Hawker's Licence in the Kaokoveld: Philip Hameva NAO 64 17/4 Trading in Kaokoveld, The Administrator to Native Commissioner Ovamboland 15/10/1952 NAO 64 17/4 Trading in Kaokoveld, Officer-in-Charge of Nat. Affairs to Nat. Com. Ondangwa 5/2/1948.
63. NA BOP N 1/15/6/1 Annual Report for the Year 1959.
64. In a consultancy report Page (1975:47/48) remarked: "Die feit dat daar geen kompetisie is teen die BBK nie wek'n groot mate van agterdog en ontevredenheid by die veeboere omdat hulle meen dat hulle veel hoer pryse sou kryn indien hulle direk kan bemark ... Blykbaar is daar ook ontevredenheid oor die bemarking van kleinvee."
65. In a later definition Glantz (1989:57) defined meteorological drought as a 25 to 50 per cent reduction in precipitation. There have been further attempts to define meteorological drought on account of the duration of rain-free periods. A clearcut definition which needs a lot of research, relates to a method of soil moisture budgeting that accounts for precipitation and temperature for a given year (Palmer Drought Severity Index, PDSI, Palmer 1965).
66. Glantz (1989:57) refers to the etymological root of drought in the Anglo-Saxon term *drugoth*, dry ground.
67. Scoones (1996:205) observes a typical pattern of losses: "Birth rates drop off dramatically to practically zero prior to the onset of significant mortality. Mortalities start slowly but then rapidly accelerate. Peak mortality periods are during the onset of rains."
68. In this chapter I will mainly concentrate on the effects of drought on livestock mortality. The effects of drought on milk production are not analysed in detail. Donaldson (1986:44ff), in a report on the Ethiopian drought of 1985, reported a decrease in milk yield of 35 per cent; the number of lactating cows was only 10 per cent of the normal level.
69. Dietz (1987:93) documents similar rates of variation for Kongelai and Kacheliba in the West Pokot lowlands. In his data no trend was detectable either.
70. If at all there is a slightly positive trend. This has also been observed for Lodwar, the centre of Turkana District (Ellis et al. 1993:36f).
71. For northern Baringo there is no data on the effect of drought on biomass production.
72. The data is skewed a bit towards younger animals because for older animals the age is frequently not remembered, but there is little question that calf mortality even during a modest drought year is still very high.
73. Rates of reproduction decrease rapidly as a result of drought. Donaldson (1986:41/42) reports that during the 1983/84 drought in southern Ethiopia the intercalving interval rose from 14.5 to at least 17.3 months. During the year following the drought he recorded a decrease in calving rates from 70-80 per cent in pre-drought times to a low of 8.7 per cent

after the drought. Homewood and Lewis (1987:628) report calving rates in the Baringo District which are reduced from 83 per cent to 69 per cent as a result of the drought of 1984.

74. One household was recorded with no losses at all. However, this household only owned 7 animals, of these only one adult cow had been with the household for a long time, two heifers had been born recently and the other four animals had been borrowed within the last few years.
75. Unfortunately the livestock census figures for these years are weak. In spite of the written statement that serious losses occurred during 1928, the figures show an increase in livestock. A major vaccination campaign in 1929 eradicated the disease for several years (KNA DC BAR 3/5, KNA DC BAR 2/11/4).
76. Enterotoxaemia (also pulpy kidney disease) is usually noted as a disease which occurs mainly in sheep (Schneider, 1994:163) but in this sample was found mainly in goats.
77. Fukui and Turton (1979) gave a first overview on the problem, representing historic cases (Jacobs 1979) and recent conflicts (Turton, 1979; Fukui, 1979; Almagor, 1979). Fukui & Markakis (1994) show how local conflicts escalated in the 1980s and early 1990s and how local conflicts were tied into larger conflicts (e.g. Turton, 1994; James, 1994; Baxter, 1994; Kurimoto, 1994).
78. Beech (1911:3) claims that the Samburu were settling the valley and that it was more the deeds of a ritual expert than of warriors which made the Samburu leave the Baringo plains. Oral accounts, however, point to the fact that Beech's Samburu must have been Laikipiak Maasai and that it was not the deeds of a wizard but the raids of warriors that made the Maasai retreat. This is also corroborated by other literature on the history of the region (see Weatherby, 1967).
79. To what extent the Rinderpest epidemic of 1891 contributed to these shifts remains unclear. Apparently both pastoral populations lost tremendous numbers of cattle during the epidemic.
80. KNA DC WP 3/43: J.W. Hym, Memorandum on the Customs of the Suk viewed in the light of the clan system, p. 60.
81. At that time the pastoral Pokot apparently owned no guns.
82. KNA DC WP 3/42: J.W. Hym, Memorandum on the Customs of the Suk viewed in the light of the clan system, p. 69.
83. ibid., p. 72.
84. KNA DC BAR 1/1 Annual Report 1917
85. KNA DC WP 3/42: J.W. Hym, Memorandum on the Customs of the Suk viewed in the light of the clan system, p. 78.
86. KNA DC BAR 3/5 Political Records for the 1918.
87. In both volumes on violent conflicts in East Africa Fukui (1979, 1994) and Turton (1979, 1994) demonstrate the effect of new turns in armament on warfare patterns in the Ethiopian Omo-Valley. Both the Bodi and Mursi in turn became victims of groups which had earlier access to automatic guns than they themselves.
88. Since about 2000, however, Pokot expanded into areas formerly held by Turkana and Samburu. This territorial expansion is not analysed here in detail. It remains to be seen whether Pokot succeed in holding on to these new pastures.
89. A closer look at the literature published in recent years on the southern Turkana, proves that at least some Pokot also gained during this period. Johnson (1990:182f) reports that about half of 22 Turkana households he worked with were raided by the Pokot during the early 1980s.
90. This comparative advantage against less courageous or less successful age mates however is levelled out at a later stage.
91. Assessing the course of violent interaction in northwestern Namibia I profited greatly from a data collection made accessible to me by Mr. Peter Reiner Windhoek who recorded events reported in local news but also in publications of SWAPO.

Chapter 4

The Perception of Droughts and Disasters

The perception of hazards and vulnerability is the foundation for risk management. Ian Scoones who looked at disasters in rural Zimbabwe stated aptly: "People must attempt to explain uncertainty in order to cope with it to offer some provisional certitude in order to provide some basis for action rather than despair. Drought has personal, emotional impacts as well as material ones." (Scoones, 1996:162). While the previous chapter described specific causes for increasing vulnerability and for disasters. This chapter will highlight the culturally embedded perceptions of individual disasters. Disasters are crucial events in every herder's life. They are part and parcel of everybody's biography and of shared memories. Manger (1996:168) describes such periods as "key events in the development of pastoral systems in an ethnographic account of the Beja". Generally the Pokot sort hazards into two fundamentally different categories. They are either *kömöy* or they are *tapan*. *Kömöy*, literally dry season, is used to denote all aspects of climatically or politically induced scarcity. *Tapan* describes dangers which arise from violent conflict. The Pokot and the Himba name years after specific droughts: *kiplelkowa* - the Drought of the White Bones, *otjiṯa* - the Year of the Dying. The generic term that the Himba use for drought, *ourumbu* (from - *rumbu*, the yellowish colour of withering grass), is used to refer to all forms of drought-related fodder scarcity and ensuing problems of livestock husbandry. The term *ondjara* - which at the same time means hunger and famine - describes, when related to a specific year, conditions of starvation and destitution. While famines (*ondjara*) frequently develop out of drought, both categories are not necessarily tied to each other. *Ondjara* also resulted from the governmental prohibition of trade and from livestock epidemics. Through the process of cultural appropriation disasters become building stones of the pastoral *Lebenswelt*[1], central in oral traditions, crucial to indigenous chronologies and turning points in life histories.

The presentation of Pokot and Himba ideas on disasters must go hand in hand with a reconstruction of single events: droughts have individual characteristics and names. Disasters are reconstructed on the basis of archival material and extensive analysis of oral accounts and personal biographies. The data on individual disasters and local perceptions is more substantial for the Himba than for the Pokot. There are several reasons for this. Archival data was richer and more accessible in the Namibian case. Then research among the Himba focused on risk management from the very beginning, while part of the three years of research among the Pokot was spent on another focus (conflict). However, there are intrinsic reasons as well which have nothing to do with the research set-up, but which reflect the treatment of disasters and history in both societies. Himba historical memory is deeper. Historical events of the 1850s are remembered well, while Pokot traditions only become more detailed when addressing processes of the early 20th century. Himba recollections on droughts are linked to their political history, to quarrels for chieftainship and colonisation. Pokot traditions on droughts are linked to the history of individual households but not to an overall historical frame.

While each drought-event is introduced with a short description of historical causes - climatic perturbations, political turmoil or livestock epidemics -leading to the disaster, the focus is on the presentation of a compact narrative of disasters for both Pokot and Himba.[2]

4.1. THE ENEMY IS US: THE SOCIAL APPROPRIATION OF DROUGHT AND DISASTER AMONG THE POKOT

Pokot oral traditions are filled with accounts of droughts and other disasters. Lineage-specific traditions frequently start with tales about droughts, earthquakes and violent conflicts which forced people to move towards the Baringo plains. The herds of the Kakirop lineage, for example, were struck by an epidemic and a short time later their people were affected by the pox. These consecutive disasters forced them to move from the hilly parts of West Pokot to the Cherengani Mountains. The Katoka, who according to tradition originated in Borana country, left the Mount Nyiru area when a disastrous drought killed all their cattle. Interethnic violence was another important cause for people moving: the Kachepunyö left the Sekerr area because of constant raids by the Karimojong, the Kaptakar migrated after many of them had been killed by the Turkana. Traditions indicate that the Pokot perceive the past as a constant struggle between different populations and between humans and a hazardous environment.

Asis – Rinderpest and the Drought of the Sun

The 1880s and 1890s are generally portrayed as a period when consecutive disasters struck pastoral and agricultural communities in Eastern Africa.[3] For the Pokot the period of disasters commenced with a CBPP epidemic (von

Höhnel 1897:768). Then Rinderpest (*kiplok*) struck Pokot herds in 1891. The panzoetic caused dreadful losses in Pokot herds. The epidemic was followed by a drought, by mange (*simpiriyon),* a contagious goats' disease, and by various human diseases. The disaster is referred to as *asis*, the sun, alluding to the scorching sun and the drought following the Rinderpest epidemic. An elder reported on the disaster.

> During *asis* people suffered a lot. There were no cattle. The Chebleng sold their women for milk and did not even ask for cattle. If Pokot men who still had one or two cows gave some milk they obtained a woman. People like the Sapiny (Sebei) were giving women for a hide. The Pokot - among themselves - gave one cow for a woman during *asis*. In bridewealth negotiations they agreed to give four cattle, but many people had to give just one cow. And if you had no cattle at all you could dig edible roots and exchange them for a woman. (Todokin, September 1993)

The Pokot had to adapt their social system rapidly to a massively decreased number of livestock. Even rich people were said not to own more than five head of cattle. Bridewealth payments were reduced to one or two head of cattle, compensation in case of murder was postponed and arrangements for livestock friendships changed. Traditions describe rather the breakdown of social standards than the disaster itself or its causes.

Former herders were reduced to foraging; they had become *pich chole mey*, poor people, the Pokot equivalent for the Maasai term *Il Torrobo*. However, neighbouring agricultural communities fared even worse. Traditions point out, that the pastoral Pokot in those days were joined by destitutes from agricultural and agro-pastoral communities. Women from other communities were married for substandard bridewealth. However, the contrary seems to have happened as well: ancestors of the Kapchepkow lost their livestock during the disaster and temporarily joined the Cheptulel Pokot in the Cherengani Mountains. Others fled to the mountains within Pokot land where they could survive on hunting and gathering. Starving people dug for tubers such as *kela* (Vatovaea triphylla) and *tapodin* (Lannea triphylla). At Korossi people dug for the *chowit* root (not identified). Traditions report that to find *chowit* roots people had to dig holes that were so deep a man could stand in them.

By the turn of the century conditions had normalised and by 1904 the Pokot had to pay taxes to the British government. Within a short period of time the Pokot had succeeded in reconstituting their depleted herds. Beech (1911:26), who administered the District in 1907, described them as wealthy herders "unintelligent but surprisingly honest people, exceptionally vain and exceptionally generous".

Pörkö Chepchok - Turkana Raids on the Pokot in 1916/1917

The Pokot did not meet colonial penetration with resistance. From the beginning they saw the British administration as an ally against the powerful

Turkana.[4] At the end of the 19th century, during a period of peace, the Turkana settled on the Baringo Plains together with Pokot herders. Since about 1910 the Turkana repeatedly raided the Pokot of Baringo District. Turkana troops, several hundred men strong, swept into the Baringo plains and the Mösol area. Loriko, an elderly informant, remembered the attack well:

> I was about 10 when the Turkana attacked the Pokot. The Turkana raided plenty of cattle and in the area of Chepchok many Pokot were killed - that is why this raid is still called *pörkö Chepchok*. Many Pokot fled to Korossi and Paka and hid there. The Turkana were so successful because they had some guns while the Pokot had no guns at all. The Turkana came in two troops. They had divided in Natan. One group went to Chepchok via Tuwöt and the other via Nalekat to Korossi. The Chepchok group started the attack. When the warning cry sounded from the Chepchok region many warriors gathered and ran towards that direction. The Turkana heading for Korossi sensed their chance and took the unprotected households by surprise. The Pokot lost many cattle and small stock that day. Later that day they tried to recover some of the animals and intercepted the Turkana on their retreat in Nasurot. They killed four Turkana but did not regain a single animal. The Turkana did not have many guns at that time. Ebei the leader of the Turkana and his son both had guns and Katodi, who raided Korossi, had a gun, in total they had four rifles. However, those Turkana were good marksmen and they killed indiscriminately. (Loriko February 1989).

The raid had gigantic dimensions. Within a day or two the Turkana stole some 13,000 cattle, 35,000 small stock and 300 donkeys from the Baringo Pokot[5]. Oral accounts and archival records describe that many women and children were killed. The Pokot perceived this raid as an attempt at their total annihilation.

Kiplelkowa - the Drought of the White Bones, 1925/27

Kiplelkowa - the Drought of the White Bones is portrayed as the major drought related disaster in colonial times. The drought is also called *kututu* which - according to Pokot informants - describes in an onomatopoeic way the trampling of cattle on a mass migration. Some older men recalled the drought from their childhood and gave lively accounts of the misery it brought. The drought probably occurred in the years 1925 to 1927.[6] While 1925 is listed as a drought year with 50-75 per cent livestock mortality rates, for 1927 mortality rates as high as 70 per cent among cattle and 45 per cent among small stock are reported in the colonial files. However, neither in the archival sources nor in oral traditions are there references to people dying from starvation in large numbers. Todokin, a senior in his seventies recounts the drought.

> I was a herdsboy in those days. My father, Kirali, a Nyongi, lived near Parpelo. We migrated to Paka to graze our cattle there. There was a

> permanent well at Tapogh. This well became so overcrowded that when you arrived at noon you had to wait until 6 in the evening to water your livestock. The herders were living on milk and *tapodin* roots. In the mountainous areas of Parpelo *tapodin, kela* and *sorich* were staple food. There were no shops and itinerant traders were rare. Occasionally an Arab trader would come along to exchange tusks for maize. Some Pokot went to Marakwet villages to exchange goats for millet and sorghum.
>
> One day an elder woke up at 5 in the morning. He cried "*Töröröt, töröröt, meyoda tupa Pokot*" (God, God return the cattle of the Pokot). They took one white oxen (however, the colour was not important T. says) and speared it. They ate it in the sacred half circle of men. They had not finished eating when it started raining. It rained continuously until a flood killed some children. The elder who said the blessing was Lowlemoi of the Nyongi generation and of the Soliongot clan and Kaparwas lineage. (Todokin, September 1993)

Two ideas are forwarded in traditions on why the drought became known as the "Drought of the White Bones". Some people say that the term refers to the fact that the ground was covered with the bleached bones of dead animals. Others say that the name was derived from the whitish marrow of cattle bones the people were living on. While there is some controversy on the etymology of the term, there is little doubt that the effect of the drought was disastrous. Cattle were dying in large numbers. As sedentary traders had not yet arrived herders could not fall back on maize. Instead they had to make use of tubers and fruits and had to slaughter starved livestock. In order to obtain food the Pokot climbed the Tugen hills to exchange skins, ropes and goats for corn. Others travelled to Marigat (about 60km away to the south) where there was a shop. Oral accounts describe wide ranging-migrations. The tradition highlights the physical strain the drought put on herders. Watering was especially tiresome because wells were overcrowded. Human control of drought is the second important theme. Only after a blessing from an elder do the rains return.

Koringring - The Drought of the Shaking Earth, 1933

A drought in 1933, which was noted with a 50 per cent mortality rate in cattle in the colonial records,[7] is remembered as the "Drought of the Shaking Earth", as an earthquake marked the climax of the drought. The alternative term for this drought, *kutiti* (migrating in full swing, everybody migrating in the same direction), is less known. Archival records mark the year as a serious drought but have little more to say on its effects. The consequences of the drought were aggravated by a further outbreak of Rinderpest[8] which occurred in the following year. However, while mortality rates of livestock soared, humans did not die in large numbers. People survived on drought food and barter trade with itinerant traders. Kanyakol, in his seventies and frail when the interview was taken, recalls the drought.

> I was still small then. We moved to Parpelo and then returned to Paka. We were together with Todokin's family. We ate those thin cows that died and oxen that were slaughtered. Afterwards the earth shook. Shortly after that year cattle were killed by Rinderpest. *Koringring* came because of the curse of an old man called Lengotum. The cattle stayed at Kokwöakalis and there was lightening constantly. An ox was slaughtered and the meat was ceremoniously eaten in the *kirket*. However, during the night a dotted hyena slept on the *eghyan* (stomach contents of the sacrificial animal placed at the centre of the ceremonial place) in the middle of the *kirket*. An old man who noticed what happened ran to spear the hyena. But he fell. When that man stood up he took a stick like that one there with three branches on top. His in-law told him 'uncle, that looks like a thing girls put on their necks'. He told him 'boy, don't you see what has happened'. After that had happened the earth started shaking and big rock boulders fell from mountains and hills. Tears were running from the eyes of cattle and when they were skinned they were fat. If the fat was fried it made a strange noise. The bones had no marrow. After that year the people had to buy cattle from the Somali. They had to give about 30 goats for just one cow". (Kanyakol, October 1993).

According to the tradition the disaster was caused by the curse of an old man, although we do not learn why he cursed the people. Kanyakol stresses the ritual desecration brought about by a hyena sleeping on the stomach contents of the sacrificial animal in the middle of the sacred ceremonial place. The negligence of men allowed this pollution to happen.

Katarngang - The Drought of Aimless Migrations, 1939

The term *Katarngang* describes the aimless migrations of herders looking for the last leftovers of pasture. In their desperate search, Pokot herders invaded white-owned farms on the Leroghi plateau. Cattle were confiscated by the police. Due to the prohibition of using pastures on the highlands, Pokot herds stayed in the bushland of the foothills. These areas are renowned for the danger of Trypanosomiasis - and in fact many cattle did succumb to the disease. While cattle died in some numbers due to the combined effect of drought and Trypanosomiasis, goats and camels were hardly affected.[9] Meanwhile one or two traders had established themselves in Nginyang. This changed the supply situation tremendously. People bartered goats and cattle for maize and sometimes obtained corn on credit. There are no longer any descriptions of wide-ranging barter expeditions into the Tugen and Marakwet highlands as described for previous droughts. The regional economy became re-orientated towards larger urban markets which supplied grain via traders in return for livestock and hides. Only those pastoralists living in the direct neighbourhood of agriculturalists still continued to barter livestock for staple foods. As in previous droughts the blessings of an elder broke the drought.

Arasusu - A Millenaristic Rebellion against the Colonial Power, 1950

Arasusu – the term for a short-lived attempt at millenaristic rebellion against the colonial state - is a disaster of a different kind. *Arasusu* is portrayed as a disaster in biographies and other traditions. The later prophet of the millenaristic movement, Lukas Pkech, joined the *Dini ya Msambwa* movement of Elijah Masinde in the 1940s (see Wipper, 1977). This movement, which preached a millenaristic renewal of the earth and the disappearance of all white people, recruited its adherents mainly among the Kitosh Luhyia in the Mt. Elgon area. When the movement was crushed by the colonial police, several adherents fled to West Pokot where they went on recruiting amongst the Pokot. There Pkech got in touch with the messages of Masinde. However, he was imprisoned soon afterwards and taken to Nakuru jail. He managed to escape in late 1949. Pkech looked for shelter in East Pokot where he started to preach his gospel to the pastoral Pokot who had not had any contact with the movement before. Pkech told his listeners that they had to go to the land of Zion, which he likened to Mount Elgon. There God would present them with many cattle, the blind would be given eyesight, and the lame would be able to walk again. The anti-colonial content of his preaching was vague and obviously just contained the message that white people would leave when the new era began. Pkech told the Pokot that bullets would turn into water and bombs into smoke. Finally a group of about 300 people left Nginyang Division to find their way to the land of Zion. After the group had marched for about three to four days, the group was intercepted in Kolloa by administrative police and some officials. The District Officer tried to convince the Pokot to give up and hand over Pkech to the police. On the 24th of April late in the afternoon a group of Pokot came dancing towards the police convoy. Whether the subsequent attack of the Pokot on the police convoy was planned or was a consequence of the first shots by the police was never really clarified. However, finally 50 Pokot, among them Pkech, died in the crossfire - and four administrative officials were killed.

That was the end of the movement. The punishment of the British administration was draconian. In the follow up to the affray some 176 men and women were imprisoned. Eight men were hanged later on and many Pokot were imprisoned for several years. To counter further disturbances the military and police was stationed in the area. The Pokot had to carry the costs through livestock sales. Additionally some 5,000 head of cattle were confiscated immediately. The entire population was forced to participate in the construction of roads and landing strips. All men had to hand in their spears. Some of these regulations were already cancelled in 1951.[10] However, there is little doubt that the *Arasusu* affair severely affected the pastoral economy of the entire Pokot division for about one and a half years. Although losses in livestock were moderate, a large part of the population was affected in one way or the other through the death of a relative or the imprisonment of a family member.[11]

The Pokot perceived *Arasusu* as a disaster - as bad as many droughts and as effective as an epidemic. Pokot lost many cattle and numerous men were killed. Many more men were taken to prisons and had to undergo forced labour. Some men I interviewed about their imprisonment were still traumatised. They were sure that they were going to be killed - or slaughtered as they said. Informants described that they were neither hanged nor shot but killed like sheep - their throats were cut. Every morning a few were taken out to be slaughtered in this way. One day the widow of a white policeman who had been killed in the affray came to visit the prison. Many feared that now the remaining men were going to be killed. However, she interrupted the executions and said that she was satisfied and revenged. It was repeatedly stressed that the widow was dressed all in white when she came to end the executions: the colour white indicates harmonious relations.

Cheruru - The Drought of Pouring out Cattle, 1955

The "Drought of Pouring out Cattle" denotes another dry period of three years (comparable to *Kiplelkowa*). The name *cheruru* describes, in an onomatopoeic way, the finishing (the running out) of cattle. The year 1955 was extremely dry (–61.6 per cent at Nginyang) and caused heavy mortality rates in cattle herds. Again, adverse climatic conditions were accompanied by livestock diseases. Famished cattle succumbed to Trypanosomiasis in large numbers. Herds migrating to the Leroghi plateau were infected with East Coast Fever. Although shops had been established at several places by then and itinerant traders were roaming the area, substitutive foods like berries and tubers were still of great importance.

> "*Cheruru* was a bitter drought. The women went to collect *makan* fruits at the river; they went to look for green *makan* which were later boiled for a long time to become soft, stirring them continuously. The cattle were finished '*lerururu*'. It was a Maina man who brought Cheruru by cursing the land. They went to kill an ox for the Maina until the Maina said 'thank you' (and revoked the curse). That is why the goats are healthy nowadays. They were brewing beer for each single Maina man. If there was a single Maina at Nakoko, young men from Merkalii and Nalekat would go there, they would go there to kill something for him and would go again and again....I killed the ox of Kangolese for an old man. It was of *ngolepurat* colour. It was very fat. The old man inquired 'who is the person who will spear this ox' and I stood up and said 'I, the son of Aremule'. He said 'yes, you shall be the one'. He gave his blessing 'the goats shall multiply'. In the year of Cheruru the people ate their cattle, they speared their oxen for their children. They killed something in this homestead and spent a day here and the next day they killed something in another homestead and spent a day there." (Wasareng, October 1993).

Once again traditions report that it was the curse of an elder which caused "The Drought of Pouring out the Cattle" and it was the blessings by elders which finished them. Beyond these generalising considerations famine first of all meant suffering and social stress. The account of Chepösait impressively stresses these sides of the drought.

> "Cheruru started in Tirioko and then came over all Pokot land. People were living on dying animals. There were no fruits. Shops were at Mukutani, Tangulbei and Nginyang. In the year of Cheruru cows had given birth to many calves but many calves died during that drought. I was not yet married in those days and lived in Amaya with my father Tamakaru. When the drought was washed by the next rains I was married to Wasareng who at that time already had two wives.
>
> The next bad year was called Chepembe. During that year I gave birth to my first child. I was very thin and had problems carrying the child. I was living with my husband near Merkutwö in those days. For the first time I felt the bitterness of life....I was so thin that I was not able to walk to my father's home anymore. I went to my father's brother Ngoreiyang. There I was given a cup of milk. I drank it little by little and after the first drop I felt my stomach aching as there was nothing in my stomach. I fell asleep with exhaustion, with my baby still tied to my back. This was the first time I found some sleep and I did not feel hungry anymore - my husband, was neglecting me and the other two wives did not share with me. While I was sleeping a big snake came and bit the child which was still tied to my back. I heard my child crying but it was as if the noise came from far away. Something in me said 'this is nothing, this is not your child'. The child kept on crying but I did not wake up. Somebody beat me with a stick and I roused from my drowsiness. I felt my child and with my backwards turned hand I felt along the body of the child. I touched the head of the snake which quickly disappeared. I went home to Wasareng quickly but the child died soon afterwards. Then I was given two lactating cows by my father. A brother of my father gave me some goats and as it rained well my small herd was doing well. Slowly I started to feel better." (Chepösait, October 1993).

Lochwer - The Drought of the Comet, 1965

Lochwer - the "Drought of the Comet" occurred in 1965. According to informants the comet, after which the drought was named, had a black head and a white tail. *Lochwer* is an onomatopoeic reference to a comet.[12] The drought was accompanied by several contagious livestock diseases which affected cattle and goats. Cattle died in large numbers. Many herds migrated towards the Leroghi plateau and once again looked for grazing on commercial farms. For the first time the Pokot received famine relief. The yellow maize which was donated by the United States is still vividly remembered and gave the drought in the following year the name *America*. Commercial farmers bought livestock from the Pokot and thereby supplied them with money to buy cereals in shops. Maize was obtained from Somali traders on credit too.

> Until the drought of *Lochwer* our animals had multiplied, the cows had become nine again and the goats were about 20. Due to a lack of rain there was no grass at all during *Lochwer*. Cattle migrated to Churo and further on to Leroghi. Out of the nine cattle we owned before *Lochwer* only five survived the drought. Goats died too during the drought but did somewhat better than cattle. During that drought we moved cattle into Luoniek ranch (a commercial farm bordering Pokot in the far south-eastern corner of their territory) during the night and drove them out in the morning not to be detected. The cattle were taking water at Luoniek dam. We lived off dying animals and from the blood we drew from cattle. Just before it started to rain a livestock market was opened at Churo. White farmers bought cattle. With the money we bought grain in Churo or Tangulbei. Some Somali traders gave out sacks of maize and collected the debts during the next rainy season. The rainy season after *Lochwer* was called *America*, as we received famine relief from America. In this rainy season cows did not give birth but goats did. There was little food. Famine relief in the form of maize, oil and powdered milk was donated. When the girls got circumcised they went to Nginyang to ask chief Kalikeno for powdered milk. This was phrased into a well known song: *onto ne chemeri? musar chekö ka America cho mi Nginyang*. "What are you doing now young girls? We will take the milk of America which is in Nginyang". (Tochil, September 1993).

Ngoroko - Interethnic Conflicts 1965-1996

There are many personal memories of the Ngoroko wars of the 1970s and 1980s. There are few people who did not lose livestock to raiders coming from Turkana District. Many lost close relatives too. While accounts of raids on the Turkana have become part of public traditions, recollections of Turkana attacks are more personal.

Between 1977 and 1983 interethnic conflicts reached a climax. Attack followed attack. Both, Turkana and Pokot were undertaking large scale raids. Nginyang, the centre of the Baringo Pokot, was raided three times in 1977. About 40 inhabitants were killed in these raids. Extreme violence and the civil war-like atmosphere of these years are echoed in the recollections of raids.

> In 1978 we were living in Nginyang. Rumours of pending raids had been many during the past weeks. Some people thought that they had seen the spoors of Turkana raiders near Kökwömogh and Orusyion. This was seen as a final proof of Turkana plans. However, nobody knew what area the Turkana raid would target. Guards (*santir,* probably from English sentry) were posted at several points. Sentries for Nginyang were posted on the hills near Nakoko, those for Tangulbei on the slopes of Mount Paka. As soon as they saw enemies they would give the warning cry *rira*. Some guards would run back to the settlement to warn people of the pending raid.
>
> The system failed. There had been no warning in advance when the Turkana attacked one morning between 5 and 6 o'clock. I woke up

when I heard shooting. First some shots were heard at Donya Sas and then near the airstrip and only a few minutes later there was shooting everywhere. Hurriedly we drove the cattle towards the small station of the Administrative Police and all the children and mothers were also taken there. The attack lasted for about three hours. The Turkana systematically attacked each and every household in Nginyang and in the vicinity. Cattle and small stock were driven off and also numerous camels were stolen. There was a lot of confusion. Dust, shooting and crying added to the chaos. More than 20 Pokot were shot dead, many of them children. (Loriko, May 1988)

korim - The Drought of Aimless Moves, 1984

korim - the "Drought of Aimless Moves" temporarily ended the period of intense raiding. Rainfall during that year was measured at 64.8 per cent below the annual average at Nginyang. The biggest problem during the drought was the combined effect of lack of pasture, violent conflict with the Turkana and army operations. The Pokot lost about 20 to 40 per cent of their cattle in one year. These losses sparked off major famine relief operations. These incorporated major programmes of several churches, the World Bank and the German NGO German Agro Action. Famished people were concentrated in several camps where famine relief was given. Many households left several people, usually children, women and older people in or near the relief camp. Able bodied people trekked large distances with the remaining stock. The case story presented here entails an individual account of drought and conflict related migrations. Campsites were changed nearly every second week, sometimes after only a few days.

"At the beginning of *korim* I lived in Cheparak, a bit down from Chepungus. I lived there with goats and cattle (autumn '83). Due to the lack of pasture I moved on to Chepchok with cattle and small stock. There I stayed for two or three weeks. When the grazing there was finished, I moved on to Sesiangimur in the Seruni plains where I joined four other camps. There the grazing was still quite good. We stayed for a month and then moved on together to Orus to use the remaining pasture there. But due to previous bushfires the grazing there was poor too.

Then the disarmament activities of the army started. For some time I and my four friends had to hide in the bush and we left our women at the camp. Shortly after we had returned from the bush Turkana raiders attacked us early in the morning before dawn. My household was the first to be raided: they started shooting at my hut as my wife's brother was sleeping outside. Right in the beginning of their attack they killed a small child and a woman. Together with my friends I tried to drive off the major part of our herds but we were immediately shot at. We had to hide in the bush. In that raid the Turkana raided some 16 Pokot households. Some of my friends had lost everything. The neighbourhood at Orus dispersed, not only that the place was unsafe, the grazing had been depleted too.

> Together with three friends I went to Napeliamajanit in Naudo, right on the border of Pokot and Turkana territory. We went into this border area to escape from another disarmament campaign by the army and to gain access to little-used pastures. (People had not grazed there for years as they feared the attacks of raiders.) Only one week later we went on to Mt. Silali, from where we could water our cattle at the Kapedo waterfalls. But we were still harassed by the army - now we feared the soldiers more than the Turkana raiders. At the same time the drought was still on, there was virtually no pasture left. Nobody was willing to hand in his gun as these guns had cost a lot of cattle and, furthermore, no disarmament of the Turkana had taken place yet. We were sure that Turkana raiders would come as soon as they heard that the Pokot were without guns. They would steal our cattle and kill our women and children. In Nakwakwa we stayed for one month. Then the soldiers caught us by surprise. We young men succeeded in fleeing but A. was caught and when he did not admit to having a gun he was terribly beaten. I fled with my small herds together with a few friends to Lomechan north of Mt. Silali - a place right within Turkana District. We watered our cattle at Lomelo together with Turkana herds. But we did not fight. I still had several Turkana friends in the area - friends I had made prior to the beginning of armed conflicts. In Nasurot warriors of both groups met - without their chiefs as they suspected the chiefs to be on the government's side. Everybody came with his weapons - just in case a fight might erupt - but finally they agreed to postpone all hostilities until the end of the drought. The Pokot slaughtered several big oxen for the Turkana and the Turkana slaughtered several oxen for the Pokot. For three or four weeks we stayed together with the Turkana. Then in autumn 1984 some rains fell on Mt. Paka and we returned. (anonymous, July 1988).

In spite of all the hardships that the Pokot took upon themselves, the 1984 drought developed into a major disaster with high rates of livestock mortality.

Summary

Personal memories of famines consist as much of remembrances of disaster-related strategies as of considerations on the origins of the catastrophe. Older informants reported that they lost major parts of their herds several times in their lives. Drought, livestock diseases and raiding were mentioned as causes. People became poor almost over night. The sole wealth of the Pokot, livestock, is portrayed as a highly vulnerable form of property. Considerations on the origins of droughts are important parts of traditions. Frequently the curse of an elder is the deeper cause of the drought. Why do elders act in such an anti-social way? The answer to this question gives a deep insight into Pokot concepts of emotions and morality. Elders are the repositories of wisdom and culture. If they are not respected, the moral code of the entire community is rejected. The society is polluted by disorder. The curses are hefty punishments on a group which goes astray. However, they do not act out of cold blood. Disrespect enrages them more than anything else and enraged people cannot be held responsible for their deeds. The curses of elders even if they push the entire group to the verge of annihilation are morally justified and finally contribute to the purity of the group. If the deeper cause of disasters lies with elders, help will be with the elders as well. Oral traditions on disasters frequently stress that the blessings of elders finished a drought.

4.2. "IN THE NECK OF A PERSON THERE IS A BONE" – TRADITIONS OF DROUGHT AND DISASTER AMONGST THE HIMBA

Knowledge on past droughts belongs to the realm of common knowledge and all elderly people have something to say about the drought of *Kaṯe* or the drought of *Kariekakambe*. Archival data contains information on permissions and prohibitions of mobility, on efforts of the government to supply famine relief food and excuses for not doing so, but unfortunately very little on the actual losses of Kaokoland's herders.

The Gun of Kaukumuha - The Kwena Wars, 1850s to 1880s

Raiders probably started to launch attacks into north-western Namibia in the 1850s. The rich pastoralists of Kaokoland offered themselves as easy prey. They were neither armed with guns nor had they any form of institutionalised military organisation. Himba traditions are full of detailed descriptions of these raids. The picture which emerges is that of continuous raiding, of resistance and bloody battles and of a final exodus to southern Angola. The landscape crossed by the refugees is mentioned in detail. Although the commandos were mainly after cattle, they also killed many people. The descriptions of attacks are full of atrocities: women were hacked to pieces only to get the copper off their arms (cf. Vedder, 1934:237), herds-boys were killed in the bush and entire households were wiped out. Himba narratives still convey the horror of these attacks. The stories are clad in metaphors which make it hard for an outsider, and sometimes even for cultural insiders, to understand the course of action. Katjira Muniombara recounts a tradition which describes the destruction of pastoral communities in northern Kaokoland. Homes are burned and people are killed.

> "I will start with the gun of Kaukumuha, the man who killed us was called Kaukumuha – an Omukuena. Now they themselves say: '!aukumuha', the man who brought a raiding party into this land, first to Ombombo of Kazukotjiuma, where they destroyed the home of the Glowing Red Iron, Kakunotjivi, our senior mother's brother, they burnt (his home) to ashes. .... The troop of Kaukumuha came here, they came to Omuzengaturundu who ate a cow of yellow-black colour with thick jaws. There they found our two men, Uatuihi of Kavange, of Ndjara of Kaoko who ate a female sheep, and

> Uambumba of Kaongo Tjimosenguaukamba of the Mwatjikaku patriline, when the elephant of Rovingu was killed there. They found them there where they burnt it (the homestead) to ashes....
>
> The people were killed by the troop, the next day they fled in a hurry: 'So, Muje the war is near, the war is near. Let us separate, they will kill us, we will be destroyed.' After they said that, immediately afterwards Ndjoue was killed, and they separated there. The one of Mbumba of the small kudu passed through the mountain pass of Otjikotoona. He drove the cow of Ngao of Kasona, of the mother of Kuziruka, and they went down the valley, they crossed the river (Kunene) at Ohamurenge. .... The Ovakuena came down to Otjinduu, they raided the cow of Tjao which had a chest, they raided the Ouoruze area. Then it (the war?) crossed there straight towards the sun. ...Those of Kaukumuha went and crossed the Kunene river.[13] It was them who followed the home of Muje and burnt the people. ....The people of our fathers fled to Angola." (Katjira Muniombara in Bollig, 1997c)

The story relates attacks of Oorlam commandos on pastoral households. People were killed and the households and their people were "burnt". In another oral account on Kwena raids, a protagonist desperately claims "*in this war there is nothing to fight; it is a fire which burns the people*". A call which desperately describes the inferiority of local people's weaponry: it was not worth fighting, the raid destroyed people like fire. The trauma and desperation of these events are well conserved in the metaphors and frozen images of these traditions.

Otjiṯa-the Rinderpest epidemic of 1897

The Rinderpest epidemic caused extremely high losses in cattle all over Eastern and Southern Africa.[14] In Himba traditions the year is called *Otjiṯa* – "The Dying". Rarely did an elder mention the name of the disease - *ombita,* the year is referred to by its devastating effects. Oral accounts do not give detailed descriptions on how the disease entered into south-western Angola and how it proceeded there, even exact accounts on the characteristics of the disease are lacking from these accounts - however, all traditions emphasise heavy losses. Generally it is claimed that only one or two cattle survived per household and that people lived on small-stock, on bush resources and to a large degree on income generating activities beyond pastoralism.

> "The death came like this. It came after the War of the Zebra (i.e. the Oorlam wars) when Muje and Mureti separated. Muje fled to Angola and collected a lot of things like wheat, millet and maize there. With these things they bartered for their cattle (and replenished their depleted herds). Those cattle multiplied until they became a large herd....But (all) those cattle ...were killed by a disease called *ombita*, that killed all the cattle in all the land. ... All those cattle which were left over from the raids died, they were finished completely....

> After that the Ovahimba survived with small calves. They milked the goats to feed a small calf. This calf drank goat's milk only until it grew up and began to eat grass. (And all this) because the mother of the calf died of the disease that we spoke about. That calf went to graze on its own or someone even went out and collected grass for it. The calf ate the grass the whole night through and the next morning the calf went out and grazed for itself.....Everything was done to ensure that it grew well and to prevent it from the emaciation that killed its mother. That calf grazed like that and after a while it became a grown-up cow. And after three months it was covered and suddenly it calved. After it had calved, all of us would drink its milk,...all of us would drink the milk of one cow only...Yes, first one person was feeding milk to the people and then later everyone drank from his own (cow)....All those cattle came from that single calf which was fed grass by hand. All those cattle came from one calf that survived when its mother died of emaciation....
>
> They died, all the cattle died....The Portuguese migrated again during the Year of the Dying, the year that killed the cattle of our father and mother, the dying of cattle and war itself....The young Himba men went to towns to look for jobs and when they found jobs there, they brought it (money, goods) home". (Vitunda Mutambo and Kandjuhu Rutjindo in Bollig, 1997c)

The effects of Rinderpest on Himba herds were disastrous. The disease hit households at random: rich herders became poor over night and many households were left with nothing. Whereas in later narratives on droughts it is always pointed out that rich people faired much better than poorer people did, here we get the idea of an overall collapse of the livestock-based economy.

Ondjara oya Katurambanda - The Year of Pounding the Leather Clothes, 1915

The 1915 drought is the first drought that is widely portrayed in Himba oral traditions. Himba herders had returned from their exile in southern Angola since about 1900. Vedder who travelled the area in 1914 met with a population which was heavily engaged in pastoralism and active in regional trade networks. Herds had increased after the 1898 Rinderpest disaster and the drought of 1915 was the first major blow to Kaokoland's livestock economy in the 20th century. While there are no rainfall figures available for that year for Kaokoland, neighbouring Tsandi in western Ovamboland recorded 153.3mm (–62.7 per cent) and Outjo on the southern expanses of Kaokoland recorded 215.6mm (–48.4 per cent) in 1914/15 and 279.2mm (–33.1 per cent) in the following year. These figures indicate that the drought was on a regional scale.

In neighbouring Ovamboland the years 1914–1916 are remembered as the Great Famine or as the *ondjala yawekomba* - the "famine that swept".[15] The effects of a severe drought were aggravated by violent conflicts between and within Ovambo polities and, most of all, by a major military campaign of the Portuguese against the Angolan Ovambo. The Portuguese added to the effects of the drought by the seizure of corn storages and livestock. On the Namibian side the South African government took the chance that the famine offered to subdue the Ovambo polities which had remained beyond the direct control of the German administration until 1915. The famine resulted in many thousands of deaths, massive internal migrations and the temporary breakdown of the social system (Hayes, 1992:191; 1997:5). Reciprocal obligations were abandoned and rules of hospitality suspended. Cannibalism and desertion of children were noted in some reports (McKittrick, 1995:75). Starved people were no longer buried and corpses littered the ground along the major migration routes. Hayes (1997:5) assumes that about 25,000 people starved to death out of a population of 156,000, i.e. some 16 per cent of the total population perished.[16] Mortality rates were probably even higher in Angola where Spiritan missionaries estimated that about 50,000 Ovambo had died during the disaster. However, losses were distributed unequally, with the western-most kingdoms being harder hit than the eastern ones and within the different Ovambo societies rich people being less affected than the other strata of society (ibid.).

While the 1915 drought is widely covered in the archival records for Ovamboland featuring missionary reports and travelogues of South African army officers, there are no written accounts on the drought for Kaokoland. The disaster fell within the interregnum between German and South African rule. Vedder, who travelled throughout the region in 1914, did not report drought conditions. After that Manning in 1917 was the next white person to journey in Kaokoland and leave a report. He, too, does not refer to a major drought. Oral traditions, however, hint at drought conditions: In Himba oral accounts the 1915 drought is called the drought of "Pounding the Leather Clothes", referring to the fact that people were forced to eat their leather clothes. While drought conditions were probably as hard as in neighbouring Ovamboland, the situation here was not aggravated by violent conflicts. Tjikumbamba Tjambiru, an elder well in his eighties, remembers the drought which he experienced as a small child.

> "... the rains did not come, the rains did not rain, they did not rain and it became a drought, from autumn until summer until winter no rain. In this year the famine began, the hunger commenced. A child that lived under good conditions was the child of a rich man. The child of a poor man, somebody like me, ate the hide(s) that we slept on, we Ovahimba, that hide stained with dirt that we slept on, that one with residues of fat and the dirt of people. Those children they were cutting, those children of poor people they were cutting them, they were cutting them with the help of others and they put them on fires. The fat of the roasting skins was making the sound 'tjetje'.....Poor weak people were completely finished, they died, they were killed by starvation. The child of somebody rich was living and the child of somebody poor tried to stay near somebody rich, in the hope that those children (when they had finished eating a goat) would give the hides and skins away. They were living off these things, it was like that in that Year of Pounding the Leather Clothes.

> Every poor adult person was roasting them (the hides). They roasted them, they beat off the ashes and put them into their mouths and the rest they gave to the children. Adults and children were living off this (food) and cutting the skin into pieces so that they could survive." (Tjikumbamba Tjambiru in Bollig, 1997c)

While in Ovamboland a quarter to a fifth of the population died, in Kaokoland there were apparently only few drought-related deaths. The evidence on death rates in Kaokoland, however, is scarce. The tradition told by Tjikumbamba Tjambiru refers to several cases of death. However, other traditions relate that only a few people died as a result of the famine. Vedder estimated the population in 1914 to be somewhat lower than 5,000, in 1917 Manning gave the same estimate which may indicate that losses of human life in Kaokoland were low. However, there is little doubt that starvation was widespread and the preparation of skins and hides was one last desperate way to overcome death by hunger.

Karasaruvyo - the Famine of Licking the Knife, 1929/30

It is the combined 1929/30 and 1932/33 drought which is remembered as *karasaruvyo* – "The Licking of the Knife" in Himba oral traditions.[17] In contrast to the previous famine and the following disaster of *Kate,* oral traditions are not as detailed on this drought. It is the first drought that the South African Administration had an impact on - in fact the records suggest that it was measures taken by the administration which turned the drought into a famine. The years 1929 to 1933 were marked by climatic perturbations[18] paralleled by activities of the colonial government to control the population of Kaokoland more firmly. Since 1925 cases of CBPP had been reported from Kaokoland and there was a widespread fear that the disease would affect settler herds further to the south. Since the establishment of a Police Post in Tjimuhaka in 1926, the colonial government had tried to control Kaokoland more firmly. The borders towards Ovamboland and along the Kunene were now heavily controlled. Internal migrations within Kaokoland were scrutinised and since the early 1930s herdsmen were required to ask for permission with the administration before they moved livestock. The most drastic step for enforcing rule over Kaokoland's population was the forced relocation of the Herero from southern Kaokoland: in the second half of 1929 1,127 people, 7,289 cattle and 22,176 goats and sheep from southern Kaokoland were forcefully relocated to the central parts of Kaokoland. This major move was carried out despite the drought conditions which prevailed throughout the year.

By the middle of 1930 serious famine conditions prevailed in major parts of Kaokoland. At the request of local leaders the administration entered into negotiations with the Portuguese authorities to allow traders to bring maize for barter to the Kunene river at Tjimuhaka. The Himba had been prohibited from bartering cattle with Portuguese traders for several years. Namibian authorities first had to apply to Portuguese officials to permit traders to trade at the river. The permission was given - but it had serious drawbacks. Kaokoland's herders were forced into a very unfavourable trade agreement.

When the rains failed again in 1932, the leading political figures of Kaokoland approached the administration to ask for traders to be allowed to trade at the river. In their appeal to the government they emphasised that they preferred to barter grain for livestock as they were short of money. At the same time a local police officer estimated that some 400 to 500 sacks of maize were needed urgently, i.e. 36,000 kilograms for about 4,300 people.[19] Native Commissioner Hahn proposed to his superiors in Windhoek that the government buy the cattle from local herders in order to provide them with money. However, the government did not respond positively to his request. Things turned really bad when the Portuguese authorities refused to allow traders to trade with the Himba at the Kunene in July 1932[20]. The Portuguese officials asked the South African government to purchase grain at Huila, but given that the administration in Windhoek did not want to spend any money from state coffers for famine relief no action was taken. In November the Administrator of South West Africa paid a hurried visit to the famine-stricken Kaokoland. He decided that famine relief should be limited to a minimum as Kaokoland's herders did not go to look for work. Only a few sacks were given symbolically as aid.[21] However, the maize was not to be given as famine relief food but was to be sold by officers, probably for "educative reasons". The gains were paid in at the Magistrates Office Outjo. Additionally, the police were authorised to shoot zebra for meat for the natives and in order to do so they were given an extra-hundred cartridges. Cynically, the ban on local hunting was kept effective and herders caught hunting were fined for poaching.

Kate - 'Go and die somewhere else', the Famine of Chasing away the Hungry, 1941

The major 1941 disaster was preceded by a minor drought in 1939. In February 1939 locals asked for permission to use the so-called neutral zone in the south. This wish was flatly denied as the government was now seriously contemplating settling Herero from the Police Zone in this vacated stretch of land. Even the further expansion of white commercial farms into southern Kaokoland was considered. In August 1939 Karuapa, the chief of an eastern Himba section, asked for permission to settle along the Kunene river because of drought conditions (settlement there had been prohibited in the 1930s in order to prevent illegal border crossings). His request was refused, too. In October and November Angolan Himba crossed the river due to the drought conditions in southern Angola and violent conflicts with the Kuvale people. In order to check on Angolan Himba crossing the Kunene, river guards were stationed at the major fords along the river. If there were any chances of crossing the river before, these were now severely curtailed.

From January 1941 onwards severe drought conditions were reported for Kaokoland. Up until the end of the rainy season 229.2mm of rain was measured in Opuwo, about 23.3 per cent below average (another figure said 194.9, i.e. 34.8 per cent below average).[22] Around the middle of the year heavy losses in livestock were reported from all over Kaokoland. Mass starvation haunted the local population. Deprived of any trading contacts and the possibility to buy or barter grains, they only had *veldfood* to rely on. A report estimated that 6,000 head of cattle (probably between 20 and 25 per cent of the total regional herd) and some 3,000 head of small stock had perished. Despite river guards threatening to shoot animals crossing the Kunene, Angolan Himba still tried to escape from their settlements in south-western Angola. The river guards and police officers knew no mercy: in March 1941 Veripaka's herd of 517 cattle was machine gunned and destroyed at Otjipemba after it had crossed "illegally" at Otjomborombonga. Some months later in September 727 sheep were culled at Enyandi. Reluctantly the government decided to donate famine relief food in order to prevent the worst. In November 1941 some 225 bags of maize were supplied as famine food. Finally, in February 1942 heavy rains broke the drought.

The drought of 1941 is widely portrayed in oral traditions. Most adult people have some remembrances of this disaster. Kozombandi Kapika is in her seventies; she was a young adult when the drought of 1941 occurred. The following passages are quotes from an interview with her.

> Kozombandi (K): "In the Year of Kate we were hungry. Then we were living here in Omuhonga. We moved because of hunger up to Outwa, to Otjinduu (a place beyond Opuwo). We ate our small stock in the homesteads. Maize had not yet come during these days - in those days we ate small stock only. Small stock were slaughtered in great numbers. People ate them during the day, the next day the hunger came back and they stayed five days, up to ten days and they still were tormented by hunger. The next day they decided to slaughter another goat - the people survived like that. People living in Ombuku ate palm nuts. But as we were far away from that place we did not find palm nuts and stayed hungry. People ate small stock, that's it. In that land where we settled in those days there were no palms. But the people ate *omboo* (pieces of wood from the *omukange* tree) and also the roots of young *omukange* trees. Yes the people ate *omboo* in the homestead of the father of Kaunahoni. When you reached the homestead it looked as if people had eaten palm kernels. That *omboo* the people chewed for a long time and then spat it away. In the morning hours...people had already started to look for *omboo* - they did not find anything else. They ate *omboo* for several days and then slaughtered a goat once again. But *omboo*, of course, does not satisfy anybody. The people chew it to get the liquid, they chewed pieces of fibre and they spat them away. How could this liquid satisfy the people?

Interviewer (I): In this year of Kaṯe, was it only the livestock which were dying from hunger or the people were dying too?

K: When people are born they have a bone in their neck. People do not die quickly. If people were things that died quickly, in this year many people would have died. They would have died in great numbers and would have been completely destroyed.

I: Why was this year called the year of Kaṯe?

K: That year was named after somebody who slaughtered a goat in his homestead. Then somebody else came and found him with the meat of that goat. This traveller said to him 'let me rest a bit, perhaps my friend will decide to give me a little bit of food'. But that owner of the homestead just said 'No, you just go away and die'. He did not want to share his meat with the traveller. He wanted to give it to his own children in order to help them to survive – that's it. The traveller walked off. He had not been given anything. That is why the year is called *Kaṯe*. In the Year of *Kaṯe* there was nobody whom you knew - only those people whom you had to feed you would know. The child of my own brother coming from Ominyandi, I would chase away and say 'go and die, here there is nothing for you'.

I: Did many people die during that drought or only cattle?

K: No people did not die. I told you about that saying 'in the neck of a person there is a bone'. That bone in the neck gives strength to somebody so that he will not die easily. That does not mean that you do not have to eat anymore because of that bone, no. That bone means that you will reach somewhere far, where you will be able to survive, or where you will be able to find food even if you are very hungry." (Kozombandi Kapika, June 94)

Kozombandi narrates how famished people tried to quench their hunger and thirst by chewing juicy bits of wood. She deplores that there were no palm trees in the central parts of Kaokoland. The wooden bits of *omukange* (Commiphora africana) were not substitute food but rather a distraction from the pain of hunger. The major way to prevent starvation was to slaughter small stock. Famine conditions became so overwhelming that rules of hospitality were discarded. Starving people were denied food and the meat from slaughtered small stock was enviously kept away from the eyes of the public.

Kariekakambe - the Famine of Eating the Horse, 1946

Another drought followed in 1946.[23] By September 1946 serious famine conditions were reported from all over Kaokoland. People driven by hunger killed a horse, which gave the drought its name. Just as in 1941 the government did not consider bringing Portuguese traders to the river in order to barter maize with herders from Kaokoland. Watundwa Kapika, today in her late sixties, recounted her personal experience of the drought of Eating the Horse.

"In the year of Eating the Horse we came from Oyotjivare and we migrated and we settled in Orotjiheke. When we left Orotjiheke we settled in Omatarara. There we slaughtered our cattle as my mother's elder brother died there, the father of Mirireko. When we came from there we returned and we settled at that small hill...There we slaughtered several

> cows and an ox because this maternal uncle had died. Then we migrated on and we went through the gorge of Nyokohe, near Ombuku. When we passed through Onyokohe we settled there in Ongorozu. It became cold season and it became very cold indeed that year. We slept like this and shivered. When the seasons changed again and the hunger started we migrated again. It was becoming dry season. We settled in Ongorozu and the hunger came in full force. Sacks of maize had not yet come. That was the year of Eating the Horse and people just survived on what was in the livestock enclosure.
>
> (During that drought) the children drank bits of soured milk - they did not know about hunger - the people who were feeling the hunger were us adults, who had given birth already. When Karenda, my youngest brother was born, we were lucky as our father slaughtered one ox for Karenda. The drought went on. Kazupotjo and Waitavera found only hunger in our homestead. There was one old goat and they said: "please slaughter that old goat". When my father heard these words he picked out one sterile sheep from the herd, which was fat indeed. He slaughtered it and people had food for some days. Later that year it rained heavily. Many small stock, which had become very thin died." (Watundwa Kapika, June 1994)

The effects of the drought were mastered by frequent migrations looking for new grazing grounds for cattle. The continued slaughter of livestock helped famished people to survive.

Droughts and Famines of the 1950s and 1960s: The Drought of Omasitu and the Drought of the Tjimbundu

The late 1940s, 1950s and 1960s generally had few drought years. At the same time the South African administration relaxed its rigid stand on cross-border migrations. Modest development efforts such as borehole development, the expansion of internal trade and the erection of the first schools and a hospital fell in these years. Himba herds grew considerably during this period: while throughout the 1930s 30,000 to 40,000 animals were herded, in 1958 about 120,000 animals were counted.

In 1958 north-western Namibia was hit yet again by another drought which lasted for at least four years and had regional dimensions. All the rainfall stations in the region recorded massive lows over a protracted period.[24] The mortality rates of cattle soared: cattle numbers dropped from about 120,000 in 1958 to 60,000 in 1960 indicating a loss of about 50 per cent. For small-stock the losses were less severe. Mortality rates were estimated to be about 10 to 15 per cent (roughly 30,000 heads). The annual report for 1959 related the massive losses to overstocking and overgrazing in most parts of Kaokoland.[25] In fact, the annual reports for 1959 and 1960 give evidence to the assumption that the administration saw the drought as a sort of mixed blessing. They had analysed massive overgrazing and interpreted the 50 percent loss of livestock as a welcome break for the environment. However, the administration did not doubt the massive impact of continued drought and high livestock mortality. A lot of communication between various administrative offices in Opuwo and Windhoek deals with efforts to prevent mass starvation. While a massive famine relief programme was set up, scarcity was due to the continued restrictive policy on livestock marketing.[26] There was hardly any opportunity to sell livestock. There was a shop at Opuwo and three officially gazetted itinerant traders occasionally came from Otjiwarongo. However, trade with Angola, although still illegal, was not prosecuted anymore. The annual report for 1959 estimated that some 10,000 head of cattle from Kaokoland were sold during that year in southern Angola and a year later another 6,000 were bartered for maize and commodities.[27] The annual report for 1960 gives some rates of exchange: a tollie worth eight to ten pounds was exchanged for one sack of maize (probably 90 kilos): this is an absolute low in exchange rates which only compares to the miserable rates Himba and Herero obtained with Portuguese herders in 1929/30.

In the Annual Report for 1959 the reporting officer had to admit that no livestock market of any size had developed in Kaokoland. The only way out was to subsidise the Tribal Trust Fund with considerable sums of money. The Tribal Trust Fund then bought food on credit from the colonial administration and distributed it.[28] In 1967 a major sum of 10,000 R was donated to the Trust Fund as a loan.[29] Most of the money was spent on famine relief.[30] The population of Kaokoland became dependent on external aid in the case of drought as livestock marketing was still hampered by numerous regulations.

In personal accounts on the year of Tjimbundu there are no indications of widespread starvation in the Himba population. The normal repertoire of emergency mechanisms sufficed to thwart off the worst effects of the drought. People migrated to the riverine Kunene forest. The ban on settling in the riverine zone had been lifted in the early 1950s. They ate palm nuts and slaughtered goats. In Himba traditions the year is remembered as a year in which a well known Tjimbundu blacksmith who produced iron beads in Okangwati died due to hunger. Again rules of hospitality were suspended - at least towards outsiders. In contrast to the massive attention given to the drought 1958-61 in public records, however, Himba traditions take astonishingly little note of the drought.

Climatic Perturbations and War - the Disasters of the 1970s and 1980s

Climatic conditions changed during the middle of the 1970s: since the rainy season of 1976/77, nine consecutive years were recorded with below average rainfall. Excluding the rainy seasons of 1985/86 and 1991/92 which were slightly above average, fourteen years in a row were recorded with below average rainfall. The prolonged drought started with two consecutive bad rainy seasons from 1976/77 (–27.9 per cent) and 1977/78 (–32.1 per cent). The drought is remembered as *ourumbu wonde - the "Drought of the Fly"*. The name of the drought is derived from obstinate flies, perhaps transmitting a disease, which were observed frequently in this year. It was reported that many cattle were still fat when they died. The following account by Muhuwa reports severe losses and shows how the Drought of the Fly merged into the centennial disaster, the "Drought of the Dying".

> "We lived in the Drought of the Fly, the Drought of the Fly finished all the cattle, all the land, everything was finished. Oh, all our cattle of Erora (a place in Angola) were finished and only oxen remained. Every person remained with oxen only, the cows died and were finished. Someone was left with goats and sheep, because small stock did not die. The cattle were the ones that died. If a cow had given birth it died and the calf was left, and it mooed in the enclosure neither eating grass nor drinking water.
>
> When the cattle died, the pregnant ones died first. When the cattle died, they died in the whole country, finish, the land of the ovaHimba. That's it; only three homes were counted whose cattle did not die. When the cattle were finished somebody who was left with heifers exchanged some with somebody else for three oxen or for two oxen for one heifer.
>
> The cattle became many again in the year following the Year of the Fly. They almost returned to their previous number. Somebody of long arms became big and somebody with short arms remained short. Two years followed in which the cattle multiplied and the years following the Year of the Fly were good, the third year the cattle which had multiplied, the cattle of Oviheke, died in a drought until they were completely finished, our cattle of the Herero. They all migrated to the land of the Ovambo and the Zemba. There we found the cattle and the oxen of the Zemba and the Ngambwe were fat. We who came with thin heifers, who came with their (the heifers' mothers) thin mothers, we finished the cattle of the Ngambwe and Zemba (meaning: we exchanged Ngambwe and Zemba heifers four Himba oxen). We exchanged (livestock). We wanted meat for food because we did not have soured milk. We exchanged our oxen, we exchanged our heifers..." (Muhuwa, April 1994)

Muhuwa's account of the drought suggests that despite adverse climatic and political conditions migrations to southern Angola across the Kunene were still possible. Himba herders coped with the effects of the drought by bartering with Zemba and Ngambwe on the Angolan side of the river and looking for grazing along the Kunene river.

Precipitation in the late 1970s did not return to normal. A climax of adverse conditions was reached in 1981 - in Himba traditions called the "Year of the Dying".[31] The drought had disastrous consequences, widespread famine, soaring livestock losses and encampment in famine relief camps. A pastoral culture seemed to be on the brink of annihilation.

Climatic perturbations were exacerbated by political conflict. The armed wing of SWAPO, PLAN, stepped up its efforts along the Kunene border. Allegedly they maintained several guerrilla camps in the areas immediately bordering the Kaokoland to the north. In order to check advances by the guerrillas, the South African administration established three major military camps in Kaokoland in the second half of the 1970s: Opuwo, Okangwati and Ehomba. Paramilitary units (Koevoet), which were placed alongside regular army units, recruited heavily among the Himba. Many young men tried to cope with stock losses by joining Koevoet units. Himba living in Angola were more affected by the war than their Namibian brethren. Places such as Chitado and Ongokwa were bombed at least twice by the South African air force (1976 and 1981). However, war-related activities were effective in Kaokoland too. In counter attacks, PLAN units occasionally took Himba as hostages or abducted young men to join their ranks. Roads all over the northern parts of Kaokoland and south-western Angola were made impassable by landmines. In order to check on the infiltration of guerrillas, Himba households in the Kunene valley were forced to resettle in the Omuhonga valley near Okangwati in 1979. Some people described their situation as dramatic: in the night guerrillas came and punished herders for working too closely together with the South Africans, and during the day South African soldiers came and, seeing the spoors of the guerrillas from the last night, punished the herders for hosting PLAN fighters. From 1978 to about 1985 mobility was frequently conditioned more by considerations for personal safety than by the needs of grazing.

Dramatic changes in southern Angola contributed to the destabilisation of the system. Since 1968 the Portuguese colonial administration had tried to bring economic development to the arid south-west of Angola. Eight huge farms were allotted to white farmers along the Kunene between Ruacana and Epupa falls. The farms were fenced off and the Himba were left with small corridors between those farms to walk their animals down to the river. Obviously violent confrontations between farmers and expropriated Himba herders were frequent. Informants claimed that several Himba were killed for entering farms unlawfully and many were flogged by the police for doing so. These conditions came to a rapid end in 1974. After the 1974 rebellion of army officers against the fascist Caetano regime in Portugal, Portuguese colonies were hurriedly released into freedom. Many white settlers and traders in southern Angola left the country almost overnight. This caused the temporary breakdown of trading in the area. In the following years south-western Angola was haunted by South African air raids. The added effects of drought, violent conflict and marginalisation led to an explosive situation which vented itself in the centennial disaster of 1981, destroying the herds of Kaokoland and bringing Kaokoland's herders to the brink of annihilation.

Traditions leave little doubt about the severity of the disaster. Kamanda, a man in his late fifties, stressed the effect of civil war on coping strategies (Kamanda Kuhange, May 1994).

> I: "In that Year of The Dying where did you try to move to, where did you go?
>
> K: No, we did not migrate, they (the cattle) were finished, we died, we died, we stayed there at the Kunene river, we were near our hills, we

stayed near the river. The land did not have any grass at all. Now the war started and the people were afraid. The cattle went into the bush and nobody could stop them. They were afraid. They feared for their lives.

I: In that Year of that Dying what did you eat?

K: Only palm nuts, the thing that is on the tree. Millies only came later and the herd was finished in those days of the war. The people were afraid, the traders hesitated to come as they were afraid to die on the way. They (the fighters) were putting landmines to kill people. Some exploded and killed people, other people were killed from planes. That's it, now how could somebody pass those things (the mines) that destroyed the cars. .. We stayed here in the Drought of the Dying that was taking away (without permission) our cattle. Then we tried to find one small cow, somebody who had one small cow and found somebody who had one small heifer and tried to exchange it.'

Vahikura recollects the drought's disastrous consequences:

"That one totally destroyed us. Some people came out of it with a few cattle and others came out with no cattle at all, and yet others came out with just one cow and that single cow was the one that gave birth again (and formed the basis of a new herd). We were eager to obtain sacks of maize until the rains came, some went to beg for livestock from paternal relatives, you asked those who still owned cattle for help....goats died too, but most of the small stock survived. The people lived off the remaining goats. Those that survived were enough for the people to live on." (Vahikura Kapika, July 1994)

The accounts leave little doubt about the severity of the disaster. Within a period of two years Kaokoland's regional herd had been diminished from 110,000 animals to 16,000 cattle - a loss of about 90 per cent. Since the great Rinderpest epidemic of 1897 no other disaster had depleted the regional herd so thoroughly. The fact of complete annihilation can be clearly seen from the informants' statements. Not only had an individual herd been erased, the "cattle of the Himba had been vanquished" and "the Himba had been destroyed". Himba were portrayed as a starving nation in the national newspapers - although, due to political circumstances, the full extent of the catastrophe was not revealed to the public. Kaokoland became an emergency zone. 13,000 R were presented to local leaders to buy famine relief food.[32] The parastatal FNDC (First National Development Corporation) donated further food aid (ENOK Bulletin 26/6/1981). In the middle of 1981, thousands of people flocked to the relief camps around Opuwo (Jacobsohn 1990:16), where government institutions and the Red Cross were handing out famine food.

Summary

Himba oral traditions on droughts and disasters are extensive and detailed. They consist of complex descriptions of events and processes and contain a wealth of accounts on coping strategies. The fact that names and events repeat themselves and are clad in metaphors apparently adds a mnemonic device. Traditions on disasters are an integral part of the overall political history of the group.

Recollections of disasters account in detail for coping strategies. Migrations are remembered vividly and drought foods are listed extensively. The availability or non-availability of drought foods is of paramount importance in oral traditions. The focus of traditions seems to be rather on coping strategies than on considerations about the causes of the drought. While Himba have some vague ideas that droughts are divine punishments this view hardly finds a place in oral traditions. Himba do not tend to look for metaphysical explanations for disasters. Even the colonial impact on disasters is not elicited in detail. The focus and pride of recollections lies in the extraordinary efforts people take to save their lives and those of their livestock.

4.3. A COMPARATIVE ACCOUNT OF POKOT AND HIMBA PERCEPTIONS OF DISASTERS

In both groups droughts feature as most important in the accounts of disasters. Violent conflicts stand out as a second source of damage in the presentations of both societies. While conflict has an ambiguous touch among the Pokot - one may win or one may loose - Himba see violent conflicts as utterly negative. They depict themselves as victims, as targets of powerful and violent adversaries. Hazards which have an immediate impact are dealt with in more detail in representations on disasters. Demographic growth outpacing economic resources, which was identified as a major cause for increased vulnerability among the Pokot, was not mentioned once as a hazard. In contrast to this degradation featured importantly in Pokot descriptions of disasters. Yet again, entitlement decline is only mentioned in passing. Forced destocking and strict regulations of trade were also not mentioned prominently. Himba do not emphasise structural hazards either. Degradation is seen as a minor problem which is largely attributable to the vicissitudes of rainfall and not to overexploitation. The severe impact of the government on trade and spatial mobility is rarely mentioned in traditions. After studying the archives for a long time on the ruthless measures the South Africans took, I was sure that the Himba would identify colonial rule as a major hazard. This was not the case: the key to this problem perhaps lies in the fact that the most ruthless acts of colonial encapsulation were perpetrated from the 1920s to the 1940s. In the 1950s the system opened up to some degree. Some amount of trade was allowed and the 1970s brought the establishment of several shops.

How do both societies interpret the causes of specific hazards? In their representations of past droughts, the Himba never deal with the causes of a drought extensively.[33] However, when asked in detail about their ideas, they forwarded more metaphysical reasoning. All informants were of the opinion that God brought rain. If he brought rain he could obviously also withhold it and cause the destruction of livestock herds: "*Mukuru wazepa ozongombe ekaeta*" – "God kills the cattle he brought" one elder voiced his opinion. He may do so if he is angry with human beings. He especially dislikes if envy (*eruru*) dominates interactions between humans. Most Himba questioned on the topic strongly believe that sorcerers of neighbouring groups had the power to stop the rain. These qualities were especially assigned to Ngambwe sorcerers. Informants were ambiguous about whether they attributed sorcerers the power to stop or prevent rains. The only factor that remained constant in their answers was that sorcerers who stop rains always come from the outside and never from the inside. Habitually the Himba voice the opinion that rainfall was unpredictable and that humans have to cope with this unpredictability.

Pokot ideas on the cause of hazards are more detailed and more consensus-based. Usually the inner causes of disasters are traced to internal conflicts. These are either generational conflicts or generally acts against solidarity. Even major detrimental changes in the environment are explained in this vein. Droughts are traced to similar incidents. Frequently accounts of droughts contain statements like: "It was a Maina man who brought *Cheruru* by cursing the land" or "*Koringring* came because of the curse of an old man". The cause for the drought is attributed to an individual elder who scorns the younger generation. Their disrespect and haughtiness pollutes the gerontocratic society and the elders' anger is depicted as justified.

Pokot seek the ultimate cause for droughts as well as for livestock epidemics and degradation in intra-societal conflicts. Thereby they appropriate hazards of an extremely unpredictable environment. (Only the danger of violent conflicts, *tapan*, originates from outside sources). Pokot ideas on the cause of disasters suggest that human action within their own society is at the bottom of the problems and that human action may show the way out of problems. Presents to elders which symbolise deferral under their gerontocratic rule reinstate harmony between generations. Prayers and blessings of seniors promise help.

ENDNOTES

1. *Lebenswelt* according to Schütz describes that part of reality "den der Erwachsene in der Einstellung des gesunden Menschenverstandes als schlicht gegeben vorfindet" (Schütz & Luckmann, cit. after Schmuck-Widman, 1997: 131). Only part of this *Lebenswelt* is based on personal experience, the major part is part of a common understanding of reality within a population and is socially communicated.
2. While the quotes from interviews with Himba are usually direct translations from taped conversations, the quotes from interviews with Pokot result from scripts made during the interview.
3. Richard Waller (1988) gave a detailed account of the catastrophe in Maasailand. There the two last decades of the century are remembered as *emutai* - from *a-mut*, to finish off (completely). The term conveys the notion of complete destruction.
4. According to Lamphear (1994) the Jie of north-western Uganda applied a very similar strategy. They too had expanded rapidly just before the advent of colonial troops. The

British were seen as natural allies who were as interested in the maintenance of a status quo as the Jie themselves.

5. KNA DC BAR 1/1 Annual Report 1918
6. KNA DC BAR 2/11/4 Handing Over Reports
7. KNA DC BAR 3/5 Political Records.
8. One informant remarked that the Rinderpest was also called *chemulunchö* (lit.: of the warthogs) because warthogs were first affected by the epidemic. As warthogs drank from the same wells as cattle, these were affected too.
9. The archival sources do not give estimates on mortality rates. In oral accounts the drought is described as having lower mortality rates than the "Drought of the White Bones".
10. The Mau Mau disturbances were casting their shadows and the British had analysed Central Kenya as the major trouble spot of the Kenyan Protectorate. They had little interest in creating another point of contention by severe punishment of the Pokot.
11. For a fuller account of the movement see Bollig 1992a: 136-140.
12. In order to explain the meaning of the term, Pokot informants pointed to the sky and made a movement with their arm which they underlined by the sound *chweeer*.
13. Chapman (n.d: 44) reports a Swartboois raid into southern Angola in 1890 which reached as far as Chibia near nowadays Lubango. The tradition probably refers to this raid. Thirty-three raiders were killed by Dorsland trekkers and forced the raiders to retreat.
14. Gewald (1996:138ff) reports on the effects of the disease in Central Namibia: "Cattle cadavers lay scattered around the settlements, in the cattle posts and along trade routes of Namibia, and as they started to rot, the scavengers being sated, the air of the territory was filled with the cloying smell of putrefying flesh". In central Namibia Rinderpest was followed by famine and a range of other diseases.
15. Hayes (1992:178-206) gives an excellent account of the advent of the disaster and its causes.
16. According to population figures given by McKittrick (1995: Appendix) which are lower than those of Hayes, mortalities were roughly a fourth of Ovamboland's population around 1915.
17. Gibson (1977) gives 1935 as the year of *karasaruvyo*. However, colonial records do not remark on a drought in 1935 while the two consecutive droughts of 1929/30 and 1932/33 are widely commented upon.
18. Tsandi in western Ovamboland recorded 143.1mm (–65.2 per cent) in 1927/1928, 163.8mm (–52.–2) two years later in 1929/30, 345.9 (–15.8 per cent) in 1930/31, 163.8 in 1931/32 (–60.1 per cent) and 274.6 in 1932/33 (–33.1 per cent). The deviations from the mean in Outjo for the same period were - 59.5 per cent, - 31.4 per cent - 45.4 per cent +0.9 per cent, –57.2 per cent, –42.3 per cent.
19. NAO 28 Native Commissioner Ovambo to the Secretary of SWA 24/6/1932
20. NAO 28 Carlos Lino da Silva, Frontier District of the Lower Kunene, Vila Pereira de Eca to Nat. Com. Ovamboland 15/7/1932.
21. A local constable commented critically "... it is of course impossible to afford relief to the whole population; 21 bags of mealie meal are being sent for distribution among the most urgent cases; 100 rounds of ammunition to shoot zebra for meat for natives." (NAO 28 Secretary for SWA Courtney Clarke to Constable Cogill 3/11/1932)
22. While Outjo had slightly higher rainfall than in normal years, Tsandi recorded minuses (from a mean of 410.9) of –11.9 per cent (1937/38), –20.0 per cent (1939/40) and –29.3 per cent (1940/41) and Kamanjab a minus of 45.2 per cent in 1940/41 (mean 303.6).
23. In Opuwo only 154.5mm of rain fell, which is 48.3 per cent below average. Kamanjab recorded rainfall 50.1 per cent below average. Outjo only had a slight minus of 7 per cent while Ombalantu booked a –40.9 per cent deviation from the mean.
24. The 1958/59 rainy season was recorded with a low of 48.9 per cent below average in Kamanjab, the lowest rainfall since the 1946 drought and in Sesfontein with even 71.1 per cent below average. In Opuwo a low of 32 per cent was noted. The effects of the drought were somewhat ameliorated in the northern parts of Namibia by the following year which recorded 12.5 per cent rainfall above average in Kamanjab and 1.3 per cent below average

in Opuwo. However, the drought went on unabated in western Kaokoland: Sesfontein measured –41.3 per cent. The two following years then recorded minuses of 32.5 per cent and 20.3 per cent in Kamanjab turning the drought into a famine. In Opuwo the rain year 1960/61 brought relief and was recorded with a 20 per cent above average rainfall while in other places of Kaokoland the drought still continued.

25. NAN BOP N 1/15/6/1.
26. In 1964 the officer in charge in Opuwo wrote to the Secretary of Southwest Africa: "Re.: Noodtoestand in Kaokoveld: Bemarking van Kleinvee", Soos u darvaan bewus is was dit baie droog die afgelope somer in die Nordelike Inboorling gebiede. Volgens verslae uit die Kaokoveld is die mielie-oeste 'n totale mislukking en sal daar nie eers 100 sal gewen word nie. Die Kaokoveld Stamfonds is hopeloos oortrokke en die inwoners skuld nog groot bedrae geld vir mielies wat gedurende 'n vorige seison aan hulle verkoop is. Daar is geen inkomste nie, en die inwoners het geen geld om of hulle belastings te betaal nie of om voedsel vir konstant mee to koop nie.Hierdie kantoor probeer om aan soveel van die inwoners as wat moontlik is werk te voorsien sodat hulle wel 'n inkomste kan he om voedsel te koop. Dit is 'n onbegonne taak om soveel werk te verskaf dat die helde bevolking tot die volgende oes daaruit kan lewe. Dit dien egter as 'n hulp. Die enigste moontlike ander uitweg waraan gedink kan word is om te probeer om 'n mark vir hulle kleinvee te vind. Soos u darvaan bewus is, is daar nog nie veel vordering gemaak met die bemarking van vee na Angola nie." (NA BOP N7/8/2 Famine Relief 1958-72, Die Administratiewbeampte, Ohopoho to Die Sekretaris van Suidwes-Afrika, 12/6/1964)
27. NA BOP N1/15/6/1 Annual Reports for 1958 and 1959.
28. NA BOP N7/8/2 Famine Relief 1958-72, Die Administratiewbeampte, Ohopoho to Die Hoofbantoesakekommissaris 21/9/1964.
29. Kantoor van die Hoofbantoesakekommissaris van SWA to Die Bantusakekommissaris 22/8/1967.
30. Bantoesakekommissaris, Ohopoho to Die Hoofbantoesakekommissaris, Windhoek 26/10/1967.
31. In Himba traditions this is the second Year of the Dying after the Rinderpest year of 1897.
32. Allgemeine Zeitung 11/6/1981: p.1
33. Crandall (1992:182ff) says that the Himba differentiate three types of misfortune: "The first category of misfortune is that which befalls a person specifically because his own actions are offensive to an ancestor....The second category of misfortune involves the ancestors in the role of protectors of their living kinsmen. Sometimes the cause of misfortune is divined to be an evil force or spell sent out to attack a person or family by another definite person or group of persons who, in my experience, remain nameless - essentially anonymous. ....The third category embraces misfortune on a very wide, collective scale. These types of misfortune are really inexplicable. Severe drought, rinderpest, diseases, etc. these are the collective misfortunes which the Himba cannot explain. In my conversations with Himba men on the subject, God is invoked as the reason why something has happened. The ancestors are perceived to be incapable of causing something on so grand a scale as this." These observations coincide with my data on Himba perceptions of dangers, especially on the last point. Considerations as to the deeper cause of major disasters remain vague.

Chapter 5

Coping Strategies during Drought and Disaster

The following section of the book describes coping strategies which are applied during drought and disaster to prevent starvation and the loss of livestock. The change in dietary habits is an important strategy which is applied early on to cope with famine (Halstead & O'Shea, 1989; Mink & Smith, 1989; Colson, 1979; de Garine & Harrison, 1988; Amborn, 1994; Watts, 1988). More animals are slaughtered than usual, food is shared more intensively, and/or everyday food is substituted by less preferred food. Herders all over the world react to drought conditions with increased sales of livestock. When milk supplies fall short, market bought cereals make up the major part of the daily food. Intensified spatial mobility is a distinct strategy of pastoral people to cope with a crisis "by taking advantage of the spatial and temporal structure of resource failure"(Halstead & O'Shea, 1989: 3). Diversification is necessary in order to exploit resources that are not affected by drought, epidemics or violence. Diversification in pastoral households ranges from multi-species herding to so-called "ten-cent-jobs" such as brewing, charcoal burning, and the sale of traditional medicine (see for example Browman, 1987; Odegi-Awuondo, 1990; Legge, 1989). However, coping strategies during a disaster are not only tied to the material world. Many ethnographers observed that during a crisis people look for explanations of the misery and hardship (Scoones, 1996; Mink & Smith, 1989; Colson, 1979). They try to reduce unpredictability by oracles and attempt to influence the course of events by rituals.

5.1. CHANGING FOOD HABITS: SLAUGHTER, SHARING, SUBSTITUTING

Milk products and cereals are major components of the diet in all African pastoral societies. Sellen (1996:109) summarises findings on this aspect: 62 per cent of dietary energy originates from milk among the Turkana, 64 per cent among the group-ranch Maasai, 66 per cent among the Ariaal and about 30 per cent among the agro-pastoral Ethiopian Borana. Meat in general only accounts for less than 10 per cent of the caloric intake. Wild plants add important vitamins to the otherwise simple diet: vitamins A, B2 and C are frequently found in berries, relishes, fruits and tubers (see Casimir 1991 for a detailed treatment of food adequacy in pastoral societies). Anthropologists have found major seasonal differences in nutrition in African pastoral groups (cf. Galvin, Coppock & Leslie, 1994:120): Turkana dietary intake relies on milk for 89 per cent during the wet season but for only 30 per cent in the dry season. Wealth-related differences in nutrition were observed too (Sellen, 1996:109). These differences can be balanced to some extent by an increased consumption of meat. During a drought the metaphor that livestock is food "stored on the hoof" (Clutton-Brock, 1989) becomes evident: as if from a larder, small stock is repeatedly taken to calm the severest hunger. Social mechanisms contribute to the distribution of available food. While flexibility in the adaptation of food-ways to seasonal constraints and social strategies prevent starvation, nutrition is not necessarily adequate all year round. Sellen (1996: 121/122) found in a comparative analysis that seasonal differences in nutrition lead to appreciable seasonal fluctuations of bodyweight (about 5 per cent), in significantly lower body weights (per height) in general and retarded growth during childhood. These are normal fluctuations which reflect a high degree of seasonality. What happens if lean periods are extended?

5.1.1. Pokot Foodways during Famines

The acquisition of food is a major topic in Pokot accounts of drought. Even in everyday conversation, food features importantly: one has been given a chunk of meat at a festivity, meat or milk has been denied to somebody, unusual food had to be tested to prevent hunger. Data on food sharing was mainly gathered during three periods of fieldwork in 1991, 1992 and 1993. An inventory of substitute food was gradually produced during the entire period of fieldwork.

5.1.1.1. Increased Slaughter

In 1992, in the midst of a serious famine, I interviewed 25 household heads from a sample of households I had worked with during the previous five years. I wanted to learn (a) to what extent they had slaughtered and sold livestock from their herds to prevent hunger in their homes and (b) whether they had appealed to other households for help or had assisted any of them. The increase in slaughtered livestock was notable.[1] During the 1990/92 drought the number of healing rituals (*tapa, kikatat, kolsyö*), which always involve the slaughter of livestock, increased and the number of meat feasts (*asiwa*) rose. The number of goats slaughtered per household in a 12 month period ranged from 2 to 21 with an average of 8 animals per household. Cattle were slaughtered much less frequently - oxen were sold rather than slaughtered. The 25 households of the survey had slaughtered 11 cattle altogether resulting in an average of 0.44 cattle slaughtered per household.

Neither meat feasts nor healing rituals are necessarily tied to drought and famine; however, the data recorded during 1992 indicates that occasions for slaughtering livestock become more frequent during periods of nutritional stress. While neighbourhood-based meat feasts tend to become less at the height of a drought due to intensive labour demands on potential male participants, simple requests of food-aid in the form of a goat or a sheep become more numerous. Each mode of support will be shortly discussed.

omisyö moning (literally: food for children): A household head lacking a sufficient number of goats for slaughter or sale can turn to a relative or a stock-friend for support. Women will turn to their kinsfolk (father, brothers, father's brothers etc), men may address friends, older brothers, father, father's brothers and occasionally affines. They ask for *omisyö moning*. This metaphoric request for food aid suggests that the adult person begging for food does so for the benefit of his children - and not for his own sake. The metaphor directly addresses the *kisyonöt* (mercy) of relatives and friends and it is very hard to deny such a request without rendering oneself deeply inhumane. It alludes to the moral self of the two persons involved in the transaction: the petitioner caring selflessly for his children and the donor giving away livestock without asking for any form of compensation. Usually a goat or a sheep is given in such transactions. The beneficiary may either slaughter or sell the animal. If slaughtered, the meat of the animal will be distributed throughout the family and neighbours will also get a fair share of the meat. If the animal is sold the money is usually used immediately to buy maize

asiwa (meat feasts of the neighbourhood): Meat feasts are initiated by the men of a neighbourhood. During their daily informal meetings at their meeting place the talk frequently circles around the obligations of the wealthier herders to supply their friends and age-mates with meat. Occasionally they will decide that one of them has not slaughtered anything for the men of the neighbourhood for quite some time. This person is then formally requested to donate an ox, or at least a goat or a sheep. He may offer a sheep or a goat in return but it is hard for him to deny the request altogether. However, he has the liberty to select the man who is going to spear his animal and thereby establishes a bond with this man, who will then have the obligation to pay back the slaughtered animal in due time. Frequently, however, the men requesting the animal have already chosen somebody from their midst to slaughter the animal - i.e. preparatory discussions of a meat feast usually specify not only the man who is to donate the animal but also the person who is to take the burden of the debt onto himself. The meat is eaten communally over a day or two by the men. Selected pieces of meat may be sent to elders who are not able to participate in the occasion due to distance or infirmity. The person who spears the animal takes the head and neck back home to his family. Intestines and other minor parts of the meat are shared in the household of the donor. Meat feasts mainly provide men with meat, women and children only participate peripherally in these meat distributions.

tapa (healing ritual): The number of various healing rituals increases during periods of stress. Rituals such as *tapa* and *kikatat* are aimed at diagnosing diseases and/or curing them. All healing rituals, on homestead and neighbourhood level, involve the slaughter of livestock. In the survey, the number of *tapa* rituals per household was recorded. Many animals used in *tapa* were donations or loans from friends and relatives. Meat is meticulously distributed in such rituals according to well spelt out standards: the intestines and some special parts of the meat, defined by the ritual specialist conducting the ritual, go to the sick person. The woman who contributed the goat (or cow) takes the head and neck of the slaughtered animal. If the animal was borrowed or presented by a relative of a household member, these parts will stay with the member receiving the animal. If it is borrowed or presented by a relative of the household head, he then decides which house in his homestead will receive the head and neck. The receiving house has the duty of repaying the debt at a later stage. The other wives of the household share the back and intestines. The neighbours take part of the rump and both hind legs. Married daughters of the household take both fore legs back to their homesteads. Boys and men of the household and neighbourhood distribute the breast among themselves. Finally the ritual specialist conducting the ritual, usually a woman, takes the skin, remaining parts of the rump and, if the animal was a sheep, the fat tail. Livestock slaughtered on other ritual occasions are distributed in a similar way.

During a drought the number of rituals increases and with them the amount of meat allotted to food-sharing increases too. These ceremonies strengthen the person undergoing the ritual and at the same time ameliorate the nutritional situation of the entire social environment during a period of food scarcity. Food is channelled through a pre-existing network of reciprocal exchanges. Table 12 shows the amount of livestock slaughtered in the drought period of 1991/92 in 25 households. The table differentiates the purpose of slaughter and the kind of animal slaughtered (small stock/large livestock). The total amount of meat procured per household is given and set in relation to the number of people (adults/children) per household. This ratio is then transformed into a figure for kcal/person/day. Households are sorted according to wealth groups into five classes, class 1 containing the richest herders and class 5 containing the poorest herders.

Meat yields per household vary tremendously. The mean of 17.5kg meat/person/year does not carry much meaning as the range is, roughly speaking, between 1 and 70 kg/person/year (Std.Dev. at 15.5). There is no pattern which explains this high rate of variation. This leads to the conclusion that the slaughter of livestock depends very much on the actual situation of the household, the proximity of trading centres and the availability of livestock for slaughter and not so much on its wealth.[2] The household of

Table 12. Livestock slaughtered between May 1991 and April 1992 (no of households 25)

House-hold Nr.	Wealth	Heal. ritual g/sh	Heal. ritual c/ca	'Meat feast' g/sh	'Meat feast' c/ca	Other g/sh	Other c/ca	Total: s/l	Meat (kg)	No a/ch	Meat/ pers/g	kcal/ pers/day
1	1	2		1		1		4/0	48	7/10	4,0	22
2	1	5				8		13/0	156	10/8	11,1	61
3	1	4		2				6/0	72	9/3	6,9	38
4	1	5				7		12/0	144	12/6	9,6	53
5	2			7	1	2		8/1	206	5/8	22,9	72
6	2	15	1	5		1		21/1	362	5/15	29,0	159
7	2	6		4		5		15/0	180	11/6	12,9	71
8	3			2				2/0	24	9/10	1,7	9
9	3	5				9	2	14/2	388	4/3	70,5	386
10	3	1		1		2		4/0	48	5/4	6,9	38
11	3	2				1		3/0	36	4/3	6,5	36
12	3	4		3				7/0	84	6/3	11,2	61
13	3	10		1		1		12/0	144	11/3	11,5	63
14	3	4		3		1		8/0	96	6/7	10,1	55
15	3	15			1		2	15/3	510	10/11	32,9	180
16	4	2	1				1	2/2	244	3/5	44,4	243
17	4	2		2		1		5/0	60	4/6	8,6	47
18	4	5		2		2		9/0	108	2/4	36,0	197
19	4	4	1	2		1		7/1	194	4/5	29,8	163
20	4	7		4		1		12/0	144	4/6	20,6	113
21	4	1		3		1		5/0	60	5/7	7,1	39
22	4	1	1	1		3		5/1	170	5/6	21,3	117
23	5			1		2		3/0	36	5/3	5,5	30
24	5	2						2/0	24	3/1	6,9	37
25	5	3		1		2		6/0	72	4/8	9,0	49

Continued

Table 12. Livestock slaughtered between May 1991 and April 1992 (no of households 25)—*cont'd*

House-hold Nr.	Wealth	Heal. ritual g/sh	Heal. ritual c/ca	'Meat feast' g/sh	'Meat feast' c/ca	Other g/sh	Other c/ca	Total: s/l	Meat (kg)	No a/ch	Meat/ pers/g	kcal/ pers/day
Average	-							-	144,4	-	17,5	93,6
Std.Dev.	-	-	-	-	-	-	-	-	120,3	-	15,5	84,7

Notes: (1) Wealth categories: 1 - very rich, 2 - wealthy, 3 - medium, 4 - poor, 5 - very poor. (2) Abbreviations: g/sh/c/ca: goat/sheep/cattle/camel. (2) Meat yields: sheep "anywhere between 10-25kg, Maasai 9.8kg" (Dahl&Hjort 1976:201), goats "meat yield 12.7kg" (Dahl & Hjort 1976:203); as a basis for accounting I took 12 kg per small stock; Cattle 117.5kg for Kenya, Tanzania and Uganda, northern Kenya, Boran 118.5kg, 107.5 Samburu, 117.7kg Maasai (Dahl & Hjort 1976:165/166); as a basis for accounting I took 110kg meat per animal. (3) For the row meat per person: children up to the age of 15 were counted as 0.5 adult (Casimir 1991:48). (4) Casimir (1991:79) assumes average energy content of meat at 200kcal/100g.

Angurareng for example who garnered some 510kg within one year resulting in an appreciable 33 kilograms per person had to slaughter livestock continuously for the healing rituals of his ailing son. Limanyang, reaching the second highest average with 44kg, had to slaughter two cows for a ceremony after his wife had health problems after giving birth. In the face of such high rates of unpredictable variation, extensive meat sharing is the most efficient solution to guarantee each household a low degree of food security.

The total average amount of meat available from slaughter during this drought period was low: only 94 kcal (about 50gr) per person per day became available as a consequence of slaughter. Only 6.3 per cent of all energy procured (from maize and meat) during the dry year of 1991/92 stemmed from slaughtered animals. Maize was of overwhelming importance for nutrition during the drought. Still, the Pokot regard meat as an important part of their drought diet. In many accounts on famines the description of starving animals (*chemosoy*) being slaughtered features importantly. However, figures suggest that the factual importance of meat as a drought food has declined over the last decades. Meat consumption did not feature very importantly in the recent drought! Drought food is characterised rather by a monotonous maize diet. There are two reasons for this: the average livestock holdings per household have declined considerably over the last decades. At the same time the exchange ratio of livestock to maize has been increasingly favourable. A slaughtered ox resulted in about 400,000 kcal (Casimir, 1991:79 assumes average energy content of meat at 200kcal/100g) in the early 1990s. If sold at the average price of 3000 KSh (1992's pricing) and if the money was used to buy maize the animal yielded 1,795,000 kcal (energy content of maize taken from Casimir, 1991). A goat slaughtered (12 kg meat) would yield 24,000 kcal. If used to buy maize the same animal would result in 236,940 kcal. A household of ten people could live well for two or three days from the meat of a slaughtered goat, but could live for 10 days on the maize bought with a goat. Hence it is perfectly rational to exchange goats for grain in the face of overall declining herd numbers. Of course, this simple juxtaposition does not account for nutritional adequacy: maize lacks fat as much as it lacks proteins and other essentials. Long lasting nutrition based on maize only leads to malnutrition and that is exactly what happened during the droughts of the 1990s and was very prominent in the drought of the year 2000 (pers.com. Sean McGovern and Rebecca Janacek, East Pokot Medical Project Project).

The little meat which is available is shared. While maize is of the greatest importance during a drought and meat only contributes to the quantity of food during a famine, meat adds important fats and proteins to the diet that maize does not have and as such remains indispensable during a drought.

5.1.1.2. The Sharing of Food

The sharing of food is intensified during a drought. There is a very general and binding idea of hospitality. The two normative concepts *tilyontön* (solidarity) and *kisyonöt* (mercy) which Pokot deem to be characteristic for their community, motivate people to give out food during periods of hardship. The imperative rule is: food must not be denied to anybody who asks for it! People who hide food away from others are quickly accused of envy (*ngatkong*) and witchcraft (*pan*), bad will (*ghöityö*) or a bad heart (*gha mikulogh*) (cf. Bollig, 1992a:154). Somebody who has been denied food will eventually respond with bitterness, curses or witchcraft resulting in disease and bad luck for the person blamed. These emotional reactions to the denial of food are interpreted as natural reactions, beyond the immediate control of the person. Withholding food from guests and neighbours does not only put internal peace at risk but also endangers personal well-being[3]. Especially in times of scarcity it is important to handle food openly and generously. The Pokot like to juxtapose their generosity and propensity to share food with the stinginess of neighbouring ethnic groups.

While standards of generosity specify general attitudes to food sharing, there are rather specific rules on what kinds of food have to be shared and how they have to be shared. Maize, which is usually bought in shops, and milk are rarely given away to non-household members. Unlike meat, there are no rules for the distribution of maize. Usually every adult woman within a household handles maize and milk very much by herself. However, in times of need there should be and factually is an exchange of milk and maize between the various houses of a homestead. In contrast to this the distribution of meat usually covers a group of people wider than the household. Standards of meat sharing are clearly spelt out. The proper sharing of meat is deemed to be an integral part of the Pokot identity (see Table 13).[4]

Indeed, the 37 households interviewed have supported each other on numerous occasions during the previous two years (see also Bollig, 1998a): men invited each other for meat feasts (*asiwa*); they gave out presents of small stock to furnish their friends' and neighbours' households with meat or capital for market exchange (*omisyö moning*); the number of healing rituals (e.g. *tapa*, *kikatat*), all connected to the slaughter of livestock and extensive meat

Table 13. The Pokot Way of distributing meat

Relation	Appropriate part of meat
Father	breast (*takat*), part of hindleg (*loyö*), tongue and respiratory organs (*ngalyap*), calf (*hagh*), liver (*koghogh*)
Sons	breast (*takat*), tongue and respiratory organs (*ngalyap*)
Wive's of father	intestines (*kwan*), stomach (*mu*), backbones (*rototIn*), feet (*kelyon*)
Married daughter	one of four legs, bit of intestines and stomach,
Pat. Uncle	upper part of hind leg (*kipes*)
Mat. Uncle	special part of breast (*kirmongö*)
Neighbours/male.	entire left hindleg (*chat*), participate in breast and tongue
Neighbours/female	part of intestines and stomach
Young children	colon (*pempö kamösöw*) part between hooves and calf (*ngoriyon*)
Woman owning the animal	participates in the sharing of meat with other women, skin,

sharing, increased during the drought. Women went to their paternal relatives to do *kibich*: they approached their kinsfolk with a lot of sentiment and made their precarious situation known. The consulted relative usually slaughtered a goat or a sheep for his sister or daughter the very same night. A major part of the meat was eaten on the spot, but the woman usually carried some parts to her homestead. Additionally she may have been given another animal to take home, to sell it at a later stage or to use it to ameliorate her milk supplies.

Transactions of livestock within the network of 37 households between 1990 and 1992 were substantial. The 37 household heads gave on average about three (3.24) heads of small stock (i.e. about 38.9 kg of meat, very few donations were cattle and none were camels) within the local network for ritual and food assistance (i.e. roughly 120 animals). Each household head additionally donated one to two goats (1.68, or 20.2kg) in meat feasts, all goats being slaughtered for public (primarily male) welfare (roughly 60 animals) (see Table 14). This means that about 180 animals were slaughtered within a period of 24 months, indicating that about every fourth day an animal procured via sharing mechanisms was slaughtered within this community of

Table 14. Transactions during the 1990-92 famine

No.	Rank	A gives B	kcal (1000) donated	A receives from B	kcal (1000) received	Total budget
11	1	3 5 6 9 12 20 21 30	192	1	24	−168
18	1	1 13 16 17 31 32	144	16 32	48	−96
32	1	9 18 25 27 31	120	1 2 5 18 20 27	144	+24
34	1	16 3 6	72	0	0	−72
1	2	2 3 4 5 6 10 11 17 20 26 31 32 33	312	18 20 24 28 31 33	144	−168
7	2	17 27	48	20	24	−24
16	2	9 17 18	72	18 22 34 36	96	+24
26	2	0	0	1	24	+24
20	2	1 4 5 6 7 17 23 27 37	216	1 4 11	72	−144
33	2	1 4	48	1	24	−24
35	2	0	0	0	0	0
36	2	9 16	48	2 3 20	72	+24
4	3	5 13 15 19 20 24 25 28	192	1 2 4 5 13 20 31 33	199	+7
6	3	9 12 17	72	1 10 11 19 20 29 34	168	+96
8	3	0	0	2 25	48	+48
10	3	3 5 6 9 12 20 21 30	192	1 37	48	−144
14	3	9 13	48	13	24	−24
19	3	3 6	48	2 3 4 10 17	120	+72
25	3	9 17 24 25 27	120	4 13 23 25 32	120	0
27	3	9 13 18 25 27 31	144	7 23 25 32 20	120	−24
29	3	17 6 21	72	21	24	−48

Continued

Table 14. Transactions during the 1990-92 famine—*cont'd*

No.	Rank	A gives B	kcal (1000) donated	A receives from B	kcal (1000) received	Total budget
2	4	4 8 19 32 36	120	1	12	−108
3	4	13 19 30 36	96	1 11 19 22 30 34 10	168	+72
5	4	4 17 32	72	1 4 11 20 10	120	+48
12	4	0	0	6	24	+24
15	4	0	0	4 17	48	+48
21	4	29	24	11 29 30	72	+48
22	4	16 3 23	72	0	0	−72
23	4	16 17 27 24 25	120	16 20 22 25	96	−24
28	4	1 24	48	4	24	−24
30	4	1 4	48	3 8 11	72	+24
31	4	1 4	48	1 18 27 32	96	+48
37	4	10 17	48	25, 20	48	0
13	5	4 14 25	72	3 4 14 18 27	120	+48
17	5	15 19	48	1 5 6 7 16 18 20 23 25 29 37	264	+216
24	5	1	24	4 23 28	76	+48
9	5	0	0	6 11 14 16 25 32 36 10	168	+168
ave.			81.1		79.8	−1.4
Std Dev			69.8		62.1	82.0

Notes: (1) Wealth categories: 1 -very rich, 2 -wealthy, 3 -medium, 4 -poor, 5 -very poor.(2) The table is to be read the following way, row 1: actor 1 gave 2,3,4 etc.; actor 1 received from 18, 20 etc., (3) Actor 17 received considerably more than the others because many had to give him a goat as payment for their circumcision.

37 households, and that about 20,000 kcal were available per week. Meat is widely distributed through the exchange network ensuring a low level but constant supply of meat during a period which is, otherwise, characterised by fat and protein deficient food. Only very rarely I observed that meat was dried for later consumption within the household. Clearly, the distribution of meat is preferred to storing it. The symbolic connotations of meat distribution are salient: not only is hunger satisfied but also is solidarity and Pokot identity stressed by extensive meat distribution. Table 14 gives an indication of the densely woven support network. All 37 actors are listed and shown with their transfers of livestock to other households during the famine of 1990/1992.

A look at the energy budgets of rich, medium and poor households gives a first generalised insight into the magnitude of assistance garnered by poor households. Very rich households gave about 528,000 kcal and only received some 216,000 kcal and wealthy households gave about 744,000 kcal and received 360,000 kcal. Poor households received 5 goats more than they gave while very poor households got 21 goats more than they had donated themselves. Medium and poor households had a more or less balanced budget while very poor households received about 556,000 kcal more than they had donated. These budgets seem enormous at first glance: expressed in goat equivalents, the relevance of these differences melts down: the four very rich households gave only 13 goats more than they had received.

Table 15 indicates that very rich and wealthy households contributed proportionately more to famine support networks than poor and medium households. The four very rich households gave 17.6 per cent of all animals donated, while the four very poor households only gave 4.8 per cent. In contrast to this the very rich households obtained only 7.2 per cent of all animals given away while the very poor households obtained 21.6 per cent of all animals given. While the very rich households gave 4.4 animals on average, wealthy households gave 3.9, medium households 4.1, poor households 2.4 and very poor households 1.5 animals. Very rich households received 2.3 animals on average, wealthy households obtained 2.4, medium households 4, poor households 2.8 and very poor households 6.8 animals.

The four very poor households of the sample profited disproportionately from support networks. Nevertheless, even the richest households from the sample addressed other, considerably poorer households for help, too. They asked for *omisyö moning* and for contributions for necessary rituals. The Pokot are not particularly critical of such a practice and the fact that rich households depend on help too is taken for granted. The Pokot maintain that all households basically deal with the same sort of subsistence problems and start off on an equal basis. Differences in wealth are not a subject of public discourse. Rich Pokot herders could very well sell surplus livestock and buy maize and other food with the money obtained. Needless to say, they would garner much more food for their households. Obviously giving away livestock as loans or donations during famine does not directly increase the food security of the donor's household. Apparently there are

Table 15. Livestock transactions during famine and wealth

Wealth category	No. of households	A gives B.	per cent	A receives from B	per cent
1	4	22	17.6	9	7.2
2	8	31	24.8	19	15.2
3	9	37	29.6	36	28.8
4	12	29	23.2	34	27.2
5	4	6	4.8	27	21.6
Total	37	125	100.0	125	100.0

Note: (1) Wealth categories: 1 -very rich, 2 -wealthy, 3 -medium, 4 -poor, 5 -very poor.(2) Correlation between wealth category and donors is at 0.18 and at 0.72 between wealth category and receivers (Pearson's Correlation Coefficient) suggesting a low predictability as to who gives but a high predictability as to who receives.

other incentives than food security to keep them convinced that giving away livestock during a drought in the long run benefits them more than selling the very same stock.

Poverty obviously does not exclude households from exchange networks, even if people donating livestock to poor households have little hope that their gift will be returned quickly. Gifts of livestock enhance the prestige of a herder and increase his symbolic capital. Likewise he adheres to a core idea of Pokot identity. He physically contributes to the maintenance of a strong sharing group in which "Pokot lick each other like calves" as one elder once told me. During the drought of 1990-92, all poor households were able to obtain some help. A year later I learnt that only massive contravention against Pokot standards of conduct give reason to exclude somebody from support networks: one very poor household head had stolen small stock from his neighbours and was subsequently forced to leave the area and another was accused of drinking away his limited wealth. Both households had belonged to the category of very poor households. Without wearing the argument down, this may indicate a general trend: very poor households are indeed endangered of slipping out of support networks.

The structure of support networks is not primarily conditioned by kinship organisation or wealth differences, but rather by gift exchanges made prior to the drought. I recorded extensive livestock exchange networks between 1987 and 1989 and had assumed that outstanding livestock debts would be "cashed in" during periods of scarcity. Observations during the 1990/92 drought clearly showed that this was not the case. Only a minor fraction of animals which were given in 1990-92 were transferred to settle a debt. It was rather the other way round. The existence of an exchange relationship seemed to pave the way for asking a partner for a goat. Frequently people giving away livestock during the drought simply said "*kïsom kile*" – "he just came and asked for it (so I gave)". Supportive exchanges during periods of stress are bedded into existing exchange relations. The exchanges implicated by bridewealth donation, bridewealth distribution, and stock-friendship and their normative and emotive connotations create a framework for co-ordination and facilitate exchange in a crisis.

5.1.1.3. Living on Meagre Resources: Substituting Food

Oral traditions on historical droughts point to the importance of less preferred foods during famines.[5] Accounts of the disaster stricken years of *asis* in the 1890s claim that many Pokot only survived on roots, tubers and berries. Table 16 gives an overview of plants which were used during droughts. Many of these plants could actually be harvested during the lean months of the year (i.e. dry season, *kömöy*, or early rainy season, *sarngatat*). While bush foods are characterised by low yields (deGarine, 1994:349), they have nevertheless contributed significantly to feeding the Pokot during famines. The roots Lannea triphylla (*tapodin*) and Vatovaea pseudolablab (*kela*) have, according to traditions, helped people through the harshest famines of the pre-colonial and early colonial past.

Nowadays, substitute foods from the bush are rarely used. Informants claimed that many of the substitute food plants which were exploited in the past have become rare over the last few decades. The work of collecting and preparing drought food is described as cumbersome and labour intensive. To rely on bush food is regarded as a sign of outright backwardness (cf. de Garine, 1994:354). The Pokot nowadays prefer to sell animals or to obtain money through "ten-cent-jobs" and to buy additional maize rather than gather food in the bush.

Substitutive food from the bush has been replaced by famine relief food donated by government agencies, churches and international NGO's. Relief food was distributed for the first time during a drought in the early 1940s. The distribution, however, was not free. Relief food was sold at subsidised

Table 16. Non-domesticated food yielding plants used by the Pokot

Fruit	Plant	Scientific Name	Season	Preparation
Sorich	*sorichon*	Boscia coriacea	*komöy*	Boiled
Lakatet	*lakatetwö*	Barleria eranthemoides	*sarngatat*	Raw
Taran	*törönwö*	Grewia tenax	*sarngatat*	Raw
Tileny	*tilingwo*	Meyna tetraphylla	*kitokot*	Raw
Tilam	*tilomwö*	Ziziphus mauritiana	*komöy*	Raw
Kakach	*kököchwö*	Premna resinosa	*sarngatat*	Raw
Adome	*adomeyon*	Cordia sinensis	*sarngatat*	Raw
Sitit	*sitöt*	Grewia bicolor	*sarngatat*	Raw
Makaw	*mokukwö*	Grewia villosa	*sarngatat*	Raw
Makany	*mokongwö*	Ficus sycamorus	*sarngatat*	Raw
Malkat	*molkötwö*	Canthium setiflorum	*sarngatat*	Raw
Riron	*rironwö*	Delonixelata	*kitokot*	Raw
Mochok	*mochökwö*	Berchemia discolor	*kitokot*	Raw
cheblis	*chebliswö*	Hibiscus sidiformis	*komöy*	Boiled
loma	*lomiyon*	Balanites orbicularis	*komöy*	Boiled
aruru	*aruruyon*	?	*komöy*	Boiled
arol	*orolwö*	Sclerocarya birrea	*komöy*	Raw
tapoye	*tapodin*	Lannea triphylla	*kitokot*	Raw
songow	*songowowö*	Zanthoxylon chalybeum	*year*	burnt bark as salt substitute
Root	Plant	Identification	Season	Preparation
tapodin	*tapodin*	Lannea triphylla	*year*	Raw
kela	*kelowö*	Vatovaea pseudolablab	*year*	Raw
kilipchö	*kilipchö*	?	*year*	Raw
kiptermam	?	?	*year*	Raw
Seeds	Plant	Idenitfication	Season	Preparation
puyun	*puyun*	Eragrostis cilianensis	*komöy*	Ground
kimökön	*kimokonö*	?	*komöy*	Ground

Note: (1) plant identifications according to Timberlake (1987), (2) *kömöy*, dry season, December to March; *sarngatat*, early rainy season, April; *pengat*, rainy season, May to August, *kitokot*, early dry season, September to November prices to local traders who then sold it at fixed prices to herders. Famine relief is frequently addressed as *omisyö nyo pö serikali* (government food). In 1965 during the drought of Lochwer, a major relief programme was started for the first time supplying yellow maize from the United States and powdered milk to the Pokot. Many people remember an intense nausea after consuming tinned fish which was distributed with the other food items. During the devastating 1984 drought famine relief was poured into the division to prevent the worst. Since then the supply of famine relief has been institutionalised. All through the 1990s food aid was given to pastoral households - no matter whether the year was good or bad. During the December 1997 elections the two candidates running were judged according to their capacity to guarantee a smooth supply of famine relief food to the constituency.

> "Critics contend that the MP also failed to call on the Government to provide the residents with relief food, given that most parts of the constituency are dry and people rely on livestock for their livelihood. ... Mr. Lotodo is expected to face a formidable challenge from a former Athi River District Officer, Mr. Asman Kamama ... Most of the constituency's elite have thrown their weight behind Mr. Kamama who is credited with using his connections within the provincial administration to provide relief food to the people during last year's drought". Daily Nation, Thursday October 2, 1997, Election Supplement III: "Lotodo walking poll tightrope"

Bearing in mind that 1997 was a very good year for rain, it shows that the delivery of drought aid has been politicised and occasionally depends more on local politics than on climatic conditions. The local political elite simply construe the necessity for food aid from the fact that "the constituency is dry and people rely on livestock for their livelihood". Clearly drought aid has become very important during the past two decades and has altered other coping strategies. Livestock poor families nowadays depend, to some degree, on regular provisions of food-aid.

5.1.2. Himba Foodways during Famines

The quest for food is a salient theme in Himba accounts of droughts. Drought food characteristically consists of maize porridge, occasional chunks of meat and various substitute foods. The increased slaughter of small stock is one major strategy to prevent starvation. Substitute foods are of great importance among the Himba. High yielding palm trees which grow in dense groves along the major river courses and several other bushes and trees are harvested in times of food stress. Food aid has become important since the 1960s and was of great importance during the centennial disaster of 1981.[6]

5.1.2.1. Increased Slaughter

Over a period of 22 months (April 94 to January 1996), all livestock slaughtered in 25 households was recorded. The number of livestock slaughtered during this period was substantial. The households observed slaughtered on average some 21.4 goats, 6.8 sheep and 5.9 cattle resulting in some 324.9 kcal per day and per person. Table 17 summarises the number of animals slaughtered in 25 households. Again the total amount of meat and kcal/person/day is calculated. Wealth differences in meat consumption are obvious in this data. While rich households slaughtered more than 10 head of cattle during this 22 month period, two households did not slaughter a single head of cattle for the sheer lack of oxen and tollies. The correlation (Pearson's Correlation Coefficient) between wealth and animals slaughtered is at 0.308 with the level of confidence at 0.095).

A look at the seasonal distribution shows that during the 22 months recorded, no clear seasonal slaughtering pattern was observed (see Figure 13). The slaughter of cattle is mainly dependent on the ritual calendar: ancestor commemoration festivities which require the slaughter of cattle usually take place between August and October. This is shown by a minor peak in the curve. Whenever a funeral takes place several heads of cattle have to be

Table 17. Monthly rates of slaughter in 25 Himba households

Household	Wealth	Goats	Sheep	Cattle	Tot. meat in kg	Adult/ children	g/person /day	kcal/ person/ day
Katjira	1	20	29	18	2568	7/6	382	765
Kandjuhu	1	9	8	10	1304	9/2	194	389
Hikuminwe	1	13	1	4	584	6/3	116	232
Mbasekama	1	21	2	9	1266	12/6	126	252
Mirireko	1	7	0	1	194	7/5	30	61
Mbangauiye	1	59	45	19	4070	8/7	527	1054
Nongava	2	12	0	15	1794	14/12	135	267
Vahenuna	2	6	2	13	1526	13/7	138	276
Mbatjanani	2	9	2	4	572	7/6	85	170
Twarekarek	2	23	1	6	948	7/3	166	332
Katuezu	2	40	2	4	944	9/5	122	245
Taku	2	20	7	4	764	5/1	207	414
Mungerinye	3	15	13	4	776	3/4	230	463
Vahikura	3	17	0	2	424	3/1	181	361
Koseve	3	24	3	4	764	11/3	91	182
Koriautuku	3	35	4	2	688	8/3	108	216
Kamanda	3	25	3	6	996	4/4	247	495
Wokumutete	4	47	17	2	988	5/6	184	368
Wakakaata	4	27	2	4	788	11/5	87	174
Huku	4	7	9	0	192	10/2	61	122
Muhatunduk	4	13	5	3	546	6/5	96	191
Kaeretire	4	29	0	3	678	5/5	135	269
Kamuhoke	5	5	3	2	316	4/4	78	157
Tjandira	5	11	8	0	228	3/1	97	194
Muhuwa	5	27	1	2	556	3/1	237	473
Average	–	20.8	6.7	5.6	979.0	–	162.4	324.9
Std Dev	–	13.2	10.1	5.2	820.1	–	104.5	209.1

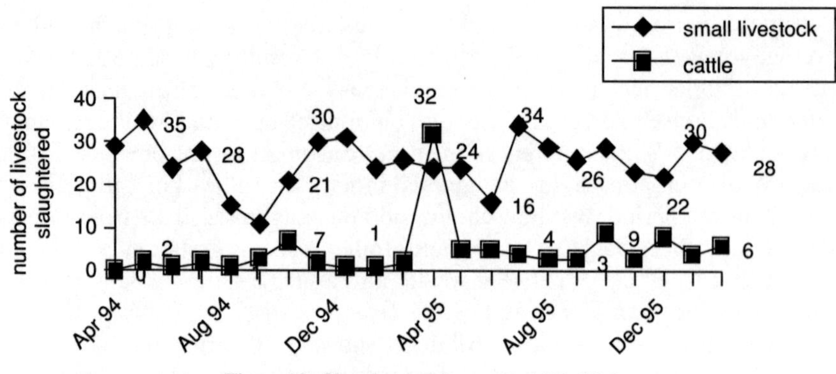

Figure 13. Slaughter of livestock, 4/94 -1/96

slaughtered. The peak in March 1995 resulted from a funeral. The slaughter of small stock is even less patterned. Oral accounts on famines suggest that the slaughter of small stock increases whenever milk supplies decrease substantially. The meat of goats and sheep is as of major importance as famine food. Figures on off-take indicate that Himba herds still have enough buffers to allow for increased slaughter.

Himba small stock herds are comparatively big: in a sample of 40 households all the herds consisted of more than 150 animals. This allows for an appreciable off-take without endangering the herd. Data from interviews suggests that Himba slaughter extensively during dry periods. A goat or a sheep may be slaughtered at least once a week. An informant claimed that in his household "goats were eaten like porridge" during periods of food stress. In this way small stock herds - much more so than cattle herds - mainly serve as buffers against periods of food scarcity. They are a reliable food source and the size of herds ensures that this "walking larder" (Clutton-Brock, 1989) will not be easily emptied.

Oral traditions leave little doubt that next to bush food the increased slaughter of small stock was and still is of tremendous importance to guarantee a steady supply of food. Mukaakaserari, a woman in her forties, emphasises the relevance of drought food in the following story. Goats and sheep were slaughtered nearly every second day. For weeks meat became the staple food.

> "And we continued our march on the road that climbs the gorge. We climbed the gorge.... Yes, we went there and settled there until we became very hungry and Mungerinyeu took out (from the goat's enclosure) a goat of my father's younger brother. My husband slaughtered it and the people ate...... We slept there and we ate that goat. The next day we stayed the whole day at the homestead and the following day we returned to the well, we scooped water, and the following day we moved once again, we descended down there. We went to Omuseravari. I was so thirsty

then, I was near the end. ...We stayed there until evening and then we went to an empty homestead. There we stayed. And I slept. That evening, when we stayed at that place, the father of Vasongonona killed another goat. The people were exhausted, they just fell asleep near the meat and we did not eat and did not drink soured milk as the cattle were dry and the calves were starving and dying.

Yes, there was some meat to make us fall asleep and to make us stay and the next day the father of Vasongonona killed yet another goat and we stayed the whole day and we ate and the next day we moved our cattle again. We were still on the way to Ongokwa..... far away, over there. ... And we walked and stayed, and stayed in the bed of one dry creek that is called Of Termites. We slept and then in the evening this man there slaughtered a goat. We slept and ate our goat. Yes, we had tied it (the stomach) with the big women's belt. This man with whom we went together,... brought us a small piece of meat from this goat that we had eaten a long time before. He brought the head. We were left behind and he said: 'I have spoken to the others there. They have lost cattle. And I have told them <I will return now and I will drive the goats to Ongokua (together with the women)>.' And the others said 'yes, and our women let them stay behind in the old homestead, we will arrive there in the evening'. And we wondered 'the men have said <leave the women behind in the old home>.' He said 'yes'. And the women wondered 'What shall we eat in the old homestead?' The man said 'I don't know'. We said 'the goats we went with we will eat'." (see Mukaakaserari Rutjindo in Bollig, 1997c).

Mukaakaserari's account gives a good idea of the hardship herders meet during periods of drought. Long treks with luggage and children on their backs in a rugged terrain demand the utmost. Thirsty and starved animals are difficult to tend. Men, women and children are near complete exhaustion. The rhythm of the day comprises daily movement, waiting and watering at overcrowded wells, slaughtering small stock and sleeping.

5.1.2.2. The Sharing of Food

There are no first hand observations on food sharing during a drought. However, observations on food sharing during normal times and interview data on sharing during a famine give a good idea about patterns of food sharing under stress. Hospitality is seen as a hallmark of Himba identity, and this implies by definition the sharing of food with guests and kin. Visitors are given milk and porridge and if a host wants to honour a visitor he will slaughter a sheep or a goat for him. The sharing of food on a neighbourhood level pertains mainly to meat. Maize and milk are not usually shared, at least not in good times. Guests will participate in eating but are rarely given unprepared food to take away with them. Maize was shared to some extent after the harvest of 1995 failed in some gardens due to heavy flooding in several river basins. I asked some women if they had acquired any maize from friends. Many of them had obtained some maize from those few women whose gardens had done well. However, in many instances the amounts of maize given away were rather symbolic: many women said that they were given a tray (*otjimbara*) of maize, which is roughly enough for one or two meals. The *otjimbara* seems to be a standard measure for distributing home-grown maize among friends. Only two women obtained a sack of maize (ca. 60kg) from friends. However, during a drought there is little chance for distributing home-grown maize as the harvest fails in most gardens and bought maize is only shared within the household.

In contrast to milk and maize, meat is distributed to a wider range of people. Whenever a ritual takes place at a homestead, e.g. marriage, initiation or commemoration, numerous guests will be present. These are recruited from the neighbourhood and from the matrilineal and patrilineal network of the person organising the celebration. Although the code of sharing meat is not as elaborate as among the Pokot there are some rudimentary rules. Elders will preferably be given meat from the breast, while the head of the homestead is given the chest of the animal. Accounts of the 1981 famine emphasise that small stock were slaughtered continuously (cattle were only rarely slaughtered) and that the meat was distributed within the group migrating together. People in one camp are usually close relatives. They belong to one extended family or have close patrilineal or matrilineal ties. Guests may be present and they will be given meat like family members. However, the typical situation is that meat is shared between close relatives.

Despite the outright positive evaluation of food sharing during periods of food stress in the Himba culture, older people clearly remembered that rules of food-sharing were discarded during serious famines. The drought of 1940/41 is remembered as the famine of *Kaṯe*, literally meaning "go away and die somewhere else". There are similar accounts for the drought of "Pounding the Leather Cloths": meat was denied to some people who were subsequently forced to eat their own leather cloths and sleeping skins. Several informants emphasised that it was only the poor who had to take to eating leather cloths, while the well-to-do households were still slaughtering livestock. In the 1962 drought, a Tjimbundu blacksmith was denied food, although he was apparently famished and near to death. Rules of food-sharing were not extended to him.

5.1.2.3. In Praise of Palm Nuts: Substituting Food

Next to slaughter and bought maize the Himba rely on a multitude of substitute foods during periods of food stress. Himba accounts on famines stress the relevance of so-called drought food which are very important to balance seasonal and annual shortfalls. While Pokot accounts on drought food stress the low yield and low density of plants harvested during a drought, the Himba emphasise the overwhelming importance of one single substitute food, the nuts of the *omirunga* palms (Hyphenae petersiana).

> "There were three things - goats, cattle and sheep - which were eaten before the maize arrived. Before these sacks arrived they ate livestock and *omarunga* nuts. For some time they ate *omarunga* nuts only and then they went to sleep (without eating anything else). That's it, only *omarunga* nuts. Only *omarunga* nuts, finish. When the people became weak in their knees (from only eating nuts), they took one goat from the goats' enclosure to slaughter it. They ate that goat until it was finished and then they returned to eating their *omarunga* nuts. When the *omarunga* nuts became too many in the stomach to climb the steep walls of a riverbed, they slaughtered a sheep, they ate it, they ate it, they ate it. And when it was finished they returned to their food of *omarunga* nuts once again. Eating *omarunga* nuts only, day after day....Sometimes they also ate palm kernels (palm starch), *omarunga* nuts, palm kernel, *omarunga* nuts, palm kernel. If they felt in their heart that it became too much, they entered the livestock enclosure and took out another sheep." (Mbatjanani Kapika in Bollig, 1997c).

Hyphenae petersiana is found in thick stands along the Kunene river between Ruacana and Epupa and its main tributaries. Especially the productivity of the stands along the Kunene are fairly stable as they rely on rainfall in humid highland Angola rather than on the rains in the semi-arid savannah of Kaokoland. With our present limited knowledge, it is hard to tell how much food is produced annually from these palm groves. First measurements show that on average 29.5 per cent (or 16.3g) of the nuts (which have an average weight of 55.4g, n = 227 nuts) are edible (i.e. edible fibre in relation to the non-edible shell and core). Some 1,000 to 2,000 nuts may grow per palm tree, this results in an average of 16.3kg to 32.6kg of food per year and per palm (rough average 24.5kg per palm tree). Palm trees yield every fourth year. Only the riverine oasis of Epupa holds an estimated 4,000 palm trees. If 50 per cent of these palms yield this would result in an annual crop of 12,250 kg of edible fibre (all figures on yields based on pers. com. A. Kuper). Table 18 indicates that the nuts of Hyphenae petersiana are a highly nutritious food source. The nuts have a high energy value and contain important vitamins, minerals and trace elements. About 30 nuts would make a good meal which contains many of the necessary important nutrients.

Oral accounts of past droughts emphasise the enormous importance of palm nuts as a substitute food. During harsh times women organised donkey caravans to the river. They stayed in the gallery forest pounding the edible fibre from the hard core. In the end the fibre was put into sacks and then transported back to the homesteads in the hinterland. While the fibre of palm nuts is usually eaten raw, there are other recipes as well: the fibre of palm nuts is occasionally mixed with soured milk or with butterfat. The hearts of young palms (*otjikora*) may be used as well during periods of scarcity. For this purpose the entire palm has to be dismantled, the palm heart is then taken out and grilled.

Table 18. Nutritious contents of Hyphenae Petersiana nuts in comparison to other food items

	Hyphenae Peters. part: fruit flesh	Maize	Wheat	Goats' Milk	RDA
Moisture, g	6.6	12.5	13.2	86.6	–
Ash, g	9.0	–	–	n.d.	–
Protein, g	4.9	9.2	11.7	3.6	56
Fat, g	0.4	3.8	2.0	4.1	
Crude fibre, g	9.6	2.1	2.0	0	n.d.
Carbohydrate, g	69.5	71.0	69.3	4.8	50-100
Energy value (kcal)	302,63	342.1	321.0	70.8	2700-3000
Ca, mg	103	15.0	43.7	123.0	800
Mg, mg	197	120.0	173	123	350
Fe, mg	2.04	2.0	3.3	0.1	10
Na, mg	–	6.0	7.8	42.0	1100-3300
K, mg	2560	330.0	502.0	177.0	1875-5625
Cu, mg	0.47	0.19	0.49	n.d.	2.0-3.0
Zn, µg	0.56	2.5	4.1	n.d.	15
Mn, µg	–	0.48	2.4	n.d.	75
Thiamin (B_1), µg	–	0.36	0,48	n.d.	1,4
P, µg	156	256.0	406.0	103.0	800
Riboflavin (B_2), mg	0,1	0.20	0,14	0.15	0.6
Nicotinic acid, mg	4,62	1.50	5.1	0.3	3 µg
Vit C	19,7	0	0	2.0	60
Carotene	0,06	0.37		n.d.	n.d.

Note: (1) The figures given usually relate to values per 100g dry matter, (2) Data on Hyphenae petersiana Klotzsch from Wehmeyer 1986, (3) Data on Maize, wheat and goat's milk from from Souci et al. 1979/82, cited in Casimir 1991:47-72, 105/107;

Himba use other bush food intensively during famines, for instance *ozongongo* nuts (Sclerocarya birrea) which are rich in fat content and the berries (*ozombe*) of Berchemia discolor (*omuve*) which are gathered in large quantities at the end of the rainy season and stored and eaten during the dry season. In the hilly areas of northern Kaokoland Adansonia digitata (baobab, *omuzu*) is frequent. Children often collect the large fruits which are a source of vitamin C, but also contain vitamin B1 and calcium in significant proportions (Malan & Owen-Smith, 1974: 165). The berries of various Grewia species are used. Bushes such as Grewia tenax and Grewia villosa are abundant in the area, as is the berry-carrying bush Salvadora persica. Cyperus fulgens (*oseu*) is recounted as a major famine food.[7] The seeds of various grasses and herbs, namely Setaria verticillata, Eragrostis porosa and Gisekia pharnaceoides are used to prepare a porridge (von Koenen & von Koenen, 1964:87) in times of scarcity. Table 19 summarises all fruits, berries, tubers and seeds which are important famine foods used by the Himba.

Despite the abundance of some drought food the Himba also had to resort to very unusual food items during past famines. During the drought of

Table 19. Plants used as substitute foods by the Himba

Fruit	Plant	Identification	Preparation
ozongongo	*omungongo, or omukongo*	Sclerocarya birrea	Fruits eaten fresh, kernels cracked and core eaten
omarunga	*omurunga*	Hyphaene petersiana	Eaten raw
omahu	*omuzu*	Adansonia digitata (baobab)	Fruits cracked open
	omboo	Commiphora africana, other Commiphoras with edible roots are referred to as *omboo* as well	Tuberous roots are chewed
ozondowa	*omutungi*	Commiphora glaucescens	Fruits eaten raw
ozongwindi	*omungwindi*	Boscia albitrunca	Fruits eaten raw, during times of food shortage roots are pounded, dried and ground and mixed with grain and made into a porridge
Ozonyandi	*omunyandi*	Diospyros mespiliformis	Berries are eaten
Ozosepa	*omusepa*	Cordia gharaf	Berries are eaten
Omakuyu	*omukuyu*	Ficus sycomorus	Figs are eaten
Ozoninga	*Omuninga*	Ximenia americana	Fruits are eaten
Ozondumise	*omundumise*	Opilia campestris	Berries are eaten
Ozombe	*omuhe*	Berchemia discolor	Fruits collected in large quantities, eaten raw or dried and stored
Ozongaru	*omukaru*	Ziziphus mucronata	Berries are eaten
Ozondjenya	*omudjenya*	Vangueria infausta	Fruits are eaten
Ozongambu	*omungambu*	Salvadora persica	Berries are eaten
Ozovapu	*omuvapu*	Grewia bicolor	Berries are eaten
Ozondjembere	*omundjembere*	Grewia flava	Fruits are eaten
Ozohore	*omuhore*	Grewia schinzii	Fruits are eaten
Ozondjendjere	*omundjendjere*	Grewia tenax	Berries are eaten
Ozohamati	*omuhamati*	Grewia villosa	Berries are eaten
Ozondape	*omundape*		

Herbs	Plant	Identification	Preparation
Omunandi	*omunandi*	Amaranthus thunbergii	Boiled, stew eaten with grain or meat, or pressed and dried as small cakes (*omavanda*) and stored
Ombowa	*ombowa*	Gynandropsis gynandra	Stem and leaves boiled to make stew, also dried in small cakes (*omavanda*)
Otjinakwi	*otjinakwi*	Ipomoea bolusiana	Small tuberous roots are roasted

Continued

Table 19. Plants used as substitute foods by the Himba—*cont'd*

Root	Plant	Identification	Preparation
	otjipia	Tylosema fassoglensis, and T. esculentum	Tuber eaten raw or roasted, Roasted parts can be stored, the bean-like seeds (*ozombanyu*) are also eaten.
	oseu	Cyperus fulgens	Small bulbs are eaten
	otjimaka	Coccinea sessilifolia	Tuber roasted and eaten
	onduvi	Lapeirousia sp.	Small tubers are eaten
	ozondungwarara		
	ozondape		
	ehwe	?	
	ehuu	?	
	ohona	Corallocarpus welwit-schii and Trochomeria macrocarpus	Tuber roasted and eaten
Seeds	**Plant**	**Idenitfication**	
	ombuma	Eragrostis biflora	Seeds ground and made into porridge
	ombuma	Eragrostis porosa	Seeds ground and made into porridge
	?	Setaria verticillata	Seeds ground and made into porridge

Note: Tab. 19 contains own data and information taken from Malan & Owen-Smith 1974.

"Pounding the Leather Cloths" starving people resorted to eating their leather cloths. In the same drought other narratives report that bones of slaughtered animals were ground and the resulting meal was mixed with cereals. Eating rock-rabbits (Heterohyrax brucei) seems to have been a normal exercise during years of famine.

Famine Relief

While substitute foods still feature importantly in Himba drought diets, famine relief has also become important over the last years. Already in the 1920s famine relief food was delivered to famished herders in Kaokoland. These had been barred from access to markets and the colonial government tried to prevent the worst by supplying food. This food was not free but had to be paid for. In 1941 during the drought of *Kaṯe* the government felt pressurised to deliver some 225 bags of maize as famine food. The 1958/62 drought sparked off major famine relief activities by the South African administration. Between 1959 and September 1961, 4,000 sacks of relief maize were brought to Kaokoland. This maize was sold to local people, to a large percentage on credit as people were short of cash: of the 900 sacks delivered in 1959 some 451 (50.1 per cent) were handed out on credit.[8] The longer the drought lasted the less capable the people were of paying for relief food: of 2,100 sacks delivered in 1960 and early 1961 some 1,782 (63.4 per cent) were sold on credit[9] and of 1,000 more sacks brought by September 1961 almost all were sold on credit. At the end of the drought a debt of 3,538.50 R had accumulated.[10] The Tribal Trust Fund was thought to act as a security. In case of debt defaulting, the administration intended to take money from the fund to pay for its costs.[11] However, income for the fund was rare and Kaokoland's Tribal Trust Fund was virtually non-existent. Finally it was decided to transfer the debts into Food for Work units, and debtors were required to work on roads and dams. In subsequent years, Food for Work became a regular institution in some parts of Kaokoland. During the 1981 disaster, the Himba relied mainly on famine relief food which was supplied by the army and the Red Cross. Through most of the 1990s Himba received drought relief handouts and sometimes even subsidies for livestock in the form of fodder (Talavera et al., 2000:28).

5.1.3. A Comparison of Pokot and Himba Foodways during Famines

While data on famine foodways for the Pokot rely on interviews and observations during a crisis, data on coping strategies amongst the Himba rely on interviews dealing with past famines and on extrapolations from a detailed survey of slaughter and sales in a wider neighbourhood.

Increased Slaughter: In both societies the increased slaughter of small stock plays a crucial role in coping with food scarcity and the frequency of slaughtering increases pointedly during a drought. The argument that pastoralists store resources "on the hoof" to buffer shortfalls is born out by both case studies. However, the potential for increased slaughter is much bigger in Himba herds than in Pokot herds. While poor Pokot herders owned 30 to 60 head of small stock, poor Himba herders owned at least 100 to 150 animals. Several Pokot households surveyed in 1992 had only a very few male goats or sheep for slaughter. This may have been different in the past when Pokot livestock herds were bigger. In contrast to this most Himba households could easily step up meat availability. While the small herds of the Pokot only allowed for an average of some 91kcal/day/person derived from meat during the drought year 1991/92, Himba on average obtained some 325kcal/day/person already during the good years of 1994/96!

Sharing of Food: There are some similarities when it comes to the sharing of food. Rules of hospitality guarantee guests in both societies an ample supply of food during periods of food stress. While meat is a major item of food sharing during periods of stress, maize and milk are not shared extensively with non-household members. There are differences as well: Himba slaughter small stock in large numbers during a drought and whoever is present at such occasions will receive meat to consume on the spot or to take home. Rules of meat sharing are rather rudimentary amongst the Himba - they mainly apply to the fact that some specific pieces of meat must not be eaten by non-kinsmen. While the frequency of slaughter during a famine is lower amongst the Pokot they have a more elaborate system which ensures the intense sharing of meat. Specific pieces of meat must go to elders, others to specified relatives, to in-laws or to mother's brothers. A widespread practice amongst the Pokot is to present needy households with goats and sheep for slaughter. This substantial support is given to related and non-related households that have close ties through the prior exchange of livestock loans and presents. In contrast, such exchanges are more restricted among the Himba. Animals for slaughter are only given away to close relatives. The intensity of meat sharing and the complexity of rules connected to it are clearly higher among the Pokot than among the Himba. These findings correspond to the hypotheses forwarded by Hames (1990) on food sharing: (1) the scarcer food items are the more intensively they are shared, (2) the scarcer the food items the less important a kin-bias is and (3) the scarcer a food item the wider the spatial scope of food sharing.

Is sharing of food unlimited and holds even during periods of severe starvation? The Himba recount at least two droughts during this century when food was no longer shared. The Pokot reported no historical evidence for food hidden away during periods of intense stress. According to several older informants meat was shared even in periods of the worst hunger. Meat sharing is much more a part of Pokot ideology and identity than it is among the Himba.

Substitute Foods: Historically both pastoral communities relied heavily on non-domesticated food collected from the bush during periods of drought. Differences in substituting food during a drought clearly stand out. The Pokot relied heavily on a variety of substitute foods (tubers, berries) during famines in the past. Nowadays, substitute foods are only used marginally. Store-bought maize and food aid - usually in the form of maize too - has replaced food from the bush. In contrast to this substitute foods are of much more importance among the Himba. Palm trees which stand in dense groves all along the Kunene and its tributaries produce nutritious nuts in large quantities. This substitute food is very reliable as palms also yield in drought years. While Pokot historically relied on a set of fairly different substitute foods widely spread over the area, the Himba rely on one localised food source which can be harvested reliably and without much labour. Parts of numerous other bushes, trees and tubers are used by the Himba as substitute foods as well. Generally, the findings on substitute foods conform to the dietary breadth model (see Winterhalder, 1981): the more intense the food stress the more frequently less preferred food is exploited. Informants in both societies could relate stories on how bones were ground to produce porridge, how skins were boiled, baboons hunted and donkeys slaughtered.

Famine-relief: Famine-relief food has been of growing importance in the drought diet of both populations over the last decades. At least since the 1940s colonial governments have been supplying relief food. Since the 1980s famine relief has become a constant in the Pokot diet. At least during droughts the population cannot be maintained by products from livestock production alone and external help is needed to prevent increased mortality. Himba had to rely heavily on drought relief in 1981. Since independence, in 1990, relief food has been more frequently donated, sometimes disregarding factual necessity.

5.2. INCREASED SALES OF LIVESTOCK

The increased sale of assets is a strategy many rural populations employ to cope with disaster. While farmers have to sell or pledge land and thereby loose access to their means of production (cf. Watts, 1983), herders sell more livestock than in normal years in order to buy more food on the market. During major droughts the exchange of livestock for grain becomes a serious problem. Frequently the value of starved stock decreases rapidly after the onset of a drought, forcing herders to sell even more animals in order to obtain an appreciable amount of maize.

5.2.1. Taking from Meagre Accounts: Pokot Livestock Sales during a Drought

The system of livestock marketing in the Baringo plains has changed tremendously over the last 100 years: from a situation without any established forms of long-distance livestock trade, to barter trade with itinerant long distance traders and lately to a full-blown market system. While during the first half of the century the livestock trader was, at the same time, a trader in grains, these two spheres became disarticulated from the 1950s onwards. Nowadays livestock traders are Tugen who just come to Nginyang for the purpose of livestock trade. The owners of retail shops selling maize and other food items are Somali and since the late 1990s more and more often Pokot. Since the late 1980s weekly livestock markets have been taking place at Nginyang, Tangulbei and Yattya. There is some degree of competition between traders, despite the fact that the Pokot frequently allege that traders have made prior arrangements to keep producer prices low. Despite interethnic conflicts and occasional attacks against single traders this market has remained operative and even extended its operations over the last ten years.

During normal years the Pokot ideally try to sell an ox at the on-set of the dry season. Oxen are in a prime state then and obtain maximum prices. The money is stored away and given to the women of the homestead on a weekly basis in order to buy maize, sugar and tea on the weekly market day. When the household runs short of money towards the end of the dry season, goats and sheep are sold to bridge the time until the next rainy season. An average household will sell one ox and three to four heads of small stock per year.

At the height of the 1990/92 drought in March and April 1992, I interviewed household heads on their selling strategies. All households surveyed during the 1992 drought were forced to sell numerous livestock - substantially more livestock than in a normal dry season. Conclusive data was obtained for a twelve month period (April 91 to March 92) for 18 households (see Table 20). One household head was able to spend his salary on maize and other goods; for the time in question he did not have to sell a single animal. He emphasised that he had taken the job as a night watchman to prevent his small herd suffering from excessive sales. All the other 18 households sold on average 5.9 goats (range 0-23, Std Dev. 5.3) and 3.4 cattle (range 1-8, Std Dev. 2.0), indicating a significant rise from livestock sales in a normal year.

Several households were forced to sell animals which had not yet reached maturity. The year 1990/91 had already been bad and many households had sold animals, which could have fetched better prices in the market at the onset of harsher conditions towards the end of 1991. While maximum prices obtained for a 70 to 80kg male goat reached about 700 KSh, some goats were sold at prices as low as 200 KSh at the height of the drought. The average amount obtained for a goat was 300 KSh (or about 50kgs of maize, 1 kilo of maize cost 6 KSh in 1992). This fairly low average price resulted from the fact that quite a number of herders sold underweight and starved animals. This price is about 100 KSh lower than the average price obtained for goats and sheep in early 1991, indicating a loss of 16kg of maize per sale of one goat. The survey (see Table 20) showed that, for example, household 15 did not have large castrated male small stock left and was forced to sell off young stock, achieving an average price of only 250 KSh per goat. The same held true for households No. 5, No. 7 and No. 19 who attained average prices of 203 KSh, 275 KSh and 260 KSh respectively. All four households belonged to the group of very poor herders.[12] Those herders who still owned at least a few big castrated males fared better. The two rich households No. 6 and No. 11 (see Table 20 for orientation) obtained average prices of 400 KSh per goat. This meant that household 6 obtained more money (and as a consequence more maize) from 15 small stock sales, than did household 14 from 23 sales. 7 out of 16 households sold goats under-priced. Out of these 7 households 5 belonged to the category of poor herders.

Poor households are forced to sell sub-optimally. They have to sell more animals than others to obtain the same amount of money. This inhibits their potential for recovery after a drought.[13] If the money obtained is transferred into maize equivalents, the differences between households become even more evident. While household 14 only obtained some 42kg of maize on average per goat and household 5 only 34 kg, households 6 and 11 obtained 67kg of maize per goat, i.e. these two households obtained almost double the amount of maize per goat than the poor household No. 5.

The average price obtained for cattle was 1,964 KSh per animal with a range of 3,000 to 1,130. Transferred into maize equivalents this meant that while household No. 4 obtained 500 kilos of maize per animal, household 16 obtained only 188 kilos and household 7 received 200 kilos, which is less than half of what household No. 4 obtained. Out of the five households selling underweight immature animals, four belong to the category of poor herders. Again household No. 1 is the exception; although being fairly rich he still had to sell livestock which did not attain high prices. Eight households sold livestock near the average price of 1,964 KSh (+/– 300 KSh) and five obtained high average prices (of these, two belonged to the group of rich herders, two were average herders and one was a poor herder). However, the relation between wealth and increased sales is not systematic. Correlation between wealth and TLU sold (Pearson's Correlation Coefficient: - 0.21) and wealth and maize available per person per day (Pearson's Correlation Coefficient: 0.04) are not conclusive. Wealth differences among the Pokot are apparently too small and too unstable to have a profound effect on selling strategies.

To figure out the amount of maize obtained per household, I assumed that about 70 per cent of all monetary income went into the purchase of maize during the drought period. This is a low estimate: figures may definitely go up to 90 per cent. The remaining money went into school fees, contributions to public funds, the purchase of tea and sugar and consumer goods such as cloth, beads and medicine. Some families had secondary incomes from trade and ten-cent jobs such as selling home brewed beer or simple cooked meals of maize and beans. However, the income from these activities was unstable and for all households covered in the sample, livestock sales were by far the major income. Maize is usually bought at a shop in Nginyang. Women go there and buy two to five kilos in one go, transport it back home, produce several meals from it and walk back again to fetch more. Only rarely are donkeys employed as beasts of burden to transport maize from the centre to the homestead. Due to the lack of transport it is rare that full sacks of maize are bought. On average the 18 households had some 0.37 kilograms available per person and per day during the drought. This translates into 1323 kcal per person per day showing that the major part of their nutritional demands was covered by maize during this drought period.[14]

If the average kcals obtained from slaughter are added to the energy obtained from maize, the picture does not change significantly. Households that have obtained only a little maize do not necessarily slaughter more stock to better their situation. The same households which were below average, looking at store-bought maize only, were still below average after adding kcals from slaughter. Hence, the situation remains that some households have considerably more food available than others.

In the previous chapter, the ways of meat sharing among the Pokot were delineated. The data of Table 20 however indicates that the sharing of maize would probably be more central to achieving adequate food supplies for all households. However, while the Pokot demonstrate symbolically through the intense sharing of meat that households share the brunt of a crisis, factually some households seem to run into severe nutritional problems. Some poor households have to sell livestock excessively and become even more impoverished during a period of drought. Household No. 7, for example, did not have the stock to sell to procure more food. At the height of the drought they settled at Lake Baringo and dropped out of mobile livestock husbandry. In 1993 household No. 7 lived from a combination of fishing and small stock herding. Households No. 10, No. 19 and No. 25, however, show that this fate is not inevitable. All three households had to cope with reduced food availability during the drought. All three households are young, small in size and very mobile. In 1993 they were still within the pastoral niche and other herders judged that No. 10 and No. 19 had done very well during the drought: they had abstained from excessive sales and had endured hunger and by doing so had protected their herd.

From the aggregated figures on energy derived from maize first conclusions on the adequacy of Pokot drought diets are warranted. While FAO recommendations for energy intake for the average adult male give a value of 2,530 (kcal) (Herren 1991:287), the measured data of energy intake among East African pastoralists indicate much lower figures: average energy intake among the Turkana was put at 1,390 kcal and among the Maasai at only 1080 kcal (Galvin, Coppock & Leslie, 1994:123). The average figures for Pokot households indicate that a simple maize-based diet furnishes most households with a sufficient amount of energy to make it through the drought. 13 households out of 18 reached a kcal level of above 1,300 kcal per person per day. Only three households attained figures significantly lower. One of these, as pointed out above, dropped out of pastoralism during that period. The two others starved themselves through the drought.

Not only does the reliance on the market rise during a drought, the very same market turns against herders during a crisis: prices for livestock decrease, and traders become more reluctant to engage in trading activities. During the 1990/92 drought, prices for livestock fell rapidly making the situation for herders, especially for poor herders, ever more precarious. In spring 1991, at the height of the drought, it became difficult to sell livestock. On the one hand there was too much livestock being offered on the market, and on the other hand, the quality of livestock was too low to fetch any appreciable prices. Furthermore the number of dead animals had increased the availability of hides and skins, lowering the prices for these products, too. While 2,000 to 3,000 KSh was the price for an ox in normal times now only 600 to 1,000 KSh was paid for a head of cattle. The price of goats plummeted from the normal 300 to 500 KSh to about 150 to 250 KSh and occasionally even less than 100 KSh (see also Saltlick 1991). This indicates that prices for cattle fell from 75 per cent to 60 per cent and for small stock from 65 per cent to 35 per cent. The prices for goats' skins dropped due to an oversupply on the market from 15 KSh to 5 to 8 KSh. Herders are forced to cope with these sudden losses of exchange entitlement with more sales and a reduction of daily food consumption. At the height of the drought in early 1992 some households could not afford to buy sugar and tea. For some weeks several households were living entirely on maize and water.

Table 20. Sales of livestock during a period of drought

Household	Rank	g/s	KSh	Av.Pr.	c/ca	KSh	Av.Pr.	do	KSh	Tot/ TLU	Tot KSh	Maize	no a/ch	Maize/ person (in g)	Kcal / per pers	+ kcal meat
6Ngorag	1	15	6000	400	2	4800	2400	1	700	4.4	11500	1341.7	7/10	0.31	1113	1135
9Limpo	1	n.d		n.d.		n.d.		n.d.		n.d.	n.d.		9/9			
12Adom	1	n.d			2	3000	n.d.			n.d.	n.d.		8/4			
21Lokirp	1	n.d			n.d.			n.d.		n.d.	n.d.		10/8			
1Lotepa	2	7	1930	276	4	5600	1400	0		4.9	7530	878.5	5/8	0.27	969	1041
11Lomir	2	9	3600	400	4	11900	2960	0		5.1	15500	1808.3	5/15	0.50	1795	1954
13Lopoi	2	n.d			4	9550	2390			n.d.	n.d.		9/8			
2Kameya	3	8	3200	400	4	8800	2200	0		5	12000	1400.0	7/12	0.30	1077	1086
3Lomurö	3	5	1410	282	3	7300	2430	1	800	4.1	9510	1109.5	4/3	0.55	1975	2361
10Domot	3	3	1050	350	1	2300	2300	0		1.4	3350	390.8	4/3	0.19	682	718
4Lokom	3	4	1050	350	1	3000	3000	0		1.5	4050	472.5	5/4	0.18	646	684
20Tepak	3	5	1500	300	5	11000	2200	0		5.6	12500	1458.3	5/8	0.44	1580	1635
18Kanya	3	n.d			n.d.			n.d.		n.d.	n.d.		10/4			
22Angur	3	0			8	15800	1980			8	15800	1843.0	7/14	0.36	1292	1472
5Awiala	3	2	610	203	6	13900	2320	1	460	6.8	14970	1746.5	5/4	0.68	2441	2502
8Limany	4	0		0		0				0	0 (inc.)		2/7			
14Teta	4	23	5760	250	4	7800	1950	0		6.9	13560	1582.0	3/7	0.67	2405	2452
15Silet	4	0			4	7200	1800	0		4	7200	840.0	2/4	0.58	2082	2279
16Kange	4	3	1000	330	6	6800	1130	0		6.4	7800	910.0	4/5	0.38	1364	1527
17Nyang	4	7	1670	240	6	9300	1550	0		6.9	10970	1279.8	3/7	0.54	1939	2052
25Lopetö	4	6	2900	480	1	2000	2000	0		1.8	4900	571.7	5/6	0.20	718	835
19Renge	4	6	1540	260	3	4460	1490	0		3.8	6000	700,0	5/7	0.23	826	865
7Tonya	5	4	1100	275	1	1200	1200	0		1.5	2300	268.3	4/8	0.09	323	372
23Chada	5	n.d			n.d.			n.d.		n.d.	n.d.		4/4			
24Tupok	5	5	1500	300	2	5150	2580	0		2.6	6650	775.8	2/2	0.53	1903	1940

Continued

Table 20. Sales of livestock during a period of drought—*cont'd*

Household	Rank	g/s	KSh	Av.Pr.	c/ca	KSh	Av.Pr.	do	KSh	Tot/ TLU	Tot KSh	Maize	no a/ch	Maize/ person (in g)	Kcal / per pers	+ kcal meat
Average	–	5.9	2,239	300	3.4	6708	1,964	0.2		4.2	8,742	1,077	–	0.37	1323	1416
Range*																
Std.Dev.	–	5.3	1596	102	2.0	4126	681	0.4		2.2	4564	533	–	0.19	684	726

Notes: (1) Row 1: Household, (2) Row 2: Wealth Rank: 1 very rich to 5 very poor, (3) Row 3: goats and sheep sold, (4) Row 4: KSh received for small stock sales (5) Row 5: average price received for small stock sales, (6) Row 6: cattle and camels sold, (7) Row 7: KSh received for large stock sales, (8) Row 8: average price received for large stock sales, (9) Row 9: donkeys sold, (10) Row 10: average price received for donkeys, (11) Row 11: total number of tropical livestock units sold, (12) Row 12: total number of KSh received, (13) Row 13: cattle and camels sold, (14) Row 14: total in maize equivalent, (15) Row 15: number of adults and children per household, (15) Row 15: Maize per person per day, (16) Row 16: kcal per day available from maize (17) Row 17: kcal per day if kcal from slaughter (see Tab. 17) is added.

5.2.2. Taking from Full Accounts: Himba Sales Strategies

While the Himba were actively trading livestock for grain and commodities at the turn of the century, this exchange was abolished by the South African administration after 1917. Border controls were strict and trade became extremely complicated. In major droughts as in 1928, 1932 and 1941, local leaders applied to the administration to be formally allowed to trade with Portuguese traders on the boundary. During the 1958 drought, Kaokoland's herders were again allowed to trade with Portuguese traders at Chitado for the first time since the early 1920s. During the drought an estimated 10,000 animals (of a total population of about 120,000) were sold there. Although trade across the border remained illegal, it was no longer prosecuted. The situation changed again only some 15 years later when the Angolan civil war and two South African invasions into southern Angola destroyed trade networks in south-western Angola. By 1976 livestock trade in southern Angola had basically come to a standstill. Ongoing civil war and a degrading infrastructure have kept livestock trade in its debilitated form in southern Angola since then. In northern Namibia trade was on the one hand hampered by civil war too, on the other hand however, the establishment of army barracks in the area encouraged traders to search for livestock for local slaughter.

Since the early 1980s itinerant traders regularly barter livestock for maize in Kaokoland. Cheap liquor is the second most important trade good after maize. Other trade goods of importance are blankets, cloth and various forms of human and veterinary medicine. Table 21 shows that all the Himba households surveyed for 22 months between April 1994 and January 1996 bought significant amounts of maize. The average was 26.2 sacks of maize (range 7 to 79, Std Dev.16.6) bought over a 22 month period (i.e. 1.2 sacks of maize per household on average per month). The average amount of maize per person per day available was ca 962 kcal. These figures do not include the substantial amounts of maize harvested in their own fields.

There are major seasonal differences in livestock sales as shown in Figure 14. The rainy season of 93/94 had not been good, and additionally many gardens in the Omuhonga basin, one of the major gardening areas, had been washed away in one of the rare floods of that year. Households already had to buy significant amounts of maize from June onwards. The number of sacks bought from itinerant traders peaked in November 1994 reaching 98 sacks (24 households, 4.1 sacks per household)[15]. A lot of maize was stored and not all 98 sacks bought in November 1994 were consumed that month. Early rains in November resulted in the sprouting of bushes and trees and an increase in lactation, so that already in December less maize was acquired. However, significant amounts of maize had to be bought well into the rainy season. From March to May the milk yields peaked so that nutrition became once again milk-based for some months. Only very little maize was bought during this period: in March, April and May only a few sacks were acquired by the 24 households (average for these months was 0.15 sack per household).

From late May onwards the maize in the fields began to be harvested, reducing the need to buy additional maize. The 1994/95 rains were good with above average precipitation and a good maize harvest was reaped in most gardens. Consequently maize purchases only picked up again in November and did not reach a peak as in the same months of 1994, as some households still had their own maize to rely on until January 1996.

Figure 14 gives us some idea about what happens during a drought year. When rains are not sufficient, milk yields drop and the maize harvest tends towards zero and therefore the purchase of maize remains at a level comparable to November 1994. In order to simulate the maize needs of the sample households during a drought, I assumed that a failure of rains, massively reduced milk yields and a complete failure of the maize harvest led to a situation in which households had to acquire almost the same amount of maize as in November 1994 (c. 90 sacks) every further month until the first rains of the subsequent year. The 24 households would have had to buy some 1,216 sacks of maize (instead of the 660 sacks they actually bought), indicating that the need for maize purchases during a drought may be almost double the amount of a normal year. This poses no problems to rich and medium households, for poor households, however, it may be more difficult to acquire the necessary amount of maize.

Of importance for the context discussed here is that of all the stock bartered during the good years of 1994-96, less than half (44 per cent of all goats and 58 per cent of all cattle) was spent on food. Most Himba households could easily invest more livestock into buying food. Livestock is rarely sold undervalued. Most households had goats or sheep to offer which were accepted by traders on a one to one basis for 60 kg sacks of maize, i.e. 1 goat fetching about 60 kilos of maize. Only one poor household had to repeatedly barter two small goats for one sack. The same holds true for cattle: while oxen with a weight of 600 kg and above can fetch up to 12 sacks, the average tollie was bartered for 5.1 sacks of maize. The 24 households bartered some 21.4 goats on average (range 6-59, Std. Dev 13.2), 6.8 sheep (range 0-45, Std.

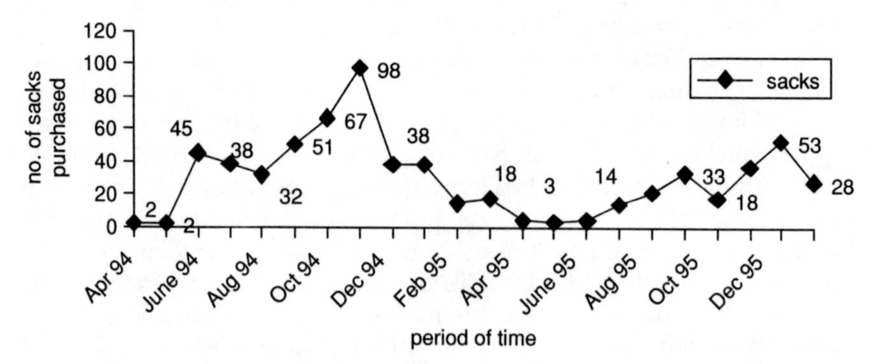

Figure 14. Purchase of maize (in sacks) in 24 Himba Households, 4/94 -1/96

Dev. 10.1) and 5.9 cattle (range 0-19, Std. Dev 5.2). There are very clear differences in marketing strategies based on wealth difference. Several households (7 households, 29 per cent) did not barter any cattle at all or only very few cattle (0-2 heads), while 5 households sold 10 and more heads of cattle during the same period. Some rich households retain good contacts with white traders who occasionally come along to buy sheep in bulk numbers and offer good cash prices.

Table 21 summarises the finds of the monthly survey on livestock sales and acquisition of food and consumer goods. The number of goats, sheep, and cattle sold over the period of 22 months is listed separately. Then the total number of sacks acquired is given. The next two rows relate to the number and percentage of small stock and cattle which was devoted to buying maize. The following two rows show the average amount of sacks obtained per head of small stock and per head of cattle for the households. Further rows list commodity goods which were bought as well. The three final rows work out the amount of maize available per day per person, transfer it into kcal available per person per day and then add up these with the average kcal obtained from slaughter. The correlation between wealth and the amount of maize bought is fairly high (Pearson's Correlation Coefficient 0.471)[16]. Rich households have more animals available for sale and they sell animals in their prime state.

The data discussed above makes it clear that Himba problems with barter trade relate to the history of trade in the region. Traders have to bear the costs of transporting animals from Kaokoland to the abattoir in Oshakati, where most of the animals finally end. Herders have to pay for these additional costs. Only the traders had any adequate information about the current pricing of livestock. Per goat they obtain just one sack of maize, i.e. a goat worth 200 N$ at the abattoir, is sold for goods worth only 60 to 70 N$. As Cashdan (1990c) rightly points out, customary prices establish security in a situation where obtaining information on actual prices is difficult and the density of customers is low. However, the Himba example shows that customary prices can also be used by traders in order to keep producer prices low over a protracted period of time.

5.2.3. A Comparison between Pokot and Himba Selling Strategies

In both societies the increased sale of livestock is a major strategy for coping with shortfalls in milk production. The rates of exchange between livestock and maize are very similar: while the average goat procured about 66 kg of maize with the Pokot (in 1988), a goat was bartered for 60 kgs of maize amongst the Himba. While the Pokot had to bring their goats to the livestock market and had to subsequently fetch the maize from a shop, Himba could immediately barter their goats for maize at the homestead with itinerant traders. Customary exchange rates apparently offer some security for producers, who may try to sell underweight animals. The exchange rates as such are bad and detrimental to the livestock producers. However, this is where the

Table 21. Himba livestock sales April 94 to January 96

Name	We	Go	She	Cat	Sac	g/c for maize	per cent for maize	Sacks/ goat	Sack/ cattle	Box	Bla	Clot	Perso	kg/per/ day	kcal	kcal + meat
1Munge	3	15	13	4	26	12/4	43/100	1,00	6.0	–	5	–	3/4	0,46	1651	2114
2Muhuw	5	27	1	2	13	16/0.3	57/15	0,59	–	8	–	–	3/1	0,33	1185	1658
3Mbatja	2	9	2	4	22	6/3	55/75	1,00	5.3	1	1	1	7/6	0,20	718	888
4Vahiku	3	17	0	2	14	9/2	53/100	0,8	3.5	–	1	1	3/1	0,36	1292	1653
5Katjira	1	20	29	18	22	13/1.2	27/6	1	–	20	3		7/1	0,26	933	1698
6Kandju	1	9	8	10	42	7/3.5	41/35	1	9	16	3	1	9/2	0,38	1364	1753
7Hikumi	1	13	1	4	28	7/1.8	50/45	1	5	10			6/3	0,33	1185	1417
8Twarek	2	23	1	6	32	7/4.5	29/75	1	5.3	2	2	6	7/3	0,34	1221	1553
9Mirirek	1	7	0	1	4	4/0	–	1	–	0	1	0	7/5	–	–	–
10Tjand	5	11	8	0	7	7	41/–	0,9	–	0	1	3	3/1	0,18	646	840
11Kama	3	25	3	6	25	5/4	18/67	0.9	5.5	15	2	0	4/4	0,37	1328	1823
12Katue	2	40	2	4	40	27/3	64/75	0,9	5.3	3	0	2	9/5	0,31	1113	1358
13Woku	4	47	17	2	23	13/2	20/100	1	3.5	18	1	2	5/6	0,26	933	1301
14Waka	4	27	2	4	17	12/2	41/50	0,9	3.5	8	3	1	11/5	0,11	395	569
15Huku	4	7	9	0	14	14/0	87/–	1	2	0	0	0	10/2	0,11	395	517

16Muha	4	13	5	3	19	7/2	39/67	1	4.5	4	0	0	6/5	0,20	718	909
17Kamu	5	5	3	2	8	4/1	50/50	1	4	0	2	0	4/4	0,12	431	588
18Kosev	3	24	3	4	21	11/1	41/25	0,9	5	1	2	2	11/3	0,15	539	721
19Taku	2	20	7	4	13	8/1	30/25	0,9	6	3	0	4	5/1	0,21	754	1168
20Mbase	1	21	2	9	46	18/6	78/67	1	4.8	10	2	1	12/6	0,27	969	1221
21Koria	3	35	4	2	29	25/1	64/50	1	4	6	3	1	8/3	0,27	969	1185
22Mban	1	59	45	19	49	26/3	25/16	1	5.3	13	3	11	8/7	0,38	1364	2418
23Kaere	4	29	0	3	15	11/1.3	38/43	0,7	3.9	6	0	0	5/5	0,18	646	915
24Non	2	12	0	15	49	6/8	50/53	1	4.6	1	0	1	14/12	0,22	790	1057
25Vahe	2	6	2	13	79	2/13	15/100	1	5.9	12	3	0	13/7	0,43	1544	1820
Totals	–	21.4	6.8	5.9	27		44/58						7/4	0,28	1005	1298
Std. Dev.	–	13.2	10.1	5.2	16.6	–	–	0.1	1,4	6.2	1.3	2.5	–	0.1	357	497

Notes: (1) Row 1: Households, (2) Row 2 : Wealth Categories, (3) Row 3 to 5: goats, sheep, cattle sold, (4) Row 6: total amount of sacks obtained, (5) Row 7 and 8: number and percentage of livestock devoted to maize purchases, read "12/4": 12 goats and 4 head of cattle sold, "43/100": 43% of the goats and 100% of cattle sold were devoted to maize purchases (6) Rows 9 and 10: average number of sacks obtained per goat and per head of cattle, (7) Rows 11 to 13: boxes of alcohol, blankets and cloth bought, (8) Row 14: Persons in household, (8) Row 15 and 16: kg and kcal available per day from maize, (9) Row 17: kcal derived from maize plus kcal derived from meat.

advantages of Himba bartering livestock for maize ends. While increased livestock sales were always an option for the Pokot throughout the 20th century, the Himba were barred from selling and bartering their livestock for grain by the South African colonial administration for several decades. The Himba had to trade illegally and to accept disadvantageous rates of exchange. However, since peace has prevailed, the barter system has developed rapidly and many itinerant traders roam the area exchanging maize and alcohol for livestock.

Potential for Increased Sales: Himba herds, being much larger in size than Pokot herds, have a larger potential for off-sales during periods of stress. While several Pokot households had to sell undersized small stock for grain, only very few Himba had to do so. The figures leave little doubt that most households could easily increase the number of small stock sold per year. In a normal year, on average, only 44 per cent of all small stock sales and 58 per cent of all cattle sales are invested into grain.

Wealth and Selling Strategies: The correlation between wealth and maize availability in both societies differs profoundly. The data suggests that there is no systematic relationship between wealth and maize availability among the Pokot. Only very poor households are really disadvantaged. In contrast, the correlation between wealth and maize availability among the Himba is clearly perceivable. Rich households obtain more food than poorer households. While among the Pokot the nutritional standard of poor and rich families does not differ much, there are clear differences among Himba households.

Among the Pokot, poor households have two options: either they sell considerable numbers of livestock and risk falling below the critical threshold of a minimal herd unit or they try to bear the hunger. There is some evidence from the households surveyed that poor Pokot households are endangered of dropping out of pastoralism altogether. Among the Himba it is regarded as normal that rich households are joined by poor relatives who then stay in the pastoral sector as dependants of larger households.

5.3. INCREASED MOBILITY

Spatial mobility is a major strategy of herders to harmonise the needs of herds with fluctuations in biomass production, to evade violent conflict and to escape from livestock epidemics. A set of factors constrains mobility: a herder may want to migrate but lacks the shepherds to do so, feels constrained by a great number of goats' kids or considers that animals in milk should stay near the people who need food. As an answer to variable options and constraints, herds are split and moved in separate units. During droughts such patterns change. The shift from rainy season to dry season settlements takes place earlier than in normal years. Herds are split up at an earlier stage and, if available labour allows, the herd is split into smaller units to increase the efficiency of herding. Remaining pastures tend to be congested and watering places are crowded. When regular dry season grazing is finished, herders have to look for as yet unused pastures beyond the normal confines of their territory. They have to migrate into areas which are only vaguely known to them, to pastures which are used by other herders and to regions which hold new problems such as new diseases and another infrastructure for obtaining food. The following paragraphs portray Pokot and Himba spatial mobility during periods of stress and contextualise them historically.

5.3.1. Erratic Moves: Pokot Mobility Patterns during a Drought

Spatial mobility is a strategy the Pokot employ to minimise various risks: starvation of livestock, epidemics, depredation by raiders and excessive control and taxation by authorities (see Map 7). The character of movements differs according to the specific danger it responds to. While moves to flee from raiders are one-time moves over considerable distances to a presumably safer place, moves responding to drought conditions are usually concatenations of many minor shifts. During a drought moves are frequent and opportunistic, i.e. each opportunity to utilise a better place for one's livestock is immediately used (before somebody else does so). Moves away from established grazing orbits towards the Leroghi plateau, Lake Baringo and towards the southern reaches of the Turkana plains between Lomelo and Kapedo follow.

Historical Development of Spatial Strategies during Drought

When eliciting the historical mobility patterns of the Pokot, two facts have to be considered. On the one hand, the Baringo plains and hills looked very different only 50 years ago and on the other hand the number of households (i.e. of separate herds units) has risen significantly while the total number of livestock has declined. In the 1920s the Pokot numbered about 6,000 people who herded roughly 90,000 cattle. This means that there were perhaps about 400 households (assuming an average of 15 people per household) managing average cattle herds of about 200 - 300 animals. Apparently there were wealthy herders who owned considerably more and poor herders who owned less, but the general picture is that households owned larger herds in the past than they do today. At the end of the 1990s there were perhaps about 4,000 households (taking an average of 15 people per household) with an average of less than 20 animals per cattle herd. While the number of herding units increased nearly sevenfold, the size of herds decreased almost tenfold. Many more separate but considerably smaller herding units have to look for sufficient fodder nowadays bringing about increasing co-ordination problems. Generally the area is much more densely settled than in the past and most dry season grazing areas are used almost all year round.

Several elders provided information on mobility patterns during the droughts in their youth in the 1920s and 1930s (see Map 8). While each homestead had its home base where goats and camels stayed for most of the

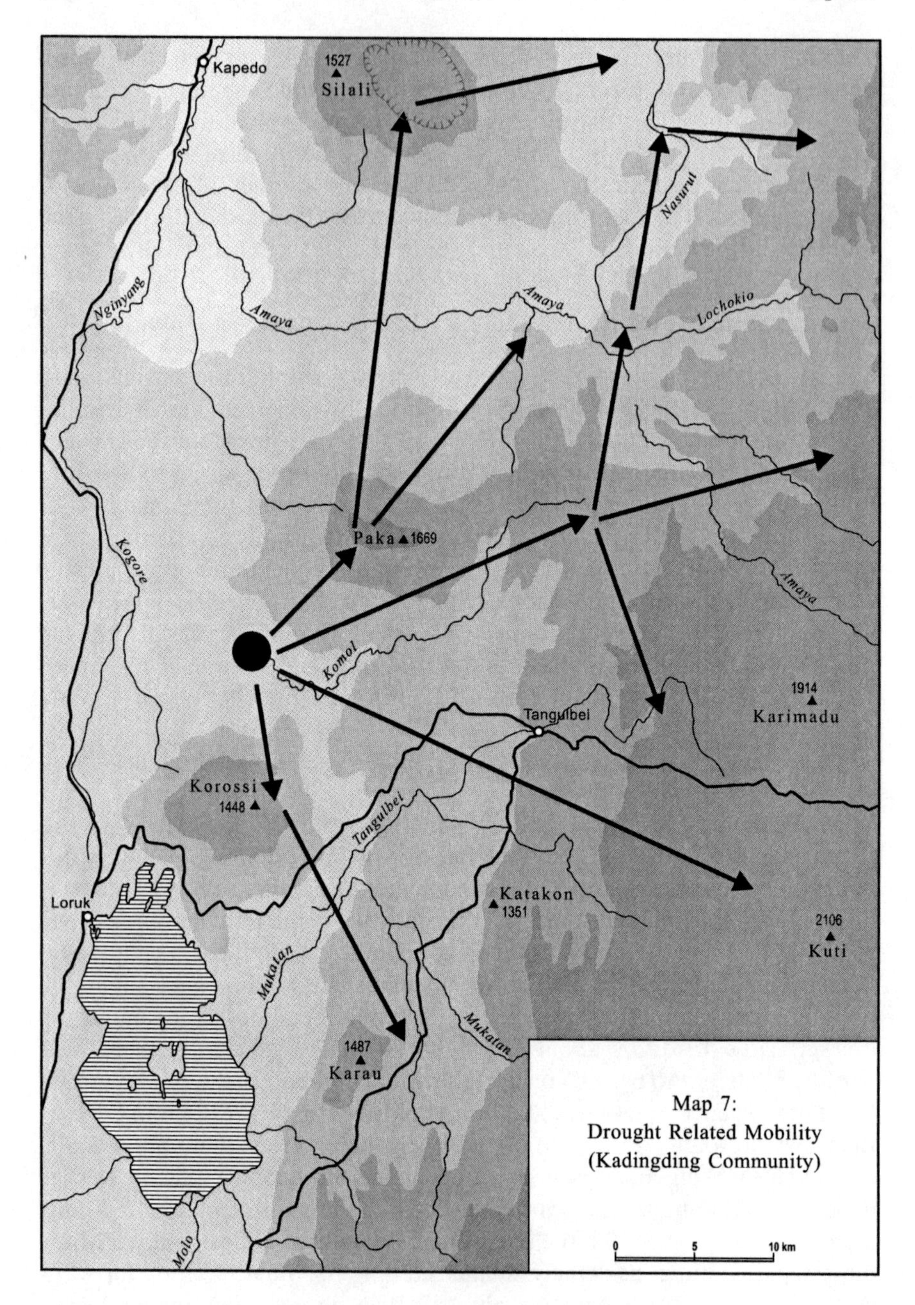

Map 7:
Drought Related Mobility
(Kadingding Community)

year, each household also had a regular dry season grazing area. During droughts cattle camps made use of the entire area that the pastoral East-Pokot called their own. During the drought of *Kiplelkowa* (1925/27) an elder remembered that he migrated with the cattle herd of his father from Parpelo to Paka, the only place where grazing was still found, covering a distance of about 50 kilometres. At Paka he met other herders coming from Parpelo and even from as far away as Luwyat in the Kerio valley. During the drought of *Koringring* in the early 1930s another informant remembered that cattle herds moved from Loruk and Nginyang to the Kerio valley. Others moved from Silali down to the swamps of Lake Baringo around Kampi ya Samaki. These pastures were highly valued as they were reliable and of a high quality. The British-instituted border committees took away most of the lakeside pastures from the Pokot in the 1930s. Occasionally Pokot herds migrated beyond their tribal boundaries. They moved into the plains between Kapedo and Lomelo (occupied by the Turkana), onto the Leroghi plateau (occupied by the Samburu and white farmers) and into the Tugen hills.

The colonial government occasionally assisted elders by making temporary arrangements for dry season grazing beyond the tribal boundaries.[17] Only when interethnic conflicts were feared, were boundaries with the Turkana and Samburu enforced by the colonial administration. In contrast to this, movements towards the commercial ranching zone were always checked energetically. On several occasions herds were confiscated or culled.[18] Some older herders reported that they occasionally herded their cattle during the night onto the pastures of private ranches and then, early in the morning, drove the herd back to communal grounds. In many other instances, Pokot herds moved towards the north-east and climbed the escarpment north of Amaya encroaching onto the tribal grazing lands of other pastoral groups.

Spatial Strategies during the 1970s and 1980s

Since the middle of the 1970s, violent conflicts have determined spatial mobility as much as the lack of rainfall. Between 1975 and 1980 most households from the Paka area moved towards Chepkalacha, Kechi and Loruk. Dry season grazing in Naudo and Silali was no longer used. Settlement areas such as Nakoko, Natan, Paraloritele and Orus were more or less given up. Households shifted southwards to get away from the threat of being raided.[19] These moves led to a complete dislocation of migration cycles. Herders who had migrated regularly from Nakoko to Silali in the dry season now moved between Loruk and Mt. Korossi. These migratory shifts led to an overcrowding in areas such as Kositei, Loruk, Chesemirion and Chepkalacha. Due to violent conflicts optimal spatial strategies were impossible during the droughts of 1984 and 1992. Herders had to weigh up between pastures they thought of as safe and those they thought of as good.

In 1984 when a major drought hit northern Kenya, the army tried to make use of the desperate situation, and forced Pokot herders to give up their guns. Herdsmen did not only have to look for the few as yet unexploited patches of pasture but also had to evade army units. The drought of 1984 was subsequently named *korim*, the “Drought of Many Aimless Moves”. The following case story gives an idea of the importance and character of mobility during this drought.

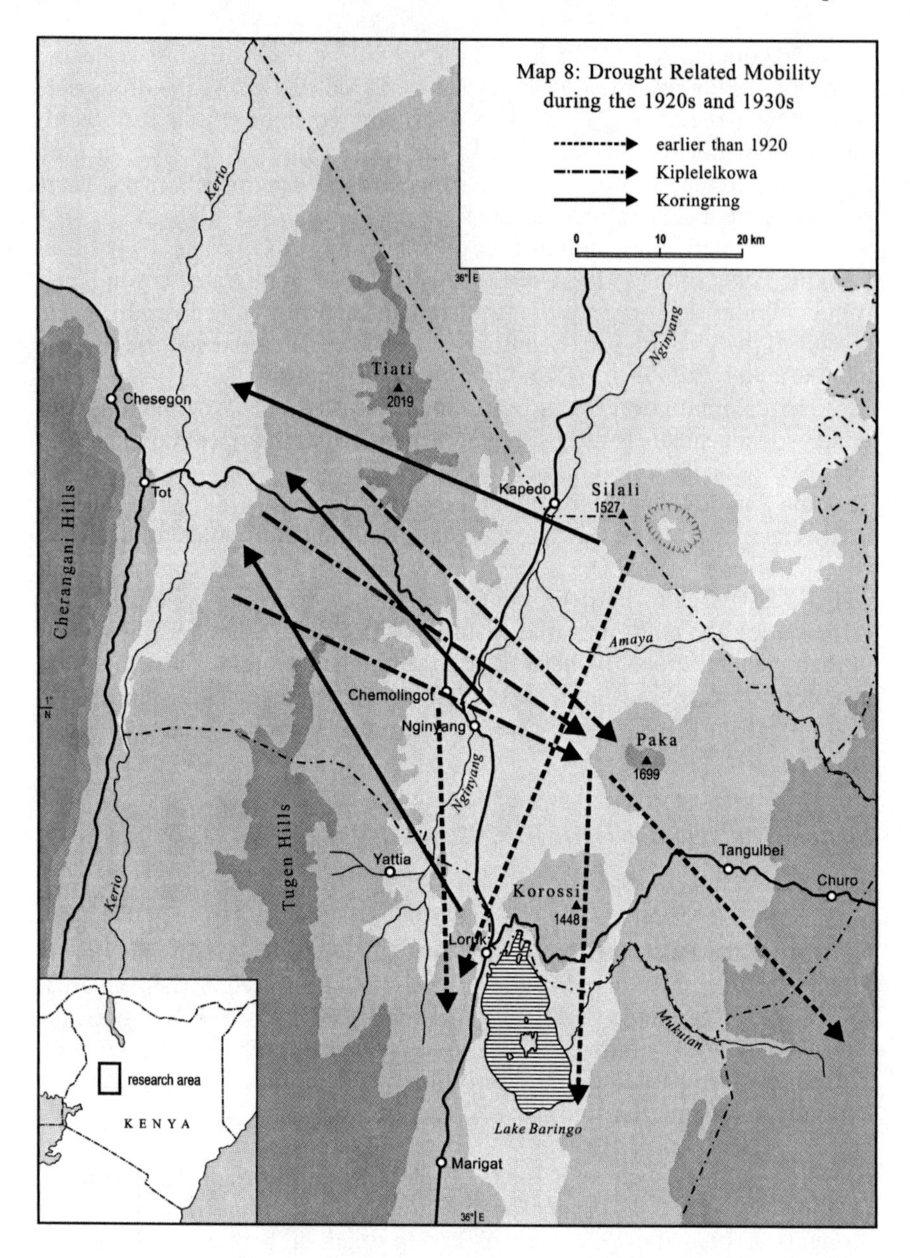

Teta's Movements during *korim* (see Map 9)

"I was living at Cheptapesyia when *korim* started. We were waiting for the early rains of *sarngatat* but only very little rain fell in Nakorete. I moved there with my two wives and my small children. Then a little rain fell in Naturkan. We moved there instantly. Although there were a few light showers, we had to water the cattle all the way down to Mukutani (about 5 hours away). At Naturkan we stayed for two months.

The May and June rains we were waiting for did not come. When I brought some of my father's small stock to my camp at Naturkan many of them died of CCPP. As there was no rain, we migrated up the Chepchok hill. There the grazing was a bit better than at our previous place. There we stayed for another month. We moved on towards Kechii to Ptare where showers had come down - we just followed the rain. Furthermore, Ptare had the advantage of being near Mukutani watering place and the cattle did not have to walk so long anymore to reach the well. However, we only stayed two weeks because the grazing was very poor and the showers had not really improved the grazing. We returned to Cheptapesyia because some rain had fallen there. But alas - after five days the water from the rains was finished. We had to move on. We moved to Chemïgh where we stayed for four days. Then the water that had been provided by a small shower was finished. Still following the rain we went on to Kasikom near Naipöchö where it had rained a bit and the grazing was a bit better. After only three days we moved on to Kaiapao (between Katengura and Mt. Paka). Now we were in the middle of the dry season. Kaiapao had received a bit of rain and there was still some grazing left. We were able to stay there for about two months. However, watering cattle was a problem there. We started moving the cattle at 2 o'clock in the night to walk them down to Mukutani and we came back the following night at about 4 o'clock; then we rested a day and then started again to trek to the well.

I decided to split the herd: we moved the cattle down to a place where the Mukutani river entered Njemps land. The goats were watered together with calves at Tangulbei borehole. Then we moved on to Sitolochel where two showers had come down. There I took on the entire cattle herd from my father. My younger brother who was herding our father's cattle had to hide in the bush as soldiers were conducting a disarmament campaign in the area. I took the cattle together with one very young herdsboy. I stayed there because Tapogh river had filled for a day or two and the flood had left behind some pools. However, after a month the water had dried up and we had to move on. First we went to Keriaw (Paka, southside) where grazing had improved a bit due to light showers....

When the soldiers had left, my younger brother joined me. We decided to split the cattle herd: Arekwen took the oxen herd via Rukus towards Lominange at the foothills of the Leroghi Plateau. I stayed behind at Mötamöt with cows and calves. Some days later I joined with the small stock herd at Cheparak, staying about 20 days, and then immediately moved on to Naipöchö. There a lion killed one of my bulls and I decided to move on to Chemïgh where I stayed about 10 days.

There I divided the herd again. I left my first wife Chemrara and her children and the young herdsboy with the goats. Together with my second wife I moved cows and calves towards Loruk. First we moved the cattle to Sengwet at the eastern side of Mt. Korossi. After three days we moved to Loruk because it had rained there. I moved on to Ngartukö on the border between Pokot and Njemps. I stayed only one day as I heard during the night that the cattle were complaining bitterly of hunger. I moved into Njemps and Tugen land, to Kampi ya Samaki. There I herded my cattle in the swamp at the northern side of Lake Baringo. Many other Pokot herders had been there already. At Kampi ya Samaki I stayed for about two weeks and then moved on to Msalabani, trekking deeper into Njemps land. I remembered that our family had some relatives there. I hoped that I could rely on their help securing me some grazing rights but I did not meet them. We Pokot now invaded Njemps land. We moved in a great number and we took the grazing by force. We did not ask the Njemps authorities nor did we ask the government. Although it never came to open fighting, bitter words were exchanged between the Pokot and the Njemps. In Msalabani we stayed for one week. Our cattle finished the grazing of the Njemps. The Njemps then decided to defend themselves and we retreated towards Kampi ya Samaki. Then the first big rains of the rainy season '85 came down. We moved back towards Loruk.

During that drought I lost about 20 head of cattle and the major part of my small stock herd succumbed to the drought."

During the drought Teta moved 21 times in a period of roughly a year covering a distance of about 180 kilometres. This does not account for the separate moves his brother Arekwen made with the cattle herd. On average Teta moved 8.5 kilometres each time. Moves over distances of more than 20 kilometres were extremely rare. For several months Teta followed the rains. Wherever he saw a shower coming down he moved towards it. Only near the end of the drought did he leave the area and move into the Njemps flats at Lake Baringo.

During my fieldwork in the late 1980s, I recorded many more case studies like these. They all tell the same story: herds were split up into smaller units and each unit migrated separately according to the needs of the animals; shifts of herds were frequent and towards the end of the drought herders moved to pastures beyond the confines of Pokot grazing lands.

Mobility patterns in the 1991-92 drought

Detailed records of migration strategies during the drought of 1990/92 clearly show that different herders adopted different mobility strategies. Some herders opted for a risk-prone strategy by moving up to the Leroghi highlands where Samburu and Turkana herders vied for the same grazing grounds and the deadly epidemic East Coast Fever ravaged, but where the grazing was good. Others chose a risk-averse strategy and stayed in the lowlands where losses were definite but calculable. The two following case studies show the different approaches to spatial options during the period of drought and epidemics between 1991 and 1992.

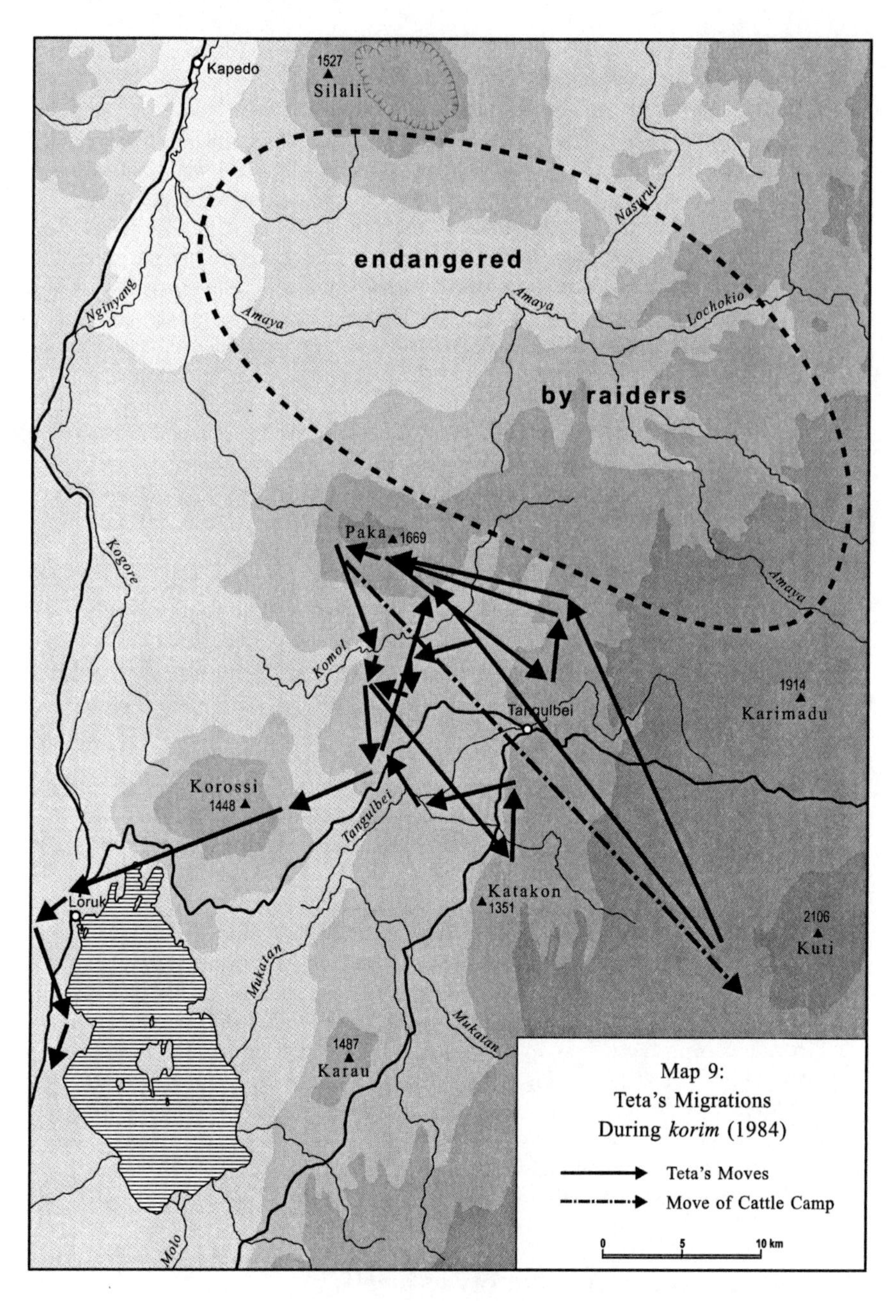

Map 9:
Teta's Migrations
During *korim* (1984)

Limpong: a risk-prone strategy of mobility

In autumn 1991, at the end of a dismal rainy season, Limpong's huge cattle herd (about 150 animals) migrated from a place near Nginyang to Naudo. That was the place where Limpong took his cattle herd most years during the dry season. But this year the rains had been so badly spaced that the grasses dried up soon after sprouting. Soon Limpong felt forced to drive his animals to the northern flanks of Mt. Paka, a place where animals were usually taken at the end of a dry season. At Paka grass growth had not been much better. Additionally there was a lot of pressure on grazing because many households had seen Paka as a last resort. Limpong felt that his animals only had a chance to survive if he moved on. He migrated to the foothills of the Leroghi plateau, to Chepïrkï, and hoped to find better pasture there in the borderland between Pokot, Samburu and Turkana. Of course he knew that the danger of an infection with Trypanosomiasis was high in the dense bush of Chepïrkï. He reckoned that the risk of a Trypanosomiasis infection was lower in a drought year than in any normal year. Furthermore, there were rumours that East Coast Fever was rampant on the plateau above Chepïrkï. Even the prophet had warned of the dreadful epidemic - but Limpong did not see any other way of saving his herd.

Limpong had bad luck. His herd was soon infected with East Coast Fever. After only a few days 25 head of cattle had died. To make things worse the grazing in Chepïrkï was not as abundant as in earlier years. When Limpong heard that it had rained on the plateau he migrated with his diseased herd to Kamuko, a place on the edge of the plateau. The Samburu and Turkana were already herding there. However, they did not cause any problems for Limpong and all the herds grazed peacefully together. The major threat was East Coast Fever. Limpong had sent a herdsboy to his eldest son, a livestock trader, to obtain money with which he could buy the expensive vaccine against East Coast Fever. As soon as he had the money he sent a young man to Maralal, the next town (about 50km away from Kamuko) to buy the medicine, Claxon. However, the drug did not have any positive effect on the health of his herd. Perhaps he had vaccinated them too late. About 40 more cattle died in Kamuko. Limpong, in fear of losing his entire herd, divided the remaining animals: all those animals which were still sturdy enough to migrate he sent back to Silali where he had come from. He hoped that the first rains which had fallen there would help these animals to make it through the drought. He remained with the sick and weak animals behind in Kamuko to see what would happen. If everything went well, some of the weak animals could recover in Kamuko within the next few weeks and then could be moved back to the lowlands.

In fact, all the animals which were taken back to Silali recovered, while none of the animals which remained in Kamuko survived.

Angurareng: a risk-averse strategy of mobility

As in many dry years before, Angurareng moved his cattle to the open grass savannahs of Serunï at the beginning of the dry season. As it was already clear after the bad rains of the rainy season that the year would turn into a drought, he did not divide the herd into lactating and non-lactating animals, but moved the entire herd to Serunï. He himself moved

> with the herd together with two of his five wives. The other members of the household remained behind at Kadingding with the camels and goats. During subsequent months the grazing progressively deteriorated at Serunï and some cattle died. However, Angurareng never seriously thought about moving up to the Leroghi plateau, like many other Pokot herders did. He had already heard about East Coast Fever being rampant on the plateau. He went towards Leroghi to have a look at the grazing and to get more details about the spread of the disease. He found that the rumours were true and thought that if the cattle were starving, he could still try and save some by migrating within the area; if they caught East Coast Fever, however, there would be not much room for manoeuvre. He hoped for some early rains and decided that he would rather track these rains than move towards Leroghi. Starved cattle recover quickly if they find fresh grass and sufficient water but animals infected with East Coast Fever are doomed, no matter how good the grazing is. Angurareng decided to stay in the lowlands and regretted that he could not divide his cattle herd into a herd of cows and calves and a herd of oxen, tollies and heifers. In the many years before he had divided his cattle into *yoswö* and *eghin* herds and had subsequently been able to adapt spatial mobility to the specific needs of these different herds. Unfortunately, just at the time he wanted to divide his herd, his eldest son got sick - and his other sons were still too young to manage a cattle herd on their own during such a harsh time of the year.
>
> Angurareng's losses over the year were great. However, a sizeable herd was retained.

It is not entirely clear what made one household opt for risk-prone or the other for risk-averse mobility strategies during the drought of 1992. The hypothesis that it was mainly young men who aimed at rapidly increasing their cattle herd and went for risk-prone mobility strategies and that old men chose risk averse-strategies was not confirmed by the data.

An important variable for determining spatial strategies is the access to information. Of course it was difficult to ascertain, post facto, what sort of information herders had had on grazing and the spread of East Coast Fever before they set off to migrate to various dry season grazing areas in January/February 1991. All the herders interviewed had heard some rumours about East Coast Fever in the highlands. This information originated from two sources: first there were some herders who had migrated to Leroghi early on and who had sent back news of the devastating disease. Secondly the prophet had warned the herders of East Coast Fever in Leroghi. Some herders claimed that they had heard the rumours and the visions of the prophet but had chosen to walk up to the plateau without any animals to see for themselves. All the herders who took the time to undertake such an exploratory tour did not move to the plateau later on, but opted to stay in the badly overgrazed but disease-free lowlands. Several informants claimed, that it was mainly herders from the Nginyang area who had dared to move to the highlands. They had been forced to leave their highly degraded home areas early on in August/September and apparently did not have the opportunity of acquiring sufficient information before they moved. Households which settled on the lower slopes of Mt. Paka stayed there with their cattle herds until October/November and apparently had more information available before they were forced to look for suitable dry season grazing.

Finally it should be noted that not all migration strategies fall clearly into one of the extremes. Out of the 26 households for which migrations were noted, 13 opted for an intermediary strategy. They moved to the foothills of the plateau between Tangulbei and Churo. The grazing there was not much better than in Seruni̇̈, Naudo or Paka. However, from there the plateau could be reached in just one day. If animals were starved they could still go from there and reach the better pastures of the plateau. With this strategy, their movements did not take them directly to better pastures but enabled them to choose between lowland and highland pastures.

Data on migrations between 1992/93 was obtained for 18 households. Map 10 and Map 11 describe the moves of three households between 1990 and 1993. While in 1990/92 spatial mobility responded to drought and livestock diseases, from 1992 onwards migrations reacted to violent conflicts with Turkana raiders. The number of moves and total kilometres covered was recorded for the period late 1990 - March/April 1992 and March/April 1992 - September/October 1993 (see Table 22).

Table 22. Nomadic moves during the 1990/92 drought and the 1992/93 period of raiding

Household	1990/92 no. of moves	1990/92 km per move	1990/92 total km covered	1992/93 no. of moves	1992/93 km per move	1992/93 total km covered
Lokolingir	–	–	–	4	13.6	54,4
Lokirpong	1	9	9.1	4	2.9	11,7
Tepakwyan	0	0	0	4	5.4	21.5
Lokomol	13	8	107	4	12.5	50
Teta	13	5	70.5	9	10.9	97.7
Silet	7	5.8	40.7	0	0	0
Wiapale	–	–	–	9	12.3	111
Awiala	–	–	–	11	11.3	123.9
Anguraren	1	9.8	9.8	4	14.1	56,5
Tochil	0	0	0	1	4	4
Kameyan	5	6.7	33.4	4	6.9	27.6
Lomurö	19	5.7	108.2	5	3.9	19.7
Chadar	3	3.8	11.3	6	7.6	45.6
Lotepamuk	1	4	4	7	8.8	61.8
Rengeruk	7	10.3	72.1	6	7.4	44.2
Lopetö	7	4.7	32.8	9	11.5	103.8
Lomirmoi	6	15.6	93.7	13	8.5	109.9
Akurtepa	–	–	–	10	3.6	35.8
Average	5.9	6.3	42.3	6.1	8.1	54.4
Av. per month	0.24	–	1.76	0,51	–	4.53

Note: "–" indicates missing data.

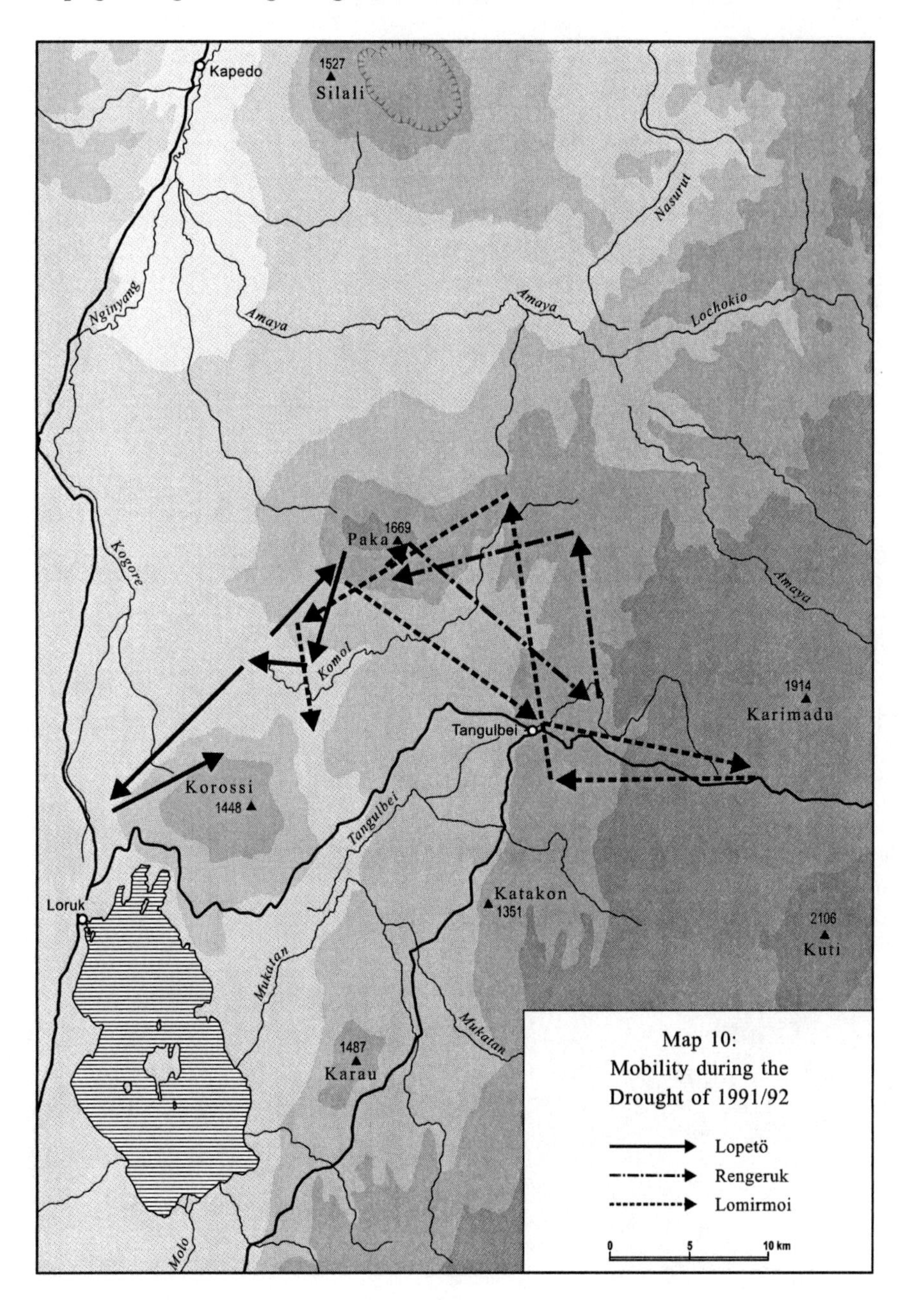

Map 10:
Mobility during the Drought of 1991/92

In the drought period of 1990/92, the 18 Pokot households migrated 5.9 times on average with an average distance of 6.3 kilometres covered per move i.e. a distance which can be covered within one day. The average total number of kilometres covered was 42.3 kilometres. While there is no doubt about the

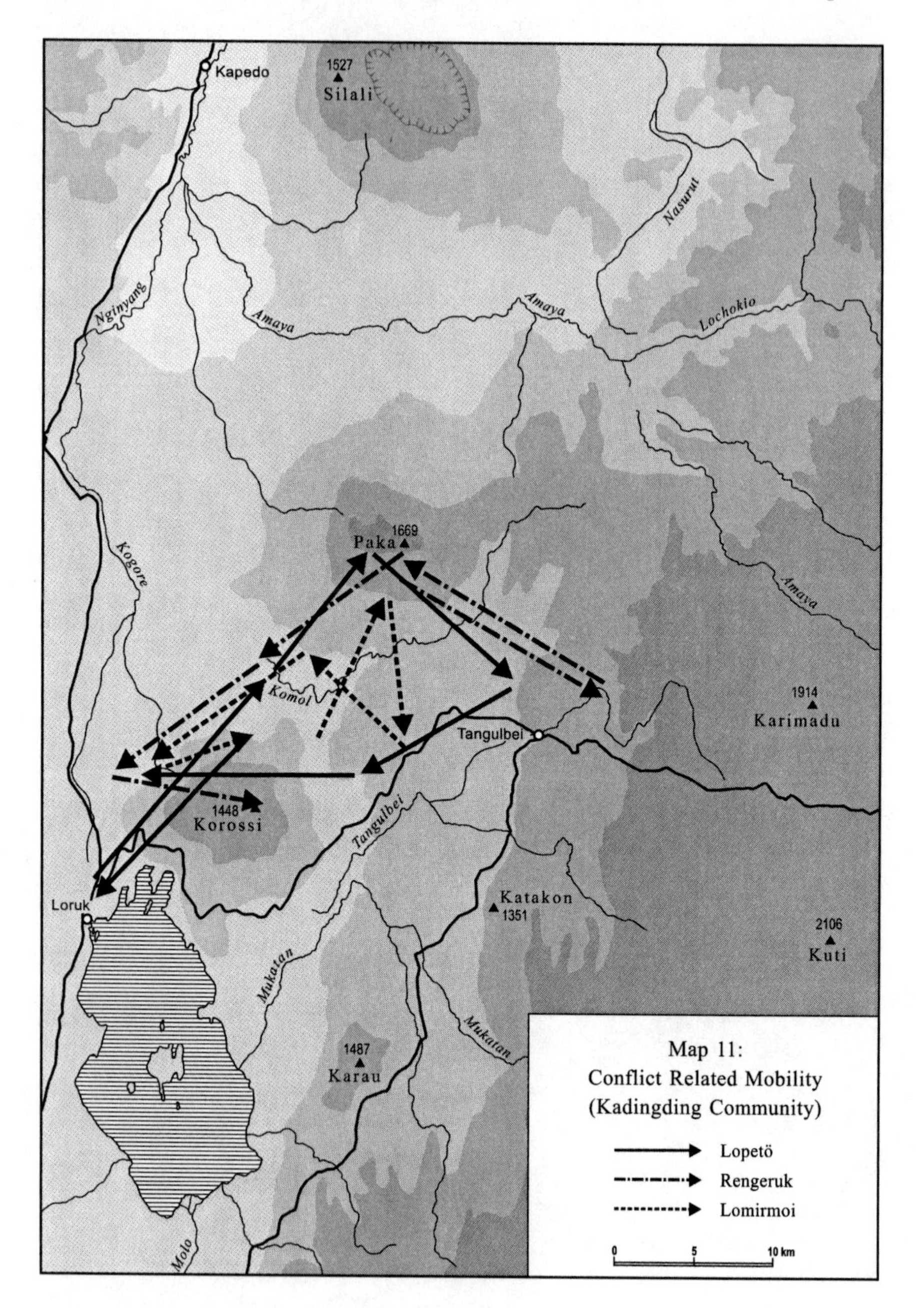

fact that the Pokot are truly nomadic, it is remarkable that they only cover minor distances even during a drought. When plotting the moves of the 18 households on a map it transpired that they had all migrated within a confined area of about 50 by 50 kilometres. It was rare that anybody moved out of this square. This indicates that during a drought remaining fodder resources in a given area are exploited initially. Only when these are completely exhausted do households move to other areas. In the subsequent year of 1992/93, it was violent conflicts and not drought which were the cause of major stress. In this period households migrated 6.1 times on average within only one year with an average distance of 8.1 kilometres per move and covered a total average of 54.4 kilometres in one year. The averages however cloud the basic difference between drought-related and conflict related migrations. While drought related mobility consists of a concatenation of moves over minor distances, migrations due to violent conflict consist of one major move into a safe area and then minor moves according to local conditions.

The 18 households differed in the way in which they applied mobile strategies. In both periods the number of moves varied between 0 and 13 and the total number of kilometres covered between 0 and 108 in one case (1990/92) and 0 and 123 (1992/93) in the other. While during the drought period households moved about every third to fourth month, during a period of intense raiding they moved about every second month. While they covered about 1.7 kilometres per month during the drought period, they covered about 4.5 kilometres per month during a period of violence. These figures suggest that violent conflict forces the Pokot to undertake more frequent and much wider moves than drought conditions do.

5.3.2. Moving to Survive: Himba Mobility Patterns During a Drought

In normal years households and herds live in a few settlement areas along rivers and near to gardening places during the rainy season. Early on in the dry season a herd of non-lactating animals is split off and taken to the dry season pastures. This herd may be split again at a later stage into an oxen herd and a herd of young stock. Towards the climax of the dry season the remaining lactating cattle and their calves may join one of these camps or establish an independent third camp. At the same time a separate small stock camp may be established. While the various livestock camps are highly mobile, the main homestead only rarely shifts its abode and usually only over short distances.

A considerable amount of pasture is set aside as a drought reserve. In normal dry seasons these areas are not utilised. In times of drought, grass from the previous year can be used in these reserves. Alternatively, herds retreat into the lush gallery forests along the Kunene river where several fodder-producing trees, notably Faidherbia albida, are a reliable resource to be used during droughts. The riverine forests at the same time hold abundant palm trees which supply humans with food. If these measures do not suffice, migrations of cattle herds to the southern Angolan highlands are a last resort.

Historical Development of Mobile Strategies under Stress

During the first thirty years of South African colonial rule (1917 to the 1950s), mobility was severely confined due to border regulations. The border to Angola was closed, as was the border to Ovamboland from the early 1920s onwards.

At the same time the riverine forest along the Kunene was declared as a no-go zone. Several times the people who had settled there during a drought had to withdraw under threat of severe punishment. Between 1926 and 1938 a police post at Tjimuhaka right at the major Kunene ford controlled herd movements. Patrols were frequently sent along the river to look for livestock spoors. If animals which had crossed over from Angola were found, they were annihilated and their owners heavily fined. But not only moves across newly instituted boundaries were prohibited: during the 1920s and 1930s moves within Kaokoland were also monitored and partially prohibited. For several years permission had to be obtained from the authorities before shifting one's homestead or cattle camp.

However, despite strict restrictions on mobility, individual herders seemingly retained a certain degree of autonomy. I found several elders who had left Kaokoland (with their fathers) for southern Angola during the 1928-32 drought and had returned later on. During the 1941 drought of *Kaṯe,* restrictions were enforced more strictly: I did not find a single household which claimed to have moved over the border during that drought. With the strict control of the Kunene riverine area by the colonial government, the gallery forests along the ephemeral tributary rivers came into focus as settlements. These forests also offered palm groves and stands of Faidherbia albida providing food for humans and animals alike.

Drought-related moves took households and livestock camps over distances of 30 to 60 kilometres and under exceptional circumstances even further than that. The total area used is at least 100 kilometres in width and length. Major migrations which are typical during droughts frequently comprise a shift of the entire migration orbit. Herders moved about 60 kilometres during a drought from Otjipupa (Angolan side) to Orui near Ongokwa (in Angola). After the drought they stayed there for about 5 years before going back to Otjipupa. In several accounts on migration histories during the 1981 drought, the same phenomenon transpired: once a homestead was dislocated beyond its regular migratory cycle it stayed in its new vicinity for several years before returning back to the area which was regarded as home. Several households which shifted to Angola during the 1981 drought stayed there for many years before returning back to Namibia.

Mobility during The Year of the Dying, 1981

In 1980/1981 an extreme drought haunted Kaokoland. About 90 per cent of the cattle died. As a first reaction to the onset of the drought, some households shifted to the Kunene river or to its tributaries despite the war. There they hoped to supplement their food with palm nuts. The goats and some lactating cattle could feed on the seeds of Faidherbia albida. However, it soon transpired that the drought was so serious that it exceeded the perturbations of a normal dry year. Especially young families then crossed the Kunene and set off to Angola. Many went as far as Otjandjou (about 100km from the Kunene, see Map 12) leaving the Himba area at its north-eastern edge. Others migrated towards the Kuvale area at the north-western rim of the Himba area. At the height of the drought some Himba herders even moved close to Lubango (personal communication Samuel Aço). The situation was aggravated by the fact that due to the Angolan civil war, very little food could be bought in south-western Angola. It was mainly young mobile families who risked the long migration taking with them the cattle of less mobile households, while the old, the children and many women stayed

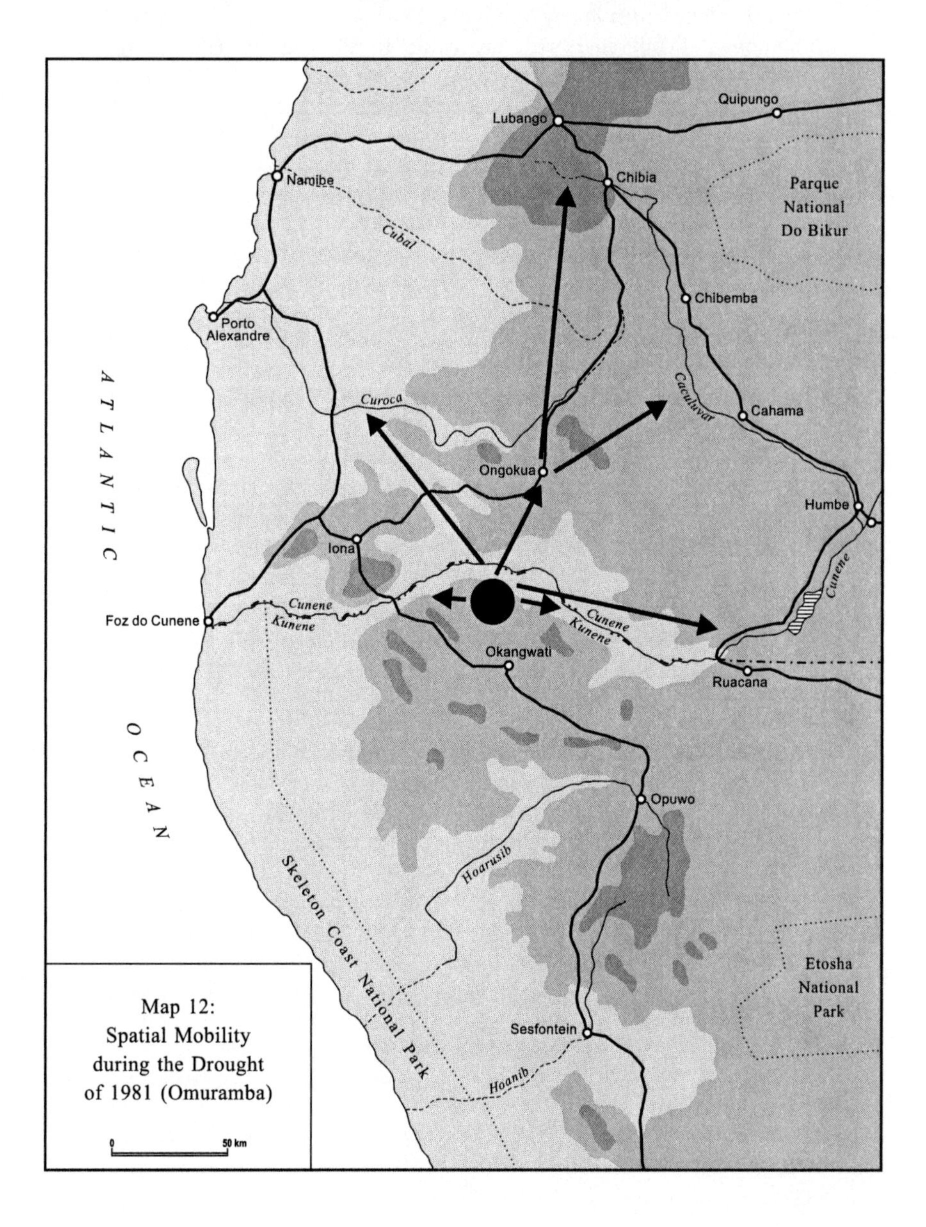

Map 12:
Spatial Mobility
during the Drought
of 1981 (Omuramba)

behind. However, herders claim that although losses in Angola were very high too, those herders who went to southern Angola were at least able to save two or three oxen.

The two following case studies give an idea of the hardship people underwent to save the lives of a few animals during the Drought of the Dying.

Tako's migrations during the 1981 drought

> "This land has seen many droughts. During one drought I lived in Omuhonga, I migrated with my cattle to the hill of Tutu in the Omavanda mountains. There I stayed. My wife remained behind at our small garden in Omuhonga. Later I asked her to come and join the cattle camp and we moved to Osia together. There I stayed with some friends for a short while and then migrated near Etengwa to a place called Okotuhama. My wife was pregnant and I brought her to her mother in Otuhama. A little time later she gave birth.
>
> I returned to my herds and drove them back to Omuhonga. There I stayed with my hunger. I rested until there was nothing left to eat anymore. I asked Vahenuna's father for a big ram which I slaughtered immediately. I ate and ate and ate, I ate with great hunger. That year was a serious drought. Then I moved again. I left my goats with my mothers younger sister, the mother of Karambongenda. I drove my cattle to Oyomiwore, the next day I set off to Okahamwanda where I settled together with some friends. When the grass was finished we moved on via Oviyere to Okayere and then to the hill of Muzema. We stayed there for some time. But the cattle became too weak to climb the hill. There was still water in the waterhole but the hole was far from where we settled. I wondered 'What shall I do?' I moved again and migrated to Otjisoko, finally I stayed in Otjitamutenya. I fetched my wife from her mother's place and brought her to Omuhonga. There our new born died in the midst of the drought. I brought my mourning wife to her father in Omuramba. Then I took off to shift my goats. I slept one night at Ombandaondu, the next day I drove them and I stayed in Ombaka. From Ombaka I went to the camps of my friends who had migrated from Omuhonga. Then it started raining. Our starved cattle died of the sudden cold. I returned to Omuhonga, but still cattle died from the cold brought by the rains. The calves died in great numbers. The women skinned them to take the hides and make them into clothing. They died and only some survived. Nearly all the calves died.
>
> When I returned to my homestead I found that many of my cattle had died. Many more had died in my herd than in the herd of the others. Yes, cattle do it that way, some cattle struggle to survive and some just give in and die. I set off again and migrated with all my livestock, goats cattle and sheep. I stayed in Oromutati. One night my calves escaped and ran back to the old homestead. One calf was killed by a hyena that night. In that place Oromutati several female cattle died. All pregnant cows had died by now, one cow after the other was skinned. The cattle just died of drought. One cow I lost there, and I never saw it

again. Then I sold one heifer and I got some millies and sugar for it on which I lived for some time.....

Now the big drought came. I stayed here in Ombuku until *oruteni* (Oct-Dec.), then I moved beyond that hill to where it is called Okomwangwei. There another cow died. Then I divided the goat herd. Some I gave to my sister's son and some I left with another herd of goats. I was all alone as my wife was still with her parents. I had to do everything by myself: I watered the cows and calves and milked the goats. At that time I heard that it had rained in Otjitanga. I tried to move there. The cattle moved well the first day. We slept at Ohamukuta. But the next morning they did not want to get up. However, in the end we arrived at Otjitanga. The rain had not been good and we had to move on, but this time the cattle were too weak and many died. I divided the cattle: the stronger ones I gave to a Tjimba herdsman, the weaker ones I took. Now the cattle started dying all over the country. Entire herds were finished. The people became sad, their hearts were red. And during this disaster the war of Ozoteri made things worse. The remaining cattle fled.

People looked for straying cattle in the bush around Okangwati. If they saw an animal they just shot it and sold the meat at Okangwati. I decided to climb the hills of Omazorowa with the remaining animals. After we arrived there all my cattle died, one by one, finished. Yes, finished, nothing was left. I said to myself 'Now I will join the army'. I sent a child with the remaining goats to my wife and told her to slaughter one for meat and sell the other ones for maize. I left to join the army. But suddenly the spirit of my mother's brother started to speak to me. And it told me not to join the army. It told me to return. And so I did. I went to Ombuku and took some of the cows I had given to the Tjimba shepherd and some animals which had been lost previously I found in the bush until I had about 10 cattle.... But the cattle still went on dying. When the cattle died in our land of the Himba, many people went to the town to look for food. A person who died was not mourned for anymore. The person was just buried and the people went back to their homes shortly afterwards. They did not stay at the funeral, not even a day." (Tako Hunga, June 1995)

The story highlights the great importance of spatial mobility. Tako moves his livestock many times in order to save a few of them. The places he visited were about 60 kilometres apart. While Tako occasionally moved together with friends, he usually moved alone. Looking at many accounts of drought migrations, I am tempted to say that this is the exception rather than the rule. Usually three or four households move together and co-ordinate their labour.

The second case study portrays the moves of three livestock camps which moved to Angola together during the 1981 drought. The account by a woman delineates the hardship of life on the move. Hunger, thirst and exceptional physical strain are characteristic for such moves. A woman has to additionally care for small children.

Mukakaserari's migrational history in the Year of the Dying

"In this Year of the Dying we migrated with our home to Okarisewandumbu, there, far away. I was hungry and at that time I had just given birth to Tjihakamo... he was still a very small child that crawled on its knees, he could not stand up yet. When we moved we stayed at Omuru wa Kaarwandu. On the journey Mungerinyeu slaughtered a goat and we slept and we ate.

Yes,... the sun was up there, and the thirst was killing us....When we settled we put our loads on the ground and we climbed the gorge of Kaarandwa with our children on our backs and our metal pots,...and we drank water there ... We went there, we settled there until we became hungry and Mungerinyeu took out a goat and my husband slaughtered it.

Sacks of maize had not come to these parts of Angola. Still there were no sacks of maize since the Portuguese had left.... We slept there and ate that goat. The next day we stayed the full day at the homestead and the following day we returned to the well, we scooped water, and the next day we moved. We went to Omuseravari. I was so thirsty then, I was near the end. We went to the well and I drank water. I looked at myself. I fell down, I lay down, my hair was in disorder. We stayed there until evening and then went to an empty homestead. There we stayed. And I slept. That evening we stayed at the place, the father of Vasongonona killed another goat. The people were finished, they just fell asleep near the meat and we did not eat and drink milk. The cattle were dry and the calves just died.

There was meat to make us sleep and to make us stay. The next day the father of Vasongonona killed another goat and we stayed the whole day, and we ate and the next day we moved our cattle again. Still we were on the way and we went to Ongokwa.

Yes, we had tied our hungry stomachs with the big women's belt. (We) drove (our) cattle, we drove, we drove, we drove. (Then we rested in an old homestead. In the evening the men wanted to slaughter a cow.) When they tried to catch the cow, they did not succeed, it just escaped, that's it, the cow just ran away each time they tried, they were very hungry. Finally Mungerinyeu succeeded in catching it. They slaughtered it, skinned it, skinned it, skinned it.

.... they said 'Let us water the cattle at the waterpoint of Ongokwa'. We were just two women. Tjikuva and the other one climbed down into the well and helped each other to bring up the water as the well was very deep. We watered the cattle. They were soon out of control and did not listen to commands, they just came towards us. We ran away and the cattle just stampeded to the shallow well. ...Yes, the two boys had to jump out of the well. They helped us to drive the cattle back some distance where they left us. They went back to the well. When they arrived there the cattle became wild again and started running towards the well with a sound like *twaaa*. The cattle did that until sunset and we wondered what shall we do now? ...We slept there, hungry, with the empty stomachs of the day. We slept there and it was cold. Look we had not gone with our blankets, as we had thought to stay there just over the day and in the evening we would

> return to (our) homes. We covered ourselves and the children with our calf-skin clothes." (Mukaakaserari in Bollig, 1997c)

Many migrations in the direction of southern Angola looked similar. These drought-related migrations were a continuous fight against adverse conditions: hunger, tiredness and heat. Many herders interviewed took at least part of their cattle herd to Angola. Many herders moved in small groups. Rough accounts of migrations were recorded for 26 households. Map 12 gives an idea of the extent of spatial mobility during the drought of 1981. Out of the 26 households, 11 took their cattle to Angola and moved partially into the Ngambwe area.

In major moves, several Himba herders took their animals about 150 kilometres away from their settlements, far into south-western Angola. Cattle are usually not driven more than twelve kilometres per day to prevent a detrimental loss of weight. A distance of 150 kilometres basically means about 15 to 20 days continuously moving with livestock. Mukaakaserari's account emphasised the terrible hardship such desperate long range moves bring with them. People and animals really show that "they have a bone in their neck" as Kozombandi Kapika said in her remembrances of the drought of *Kaṯe*. However, it is remarkable that the Himba have the opportunity of covering such a wide space if necessary. A migration radius of about 150 kilometres makes very different eco-zones accessible to herders. It takes herders from northern Kaokoland well into the southern Angolan highlands and far into the south of Kaokoland. The probability of finding areas less affected by drought increases with the distances covered by migrations.

5.3.3. A Comparative Account of Mobility Patterns during a Drought

Spatial mobility is a major strategy to cope with adverse conditions in both populations. In both cases herders make extensive use of the opportunity to create several smaller and more homogenous herding units. The availability of labour is of crucial importance for Himba and Pokot to apply mobile strategies optimally. However, drought-related mobility is clearly tied to local conditions and at several points Pokot and Himba mobility patterns during droughts differ sharply.

Drought Grazing Reserves: The Himba in northern Kaokoland maintain drought-grazing reserves. These areas are not grazed (or hardly grazed) during normal dry seasons. The major drought reserves are the ranges of the Baynes Mountains. The mountains are covered with a superb pasture of perennials interspersed with annuals. These outstanding grazing qualities have led to the metaphorical name "the mountain that never ends", meaning that the pasture on these mountains has never been exhausted in living memory. The Kunene valley between Ruacana and Epupa with its lush riverine forest acts as a buffer as well. The riverine forest with its thick stands of palms and Faidherbia albida trees is a most important fodder resource during the drought. Obviously the value of this type of pasture is particularly high because there is water in abundance nearby.

The Pokot do not maintain drought-grazing reserves. They even find it difficult to guarantee a restrictive use of dry season grazing. It is doubtful if the Pokot have ever maintained large scale drought-grazing reserves. It seems that constant warfare brought about the existence of such areas only as a by-product. Areas on the border between Turkana and Pokot are sometimes not used for years, sometimes even decades. However, this type of non-management of drought resort pasture is highly risky. What if drought and warfare coincide? This has happened several times over the last few decades with disastrous results.

Mobile Strategies during Drought: The way the Pokot and Himba employ spatial mobility during a drought differs profoundly. Pokot spatial mobility answers two basic problems: drought and security. Pokot herders try to use remaining pastures thoroughly. Herds are moved quickly within a particular area. The average distance covered per move was about 6 kilometres, a days walk with animals. Frequently herders just followed the first rains of the rainy season. These moves were not coordinated. Historical accounts of Pokot mobility lead to the assumption that Pokot mobility has been confined by the growing population density within the area. Long distance moves have been replaced by a concatenation of erratic moves within a restricted area. There is ample evidence that the Pokot are not an exception: in many East African herder societies spatial mobility has decreased massively during the last decades (see Sobania, 1988).

During a drought the Himba first of all make use of dry season grazing reserves in the hope that rains may still come. When these ranges are exhausted they turn to extra-territorial pastures. The moves towards dry season grazing areas, moves within these areas and shifts towards drought-resort grazing are co-ordinated. Leaders of livestock camps try to find a common strategy for their moves. During a drought Himba cattle herds cover large distances. Frequently herds cover 20 or 30 kilometres in one go.

Moves out of the Tribal Grazing Area: Both the Himba and Pokot report on drought-related migrations beyond the territory they usually use. The Himba however were seriously hampered from making such moves towards the foothills of the Huila highlands. Since the 1960s these pastures have become accessible once again. The Himba made excessive use of extra-territorial pastures during the 1981 drought despite the civil war conditions on both sides of the boundary. There are no reports that the Himba moves into the area brought about conflicts with local people. Frequently the Himba gained permission to access extra-territorial pastures via cross-cutting clan ties.

Pokot herders maintain close ties with Turkana herders as well. While several Pokot clans claim Turkana descent, they do not exploit these cross-cutting ties systematically. Occasionally Pokot herders made use of such friendship ties in order to gain access to Turkana land during the drought of 1984. In general the Pokot rely on their military strength when invading the pastures of other groups. They move to these areas heavily armed and guard their animals intensively.

5.4. DIVERSIFYING INCOME GENERATING AND FOOD PRODUCING STRATEGIES DURING A CRISIS

During a drought pastoralists and agriculturalists took to gathering and hunting in the past and opt for ten-cent jobs such as producing handicrafts, charcoal burning, brewing liquor etc. nowadays. Diversification during and immediately after a drought frequently comprises activities that need little capital input and are mainly based on labour. There are two principal ways of diversification: either to exploit food resources that are not affected by the crisis or to engage in activities that generate money and/or exchangeable goods that enable the purchase of food. There is a transitional field between short term forms of diversification and more permanent forms of diversification. In this chapter measures of diversification will be discussed that respond to problems brought about by the immediate effects of drought.[20]

5.4.1. Ten Cent Jobs and New Niches: Pokot Attempts at Diversifying their Economy

During the crisis of the early 1990s many herders developed strategies which opened up incomes outside the pastoral sector. Obviously such chances for alternative income are rare in the marginal areas of Eastern Africa. Wildlife in the Baringo plains is few to non-existent, non-domesticated substitute foods are rare and agricultural land is scarce. Neither state nor development institutions supply many jobs or alternative incomes. Opportunities for procuring additional food and alternative income during a crisis are highly gendered: while women explore ways of how they can commoditise their activities (e.g. cooking, brewing, selling traditional medicine) men look for ways of finding alternative incomes which they regard as appropriate for their aspirations (e.g. livestock trade).

Male Income Generating Activities

While from 1987 to 1989 the number of Pokot men trading livestock was minimal, in 1996 many young men had gained some experience in livestock trading and/or selling veterinary drugs. Until 1989 herders who wanted to sell an animal brought it to the weekly markets in Nginyang, Tangulbei or Yattia. There they sold animals directly to Tugen traders. Frequently herders voiced their uneasiness at selling livestock to people whom they did not know and whom they accused of all sorts of mischievous tricks to lower the prices. In 1996 most of the transactions at the weekly markets took place between Pokot intermediary traders and Tugen livestock traders. While small scale Pokot traders bought goats and cattle in the hinterland and brought them to the market, Tugen traders transported the livestock to the abattoirs and butcheries of the urban centres. The inclusion of intermittent traders lowered the transaction costs of livestock trading in the area considerably. The better capitalised Tugen traders no longer had to bargain with numerous herders in order to buy single animals. Small scale Pokot traders had created a niche for themselves. By 1996 ten men out of 47 from the sample I had worked with for almost a decade now traded livestock. Three more men had done so in the past few years. These young men mainly operated in the wider neighbourhood of their own homestead. Apparently their business was made somewhat easier by the herders' increased need to sell livestock during a period of stress. Nearly all Pokot traders were young: nine of the ten traders belonged to the youngest generation set and were aged between 25 and 40 years. The explicit reason given as to why livestock trading was taken up was to generate money for food acquisition and to relieve the household from excessive livestock sales. Whenever the business ran well and the food supplies of the family were sufficient, additional livestock was bought with the surplus money. The capital to enter the livestock business was usually procured by selling one, rarely two oxen. Hence, most Pokot traders started their business on a fairly limited capital basis. With the money goats were purchased from neighbouring homesteads. The goats were then driven to Nginyang's or Tangulbei's weekly market. The income from small scale livestock trading is unstable. At the height of a drought (e.g. in 1992) or during a period of intense spatial mobility due to raiding (e.g. in 1993 and 1996) the small businesses have to close down. During these periods labour demands on herders are high and there is no spare time to search for goats and to drive them to the market. Other small scale traders had problems to save enough money from a sale at the weekly market to reinvest into further acquisitions. During a drought the need for money to buy maize increases whereas the livestock trade becomes more risky. Two of the three traders who went out of business during the drought claimed that their capital was depleted by buying food for their family. Time input into livestock trading is relatively high but potential gains are good too. About 20 to 50 KSh can be gained per goat, with oxen the profit ranges from 200 to 400 KSh. This corresponds to a gain of 3 to 8 kilos of maize per goat and between 30 and 60 kilos per cow (prices for 1992).

Trade with veterinary medicine needs less financial input. The sale of three goats is enough to buy a one litre bottle of broad spectre antibiotics (Terramycin, Adamycin or comparable products). Although the sale of broad spectre antibiotics to lay persons is illegal for understandable reasons, it is rarely prosecuted. Traders buy the medicine in Nakuru in a pharmacy. The drug is then sold in minimal quantities when traders move from homestead to homestead. The trader takes out the required quantity with a syringe and injects it into one of the small bottles for veterinary medicine every household possesses. It is not rare for a homestead to buy as little as 2 millilitres.

Of course, under-dosing antibiotics is frequent and over the course of the years there is the danger of creating resistance in livestock. While buying one millilitre cost about 1 KSh in 1993, selling the drug generated about 1.50 KSh, this means that per bottle of 500 KSh a trader can make a profit of about 250 KSh. This is a considerable gain compared to the gains from livestock sales. However, the work is very time intensive as homesteads have to be visited one by one. In order to contact more clients dealers in veterinary drugs visit the weekly markets in the area in order to sell there too. Most traders of veterinary drugs are young men who live with an older brother or with their father and who are not immediately needed for herding a household's livestock.

Female Ten Cent Jobs

During the drought of 1990/92 the number of women who sold home-brewed beer rose quickly. Previously during the major period of field work between 1987 and 1989 I did not observe any women brewing beer in the pastoral areas. Earlier documents on northern Baringo do not refer to any brewing activities and it seems warranted to assume that beer brewing is a relatively recent phenomenon among the pastoral Pokot. In the 1980s brewing was confined to the small centres where there was a limited market for home brewed beer. Even there it was limited as it had been declared illegal in the early 1980s and was prosecuted by chiefs and administrative police. During the crisis of 1992 many women took up the brewing business. Remarkably it was only women who did so, whereas the clients were predominantly men. Characteristically they focussed their activities around wells and dams, where many men go on a daily basis. The sale of alcohol around the centres increased too and, for most of the time, was not prosecuted. Brewing was not confined to women from poor families. Frequently, these even lacked the income to handle the basic investments necessary to start brewing, although the money needed was comparatively little: 6 to 10 kg of maize (6 KSh per 1 kg), and 2 kg of millet (7.50 KSh per 1kg) are needed. With this quantity about 50 litres of beer can be produced. One litre of beer is sold at 4 KSh, i.e. a maximum profit of 125 KSh per drum can be achieved. Of these again 75 KSh have to be reinvested into the next brew, reducing the netto profit to 50 KSh (ca. 8.3 kg of maize).[21] Maize, millet and sugar has to be brought from further afield. The grinding of millet takes time and the dishing out of the brew is time consuming too. Given a labour input which amounts to at least an estimated ten hours the income is low - but it is the only way women can make any income. Women face yet another problem. They are serving male clients only, usually relatives, friends of their husbands and frequently their husbands themselves. Men often try to convince women to give them brew on credit. But the moral of paying back these credits is low. Accordingly women have to press for immediate payment. This may be one reason why brewing attracts mainly middle aged and older women. They can assert more authority towards their male clients than young women can. Although there are several constraints brewing seems to be profitable. Several women claimed that they would like to start brewing if they had the necessary money. The reason for earning money is the wish to safeguard the supply of essential food items. There is some evidence that beer brewing is a way to redistribute money within the community. Men give away their money which results from selling livestock. The money goes to women from the same local community and is reinvested into maize, sugar and tea.[22]

During the drought the number of small restaurants (*hoteli*) near dams and centres rose significantly. These *hoteli* are pretty basic. A rudimentary shelter from branches, one or two benches from raw wood (two forking branches and thicker pole put across) a number of plates and cups, some spoons, one or two cooking vessels. Many *hoteli* just offer maize and beans, *chapatis* and tea. In 1992 a portion of maize and beans was sold for 4 KSh and a cup of tea cost between 2 and 2.50 KSh. Capital and labour input in the *hoteli* business are higher than in the brewing business. As with brewing two or three women co-operate to run one business. A small survey of hoteli's showed that women co-operating in one restaurant were usually closely related.[23] Given that in better years only a few *hoteli* operate this number is considerable. At dam sites there were usually between three to six *hoteli* operating. *Hoteli* operated mainly during the drought, as soon as the rains started and customers dispersed the open-air restaurants closed their businesses too.

Several activities which are mentioned in other reports on diversification in herder societies are conspicuously absent among the Pokot of northern Baringo.[24] Charcoaling is rare in the Nginyang division. In 1992/93 there were only three or four men living on charcoaling in the wider region, at least two of them were Turkana. However, during the field trip in 2004 I noticed that in the Kositei area charcoal production was rampant and also in other areas near streets the production of charcoal had apparently picked up. Handicrafts are not of importance among the Pokot. On the one hand a major urban or tourism market is lacking to make charcoaling or handicrafts a profitable business on the other hand cultural preferences inhibit investments into craftsmanship.

5.4.2. The Failure to Diversify? The Himba Approach to Diversification

Many Himba households maintain gardens and gathering becomes important during droughts. Beyond small scale agriculture and gathering, however, there is very little the Himba do to diversify their economy during a crisis. Several odd jobs are left to members of other ethnic groups. Himba regard the Hakaona, Thwa and Kuroka as very efficient craftsmen and healers.

Jewellery is only made by Thwa smiths and healing rituals are frequently conducted by the Hakaona, Thwa and Kuroka. Menial work such as building livestock enclosures, constructing huts and digging wells is done by casual workers for rich Himba families - usually non-Himba who come from Angola and are regarded as good workers. In this way many activities, which serve as supplementary income generating activities in other rural societies, are only deemed to be appropriate work for non-Himba.

Three reasons speak against the involvement of Himba in "ten-cent-jobs" such as brewing and charcoaling. Population density in Kaokoland is very low. The market is limited for any economic activity in the service sector as potential clients live far apart. The scattered options for "ten-cent-jobs" are exploited by young Angolan men who do not have any obligations in the livestock economy and who are free to settle wherever there is work. Furthermore Himba would gain little from small additional incomes. As long as transactions between Himba livestock producers and traders run on a barter mode of exchange, there is little use in obtaining small amounts of money - it cannot be used to buy smaller amounts of maize as maize is only bartered in sacks. In order to buy food livestock is needed – "ten cents" and other smaller units of currency are of little use.

The size of herds, the availability of substitute foods and agriculture apparently lower the need to procure alternative income to buy food. If herds are depleted beyond the possibility of recuperation several options arise: Frequently impoverished households break up and members settle with different other households. Other households opt for a temporary change away from pastoralism.

> "Yes, those cattle of people with short arms all died and they stood there without anything. When the cattle died I asked there in the land of the Ngambwe. I thought that when the cattle had died, all I could do was to plant maize in the land of the Ngambwe and I planted maize. I knew how to plant maize but I thought that I was not born for maize. That's it. 'Let me go and return to our land even if I will be hungry there', I said to myself. I left that maize there. It could not help me because you cannot drive the maize. It is eaten only and put in the *otjindu*-basket, that's it.
>
> I saw that it cannot be driven, it cannot be killed for somebody who died. If somebody died you could not mourn after the funeral. Yes, maize cannot be inherited it cannot be inherited to your family. It is only inherited to those very near, who are in the place of the dead person those who are near where this person planted his garden. Somebody living in the land of the Ngambwe he will not come from Ngambwe only to inherit maize here and a person from here will not go to Angola only to inherit maize there. Sombebody coming from Angola, somebody from Namibia wants to inherit cattle from Angola or somebody from Angola wants to inherit cattle in Namibia. Maize cannot be inherited. I left it, and said to myself 'let me move even if I find hunger – that's it'. (Muhuwa, April 1994).

A third possibility which was chosen in 1897 after the dreadful Rinderpest epidemic and in 1981 after a devastating drought was to recuperate losses as a mercenary or soldier in a foreign army. After the disastrous Rinderpest epidemic of 1897 Himba and Herero offered themselves as auxiliaries. From 1897 to 1906 several punitive expeditions took place every year. In each expedition several hundred Himba men were engaged. They were paid in loot, usually cattle which had been raided from so-called "rebellious" natives. In 1981 a similar situation arose. Himba herds were severely diminished by a disastrous drought and again many young Himba men decided to become soldiers. While the South Africans did not pay in livestock they paid sizeable salaries which allowed restocking to take place.

5.4.3. A Comparative View on Diversification during Periods of Stress

Strategies of diversification are conditioned by constraints and options of the wider regional system and depend on the inclusion into markets and the specialisation of neighbouring groups. Pokot strategies of diversification tend to be directed towards the service sector. Intermediate trading and

Photograph 10. Himba Mercenaries (from de Almeida, 1936)

Photograph 11. Vita Tom and his Troup of Mercenaries (from de Almeida, 1936)

brewing are perhaps the most prominent activities which the Pokot adopted. Livestock trade and the selling of veterinary drugs require a large amount of time as customers have to be visited one by one. At the same time livestock trading during a drought is beset with specific risks. Animals may die before being sold at the market and prices are low for underfed animals. Brewing and running smallscale restaurants are highly seasonal activities.

Himba face different constraints. On the one hand Himba ways of diversification are more stable as they invest labour in agriculture on a permanent basis. Additionally they make use of riverine gallery forests along the Kunene and its tributaries to harvest the nuts of Hyphenae petersiana during a drought. On the other hand there are few prospects for further diversification in response to drought. There is no brewing, charcoaling or increase of handicraft work. The market for these activities is minimal as the population density is low and the demand for extra-services is very limited. Diversification becomes important in a situation of (nearly) total stock loss. This happened twice to the Himba during the course of this century: both times stockless people opted for a temporary shift away from pastoralism and into agriculture and to commercial soldiering to recoup losses. Remarkably the Pokot do not report of such major changes in economic orientation.

5.5. CRISIS MANAGEMENT THROUGH RITUAL

In many societies magic and ritual is used to counter personal trouble and disasters and/or to prevent their occurrence. Minor magical devices, for example an amulet, a prayer or a magical formula may protect an individual. On a family level, more encompassing ritual devices are chosen, engaging family members and neighbours. Communal rituals involving many related and non-related people of various localities are generally conducted to counter major problems such as epidemics, climatic perturbations and war. Both societies use magic to counter disasters: both the Pokot and Himba frequently wear amulets, honour blessings and believe in various forms of sooth-saying. The Pokot think of communal rituals as an essential means of preventing mischief. Himba crisis rituals are smaller household or descent-group based affairs. Ritual activity and symbolic exchanges are not directed towards the group but towards the ancestors. The analysis of crisis rituals does not only add a further perspective to risk management, it also allows an important insight into risk perceptions.

5.5.1. Reducing uncertainty and fighting hazards through the use of oracles and ritual among the Pokot

Disasters have to be fought physically and spiritually, just as human diseases (Nyamwaya, 1987) and animal diseases (Bollig, 1995a). Pokot beliefs in the aetiology of disasters are adamant about seeking the origins of negative events in interpersonal and intra-societal conflicts. This pertains to a crisis befalling a household (e.g. a livestock epidemic) as well as to serious problems affecting the entire community. While interpersonal conflicts resulting in intense negative emotions are seen as the main cause for sudden herd decline, bad luck and other mischief, the cause for disasters with a regional scope is attributed more vaguely to continued sinning and the transgressing of taboos, disrespect and a lack of solidarity among people. Disasters result directly from accumulated sin, adultery, illegal sexual liaisons and lack of mutual solidarity.

The Pokot diagnosis of problems and crisis management is characterised by a wealth of household and community-based rituals. Whenever an animal is slaughtered, a specialist (*kipkwan*) is called upon to read the intestines and identify future problems. Prophets have dreams about forthcoming problems, e.g. raids that are threatening and pending epidemics. Rituals on the one hand can be based on the direct observation that people are sick and/or livestock are dying and on the other hand may be instigated by a worrying forecast in an oracle. On the household level, rituals may be a short one or two hour affair. More often rituals conducted for the welfare of the household will take the form of a *kikatat*, a ritual which engages all the men of a neighbourhood who, so to say, unite spiritual forces to thwart off negative emotions, curses, or witchcraft directed against one of them. Disasters with a regional scope demand activities by the entire community. Senior men from different neighbourhoods meet, frequently in special ritual places, stay together for several days, purify themselves and the country and conduct prayers to bring about rains (i.e. to end a drought).

5.5.1.1. Oracles: From Reading Intestines to Prophetic Visions

The art of throwing sandals in such a way so that they produce reliable forecasts about pending problems or give causes for existing ones was learnt from the neighbouring Turkana (see also Best, 1978:160ff). According to Pokot informants, the best Pokot *kipkwegh* (the specialist handling the sandals) are still found in the Tiati and Silali regions in close proximity to the Turkana. The capacity to conduct a sandal divination is not tied to a specific clan and can be learnt informally by frequently watching such prognostications. The art of reading the sandals is more a matter of experience and technique than of magical power. The divination is simple and involves little input and as such is suitable for everyday use. A *kipkwegh* will first ask his client about any present problems. Then he blesses his sandals by spitting on them and throwing them up in the air. While throwing them up he asks questions such as "was X cursed?" or "is the goats' disease in X's herd the result of quarrelling between X and Y?" The position in which the sandals fall back holds the answers to these questions. Obviously it needs expert knowledge to read the sandals adequately. The sandals can give information about household problems and broader social problems. A *kipkwegh* may forecast a raid or a livestock epidemic as well as human diseases or individual misfortune.

Reading intestines is mainly used to diagnose the spiritual base of problems or to forecast the future. It is a form of divination that is used frequently and every elder has some basic knowledge in reading the intestines. Again the capacity to read intestines is attributed more to experience than to spiritual power. The intestines are read like maps. An entire cultural landscape is projected onto the intestines: there are rivers, mountains and households, there is a home range, a peripheral area and the land of enemies. The specialised vocabulary of reading intestines is complex. Almost every animal which is slaughtered at home will be read and certainly every animal slaughtered in a communal ceremony will be analysed. Immediately after an animal is slaughtered the stomach and intestines are taken out of the animal. The intestines are then ordered in a special way. Each part of the intestines represents a specific part of the Pokot territory. One sector of the intestines provides information about the future fate of the household, from which the slaughtered animal originates. The outer wall of the intestines has veins which may either be swollen with blood and red or without blood and whitish. Furthermore dark and reddish spots are frequently found in intestines. The colour and course of veins, their position in relation to each other and to these spots is what is read. If for example the veins in the part of the intestines representing the Amaya river are reddish, this indicates a pending attack by enemies. The part of the intestines representing the household can be read in a very differentiated way. Each and every person in the household is likened to specific dots. If one of these spots is red, it is said that a specific household is in danger. Further questions may be asked to the oracle to clarify exactly where the danger is coming from. The following text gives an impression of the way intestines are read.

Reading intestines during a sapana ceremony in Tilam, October 1988

First Speaker: "It (the blood vessel) reached there in the bordering area, and the stomach asks 'what is this thing, Trypanosomiasis, what kind of disease is this?' This ox which was slaughtered here says 'From over there in Naudo down to Nakoko, covering Mt. Silali and covering Kapedo'. Well, it is difficult to identify clearly, but the stomach says 'there is danger'. What kind of danger is this? The cow says 'now my friends, from Kapedo down to where I am, there is danger, danger, the danger of coughing'. This ox says that way. Up there on Leroghi, covering the whole area, it is as if it is coming this way towards the *mbochö* (a part of the intestines). The stomach asks 'are the cattle frightened or what?' No, it is not that way. The *mbochö* is twisted and goes beyond as if it is coming towards this second *mbochö*. It becomes the *alepit* (other part of intestines) and this *alepit* says: 'I have quarrels only, there is coughing.'" ....

Second Speaker: "Bring all words together now so that this ox will give us answers to our questions by reading his stomach. I heard that this ox came from down this side, you should send a boy soon to slaughter a small goat in that house the ox came from (in order to get clarity). However, there is nothing (bad) for you to fear. Therefore the words of this ox have come to an end. The other ox says: 'I, this ox, talk about the whole country'. That ox covers all homesteads. The intestines are filled with *asyokantin* (small reddish dots in the intestines). Nobody knows if they stand for cars or something else. What are these things covering all the intestines indicating? Cars have gone up to that place where the homestead is. But I doubt that ox over there showed something different. And that ox now, it does not show any problem. So what do these cars mean? What are the roads for? And what are the ropes for? What is this all for, my friends? These cars are like the ones the government came with last time when they came up to Chesitöt. We have only to follow those words which we have seen last time, and what are these cars now, are they not those which were going round in the area? Look, now those cars went down this way and then all came up here. The vehicles passed everywhere. Those two oxen said these things. Therefore we brought a goat (to clarify these issues by reading the intestines of the goat). This goat then swept away all bad omens which were suggested by those cattle. This goat said 'I am good luck, good luck all over, covering the entire country'. My friends, these are the ways the intestines of cattle are sometimes confusing people.

The problems discussed via the intestines are livestock diseases and governmental taxation. The two specialists are responsible for reading one ox each (two were slaughtered at the celebration). Results are discussed and as questions remain unresolved and the forecasts from both oxen are to some extent contradictory, a small goat is slaughtered. The reading of the goat's intestines finally brings clarity.

The *liokin* is a third ritual specialist who is specialised in exposing witchcraft and bad words (*kutïchi*) - a generic term subsuming slander, curses and witchcraft. In contrast to specialists conducting divinations, the *liokin* can diagnose a problem as well as heal a disease. He is working with sacred coloured earth (*muntin*) and other magical devices. While sandal divinations and inspecting intestines are everyday rituals, the work of a *liokin* is highly specialised and has to be paid for.[25]

The prophet is yet another expert who forecasts problems, and probably the most potent one. In many Nilotic groups, prophets are in prominent positions to safeguard their people and to warn them of dangers ahead. Occasionally such prophets became important as political leaders, in organising raids against enemies or co-ordinating resistance against colonial troops. The Maasai *laiboni* were essential in unifying forces in interethnic conflicts (Berntsen, 1979; Jacobs, 1979), the Nandi *orkoiyot* organised a drawn-out war against the British (Matson, 1972), and the Turkana prophets (*emuron*) sent armies of several thousand men against British forces (Bollig, 1987a). The Pokot prophet is less involved in day to day politics and in the actual organisation of raids. His name is not given to outsiders and everything dealing with him is *kablawach*, secret. If people talk about him in public they give him cover names like "the child". He rarely, if ever, speaks up in public himself in his function as augur. Usually men are sent to his homestead to inquire about his dreams. These men, whom he then informs about his ideas, take the message back to their communities. Nevertheless, his words are listened to carefully: if the prophet says that an area must be evacuated because a raid is pending, usually this will be done. In September 1988 he was asked if preparations for the massive circumcision ceremony (*tum*), in which some 11,000 men and boys were to be circumcised all over the country, should go ahead, or if the Turkana would take advantage of the situation and raid as soon as they found that all the Pokot young men were disabled. He sent them back with the message that no raids were to come in the near future and that as a precaution a grey buck should be slaughtered in a place right on the border of Pokot and Turkana. The person slaughtering the animal should be a young, as yet uninitiated boy from the Kiptinkö clan who would have his initiation ceremony later that year. Every major communal activity is presented to him before it takes place: raids, age set celebrations, circumcision.[26]

5.5.1.2. Individual and Household-based Rituals

Individual magic is used frequently to counter threats to personal health and wealth. Whenever something falls or is broken unintentionally people will exclaim the name of a prominent ancestor. It is believed that disgruntled ancestors may cause all kinds of minor trouble. However, ancestors are usually thought of as benevolent spirits which roam the vicinity of their earlier homesteads. They usually just need a sip of milk or a crumb of tobacco to be satisfied again. There are no major rituals which are devoted to ancestors. There are ritual specialists who produce amulets that protect against witchcraft and the evil eye. Protective magic, however, mainly originates not from objects but from the power generated by communal, family or community-based rituals. There are a variety of rituals which are conducted by the household in order to fight problems such as human and animal diseases. Minor rituals just involve the household members while larger rituals engage the entire neighbourhood.

kïrïset: a household-based ritual to thwart off livestock diseases

The *kïrïset* ritual is applied as an ad hoc aid to cure sick livestock - before giving any medicine to diseased animals or taking care of them otherwise. These ad hoc rituals involve just a few people. They last less than an hour and are thought of as a kind of initial help. They may be followed by other rituals involving more people. The following ceremony (*kïrïset*) gives an example of such a household based ritual:

> Wasareng's goats were *kitongu*, i.e. they were generally in a bad state and goats' kids were dying in numbers. Some goats were sick with severe diarrhoea. Wasareng then decided to conduct the *kïrïset* ritual. The ritual is aimed at getting rid of diseases and is conducted whenever a disease is ravaging the herd. It is also aimed at every disease brought about by envy (*ngatkong*).
>
> Wasareng first collected branches of various bushes and trees, specific for the *kïrïset* ritual: *sitöt* (Grewia bicolor), *tontolwö* (Hermania kirkii)[27], *kelkeleyan* (Barleria acanthoides)[28], *lopot* (Solanum dubium, Solanum incanum), *adomeyon* (Cordia sinensis), *kïkïchwö* (Premna resinosa) and the grass *chemnganya* (Cymbopogon caesius).
>
> Then a circle in the middle of the enclosure was cleared. The dung was removed for practical reasons: the fire which was lit could otherwise easily set the dry dung on fire and destroy the entire enclosure. Just before the fire was lit, Wasareng went into the enclosure with his two spears. Clinking the spear blades against each other and symbolically thrusting them against the diseased animals, Wasareng moved through the enclosure saying words to ensure that the "bad words were cut".
>
> When this part of the ceremony called *kilyokat* was finished, Wasareng started the *kïrïset* proper together with his son Teta. A fire was lit with all the branches of different trees as collected before. The goats were then driven around the fire many times, until the fire burned out. Nothing is said while driving the animals around. It is the smoke of this purifying fire which cures the animals.

There are several other rituals similar to the *kïrïset* described above. They are all based on similar combinations of protective magic. First the problem has to be identified: In the example above the goats became sick without any obvious cause. There was plenty of pasture, water was near and herding took place at a regular pace. According to Pokot ideas on disease and mischief the problem is sought in some past conflict between a member of the household and somebody else. The actual perceptions of the origin of the problem may still be vague. While the Pokot like to explore the possibility of various past conflicts having an impact on their household's welfare, they shy away from naming persons or from giving a clear-cut story where a culprit is identified. In the ceremony they may talk about the bad words of a man or the envy of an old woman. The *kilyokat*, the clinking and jolting of spears towards the object to be protected, symbolically cuts off the power of bad words and envy. The *kilyokat* precedes other parts of the ceremony which will only have an effect if the evil impact of spiritual aggressions against the household is eliminated. The symbolism inherent in the *kilyokat* is clear. The "bad words" are "threatened" by the determined community of men and are "chased away by sharp weapons". The part of the ceremony following the *kilyokat*, the *kïrïset* proper, is thought of as curative: various trees are burnt which are said to bring good health to the herd. The smoke purifies the animals from the spiritual pollution originating from "bad words". A ritual like the *kïrïset* is only thought to have limited power. If the ritual does not show any effects, a larger neighbourhood ritual has to be organised.

Kikatat: A neighbourhood-based ritual curing human and livestock diseases

When reasoning about diseases a culprit will usually be named (not by a personal name but in allusion) who has had bad feelings towards the owner of the animal. The accusations vary in intensity and openness. There is rarely much reasoning about a single goat affected by foot-rot; however if the entire herd is affected rumours will spread quickly as to who is thought to have caused the problem. Serious diseases are treated both physically and magically. Frequently a herder will ask the men of the neighbourhood to help him conduct a *kikatat* ritual and the elders of his community to say a blessing for his sick goats, cattle, sheep or camels. The *kikatat* is a ritual especially designed to ward off the malevolent effects of personal enemies' curses and witchcraft. The herder whose livestock is suffering calls upon the men of the neighbourhood. He invites them to come to his home for a ritual to defend his sick animals. The animals are then placed in the middle of a semi-circle of men who are ordered according to their generation-set. Again the ceremony starts off with a *kilyokat*: Each man brings with him two spears. Both spears are then jolted towards the sick animals in the centre of the semi-circle. Whilst clinking the spear-points rhythmically against each other and towards the sick animal, the *kikatat* verses are said. The spear points symbolise the collective defensive power of men. Their strength and their will are dramatically enacted in the *kikatat* ritual (cf. Bollig, 1995a).

The blessing cited below was spoken by one of the eldest men present on such an occasion. His blessing was meant to chase away foot-rot (or perhaps rather the metaphysical causes for it) from the herd of Ptïkom, a middle-aged herder.

Apuriang's Blessing at a kikatat

(The slashes in the first lines indicate where the main speaker stops and the chorus of men gathered in the *kirket* repeats what has been said. The entire text is spoken in that way.)

"A white cow/ What did she say?/ How did she chase away the bad words?/ Gather all here./ Why do you gather here?/ For the goats' wellbeing./ Let us ask God,/ to give us many goats/ and many children. We have heard/ that foot-rot said/ 'I am going,/I will leave these goats for you/. I am going/ I will leave these goats to stay in peace./'

We ordered the goats of Nginyang, to leave foot-rot in the bush. Who said that? The circle of men said that with much emphasis. The council of men said, that the land up and down shall be anointed. The smell of fried oxen flesh fills the air, so much, so much, so much.

These bad words are gone; there is no foot-rot anymore. These men here chased the disease. Yes, it is gone. When it was ordered to leave the homestead, it fled immediately.

We bless the goats of Ptïkom, who told us yesterday, that his goats are sick because of an ox. We told that disease: 'Go away! Come to an end! Disease, which befell Ptïkom's goats, come to an end!' Even if this man hid away his oxen, when men came to ask for it, these goats shall not suffer because of this. Whose goats? Whose goats? Ptïkom's goats. ... Foot-rot has been chased away, hence Ptïkom's goats got rid of the disease. We have heard that this thing went, this thing that caused whose disease? That caused the disease of the goats of this homestead. When the council of men said 'go away' and 'leave this place' it was chased away.

We have said, that Ptïkom's goats shall be free from the disease, as the council of men said 'Go away'. Where did these goats go to? They went into the bush! This night rain will wash away the disease of the goats. Wash away from whose goats! From Ptïkom's! And from whose camels? From Ptïkom's camels!

We told the disease of the sick camels 'This disease shall come to an end. Sneeze out that disease. Sneeze it out. Sneeze it out!' Even if the foam at the mouth gets cool, disease 'Come to an end!' The bad foam making camels sick was left behind in the bush. This thing shall stop. The council of men says 'Hey! hey! hey! This disease affecting the nostrils of camels.'

My friends, we tell this disease that affected the camels' eyes 'Come to an end! Finish at once!' Shall Ptïkom, who talked about the disease of his camels, be rendered a liar? The council of men chased the disease of Ptïkom's camels.

We told Ptïkom who was so scared because of his sick livestock, to go home and to come back tomorrow with better news. Yes, we, the

> council of men, say that this disease befalling the livestock shall come to an end. Where shall this disease be chased to? To the far end! To the far end! To the far end! Go away, go far away!
>
> We order Ptïkom's camels 'Be cured, be cured, be cured through and through!' They shall be blessed, shan't they? Shall they not be blessed? Shall they not be blessed? Ptïkom, if you go home now, you will see that your animals have been cured.
>
> Be blessed! Be blessed! Be blessed! This house! Whose house? Ptïkom's! We have defeated the disease! The disease, who told us of it? The disease Ptïkom told us of. Be blessed! Be blessed! Be blessed!"
>
> (Until here all people present repeat the words of Apuriang, the following words Apuriang speaks alone.)
>
> We say now 'Stop now!', these words, the words of the council of men are powerful, this council which has been purified which is sitting on clean ground. We chased the bad things which were in our round. Which are where? Which are in the council of men. Whose council? The council of the Pokot. We are powerful and chase this thing. And what was said? It was said 'Vanish to the far end side'.
>
> We say now! We say 'young men, stay in a safe place, and herd the cattle, herd the cattle, herd the cattle'. All bad words shall be forgotten. And it will be only the herding of animals which matters. Herding! And what else? The *chepkïliny* bird shall fly in front of the herds. The council of men tells you young men: Be clean! Be clean! Be clean! Be without any bad words! Be blessed! Be blessed! Be blessed!

In order to understand the structure of the blessing, it is necessary to summarise Ptïkom's problems: many goats of his herd suffered from foot-rot and quite a number had become very weak. Ptïkom's camels suffered, too. No specific disease is named but we learn from the blessing that they had a 'bad foam' on their mouth. Ptïkom just had a vague idea about who heeded bad feelings against him: in the dry season men from his neighbourhood had approached him to offer an ox for a celebration. But he had refused to offer them anything. The men went home hungry and, probably, in an extremely bad mood. Perhaps one of these men had caused Ptïkom's problem.

The senior saying the blessing mentions all aspects that are related in one way or the other to Ptïkom's problems. The rhetoric of the blessing is characteristic for this type of oral literature. The disease is personified and hence can be spoken to directly. Several times it is ordered to leave the body of the sick goats. Occasionally Apuriang states that the disease has left the body of the animals already, at other times he still orders it to do so. In this magical context such statements are not contradictory. Both sentences imply that the animals are without disease - right now and in the near future. The emphasis on a disease-free future physical state is thought to have direct impact on the factual state. The council of men is the agency invoked. This institution has the spiritual power to counter the disease forcefully by choosing powerful words for the blessing. The *kirket*, the council of adult men, is taken physically and socially as a moral space, from which blessings aptly develop their impact.

A *kikatat* is frequently held for sick people as well (cf. Bollig, 1992a:188f). The ceremony does not change much in content and symbolism. Instead of animals the sick person or persons are seated in the semi-circle of men. Spears are jolted towards them and all men present join in the staccato of the words said by an elder.

5.5.1.3. Community-Based Rituals

The Pokot believe that neither family-based nor neighbourhood-based rituals can counter droughts or livestock epidemics of a wider scope. Such problems have to be handled by the community at large, represented by their elders. The following two case studies of rituals represent two community-based efforts to counter threats to the Pokot nation - in the first case a livestock disease and in the second case a drought.

A ritual to ward off a major livestock epidemic - the putyon ritual in Kadingding

The Pokot prophet had dreamt about the outbreak of a serious goat's epidemic. In order to counter this threat he ordered that three goats with a specific coloured hide had to be slaughtered. On the 13th of October 1993, many senior men gathered in the main meeting place at the dam of Kadingding. For a long time they discussed who should donate a goat and who should spear it. The first person mentioned as a potential donor initially refused and maintained that he did not own a goat of that specific colour. After the group put more pressure on him he relented and agreed to bring the goat, but he requested that whoever speared his goat should return a young female sheep the same day (a very unusual request). Several men denied that they had any young female sheep - as unbelievable as the initial request was unusual. To solve this deadlock in discussions one senior man who had spoken a lot in favour of the ritual finally agreed to return a sheep for the sacrificial goat. The question who should contribute the second and the third goat was as difficult as the solution for the first one. Important in all three cases was the issue as to who was to spear the goat and thereby accept the obligation to repay an animal to the owner. All three men who finally accepted stood up in turn and made a small speech commenting on the reasons for their apparent generosity. However, due to several other difficulties the ritual only started eight days later and finally took place in Kadingding from the 21st to the 22nd of October 1993.

Around noon, it looked as if the ritual had to be postponed once again. But after some elders put pressure on the group, things developed swiftly. At about 2 p.m. the group gathered at the meeting place in Kadingding. At about 2.30, the group was growing rapidly in size and an elder asked one of the youths to get a bird's nest for kindling the fire. Some boys were sent to gather branches of *lopotwö* (Solanum dubium) and *kisipö* (probably Endostemon camporum)[29]. A few elders set about making sacred fire-sticks (*pighun*). Then the ritualised fire-making began. The sound of the crowd suddenly vanished into a whisper. People were moving about quietly and solemnly. A man from the Koimö clan and somebody from the Tul clan began making the fire with the sacred fire-sticks. Tudokomol, whose mother was from the Koimö clan and who accordingly had the right to light sacred fires, helped them. Each of the three men made his own fire, drilling the fire-sticks long enough to kindle the grass of the bird's nest. Then the three small fires were united. The order in which certain wood is put on the fire is fixed: first the rest of the bird's nest are burned, then wood from *kïkïchwö* (Premna resinosa) is put on top, then wood from *adomeyon* (Cordia sinensis), then branches of *sitöt* (Grewia bicolor), and at the end a pile of branches from *lopotwö* (Solanum dubium) and *kisipö* (probably Endostemon camporum).

When the fire was set alight some of the elders were still missing. Boys were sent to fetch them. At the same time the elders realised that the sacred white and grey earth (*munyan nyo rel* and *munyan nyo miryonto*) were missing as were the two bowls (*otupö*) to mix the substances. Again young men were sent to fetch bowls and sacred earth colours.

At about 3.20 the sacred earth arrived, packed in bundles of grass. The men immediately started with their ritual work. Ten minutes later the first goat was speared. Spearing was done as in other rituals: the goat's right hind leg was held with the left hand. The spear, held in the right hand, was thrust into the right side of the animal. The difference to other rituals was the way the spear was thrust into the animal: the spear was first thrust in such a way that it went right through the ribs into the body of the animal; the spear was then pulled out only to insert it a second time through the gaping wound. Then the blade of the spear was put on the wound as if one wanted to suppress the flow of blood. After doing so the blade was inserted a third time into the wound. Finally the dying animal was laid down in such a way so that it was facing the sacred mountain Mtelo. At the same time the leader of the ritual, W., blessed the place and all participants with water. Slowly and in an anti-clockwise direction he moved in the sacred semi-circle (*kirket*) of men, spraying water with his hands. An assistant went behind him and continued to pour water into his hands. This initial blessing took several minutes (probably because W. was slightly drunk). Apuriang, the oldest man present, stood at the upper apex of the *kirket* and blessed the place and participants.

While the goat was being speared, the mixing of the sacred earth was interrupted for a short time. At about 3.40 the sacred semi-circle (*kirket* and *kwen*) of men was prepared (see Photographs 12 a,b,c). The *kwen*, the centre of the sacred semi-circle and the semi-circle itself were covered with branches of *tuwöt* (Diospyros scabra). As in other important rituals, the two main ritual leaders, who were later to say the main blessings, started by blessing each other. Both of them took contents from the goat's stomach (*eghyan*) and smeared it on their breast. Only then did they start blessing the others. Spitting into their hands, they first took from the greyish earth and then from the whitish earth. Everybody was blessed first with *munyan nyo miryonto* (greyish earth) and then with *munyan nyo rel* (whitish earth). The order in which body parts were painted with earth was always the same: first the breast, then the head and finally the legs were anointed (this was the same for both colours, grey and white). The greyish earth which represented the disease was then symbolically washed away by white coloured earth. W. who was the main figure in the set-up and who had made his blessings first, stood directly behind the two elders giving out the blessings. He supported their words with a murmured blessing which, freely translated, means 'be blessed' and he uttered the name of a prominent ancestor. Immediately after this blessing the distribution of meat began according to the rules of meat distribution in the sacred semi-circle.

At 5.00 p.m. a non-castrated young ram (which was brought about an hour earlier and was then tied to a tree), was slaughtered. While eating the meat some speeches were given in the sacred semi-circle. They were not fiery. In contrast to other occasions, e.g. the typical age-set celebrations, the speeches emphasised harmony and common goals, rather than conflicts between different sets within the society. Nobody talked, held or swayed his spear - a gesture which in other cases underlines determinacy and power and is normal in such a context. Meanwhile about 40 men from different age-sets were present.

Photograph 12. a, b, c. The *putyon*-Ritual at Kadingding against a Goats' Disease

Photograph 12. cont'd

At about 5.40 the meat of the ram was distributed and A. spoke the first major blessing. While speaking his blessing, A. first moved around and then out of the semi-circle, walking about fifty metres away while constantly continuing his blessing in a loud voice. Symbolically he was taking the disease away from the group gathered in the semi-circle and, as it were, walking it out of their personal sphere. The blessing corresponded in content and form to the blessings recorded at other instances when diseases were “chased away”.

At about 6 p.m. the group slowly disintegrated. Some men went back home but the elders, stayed behind and slept that night at the meeting place. Tea was prepared for them by two youths. Then the men talked in small groups.

In the morning the first men woke up at about 4.30 a.m. and started talking. At 6.00 in the morning the third goat, a small white non-castrated goat, was slaughtered. Some fresh leaves from *tuwöt* (Diospyros scabra) were used to “renovate” the sacred semi-circle. This last goat was not thrown onto the fire still in its skin, but was carefully skinned beforehand. The skin on both hind legs, on one foreleg and a strip of skin on the stomach was left. A. who speared the goat himself (in the same way as the goat and the ram the day before) opened the stomach of the goat and blessed himself with the stomach contents. At about 7 a.m. the blessing with the sacred grey and white earth began anew. As on the previous afternoon, first the grey earth was applied and then the white earth. The blessing was conducted in exactly the same way as the day before: first the grey earth was painted on the breast, forehead and legs and then the white earth was applied in the same way. Then each man was given a bit of white earth to take home with him and it was seen to that at least everybody received some of the sacred earth. Only a few elders took some of the “dangerous” grey earth. Everyone took some white earth back home so that they could bless their entire family with the coloured earth later in the day. Only four days later the men met again and washed off the earth together.

At 8.00 the skin of the white goat was cut into strips (*putyon*) and everybody was given one of them. These strips were tied around the neck, on the upper arm or around the wrist. Some men took strips back home and tied these onto the neck or leg of the goat which usually leads the herd and the one which habitually strolls at the back of the herd. A short time later the gathering disintegrated.[30]

The ritual involved about 40 men. It was clearly more extensive than any neighbourhood-based ritual. Within the ceremony the gerontocratic order of Pokot society was re-enacted. The sacred half circle of men was prepared with a lot of care to provide a purified ground. The elders sat at the apex according to their age while younger men sat on the sides of the semi-circle. The left side of the semi-circle was taken by men of the Ngimur alternation while the right side was occupied by the Ngetei alternation. The ritual symbolically emphasised peace between age and generation sets and enacted an ordered world. The oldest men of the community said powerful blessings on behalf of the community of men. The blessing with grey and then white coloured earth symbolically removed the danger of disease: white, symbolising peace, health and harmony, displaces grey, symbolising disease, sin and bad luck. The order in which the goats and sheep were slaughtered adhered to a specific colour symbolism: the first goat that was slaughtered was grey (*miryonto*), the colour symbolising ticks and diseases and relating to ticks and *ngwur* (some sort larvae nesting in the fleece). It was followed by a white goat, from whose skin strips were cut. The white colour removes disease and disharmony and protective amulets are produced from the white skin. Each man receives a few of these strips to wear himself or to fix to one of this animals. The ritual produces the magical means to protect the small livestock herds of the Pokot and at the same time emphasises solidarity and common identity.

Rituals to ward off a major drought - the saghat and mwata rituals

Next to major communal rituals directed against diseases there are rituals that work against drought and famine. These may be instigated by the elders of a community or by the prophet.[31] The community of men, led by their most senior elders, prays communally for rain. According to Pokot concepts of the cause of disasters, drought is a consequence of social disorder. Consequently, rain-making rituals first aim at the identification of social conflicts and then at the recreation of peaceful relations between individuals and social groups.

The rainy season of 1993 had not been good and by September it was feared that the next rainy season could be worse. At that stage elders decided to conduct a ritual to ensure early rains. The ritual itself consisted of two parts: the *mwata* (purification) ceremony which took about two days and the *saghat* (prayer) ceremony which took place about ten days later. The *mwata* ceremony is meant to purify all people taking part in the ritual, and to transfer them into a state of spiritual cleanness (*tïlïl*) before the actual prayers for rain start. Again purification identifies sins and conflicts and harmonises social disorder. Prayers for rain are already said at this occasion. The *saghat* ceremony is devoted solely to prayers for rain. The ceremonies were very similar in style and content.[32] For both ceremonies the same place, Nakurkur, was chosen. Nakurkur is a small (only about 100m in diameter) green spot in a landscape of solidified black lava at the foot of the extinguished volcano Mt. Paka. The sharp dark silhouettes of the rock stand out in magnificent contrast to the green grass and the high trees of Nakurkur itself. The elders stressed during their speeches that the ground and the rocks of Nakurkur had often been soaked with the blood of sacrificial oxen.

The *mwata* ceremony - purification and harmonisation: Ritual activities for the *mwata* ceremony started on the 23rd of September in the afternoon. About fifteen men, among them several senior elders of the Chumwö generation set and senior persons of the Koronkoro generation set and eight fellow Koronkoro gathered together. Wasareng, who had already led the *putyon* ritual in Kadingding, was chosen as the person to guide the ceremony. Additionally the seniors had selected four young men to accompany them and to help them with the practicalities of living in a bush place for some days (i.e. fetching water, fetching firewood). On the evening of the 23rd the first goat was slaughtered, speeches were made and a few blessings were given which were meant to ensure the welfare of the people present.

Speeches throughout the first and second day did not deal with the drought at all. Every speaker focussed on identifying social conflicts. One elder spoke about the latent conflicts between the three present generation sets. He criticised the members of the junior generation set Kaplelach of "isolating" themselves from the elders and accused them of haughty behaviour. Another elder emphasised the problems adultery brings about, especially when adultery is committed by young men with the young women of senior men. Yet other speakers challenged specific persons or criticised unfavourable political conditions in their speeches. Blessings were said to ensure good luck for the men donating goats to the gathering and to guarantee a smooth conduct during the ceremony.

Overnight some younger men went back to their respective homesteads and in the morning only a few elders were present when the sacred *amurö* meat was distributed and eaten. Otherwise there were few activities related to the ritual that morning. Some young men were sent to fetch millet beer. The men then spent the rest of the morning carving sticks and sipping beer. The remaining parts of the sacred *amurö* meat were distributed and alongside the distribution some blessings were made. Then everybody had a prolonged siesta.

In the late afternoon a junior man was picked to supply a goat for the evening. The goat was later driven to the place within a small herd as it was prohibited to bring sacrificial animals alone to the ceremonial place. The goat was then swiftly taken from the herd and speared by an elder. He held the goat with his left hand and thrust the spear into the right side of the animal. The corpse of the dying animal was, as usual, turned westwards towards the sacred mountain Mtelo. The corpse was then carried into the middle of the sacred semi-circle (*kirket*) and opened up. The entrails were carefully cut out and experienced haruspexes started reading the intestines. Fortunately they did not find anything disturbing and the ceremony was allowed to continue. The goat was then thrown onto the fire by two junior men with the skin still on. After a couple of minutes the sacred *alamachar* meat was cut out and given to the two most senior men. Parts of the intestines were given to the elders according to their rank in seniority. The Chumwö seniors called upon some Koronkoro to assist them with the preparation of their part of the intestines for the fire. Carefully the contents of the intestines were pressed out. The well prepared parts were then given to the junior men in charge of the fire, whilst reprimanding them not to confuse the single parts and to ensure that they go back to the right senior after roasting them. The preparation of the intestines reflected the re-enactment of the social order: from the sacred centre of the meeting place the intestines were given to the elders of the Chumwö generation. They called upon the next junior set to assist them in preparing the intestines. After cleaning the parts they in turn called upon members of the youngest generation set to roast the intestines.

In the meantime and parallel to the distribution of meat, Wasareng started to bless all the participants at sunset. A pot with water was put on the stomach contents in the middle (*kwen*) of the sacred semi-circle. The area was prepared with a layer of branches on which the stomach contents, intestines and the water container were placed. The corpse of the goat was lying at the ritual centre of the semi-circle, the *kwen*. A spear with a sharp blade pointed at the goat. Four Koronkoro men placed themselves in front of the water container. With a bundle of *adomeyon* (Cordia sinensis) twigs Wasareng, the ceremonial leader, sprayed water on the chest, forehead and legs of the four men. Then he took his spear and led the men once around the *kwen*. Walking, or rather marching, they sang hymns glorifying the deeds of past generations and exploits of recent raids and idealised the herding of livestock. All the men joined in these hymns and stood up to join the five elders walking around the *kwen*, holding their spears in their right hand. In front of the entrails each man stopped short, took his spear into the left hand for a moment, took up a bit of the stomach contents with his right hand and smeared it onto his chest, rubbing the stomach contents from the upper part of the chest to the lower part. After all the men had anointed themselves with the stomach contents and had thereby established a purified community, they sat down again and the preparation and distribution of meat started. While they were eating, each participant saw to it that each and every bone was broken, as non-broken bones would inevitably bring trouble and bad luck in the future.

More blessings were said after eating. The person who donated the goat and the person who speared it were placed on the stomach of the goat at the centre of the ritual place and blessed. A senior elder said the blessing for them and led them once around the ritual centre of the semi-circle. Another elder said the central blessing. He implored past sacrifices which had brought about rain and implored water and lightening to arise from the nearby lake to bring rain. He ordered the land "to become soft".

Todokin's blessing spoken on the 23rd of September 1993 in Nakurkur

"I order the goats now,
you shall find, you shall find, you shall find
the goats shall buy a cow
and the cattle shall buy camels
and the camels shall buy donkeys and that is four
and the cattle shall fetch wives and that is five
(part where the two men who donated and speared the goat are blessed)
I order Woywoy (Lake Baringo) now,
produce lightening
this is what I say, I, the son of an elder
what do you want?
you, lightening, arise from the island of Kokwö,
arise and look around,
look at the stars, look at the clouds,
bring down the clouds, bring them near
I speared the buck Kipköröy in Njemps country

where you, lightening, are so near
and the Njemps elders blessed me in Lokeriyo when they still wore beads on their necks
I order you, water of Lake Baringo, arise, arise, arise, arise
arise because of that buck Kipkoröy,
which I speared for whom?
for the Njemps in Lokiriyo and in Natir
all elders, including Naur, all elders living in Rukus
facing what?
the sacred mountain the Njemps pray to,
that sacred mountain shall help the cattle
I order that mountain, be clean, be clean and come together with what?
with the clouds and what else?
at Mt. Kenya and where else?
at Tiati and where else?
in Kisima,
arise moutain and open the armpits, the armpits, the armpits
open them now as
I slaughtered an ox right here in Tilam
the ox of Pumpïr when all those Maina where still alive,
they blessed me
elders like Kangolese and Rwatale blessed me, blessed me.
those who were responsible for the land Tamakaru and Karanja ...
I tell you ancestors, I tell you, be happy,
all those who ate that ox, you are still in this land
we were given the sacred stomach contents
those stomach contents which are still in the waters of Tilam,
the land shall be clean, the land shall be clean, the land shall be soft ....
I recently asked a Turkana to bless these two rivers,
to bless for me Napeliamachanit and Tapogh
and that man blessed both rivers
he was a Turkana whom we gave all the abolatin and the alamachar meat
I tell that Turkana now:
be sweet (let what you implied in the blessing happen)
what shall be sweet? your mouth
and who shall help? God
Let me slaughter a buck.
Let the blood flow on the land down to where? to Suguta
I tell you, you old Turkana, help us now, help us,
become soft (i.e. have more grass) mountain Silali,
have more grass mountain Paka
have more grass mountain Korossi
have more grass mountain Tiati
I say 'land become soft'.
Be blessed! Be blessed! Be blessed!"

Subsequently many more speeches were made: again the drought did not feature prominently. Everybody tried to identify potential conflicts or talked about problems he was involved in. Social problems featured as prominently as intra-familial conflicts. Everybody had the chance of saying what was on his mind and what he thought was contradicting a general situation of peace: somebody complained about his adulterous wife, another person about his sick goats and a third man claimed that he had been bewitched. In the evening at about 10 p.m., I fell asleep and when I woke up again around 4 a.m. men were still sitting around the fire discussing the problems of the land.

At 7 a.m. in the morning meat was distributed anew. At about 7.30 Wasareng was busy preparing the major part of the ritual. One junior had fetched a bowl with sacred white earth (*munyan nyo rel*) and a calabash with milk from a blessed *roriyon* cow. The *roriyon* cow is a young cow which has calved only once and never had an abortion. The bowl and the calabash were put on the left side of the stomach contents of the goat slaughtered the evening before. Wasareng began to mix the white earth with water and milk, making sure that he only made use of his right hand. Pouring milk slowly into the bowl, he kneaded the earth carefully for about 20 minutes. In the end he poured some water into the concoction to increase the quantity. A junior assisted him holding the bowl and waving off flies with a fan. After Wasareng had decided that the sacred white earth was mixed well enough, he anointed the calabash first. Then he tied a number of *adomeyon* (Cordia sinensis) twigs into two bundles which he intended to use in the subsequent blessing (which finally began at about 8 a.m.). With the bundle of twigs he sprayed white earth on the chest, forehead and legs of the participants. He started off with the eldest men, the Chumwö seniors. The oldest participant, Lotepamuk, blessed Wasareng. Both men, Wasareng and Lotepamuk then together blessed all the others. First Lotepamuk blessed everybody and then Wasareng did the same. The men then walked solemnly around the *kwen,* the ceremonial centre, holding their spears in their right hand and each person was blessed twice. After the second blessing another hymn (*akiwiyar*) was sung. Under the guidance of the two blessing seniors, the men walked off about 15 metres towards the east where the morning sun was gradually rising. They marched up to the boulders of solidified lava where Lotepamuk said another blessing. Then they all moved back and sat down again in the sacred half circle. More speeches were made imploring the unity of the group. The gathering dispersed at around noon.

The saghat - Prayer for Rain

On the 2nd of October, about ten days after the indispensable ritual of purification (*mwata*), the actual prayer for rains (*saghat*) followed. One elder donated a camel and an ox to furnish the gathering with sufficient meat as slightly more men had gathered than the last time. While the focus of the last ritual had been to purify all men from sin and bad feelings towards each other, the *saghat* was meant expressly as a prayer for rain. There had also been prayers for rain during the *mwata* ceremony. In substance and style these were similar to the prayers spoken during the *saghat*.

On the first evening the obligatory blessings with stomach contents took place. Additionally all the men were blessed with white earth (*munyan nyo rel*) which was mixed with some bark from *lekititwö* (Barleria eranthemoides)[33], *cheptuya* (Euclea divinorum, E. racemosa) and parts of *siyoyowö* (Ficus dekdekana), *mokongwö* (Ficus sycamorus) and *kowontö* (Cymbopogon caesius). When all the elders had arrived in the evening on the 2nd October, the meat distribution followed, but most of the meat was eaten the following day. Part of the meat (parts of the breast) was taken to *serim*, to the borders of the Pokot territory, to places which were constantly endangered by enemy raiders. The sending of meat to these areas symbolically emphasises the ties between the central parts of Pokot land, where elders live and homesteads are situated, and the endangered fringes of the territory, the prototypical sites warriors live in. A short while later another ritual was to be conducted at *serim*, at a place called Kokwönyongi. Three oxen, one for each generation, were going to be slaughtered there. It had to be elders from the Kiptinkö and the Talai clan who were in charge of that ritual. Elders from these clans conduct rituals in border areas, which are deemed as dangerous and beyond the control of elders: Kiptinkö pray in times of danger related to violent conflicts; Talai prayers are directed against enemies and drought. Preferably it is animals from the Kiptinkö which are to be slaughtered, as enemies can never cross a place where a Kiptinkö ox was slaughtered in a ritual.

Very early the next morning all the men met in the semi-circle again. In an interview a few days later Wasareng gave the text of a song which was sung first just before the main prayer asking for rain was said:

lekititwö, lekititwö, aye lekititwö kiropan ilat
oh lekititwö (Barleria eranthemoides), lekititwö, aye lekititwö let it rain

At about sunrise the leader of the ceremony, Wasareng, started the prayers. He stood in the centre of the sacred half circle facing east, towards the sunrise, as he prayed.

oh yo oh yo /paponcha/	oh, yo oh yo our father
oh yoho /ilat/	oh yoho rain
ikonu peruru	give us blessings
ingara kech	help us
ikonu kalya	give us peace
ikonu arupe	give us showers of rain
ingarach tuka	help our cattle
ingarach nekö	help our goats
ingarach poyï	help our elders
ingarach kore	help our land

The other men walked around him four times and the song was repeated several times. In a chorus the men present repeated the words of the person leading the prayer. The prayer was conducted, as Wasareng explained later, in order to ensure that God would return the cattle which were lost in the past due to drought and disease and to prevent future droughts. While saying the prayer all the men were blessed with the concoction of white earth, milk and various plants.

After saying these prayers and making the blessings, the old men returned to their places in the semi-circle. Over the day more prayers were said.

Meinberg, a missionary, observed a similar ritual in 1977 or 1978.[34] He described the content and style of the central blessing in a *saghat* in detail: "At 7 a.m. a vessel, called *otupö*, is brought, some water is poured into it. The entrails of an ox (killed the same morning) are added to it and two elders sprinkle the mixture towards heaven in a gesture of bringing down the rain. Then all (male) participants are anointed with the stomach contents on shoulders, breast and stomach. The same happens to the stem of the tree in the centre of the meeting place. Afterwards a second vessel is brought with white earth from Mt. Paka. Again all participants facing east are whitewashed, face, shoulders, breast and stomach. The following prayer is said:

The cows moo, the calf is in the warriors enclosure and sleeps with the warriors
We implore you elders: you Lengotum, you Akanichum, you Luwyalan and you Kamoto
You, obtain for us, God
Pray for us to, God
You elders, who have been living in this sand (on this land),
The stomach contents of the cows,
Which are here in Kositei
Which are here in Murgurr, and where else?
In Kaduguyang (near Maron), and where else?
In Parpelo
Yes, those, which are in Nginyang
Which are in all the land
All obtain rain for us,
God, obtain rain for us,
(Obtain) blessings, something to eat, livestock, fresh pasture
Save us, livestock, save us
God, Father, save us
Livestock, I have obtained it
I have hindered, now, now, now,
It will rain, God, in the evening,
The day after tomorrow, forever
Be (our) father, who looks after us

God
Where?
Here
Be blessed! Be blessed! Be blessed!'

The ritual of purification, as well as the actual prayers for rain, were conducted by the entire community, represented by their elders. Both ceremonies are very similar in content and form. The gerontocratic order is enacted and played out dramatically. Livestock is slaughtered ritually. Blessings as well as prayers are said to ensure rain. The *mwata* ceremony gives room for discussing possible causes for the drought. According to Pokot standards, all evil is traced back to social conflicts, to adultery, to disrespect against elders and to the neglect of duties. Symbolically social order and purity is restored. White earth representing peace and harmony washes away grey earth symbolising sin and disorder. Only when these disharmonies are harmonised will blessings have an effect. The ultimate power to say prayers lies with the elders and all blessings and prayers are based on and derive their moral power from the consent and harmony within the group of men seated in the sacred semi-circle.

The symbols used in both rituals are similar. Plants are frequently imbued with a specific symbolic meaning: *lekititwö* (Barleria eranthemoides) is found in highlands, where there is more rain than in the plains. Its fruits have milk which is associated with the abundance of milk in the rainy season. *Lekititwö* is found near river beds, another association with water. *Cheptuya* (Euclea divinorum) is only found in the highlands, again indicating its close relation to rain. It has no thorns and symbolises softness and the absence of danger. Therefore people use it when they carry out purification ceremonies (indicating rain) and before raids. As there are no thorns, nothing sharp or dangerous will rise up against you. *Siyoyowö* (Ficus dekdekana) produces milk and is a highland plant too. *Mokongwö* (Ficus sycamorus) trees stand along river courses. These huge trees only grow where there is an abundance of water. *Kowontö* (Cymbopogon caesius)[35] has a good scent (like *siyoyowö*).

Colour symbolism plays a very important role in Pokot crisis rituals. Various types of coloured earth are regarded as sacred. They are dug in ritually important places. The most important place to find white coloured earth is the sacred mountain Paka, an extinct volcano. There, inside the crater men dig out white coloured earth. Only elders are allowed to possess coloured earth. *kapolok,* ritual specialists, known for their capacity for dealing with coloured earth and blessings, usually have several small sacks of coloured earth stored in a secret place in their main hut. Grey and white coloured earth (*muntin*) represent different states: grey is used to denote disease and conflict, whilst white carries the notion of peace, harmony and health. In the crisis rituals depicted above, attendants were first painted with the grey colour and immediately afterwards with the white colour.

Symbolically disease and disharmony are washed away by solidarity and peace.

5.5.2. Co-Opting the Ancestors: Himba Attempts at Reducing Uncertainty and Reducing Hazards

Himba ideas on the supernatural causation of disasters are more vague than those of the Pokot. There are no specific rituals to make rain or to prevent livestock diseases. Personal trouble or disasters that befall only one household are usually connected to strained relations with ancestors (both paternal and maternal). In contrast, the causes for disasters befalling the entire community, such as droughts and major livestock epidemics, are not easy to identify. Having asked informants on the causes of droughts, I received contradictory answers. Informants repeatedly emphasised that the human capacity for locating the origins of such disasters are very limited. Some thought that it was God who brought the rain and, accordingly it was God who then withheld it. While there was not a grain of doubt that outraged ancestors cause havoc among their descendants and their herds when disgruntled, there was nobody who forwarded the argument that it was ancestors who were responsible for droughts. Some people thought that powerful sorcerers could work to stop the rains. However, this was disputed by others. There is no homogenous frame for the interpretation of disasters. Problems befalling the individual and the household are habitually cured at the ancestral fire where ancestors are addressed ritually and asked for help (*okuhuhura*). Himba elders usually plea for rain in their prayers to ancestors. However, many claimed that they do not think that ancestors render much help in preventing a drought. Droughts come and go and humans cannot do much about them. The possibility of making rain themselves was denied outright by the Himba. In the past the Himba contacted Ngambwe rainmakers across the Kunene in southern Angola to make rain. Until today they regard Ngambwe magic as more powerful than their own and say that only Ngambwe sorcerers can actually "make rain".

5.5.2.1. Oracles

Himba forecasting of the future is based mainly on the inspection of intestines. Ritual specialists have other more secret and specialised ways to identify witchcraft (*okuvetisa*) and spirit possession (*okutumbika*). Divination based on haruspexing is the only reliable way to forecast the future for a wider public. Whenever a goat, sheep or cow is slaughtered the intestines are read (*okurora oura*, lit.: to taste the intestines). Every senior man commands some knowledge on how to read intestines. When an animal is opened up two, three or four men sit around the entrails and read them carefully. They discuss the signs on the entrails until they reach a consensus (see Figure 15 and Photograph 13). This may take a considerable time, one or two hours is not unusual to complete the

Photograph 13. Reading the Intestines

job. The haruspexes later report the results back to the larger meeting; informally, they chat about what the intestines have said.

Reading the Intestines (okuroora oura): There are different ways intestines can be read. It is a simple version to find out about the future of mem-

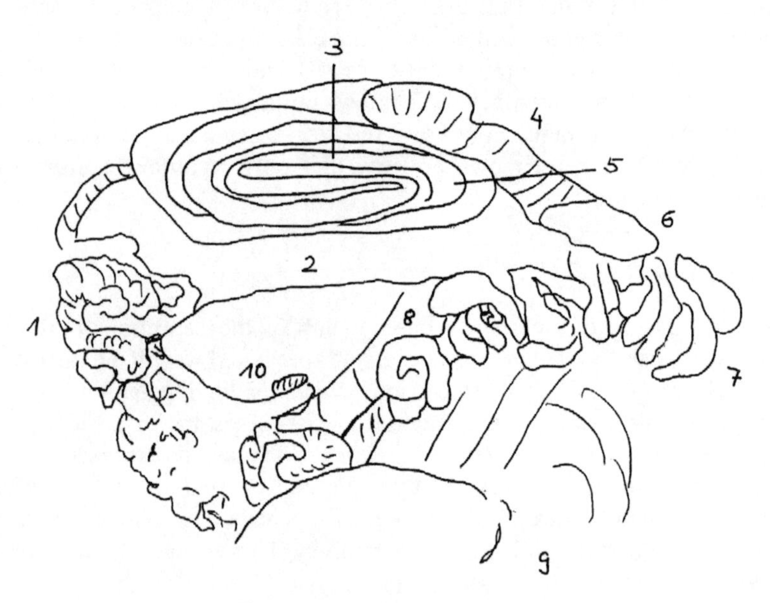

Figure 15. Reading the intestines

bers of the community and their livestock. There are other versions which are geographically more specific, showing topographical features such as hills and rivers.

(1) *etako roura*: lit. Back of the intestines, shows future events happening in the east
(2) unnamed part showing the fate of the cattle in homestead and cattle camp
(3) *otjikata*: shows future events in the place where the animal was slaughtered
(4) *otjinyuku*: shows the benevolence or non-benevolence of ancestors; if there are dots there is danger
(5) *orutwe rwozongombe*: shows the fate of the oxen of the homestead
(6) *ohongoze*: shows the fate of community leaders and elders
(7) *otjiuru tjooura*: lit. Head of the intestines, shows future events happening in the east
(8) *omusepa*: shows if people will die soon
(9) *ombumbi yokoura*: dots on this part show the fate of the household head
(10) *ombumbi yomonganda*: dots on this part also tell about the fate of the ownder of the homestead

Intestines are usually inspected in four steps: (1) After the stomach and the intestines are loosened a bit from the cavity, intestines are read for the first time, (2) then after the stomach is detached from the cavity, the *oruehe* (the membrane surrounding the intestines) is stretched over the stomach and read, (3) then the intestines are put on top of the stomach and read again and (4) finally the intestines are turned and read from the back. At all four stages there is information on the homestead (*onganda*), the wider neighbourhood (*ehi*), and the larger region (*ouye*). The part called *otjinyuku*, for example, shows if the ancestors have been worshipped appropriately. Another part called *orutwe rwozongombe* gives information about the dangers for the oxen of the household and cattle camp. Yet another part called *ohongoze*, shows the dangers awaiting the elders of the community. Major landmarks such as the Kunene and the southern Angolan highlands are discernible on all four layers. Blood red vessels in the part representing southern Angola were read as an indication of continued violence in Angola. The careful inspection of intestines allows the prediction of where it will rain, where livestock diseases may cause harm and which areas are safe.[36]

There are two further oracles which are limited to identifying personal problems: *okuvetisa* is the ritual applied to identify witches. Specialists conducting this ritual are frequently from other ethnic communities. People, who think of themselves of having been bewitched, often consult Ngambwe or Hakaona specialists in Angola. *Okutumbika* is a communal ritual which involves a specialist and a group of people for clapping and singing. It is used to identify spirit possession and to find out more about the spirit.[37]

5.5.2.2. Protective Magic (okuvindika)

Most individuals, irrespective of their sex and age, own amulets which have been produced by sorcerers (*otjimbanda*) (see also Estermann, 1981:156). The more powerful sorcerers the Himba consult are generally members of other ethnic groups, usually originating in southern Angola. These amulets protect against several problems: witchcraft, disease, sudden losses of livestock, miscarriages etc. Another form of protective magic is *okuvindika*[38]: many adult men wear the two horns of the duiker as a magical weapon (*ohiva*). Others wear pieces of the magic root *ondao* (not identified) attached to a string. Whenever danger threatens, one may whistle the *ohiva* horns and recite words to assuage with it. This will influence the source of danger, be it an enemy or a wild animal. It was said that under the influence of an *ohiva* spell enemies forget the purpose of their mission and they returned home disoriented. Lions and elephants just seek a different way and no longer endanger the person. In an oral tradition, an elder described how an ancient pre-colonial big man influenced enemies by performing *okuvindika*. Nama raiders were in the vicinity of the old man's homestead and were threatening to raid the place. The Himba were determined to resist the enemy but feared annihilation as their weapons were inferior. In a situation of utmost anxiety and terror the old man blew his magic whistle and told the enemies to be weak and his sons to be strong. The magic incantation, according to the tradition, worked long enough to enable the homestead to flee across the Kunene river into southern Angola (Tjambiru Tjikumbamba & Kozongombe Tjingee in Bollig 1997c:84, 94). When using the magic root *ondao*, the owner of this protective amulet chews bits of the root and spits it in the direction the perceived danger is coming from, while uttering a magic spell. The magic based on the *ondao* root works in the same way as the magic based on the *ohiva* horns.

Not all the Himba use protective magic to the same degree. While all the Himba wear amulets of a certain type, only the western Himba are deemed to be capable of doing *okuvindika*. The amulets produced from the magic root *ondao* are especially powerful if they come from the western Himba, who maintain that they obtained this protective magic from the Kuroka, a small pastro-forager group in south-western Angola. When fleeing from Nama raiders in the second half of the last century, prominent western Himba families settled with the Kuroka and, according to oral traditions, learnt the art of protective magic from them, asserting once again, that magical power derived from southern Angolan groups is stronger than Himba magic. The western Himba are still closely connected with the Kuroka. Occasionally Kuroka healers come over and treat sick people. A recent form of *okuvindika* imported from the Kuroka is the Kuroka protective fire: apparently this fire was kindled outside the homestead of leaders of the western Himba to protect them against guerrillas during the civil war in the 1970s and 1980s (personal communication Christopher Wärnlof).

5.5.2.3. Family-Based Magic

All the major rituals are conducted as clan-based rituals addressing the ancestors of a specific descent group. There are two major forms of ancestral rituals. The *okuyambera,* the ancestor commemoration, involves a visit to the graveside of an ancestor. At the graveside the ancestral fire is rekindled and ashes are taken from there to the ancestral fire in the homestead, symbolically connecting ancestral gravesides and ancestral fire. Frequently a heifer is consecrated during such a ritual. A lesser form of ancestral ritual is the *ongumbiru.* In this ritual, ancestors are only commemorated at the ancestral fire within the homestead. Both rituals are not necessarily crisis rituals. A first commemoration ritual is conducted about one year after the death of an adult person. While this ritual is a big affair if the deceased was the owner of a household, it is only a minor ritual if the deceased person was female or a young man. Afterwards commemoration rituals are conducted at irregular sequences. The first ritual is compulsory but those following are fixed according to the needs of the living and are frequently conducted when the present household head feels that his home is beset with bad luck.

The *ongumbiru* is more clearly connected to crisis prevention. The owner of an ancestral fire may decide at any time that he wants to turn to his ancestors with a specific problem. The *ongumbiru* contains large parts of the commemoration ritual but not all. The graves are not visited and the entire ceremony is centred around the ancestral fire in the homestead. Early in the morning the Mopane tree branches of the ancestral fire (*oviso*) are renewed and a shelter (*otjoto*) is erected at the side of the cattle enclosure. In the morning guests are introduced to the ancestors at the ancestral fire. Over the day the meat of an ox or of a sheep is eaten. Late in the afternoon the formal evocation of the ancestors represents the main part of the ceremony. These ceremonies do not have the mourning character of commemoration rituals. Usually the *ongumbiru* is connected to some sort of trouble observed or feared by the person conducting the ritual. It is not unusual that the people participating in the ceremony do not even know about the deeper reason for the ritual. Only very few members of the homestead will know exactly if the ritual is undertaken to thwart off witchcraft, to ask for assistance in healing sick sheep or just to help with a general feeling of uneasiness. The *ongumbiru* ritual is communal only in the sense that people stay together for a day and share the meat of a sacrificial animal. The ritual act proper however is conducted by a single person. He directs his words towards the ancestors. It is from them that help may come.

5.5.2.4. Community-Based Magic

There are no rituals where representatives of the entire Himba community convene to conduct a ceremony in the name of the entire group.[39] Rituals are characteristically tied to descent groups. The leaders of specific clans run the ritual and are in charge of the welfare of their descent group. Drought and livestock epidemics, disasters with a community-wide scope, are not countered with rituals encompassing the community.

In earlier years the Himba contacted Ngambwe chiefs across the border in southern Angola for rainmaking. In 1949 or 1950 the Himba paid cattle to the Ngambwe chief Tjisunga to perform rainmaking rites. Tjisunga was a renowned rainmaker (Gibson 1977:109).[40] Allegedly it rained heavily after these rituals.[41] The following year the Ngambwe chief again conducted rituals to induce rain. Already in 1932 or 1933 a lesser known Ngambwe sorcerer, known as Mbumba (after the large multicoloured blanket he habitually wore), made rain after being paid with cattle. Himba emphasise that they do not have any power to make rain themselves. In recent decades no rainmaking rituals have taken place. Most Himba do not believe that humans have the capacity to make rain. In general, informants referred to the utter unpredictability of the phenomenon. Ideas on a divine hand in matters of rain and drought are vague.

5.5.3. A Comparative Account of Ritual Approaches to Crisis Management

Pokot and Himba perceptions of causes of disasters differ profoundly. Pokot locate the deeper, metaphysical origins of a disaster within their society. They see social disharmony as one major factor leading to different forms of crisis. Himba do not look for metaphysical causes of major disasters. They are not caused by humans and they are not amenable to human action. Crises befalling just a single household, however, are connected to ancestral wrath.

Reducing Uncertainty: Divination plays a role in both societies. While among the Pokot numerous specialists forecast the future using different techniques and working at different levels, among the Himba reading intestines is the major way of predicting the future. In a highly variable environment, plagued by high unpredictability of rainfall, frequent livestock diseases, raids from enemies and constant anxiety of witchcraft, envy and evil eye, the Pokot seem to be obsessed about predicting the future. Haruspexing is done as a routine whenever an animal is slaughtered. There are plenty of ritual specialists conducting various forms of divination. Prophets dream about the future fate of the group as a whole. They are able to forecast major dangers ahead: raids, livestock epidemics, droughts. The techniques of haruspexing among Himba and Pokot are similar: the regional landscape is cast onto the intestines so that rivers, mountains and roads become discernible.

The range of problems the Pokot and Himba detect in intestines is quite similar, too. On the whole, however, oracles play a less important role among the Himba.

One may justifiably ask what the use of predicting a pending raid or a livestock epidemic from the intestines is. Beyond the immediate advantage everybody has from knowing what is ahead, communal divination is a way of creating consent. The Himba and Pokot take a long time to discuss what intestines have to say. The discussions are public and egalitarian. Consent is created on a potential hazard in public discussions. Especially in Pokot society, which is radically egalitarian and in which authorities are not easily accepted, divination is a way to base decision making on a common understanding of future disasters.

Protective Magic: Both societies use individual protective magic extensively. Amulets are used to thwart off witchcraft and the envy of others. Himba beliefs in the power of personal protection through amulets go further than those of the Pokot. At least the western section of the Himba hold that enemies and other disasters can be cast away by using individual magic. The Pokot would relegate activities against these dangers to the realm of communal rituals.

Communal Rituals to Ward off Disasters: There are major differences considering communal rituals to ward off disasters. Pokot rituals include numerous, non-related people. Their rituals of crisis management are communal acts. The ritual process takes place within the group. Among the Himba communal rituals are kinship based and the major ritual activities are conducted by few closely related elders. The ritual process is directed towards the ancestors. The major ritual power is created in a communal act. As it is thought that social disharmony is a root cause of any crisis befalling the community, discussions on internal conflicts take a lot of time in the preparatory phase of a ritual. The purification of social disorder is ritually sanctioned. Only this joined act of purification creates the basis for a successful ritual. The power of communal rituals is created by blessings spoken within the sacred semi-circle. The joint effort is emphasised in the way the blessings are spoken: the elder speaks one line of the blessing which is then repeated by the community of men. Word by word the blessing is a communal effort. Ritual power emanates from the group and not from a single expert. Women do not play any important part in Pokot communal blessings. The ritual community is constituted only by men.

Himba rituals are very different. The ancestral rituals are kinship affairs and non-related neighbours and guests become involved only as spectators. The only joint activity is that the senior man conducting the ritual introduces the attendants, men and women, to the ancestor to be commemorated. Women and men participate in the ritual. However, the power of a ritual is not created in a communal act but in the communion between one senior elder and the ancestors.

ENDNOTES

1. Although it is hard to determine an annual mean of animals slaughtered on the basis of data obtained between 1987 and 1989, it seems warranted to assume that the average household slaughters about 5 goats or sheep per year and 1 cow every third or fourth year. I did not observe significant differences between rich and poor households in this respect.
2. The correlation between wealth and kcal/pers/day is low 0.12 and at a level of 0.014 (Pearson's Correlation Coefficient) insignificant. The figures suggest that there is no correlation between wealth and animals slaughtered.
3. In another publication (Bollig, 1992a:340) I argued that the constant talk about *ngatkong* (envy) and its dangers probably contributes to increased food distribution and the exchange of goods. Everybody is afraid of being accused of envy and stinginess; in order to prevent such accusations he/she will steer a course which satisfies the social world around.
4. In 1988 I recorded a tradition where the transition of agro-pastoral Pokot to pure livestock-based nomadism is described. One major point of the tale is that the Pokot are taught how to slaughter an animal in an appropriate way and how to distribute the meat subsequently in a true herder way.
5. DeGarine (1994:344) rightly warns not to solely rely on interview data when collecting ethnographic data on famine foods. Interview data provides insight into the knowledge of wild foods within a society but does not reveal reliable information on the actual use of these plants during a drought. However, for the following presentation of substituting foods that the Pokot make use of, I had to rely mainly on interviews. Many of the species mentioned, are if at all, only rarely used in the present day.
6. Information on Himba foodways during famines relies on interviews and extensive surveys of household related herd off-takes (slaughter, sales) and food sharing practices. The data were not gathered during a famine. However, quantitative data on slaughter, for example, allows for an extrapolation of potential off-take during a crisis.
7. DeGarine (1994:346) points out the high nutritional value of Cyperus rotundus and gives 302 kcal/100gr as the mean energy content.
8. NA BOP N7/8/2 Famine Relief 1958-72, Toesighoudendebeampte, Ohopoho to Die Hoofbantoesake-kommisaris 11/1/1960.
9. NA BOP N7/8/2 Famine Relief 1958-72, Toesighoudendebeampte, Ohopoho to Die Hoofbantoesake-kommisaris 27/4/1961.
10. NA BOP N7/8/2 Famine Relief 1958-72, Kantoor van die Hoofbantoesakekommissaris va'n SWA to Die Administratiewe Beampte, Ohopoho 3/1/1962.
11. NA BOP Annual Report 1960.
12. Household No. 5 was still ranked among the group of medium herders. However, a major compensation had to be paid by the household in 1992, leaving it with hardly any livestock at all. In 1993 the household was ranked together with the very poor households.
13. That this trend has to bear with exceptions is shown by household 1, a rich household, which only attained an average price of 276 KSh, and household 25, a comparatively poor household, which obtained the highest average with 480 KSh per goat. However, No. 25 is a livestock trader and it may well be that he sold animals which he had bought from others and not from his own stock.
14. Six households were massively below the average (more than 0.1 kg below average). Of these six households, three belonged to the category of poor herders, two were medium herders, and one was a rich herder. Seven households were well above average; of these only one belonged to the category of rich herders, two belonged to the category medium herders and four were classified as poor herders indicating perhaps that these herders had to cover their entire food needs via maize when there was no supply of milk. As pointed out before there is no statistical correlation between wealth and maize availability.
15. Household No. 9 was not accounted for. The household head of No.9 is a community game warden. He gets food every month from the government.

16. The significance value is fairly low at 0.222. This however is mainly due to the small sample size.
17. In 1943 a district official claimed "The Suk and Turkana are good neighbours now-a-days and both tribes agree to a considerable degree of interpenetration and help each other with grazing and water." (Annual Report 1942 KNA DC BAR 1/3) and in 1943 the Annual Report said: "Relations between the Suk and Turkana appeared quite friendly ... At a big joint baraza at Karpeddo in November ... agreement was eventually reached and the Suk have allowed these particular Turkana herds to move into their territory until the rains break". (Annual Report 1943 KNA DC BAR 1/3).
18. KNA DC BAR 1/2 Annual Reports 1924-1933.
19. The Annual Report 1976 states "Ngoroko stock rustlers terrorised East Pokot all the year round and inflicted much damage and loss of property and life. Consequently people deserted the northern most locations completely and headed for the relatively peaceful grazing areas to the south of Pokot country".
20. Gathering, which is one alternative to herding and may add important food supplies during a drought (as in the Himba case) has been discussed in chapter 5.1. In chapter 6.1 more permanent forms of diversification will be discussed.
21. For another, more sophisticated beer called *kangara* 20 litres of ordinary beer are mixed with two kilos of sugar and then left for 4 to 5 days to ferment. Distilled liquor is still rarely produced. Only the main centres have several distilling businesses. Distilled alcohol is sold in Tree Top bottles at 20 KSh per bottle. Beyond the costs for maize, millet and sugar there are, of course, further costs: a metal drum is needed and several tins in which the drinks are dished out.
22. During a short field stay in 2004 I found a rapid expansion of brewing. Beer was produced even in remote places.
23. At the height of the drought there were about 31 informal *hoteli* in Nginyang, 27 operating the entire week and 4 only operating on Monday, the market day .
24. Similar observations are made in other marginal areas of East Africa. Odegi Awuondo (1990:83) reports on brewing, charcoaling, menial jobs in local restaurants among the Turkana; Hjort & Dahl (1991) reports that many Amar'ar Beja live on charcoaling after the loss of their herds, Kerner & Cook (1991) describe that beer brewing has become an important strategy to cope with low household budgets in northern Tanzania.
25. A *liokin's* work demands practical skills as well as a command of magical capacities. When visiting such an expert one is advised to bring him presents. He may demand that a goat be slaughtered. He works with coloured earth, white, red, black and grey which he mixes with several other substances to create powerful potions. Over several days the diseased person is anointed with these substances. Additionally the *liokin* chews the roots of the *moikut* bush (Cyperus rotundus) and spits it on the shoulders, head and breast of his client. While doing so he utters the name of the person who has bewitched or cursed his client. He orders the "bad words" to leave. The *liokin* then stands up and walks over his crouched client with open legs from behind. Alternatively the client creeps through the open legs of the *liokin*. Finally the patient is painted from head to foot with white coloured earth which symbolises the regained harmony between him and his social environment.
26. In all Nilotic groups, the legitimisation of prophets is similar. They usually stem from a specific clan (among the Pokot the Talai Clan) and they learn about the future in dreams. The prophet is not necessarily a wealthy person or a man who is known as a courageous warrior. While with other ritual specialists techniques play a major role, the prophet's power is apparently based on spiritual power only. For a detailed description of the institution of the prophet among the Pokot see Peristiany 1975.
27. Timberlake (1987:101) gives three further identifications: Hibiscus aponeurus, Hibiscus micranthus, Triumfetta pentandra.
28. Timberlake (1987:88) gives Barleria eranthemoides, Dicoma tomentosa, Ecbolium revolutum as further identifications.

29. Timberlake (1987:90) gives Ocimum basilicum – a strong smelling tree - as further possible qualifications.
30. The last thing that happened in the sacred half circle before the meeting disintegrated was that a young man's gun was blessed with sacred white earth (but significantly not with grey earth).
31. The rain rituals I observed and heard about were all instigated by the elders. However, the fact that the prophet may initiate such a ritual was confirmed by several informants.
32. I took part in the *mwata* ceremony and took many interviews on the *saghat* ceremony.
33. Timberlake (1987:92) gives Crassia edulis as a second possible identification.
34. Meinberg, the catholic missionary, who founded Kositei mission, described the ritual in a manuscript. The script is stored at Kositei Mission.
35. It seems odd that Cymbopogon caesius is used in this context, as it is hardly a palatable grass.
36. Minor diseases of individuals are not found in the intestines while the death of community members is indicated. There is some disagreement as to if witches can be identified from the inspection of intestines.
37. Estermann (1981: 155) gives a short outline on how the *okuvetisa* is done in southern Angola.
38. The overall impression from several interviews on the topic is that amulets work against hidden aggressions such as witchcraft. *Okuvindika*, on the contrary, is used to prevent open violent transgressions.
39. In recent years the commemoration ceremonies for the pre-colonial hero Mureti which are conducted at his place of birth at Omuhonga aim at such a scope. The political elite of Kaokoland are clearly looking for unifying symbols to enforce their say in national matters.
40. Ribeiro, Cabral, Ramiro C. 1948. O soba Chifunga, senhor das chuvas. Mensário Administrativo. No. 6 Feb 1948: 29-30, contains a short account on this rainmaker.
41. Estermann (1981:135/136) described how the power to make rains was vested in the chief among the Nkumbi, a group closely related to the Ngambwe: "... the sacred and indispensable intermediary between God and the earth in need of rain is the chief or a member of his family. ... Among the Nkumbi there was also the belief that rain-making power was connected with the marrow of the shoulder and shin bone of the reigning chief. After the death of a chief they therefore would wait until the body decomposed so that they could separate the bones of the right arm and leg and extract the marrow from them. This, mixed with butter, afterwards served ... for anointing the body of the heir and transmitting the aforesaid power, so that he would be capable of producing the beneficial rainfall."

Chapter 6

Buffering Mechanisms: Minimising Vulnerability

While individual strategies for coping with disasters (e.g. mobility and substituting food) are necessary to prevent human mortality and excessive losses of property, buffering institutions (e.g. social networks, resource protection) lower the vulnerability of communities and enhance their potential for quick recovery. In contrast to strategies employed during a crisis these institutions are of a more complex organisation and demand more co-ordination (Forbes, 1989:89). They are firmly integrated into the socio-economic system and belong to a standard repertoire of strategies. Typically individual benefits are sacrificed in order to create security for the common good. Pasture protection is a good case in point: herders give up their chances to exploit virgin grazing for the common good of a regulated and predictable range management.

While economic diversification is rated as a major risk-reducing mechanism in many studies (Colson, 1979; Halstead & O'Shea, 1989; Shipton, 1990), the flexible arrangement of property rights has been rarely discussed as a risk buffering institution. Risk minimisation through social exchange has been widely reported in hunter and gatherer societies (Wiessner, 1977, 1982; Kent, 1993). Halstead & O'Shea (1989:4) point out that social exchange "functions in a fashion similar to storage, in that present abundance is converted, this time via social transactions, into future obligations in time of need". Institutionalised resource protection is rarely discussed in the field of risk minimisation. However, in many populations intricate forms of communal management are employed to lower vulnerability. At the same time a moral economy, i.e. an ideology that ensures the provisioning of the poor, must be regarded as an institutionalised way to lower vulnerability.

6.1. THE DIVERSIFICATION OF THE ECONOMY AND FLEXIBLE PROPERTY RIGHTS

Rural societies diversify their productive strategies in order to cope with the probability of shortfalls in one sector of the economy (Colson, 1979; Shipton, 1990; Fleuret, 1986; Watts, 1988). While ad hoc forms of diversification characteristically demand little capital/labour input and are fairly individualistic, institutionalised forms of diversification demand continued and co-ordinated efforts. Next to the flexible handling of production strategies, the arrangement of property rights is essential in the context of risk minimisation. How do people gain access to property and usufruct rights in livestock? These issues become tremendously important in a period of restocking when many people aim at obtaining property and/or usufruct rights in livestock.

6.1.1. Diversification at the Margins: Pokot Attempts at Herd Diversification and Agriculture

The Pokot attempts at diversification of their economy are hampered by several constraints. Since Pokot herds tend to be small, there is little buffer to sell off animals in order to procure capital for input into other income-generating fields. The lack of capital constantly hinders the efforts of Pokot livestock traders to invest in stores or cars. Options for agriculture are restricted. Rainfall is highly unreliable and planting maize has high risks. However, the scarcity of arable land alone cannot explain the reluctance of the Pokot to engage in cultivation - even the little arable land which is available is not cultivated.

Pokot herders do not actively seek employment in urban areas or on farms in great numbers. Unlike the Maasai, Borana, Samburu and Turkana they rarely go for jobs in nearby towns. In a survey of 378 adult men from Loyamoruk and Ribkwo locations, it turned out that only 92 of them had worked in such jobs before. Only 26 (6.9 per cent) had actually migrated to seek labour as employed shepherds, watchmen or as workers on road construction sites. The others had worked locally as casual workers for local missionary stations or development projects. In recent years more children have been sent to school. Even conservative herders think it appropriate that one son of the family should enter school. All parents sending their children to school hope that they will find employment in the formal sector later on, and then help the family to buy food and pay for other expenses with their monthly salaries. However, at this stage the only significant forms of diversification are the expansion of camel husbandry, agriculture on plots in the highlands and an incipient engagement in trade. Dyson-Hudson & Meekers (1999) observed amongst the neighbouring Turkana that a great number of them left their district in order to search for labour in down-country Kenya. These migrants usually originate from the poorest sector of the society. Frequently they leave pastoralism for good as ties between migrants and their families remaining in the district are weak to non-existent. Such a degree of out-migration has not been observed for the pastoral Pokot.

Expansion of Camel Husbandry[1]

The pastoral Pokot of the Baringo Basin are not classic camel herders as are the Somali and Rendille. Up until today only a third of all households own camels. Usually these are the richer households. Pokot camel herds are generally small and most of them consist of less than 10 animals. Nevertheless, those households which do own camels profit from them. Camels have a longer lactation period than cattle, they yield more milk than any other species and they can be milked throughout the dry season. In the pre-colonial past camel husbandry was of only limited significance for the Pokot. Lineages claiming ancestry in either the Rendille (such as the Oro) or the Turkana (such as the Kamadewa) report that their ancestors owned camels before the advent of the colonial administration. However, early written evidence, corroborated by oral traditions, asserts that camel husbandry was of no importance among the Pokot in the early 20th century.[2]

Nevertheless, the Pokot have been very keen on extending their camel herds since the early decades of this century. This may reflect an adaptation to a constantly deteriorating environment in which Acacia bushes have become progressively dominant. In 1917 when Pokot mercenaries took part in British punitive expeditions against the Turkana, they were partially paid with camels confiscated from the Turkana. The Pokot were given several hundred camels out of the loot. Since the 1920s the Pokot tried to trade in camels with itinerant Somali traders. In the 1950s they paid up to three oxen for one camel heifer. This trade continued well into the 1970s until the onset of interethnic violence. Out of 328 adult female camels counted in a survey in 1992, 82 (23 per cent) were acquired from outside northern Baringo (Bollig, 1992b). Of these about half were of Turkana origin and the other half of Somali origin. Over the decades Pokot camel holdings have grown from several hundred in the early 20th century to 3000 - 3500 animals in the 1990s. In recent years a Kenyan NGO and a mission-based project have fostered camel breeding among the Pokot.

A survey in 1992 resulted in an overview of the present state of camel husbandry among the Pokot (Bollig, 1992b). A total of 134 herds with altogether 1005 animals were surveyed. The average camel herd had 7.3 animals. More than 70 per cent of all herds consisted of less than ten animals. Only 12 per cent of the herds were larger than 14 animals (see Table 23). Despite the fact that herds are small, the Pokot usually decide to keep their herds at the homestead as long as there is one lactating camel in the herd. If a herd of three or four animals only consists of males or dry females, the herd is joined together with the herd of a friend or relative. Only if labour is very short are camels herded together with the small stock.

Table 23. Average sizes of camel herds

No. of animals in herd	No. of herds	Per cent
1-4 animals	57	41.3
5-9 animals	45	32.6
10-14 animals	19	13.8
over 14 animals	17	12.3
Total	138	100.0

The Pokot regard camels as the livestock species which is the most difficult to herd. Camels tend to stray and a lot of time is spent on searching for lost camels. Frequently camels have to be tied individually in the evening in order to prevent them from walking about in the night. While camel fodder is not scarce in East Pokot, camels have to be taken to salt pastures every few months for several weeks to ensure the good health of the herd. This implies that for several weeks two herders have to accompany the herd. Additional mobility is also required for watering the camels. The Pokot water their camels more frequently than other camel keepers do. Camels are taken to a well at least every third day. Even during a drought the Pokot rarely wait more than four days before bringing their camels to a well. They argued that their camels would suffer if they were not watered frequently.[3]

However camels do not only bring extra labour requirements. They are also an additional asset. The milk yield is reliable and a lactating camel can help to bridge a bottleneck of milk supply during the dry season. Camels yield from 3.5 to 4.0 litres of milk per day. Occasionally, just after calving, the yield may even go up to six litres per day (Fratkin & Smith, 1994:94). Beyond that, camels are slaughtered for ceremonial purposes. Sometimes Pokot sell camels to local Somali traders and only very rarely are camels sold at the livestock market as there is no market for them down-country. Local Somali traders keep the camels and butcher them to supply their small restaurants with meat. If a groom owns some camels, one or two camels will surely be demanded as part of the bridewealth. Other exchanges frequently take on the character of more balanced transactions. A herder owning camels but lacking cattle or small stock, may decide to exchange a camel ox for eight to ten goats or for a bovine heifer.

In a survey of 1300 camel progeny histories in 1988, I found that most camels leaving the herd are sold. Altogether 124 camels were sold, 103 males and 21 females. It is mainly male animals that are slaughtered, exchanged for other livestock or sold. Only in social exchanges do female camels leave the herd in larger numbers. In total 100 animals were given away in social exchanges, 78 females and 22 males. Out of 66 camels exchanged directly for other livestock, 65 were males and only one was female. Of these 48 (72.7%) were exchanged for bovine heifers - in the case of losses in cattle herds, camels are a convenient medium to reconstitute them via exchange; 8 camels (12.4)

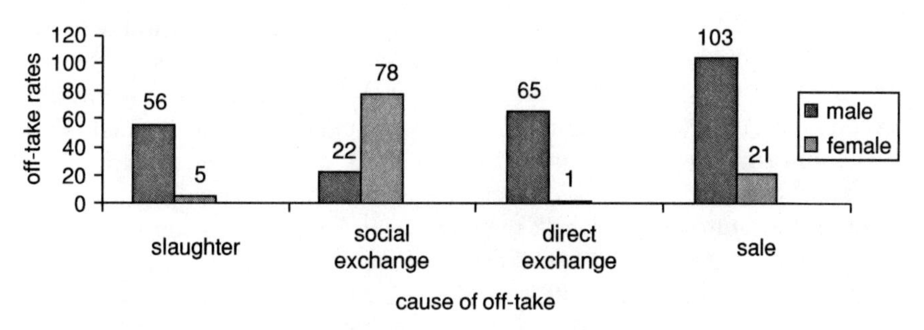

Figure 16. An overview of the off-take rates of camels

were exchanged for camel heifers and 10 (15.1%) for mainly female goats (see Figure 16).

Are there any other ways of diversifying pastoral herds? Good market prices for small stock, especially for goats, induced many Kenyan herders to expand their small stock holdings (Grandin, 1987; Ensminger, 1992; Herren, 1991). Among the Pokot a trend towards larger small stock herds was not observable: while the ratio cattle/small stock was at 1:2.0 in the 1920s and 1930s, it went up to 1:2.8 in 1950, but fell again to 1:2.4 in recent years. The total number of small stock declined over the course of the century, just as the number of cattle did. While about 175,000 small stock was the rough average for the 1920s and 1930s, nowadays 120,000 head of small stock are herded in northern Baringo. Hence, the expansion of camel husbandry is the only significant form of diversification in livestock husbandry in recent decades.

Agriculture

In the lowlands there is little opportunity for rain-fed agriculture. Good alluvial soils along the river basins are restricted. There are only a few places where households plant gardens irregularly in the lowlands. The total amount of land used for agriculture is negligible. It was only in 1996 after a good rainy season that I observed many households planting small gardens. In 1988, when similar good rains had turned the country green, nobody seemed to see the abundant precipitation as a chance for agriculture - despite constant encouragement by various development projects. A short field stay in March 2004 showed that the trend towards more agricultural activities has accelerated over the last eight years. Agriculture is practised in areas where in the 1980s nobody even considered the option of agriculture. Nowadays, in the Kadingding area where most of the households I worked with are settled, more than 50% of all households have gardens; most of which are small. In other areas, however, large fields are cultivated apparently aiming at producing a surplus for sale.

For years there has been a constant expansion of agricultural activities in the Churo area. Churo is near the Leroghi plateau and receives higher rainfall than the rest of the Nginyang Division. It is especially young educated Pokot men who seek to diversify their income by taking up gardening in these areas. Information on agriculture in the Churo area is meagre. None of the households in my sample was directly engaged in it. The usual way to obtain land in the Churo area is to ask the local chief for the allocation of a plot of arable land. Minor bribes to the local chief were not unusual in the 1980s. As soon as the land is fenced off and planted, the owner went to the District headquarters Kabarnet and registered his plot. Until the early 1990s there was no legal basis for registering private property on communal lands in Baringo District but the relevant ministerial offices apparently maintained files for pre-registration entries.

People from the lowlands who own land in the Churo area frequently have it cultivated by an employed worker or a poor relative. He will either be paid with a share of the harvest or with money. The major part of the harvest is not consumed in the highlands but transported to the main household in the lowlands. The maize is consumed mainly within the household. Many field owners voiced the opinion that maize fields helped them to abstain from excessive livestock sales. All efforts at diversification finally aim at the consolidation of the household herd.

6.1.2. Sharing Meagre Resources: Pokot Inheritance and the Splitting of Property Rights

Pokot property rights in livestock are highly complex. Schneider (1953:256) argued that "the obligations involved show the absence of any ideas of absolute ownership and the difficulties of any attempt to accumulate a surplus". While at face value all the livestock of the homestead is owned by the household head, a closer analysis shows that livestock ownership and usufruct rights are transient categories and amenable to constant negotiations. The following case story - a fictitious story invented by one of my informants to explain the intricacies of livestock ownership and inheritance - shows some of the complexities of Pokot concepts of livestock property.

> L., a young man, owned 15 female goats and 7 cows. He had married once and all his livestock were managed by his wife. He then married a second time. From the 15 goats he took 6 and from the 7 cattle 3 and allotted them to his second wife. Both women had too little stock to feed themselves and their children. L. went to his father's homestead in order to ask him for part of his inheritance. As he was the first son of his mother, he had a claim to a rather considerable part of the herd. First he addressed his mother and his younger brothers in order to reach an agreement with them before turning to his father. They decided not to take off the entire inheritance as the woman and her non-adult children intended to stay with their husband and father. The young man was

> finally given two cows and four goats. These were taken entirely from his mother's share of the household herd. Arriving back home he decided to share the animals equally between his wives - each wife got one cow and two goats. L. busied himself and went to collect outstanding livestock debts from his father's friends that he had a claim to. He succeeded in obtaining four goats from three debtors. As his first wife already had five children, he gave three goats to her and only one to his second wife who had just one child. A short time later L. took part in a raid and brought back one huge oxen. He exchanged it for a camel heifer with another Pokot man. He allotted the camel to his first wife. At the same time he promised the second wife that the first female offspring of that camel would be hers so that she would not be envious.
>
> Gradually L.'s sons were growing up. They were having an ever more important say in the management of cattle. More and more they learnt to regard the animals in the herd of their mothers as their own animals. Ownership rights in the household herd became even more complex as both wives obtained animals from their paternal families. These animals were solely their own property and L. had no say whatsoever about them.
>
> A few years later L.'s oldest son M. got married. L. had arranged the marriage. 12 cattle and some 25 goats and sheep had to be paid. It was agreed upon that roughly half of the bridewealth payment should be paid straight away, when M. came to take his wife. Out of the six cattle, five were taken from M.'s mother's share in the household herd and one cow was donated by the second wife. Out of the 15 goats which were given, 10 were given by M.'s mother, one by the second wife, three by friends of M. and one by a maternal uncle. These losses were partially compensated for when L.'s first daughter married. As a first payment of the bridewealth, L. received 8 cattle and 20 goats. He had invited two of his brothers and all his sons for the day the bridewealth arrived at his homestead. The bridewealth was shared into many minor parts. L. himself was only left with three cows and 8 goats.

The case study was continued here. L. was still confronted with more claims and counter-claims on animals herded in his household. Sons married and daughters were married, friends brought livestock presents and others asked for the payment of livestock debts etc. The story makes clear that livestock transactions feature importantly throughout L.'s life. There is a constant swapping of ownership rights in animals from one person to the other. It is remarkable that most transactions only include two or three animals at a time. Property rights in animals are continuously divided and redistributed. It is typical for the Pokot to transfer livestock-based payments in instalments. Bridewealth payments are paid over a period of ten years and more. A young man will rarely take all the animals which belong to him out of the paternal herd at once. He is perfectly satisfied with taking out a few animals and to know that he has a claim to further animals in the paternal herd. Next to the sharing of ownership rights in existing animals, promises of livestock are handled in a similar way. In the case study L. receives a camel from one transaction. He gives this camel to his first wife but promises his second wife the first female offspring of that camel. In a similar vein, fathers promise livestock to their sons. Due to this constant redistribution of ownership rights in livestock, it is not difficult for a man to obtain ownership rights in a few animals. However, it is difficult to accumulate large herds. There is a constant pressure on those owning livestock to distribute their property.

6.1.3. Standing on Two Legs: Himba Herd Diversification and Small-Scale Agriculture

Himba herds are less diversified than herds of pastoral nomads in East Africa. The Himba are first and foremost cattle herders. Nevertheless, small stock are important amongst the Himba too. Most Himba households are committed to some sort of agriculture. The opportunities for a further diversification of the pastoral economy were severely curtailed by the colonial encapsulation of Kaokoland during the colonial period.

Herd Diversification

All Himba herds consist of cattle and small stock, whereby goats are usually more numerous than sheep. In the middle of the 1990s there were about 140,000 head of cattle and about 280,000 head of small stock resulting in a ratio of 1:2 of cattle to small stock. Nevertheless, large small stock holdings offer securities during droughts: small stock herds declined to a much lesser extent than cattle herds did during the major droughts of the late 1950s and the early 1980s. Between 1958 and 1960 cattle numbers declined from 120,000 to 60,000, by about 50 per cent, while mortalities in small stock were moderate and around 10 to 15 per cent. Between 1980 and 1982 about 90 per cent of all the cattle perished in a drought. The losses of small stock were high too, but small stock herds recovered swiftly (see Figure 17). Lately small stock herds have expanded and the ratio of cattle to small stock was 1:2.7 in 1999.

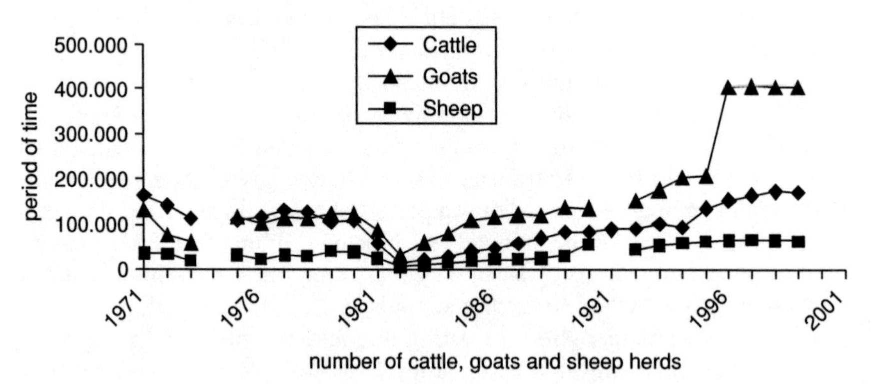

Figure 17. The growth and decline of cattle, goats and sheep herds

Some insights can be derived from the figures on the growth and decline of various herd species: cattle herds required a period of about 15 years to recover from the 90 per cent losses of the centennial drought of 1980/82, while goat herds needed only five years to recover. The total number of sheep is less variable; there are neither steep declines nor steep increases. During droughts sheep are apparently the most resilient livestock species.

Small-Scale Agriculture

The expansion of agricultural activities was a major long term strategy of Himba households to cope with the restrictions imposed by the South African government on the livestock/grain trade. Nowadays most Himba households have gardens. Himba gardens along the Kunene between Swartbooisdrift (Tjimuhaka) and Enyandi have an average size of c. 7000m^2 (75 gardens) and 9,000m^2 (10 gardens)[4] in the Omuhonga Basin (see Map 13).The range of garden size is shown in Table 24.

The size of gardens does not correspond to wealth. Gardens are sometimes joint efforts of the women from a neighbourhood and sometimes purely family gardens. Some belong to households settling immediately at the river (the Kunene or the ephemeral Omuhonga river) and some are worked on by people coming to the river just for gardening.[5]

It is very hard to measure or estimate the yields of Himba gardens as the harvest takes place over a protracted period of time. Coppock (1993:123) reports that Borana agro-pastoralists in southern Ethiopia harvested some 1,100 kg/ha of maize in their gardens. Southern Ethiopia receives about double the amount of rainfall of northern Namibia and the agro-pastoral Borana devote more time to agriculture than the Himba do. Hence, it seems safe to assume that Himba gardens yield about half the amount. The figures above have shown that the majority of gardens are about half a hectare in size resulting in a potential yield of 500 to 600 kilos of maize. These data, which is just a very rough estimate, suggest that only very small households are able to live for more than half a year on home-produced maize. Almost all households have to buy additional maize. However, it is significant that a yield of about 600 kilos, equivalent to about 10 sacks of maize, makes it unnecessary to sell 10 goats or an ox. Most Himba informants suggested that the cultivation of gardens prevents the excessive sale of livestock in the dry season.[6]

Table 24. The size of gardens at the Kunene and Omuhonga

	< 0,1 ha	0,1-0,4 ha	0,5-0,7 ha	0,8-1 ha	1,1-1,5 ha	>1,5ha
Kunene	2	20	20	14	12	7
Omuhonga	0	1	3	1	5	0

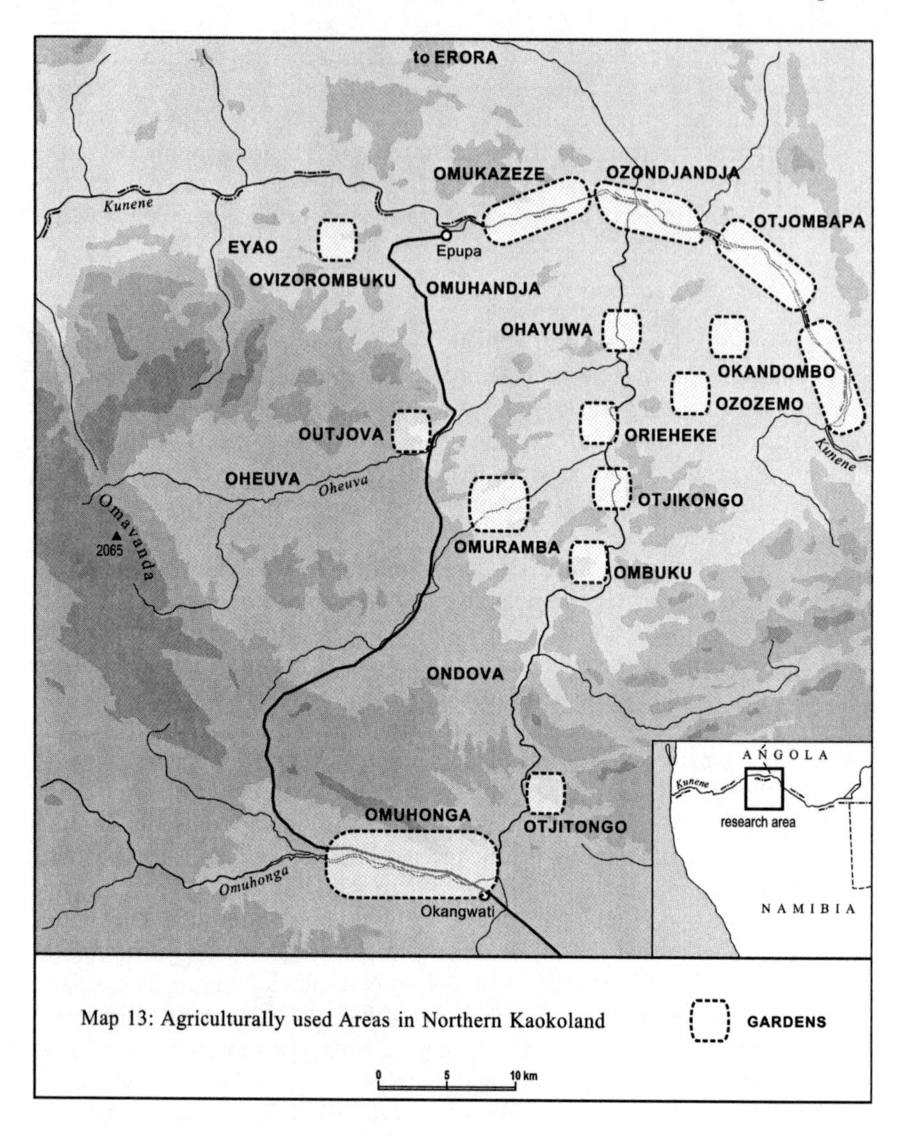

Map 13: Agriculturally used Areas in Northern Kaokoland

The fact that most fields are located on plots which are inundated annually is at the same time a major hazard and a major opportunity. On the one hand gardens do not have to be frequently relocated due to the annual inundation of plots. Essential nutrients are replenished with each flood. On the other hand inundation may occur after the maize has been planted. In 1994 many gardens in the Omuhonga valley were destroyed by a flood in late February.

6.1.4. Concentrating Resources: Himba Livestock Ownership Rights and Inheritance

Property rights in cattle are held by the head of a homestead. The household herd is only divided up at his death. Herd owners, however, may give out livestock on a loan basis. A look at the distribution of ownership rights within single herds shows the significance of livestock redistribution via loans. The distribution of livestock ownership correlates fairly well with the age of the household heads (Figure 18). While older men possess up to 100 per cent of their animals themselves, younger household heads borrow up to 98 per cent of their cattle.[7]

While use rights in livestock are transferred frequently, ownership rights are predominantly transferred after the death of a herd owner. Rules of inheritance guarantee (a) a co-ordinated and unambiguous transmission of property from the testator to one heir and (b) the concentration of property rights in livestock in the hands of a few. Households are inherited within the matriline (*eanda*), first from the deceased to a brother (of one mother) and then to a sisters' son (ZS), or any equivalent in the matriline (like ZDS or ZDDS), if a sister's son is non-existent.[8] As matrilineal relatives tend not to live together, inheritance usually implies a major shift of the entire household. As it is well known who has inherited from whom in the past and who will do so in the future, one may represent inheritance transfers as chains. One such inheritance chain may serve as an example for the spatial scope of inheritance transfers. The example highlights the rules of inheritance as well as the spatial character of transactions.

> Tjandero, an extremely wealthy herder owning about 800 to 1000 cattle, will pass on the entire herd to his sister's son (*omusyia*) Kasorere. Kasorere, who is already in his sixties and not much younger than his uncle, will in turn pass on the herd to his next younger brother Mbatjanani. Mbatjanani will once again pass it on to a younger brother, Karenda. Then the herd will be transferred to their sisters' sons' line. The first heir of the herd will be Wezuzura,

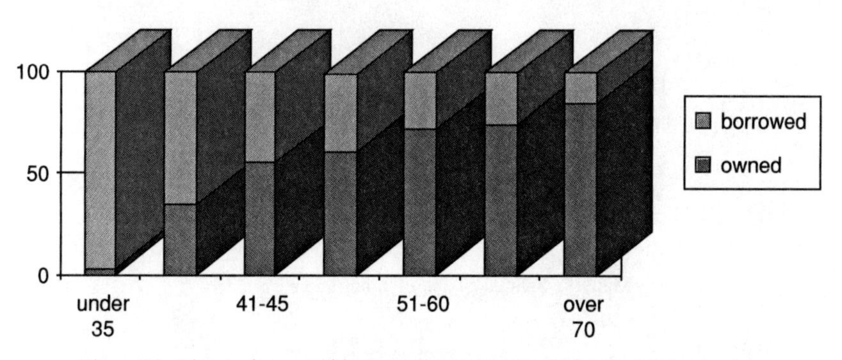

Figure 18. Livestock ownership according to age (n=36 herds, 4482 animals)

> the eldest son of Kasorere's, Mbatjanani's and Karenda's oldest sister, Watundwa. Wezuzura will hand on the herd to his younger brother Maaesuva. If Maaesuva should die, then the sons of Kasorere's, Mbatjanani's and Karenda's second eldest sister (Mungerinyeu and Manuele) will inherit and when the herd has been handed down the brotherly line, the herd will fall to the son's of Kasorere's, Mbatjanani's and Karenda's third and youngest sister.

The household herd in question will cross the Namibian/Angolan border twice. Within a period of about 30 to 50 years residential changes of the household may take place eight times (see Map 14).

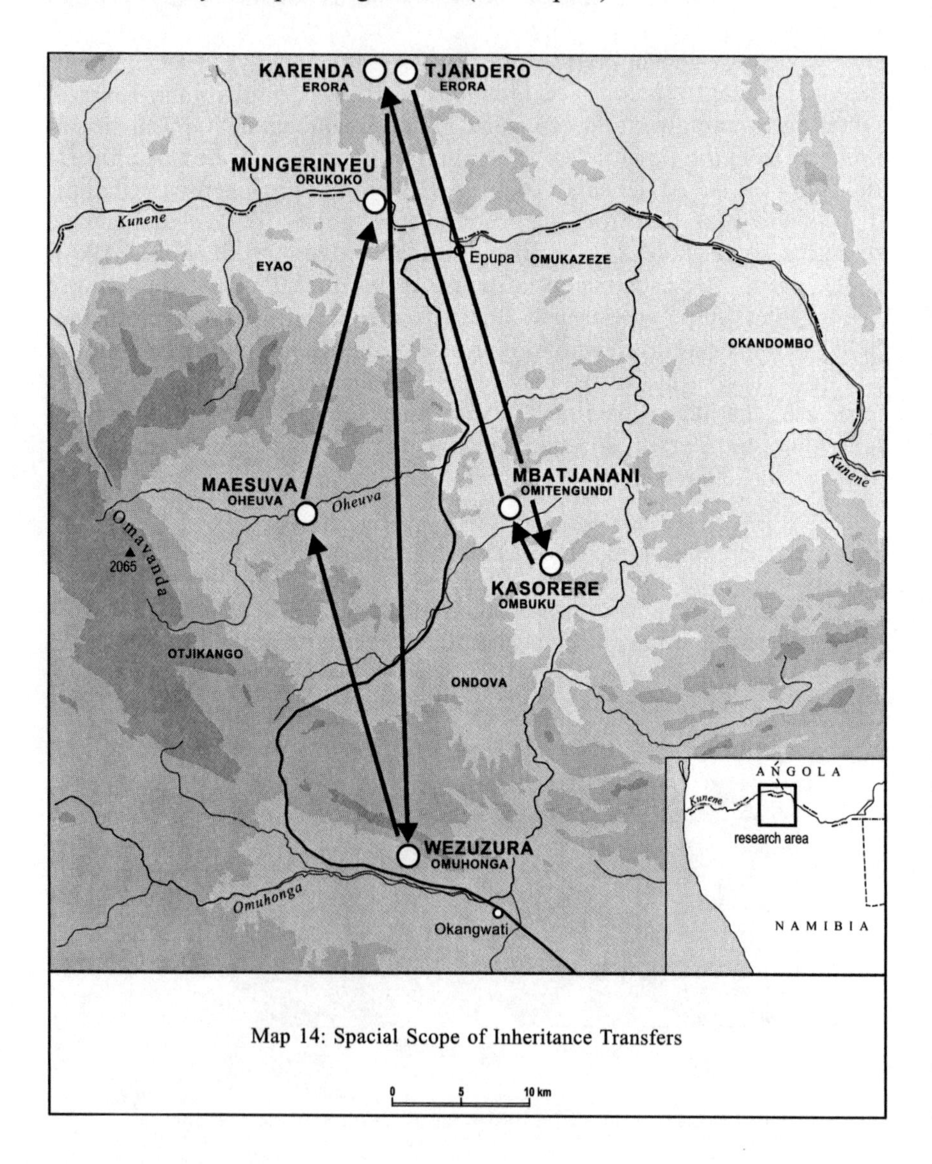

Map 14: Spacial Scope of Inheritance Transfers

Inheritance leads to a continuous redistribution of the regional herd. At each inheritance transfer, several herders, closely related to the deceased matrilineally, may take out a few head of livestock from the herd. This takes the form of tolerated theft (*okuramberia*). The night before the inheritance is finally allotted to the heir by the elders of the clan, they pick out single animals and leave the scene. Referring to this rather clandestine way of livestock appropriation, these secondary heirs are referred to as *ovarumata*, the biters.

While inheritance on the one hand guarantees a concentration of property rights it also leads to a continuous redistribution of use rights in livestock. The case presented above entails the ideal line of inheritance and it is hard to imagine that the inheritance of Tjandero's herd will actually run on the lines prescribed by several Himba informants. It is obvious that this presentation of inheritance glances over several incongruities. First of all it is highly unlikely that all the people on the list will be still alive when it is their turn for inheriting livestock. Tjandero, Kasorere and Mbatjanani are roughly the same age (the oldest perhaps seventy and the youngest sixty), and Wezuzura, Maesuwa, Mungerinyeu, Manuele are all between 25 and 40.

There is something very special about the Himba inheritance system which combines a concentration of property rights and, at the same time, facilitates the distribution of rights. Many inheritance transfers listed above include a transfer of property rights from a deceased to his brother. Frequently the brother is roughly the same age as the deceased person. Even transfers from a deceased to his sister's son do not always bridge a generation. Many sisters' sons who were enlisted as heirs were only slightly younger than their mothers' brother. In fact, cases in which an entire herd is transferred from a deceased elder to a very young man are rare. Most men who have inherited a herd have done so when they were older than 45. While only 1.5 per cent of all the animals obtained via exchange by herders under the age of 45 originated from inheritance transfers, herders between 45 and 59 obtained 16.5 per cent of all animals through inheritance, and herders of 60 and above even obtained 27.9 per cent through inheritance transfers. Of all the animals transferred in inheritance, herders of 60 and above obtained 52 per cent, people between 45 and 59 obtained 44.9 per cent while herders under the age of 45 received just 2.6 per cent of all the inheritance transfers.

Table 25 shows the probable life expectancy of elders between the ages of 60 to 70. Several models are tested. The life expectancy lies between 13.1 to 13.9 years for a 60 year-old senior and between 8.1 and 8.5 for a 70 year old man. Hence, a senior man inheriting a herd will own that herd roughly between eight and fifteen years. Then the herd will be handed on to the next heir.

Himba rules of inheritance constitute a system in which many men have a vague chance of becoming owners of larger herds at some indefinite time in the future. The practise of inheritance guarantees that most herders will only become wealthy herd owners when they are already quite aged. Property is moved within the matriline from "station to station" through generations.

Table 25. Life expectancy of seniors according to different models

Mortality-Table/Model West	Age	Life expectancy in years
Level 9, e0=40	60	13,1
	65	10,4
	70	8,1
Level 10, e0=42,5	60	13,5
	65	10,8
	70	8,3
Level 11, e0=45	60	13,9
	65	11,1
	70	8,5

Note: Based on comparative data on mortality rates in sub-Saharan Africa (Coale & Demeny 1966/83) it is fairly sure that the average life expectancy does not lie above 45 years (e0 = 45). A life expectancy of 42.5 or 40 seems to be more likely.

The "rotating pot" works like a trans-generational, matriline-bound savings association: everybody contributes (livestock and/or labour) and has the hope of being the sole owner of the herd one day.

6.1.5. Comparing Property Rights and Diversification

While herders in both pastoral economies attempt to diversify their herds there is a clear trend among pastoral Pokot to look for more diversification in the 1990s. Young men were opting for trade and a trend towards more small scale agriculture was observable. In contrast to that, Himba agriculture has been part and parcel of the pastoral economy since the beginning of this century. Three quarters of all Himba households may count on a harvest which feeds them for at least three to six months. A good harvest allows a herder to sell only a few animals. In contrast, in the mid 1990s only very few Pokot had the chance of desisting from livestock sales on account of home-produced maize.

Himba and Pokot systems of property rights in livestock are organised in very different ways. The Pokot opt for a constant redistribution of livestock capital and prevent the emergence of large accumulations of livestock. The moral obligation to redistribute property is a cornerstone of Pokot ethics and underscores their egalitarian political system. Young Pokot have the opportunity of gaining property rights in livestock at a very early stage. Impoverished herders can rely on an extended network of relatives and stock friends. In contrast, young Himba men find it hard to obtain property rights in animals. Most men between the ages of 25 to 40 are still largely dependent on livestock loans. Inheritance rules among the Himba ensure the concentration and distribution of property rights in livestock at the same time. Through matrilineal inheritance everybody has at least a vague chance of inheriting a herd one day. However, due to the fact that livestock inheritance goes first of all from brother to brother, most men are already seniors when they inherit a herd. There is a good chance of them owning the herd for only a decade or two before they die themselves. The comparative analysis hints at the overall importance of livestock exchange networks. These will be the topic of the next chapter.

6.2. NETWORKS OF SECURITY - NETWORKS OF DOMINANCE

Social networks are of major importance as risk-reducing institutions in herder societies (Bollig, 1998a; McCabe, 1990:151). These networks are instrumental for the distribution of property and use rights. In cases of disasters, herders can rely on exchange partners – their social capital - for immediate help and later for help in restocking depleted herds. In both societies social relations are expressed through the exchange of livestock. The contents and structure of this social capital however differs: in the Pokot case the exchange of livestock is shaped by and eventually reinforces egalitarian conditions; in the Himba case the exchange of livestock establishes patron-client relations. Livestock transactions mainly take place between men in both societies. Women are excluded from owning and transferring stock amid the Pokot. Among the Himba they own a moderate share in herds and especially older women loan animals, usually to their sons or grandsons. The data for exchange in both societies is based upon in-depth interviews and extended quantitative surveys. In both cases progeny histories were recorded in order to document livestock exchanges quantitatively. At the same time partial networks of roughly 40 households were noted and all exchanges between the sampled households were described in a detailed way in both societies. Herders in both samples commented at length on the moral and economic aspects of these exchanges.

6.2.1. Kinship, Friendship and Exchange among the Pokot

Among the pastoral Pokot all transfers of livestock are accomplished in order to initiate long term relations between two herders. On the one hand this includes the promise of further exchanges of livestock and on the other hand entails strong emotional ties and social support. Bridewealth exchange, bridewealth distribution, stock-friendships, distribution within the descent group, and exchanges between two fixed descent groups stand out as the major institutions of reciprocal exchange (see Bollig, 1998a).

Bridewealth Exchange: Marriages can only be legitimised through the exchange of animals (*kanyoy*). Up until now money has only been rarely accepted as an alternative to payments made in livestock.[9] In contrast to the neighbouring Turkana (Gulliver, 1955:228ff), bridewealth payments are protracted and comparatively low. At a formal meeting (*aloto*), the bride's and husband's male kin fix the amount of bridewealth to be paid. While the first marriage is usually arranged, the arrangements of further marriages fall to the actual bridegroom and the bride's father. The first part, usually about half of the total bridewealth, is paid before the bride takes her place in her husband's homestead. Over the following years the remaining part of the bridewealth is paid in smaller instalments.[10] The last instalment is marked by a ritual (*koyogh*), which removes the bride-taker from a situation in which he is indebted to the bride-giver. The relation is now regarded as "heavy" (*nïkïs*) and both men involved, the bride's father (plus his closest agnatic relatives) and the son-in-law (plus his closest agnatic relatives), may now exchange livestock voluntarily.

Bridewealth payments are not rigidly fixed; the average is about 12 heads of cattle, 2 to 3 camels and 30 goats plus some sheep. But if a rich herd owner (or the son of a rich herd owner) wants to marry, discussions on bridewealth may drag on for a long time as the bride's kin try to claim more animals. On the other hand, bridewealth requirements may be reduced considerably for poor bridegrooms.[11] Despite wealth differences it is very rare that more than 15 cattle have to be paid. The maximum number of cattle paid in a bridewealth was 17 and the lowest number paid was 7.

Bridewealth payments are the only transaction in which large numbers of female stock are given away. Next to a lactating cow (*tepa pokïr*, the cow of the mother-in-law) and a huge ox (*egh pö papo*, ox of the father), most cattle included in a bridewealth payment are heifers (46 per cent) and young steers (33 per cent). The sex and age distribution of camels paid in bridewealth exchanges is similar. 28 per cent of all goats paid in bridewealth were lactating ewes and lambs, 36 per cent were immature female goats, 20 per cent immature males and 10 per cent castrated bucks. Sheep did not feature importantly in bridewealth exchanges (see Figure 19).

The entire bridewealth comes from the herd of one family. Major contributions to bridewealth payments from outside are rare. Out of the 36 bridewealth payments recorded in detail, 22 were paid without any contribution from outside the household. If people from outside contributed to a bridewealth payment they were generally brothers.

There is intense pressure on wealthy herders to exchange surplus livestock for women. In fact, the Pokot regard it as natural that large herds are depleted in exchange for women. In 2004 the two wealthiest herders of my sample had eight and six wives respectively. During initiation ceremonies of girls and boys, rich men who have not married for some time are openly ridiculed casting doubt on their virility. A specific tree is pointed out where women sing mocking songs about these avaricious men. They are thought of as selfish and preoccupied with amassing cattle instead of investing them into the culturally-defined circuit of exchange: cattle for women.

Bridewealth Distribution: While bridewealth payments have to be borne mainly by the bridegroom or by his father (for the first marriage), incoming bridewealth payments are distributed throughout the personal network of the recipient. There is a moral obligation not to keep large numbers of livestock

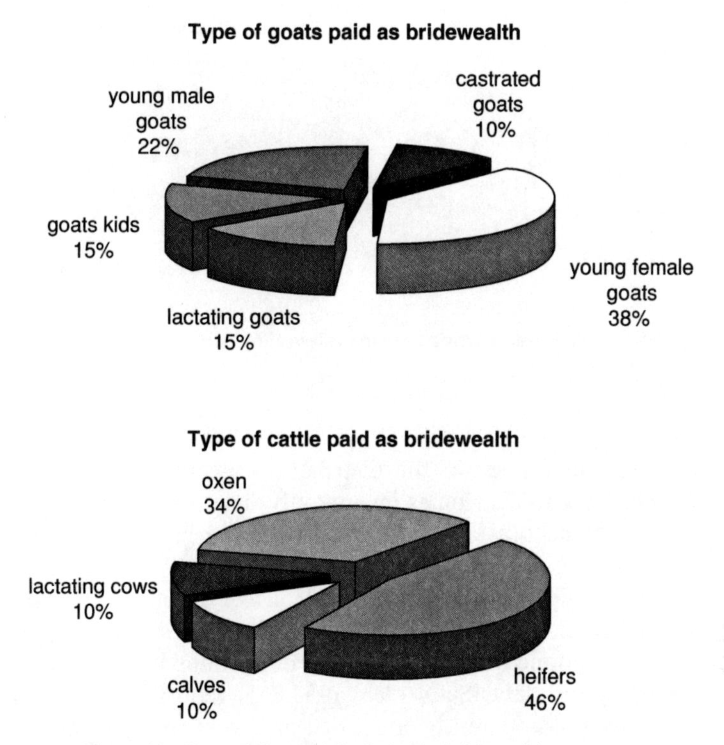

Figure 19. Type of livestock included in bridewealth payments

received as bridewealth payment for one's daughters or sisters. Instead bridewealth animals should be distributed to relatives and friends generously. In the 36 bridewealth distributions recorded there were always at least eight people participating in the distribution. About 70 per cent (n = 165) of all bridewealth cattle recorded were given away to members of the patrilineage; of these 19.0 per cent remained with the bride's father and 23.7 per cent were given to his brothers while 17.2 per cent went to the bride's brothers and another 10.7 per cent to other paternal relatives (especially the bride's father's father's brothers and their sons). Affinal relatives and friends participated in the distribution too, each group receiving 14.7 per cent of all the cattle given away. Goats were distributed to a lesser extent: 32.6 per cent stayed with the father of the bride, 29.1 per cent to her brothers, 17.4 per cent to the father's brothers or father's brother's sons, 4.6 per cent to the sisters of the bride's father and only 6.6 per cent went to affinal relatives and friends (see Figure 20).

The distribution of bridewealth payments is essential for reinforcing kinship networks of mutual obligation. Although the donor cannot directly ask his relatives or friends for equivalent gifts in return, he can well expect the recipient of his present to invite him to similar distributions. Hence, on the day in question, the person receiving a bridewealth payment is eager to include

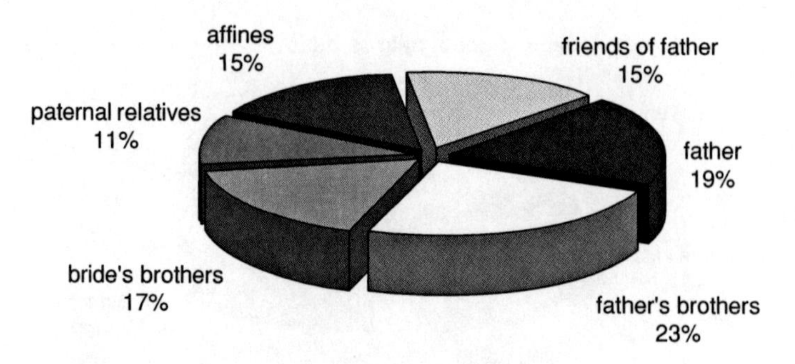

Figure 20. People participating in bridewealth distributions of cattle

those relatives and friends in the distribution whom he trusts as reliable herders. A herder who distributes incoming bridewealth generously is highly regarded, while somebody keeping bridewealth payments selfishly for himself is ridiculed in public meetings.

Stock-friendships: Schneider (1953:255f, 263) referred to the great importance of stock friendships among the Pokot. Schneider's emphasis on the fusion of economic interest and social obligations in these exchanges is corroborated by my data. Schneider (1953:255) reports that "a true best-friend relationship is apparently not sealed unless one presents an ox or cow to the other without immediate repayment. This indicates a trust in the other which he reciprocates when the need arises ...". If bridewealth exchanges and bridewealth distributions mainly include and establish kin relations, stock-friendships create bonds between non-related or distantly related actors.

Neighbourhood-based rituals and celebrations of the age and generation - sets, are frequently the arena in which stockfriendships are initiated. The person donating an animal for slaughter at such a ritual selects one of his guests to spear the goat or ox he is offering. He will look carefully for somebody who is reliable and has a lucky hand as a herder. Schneider, too, emphasises the great care and economic interest with which stockfriends are selected: "... a man does not make best friends indiscriminately ... There is little doubt that exchange between best friends, as with other ties in which stock are transferred, involves an exact accounting on what has been given and how much one is in debt to the other". (Schneider, 1953:256). Donating and spearing an animal at a neighbourhood ritual or a huge age-set celebration is highly prestigious. In the political discourse one is frequently treated to an enumeration of where and when the speaker speared an ox. The spearing of a famous ox may be incorporated into one's own songs of personal aggrandisement. It is mentioned in the same breath as more ferocious deeds such as killing enemies and stealing livestock from neighbouring pastoral groups. Somebody who has speared an ox or a ram at a celebration is obliged to return the equivalent of the animal he slaughtered some time later (*ghosyö*). However, it is the donor's right to stipulate the terms of the exchange. He will state if he wants an ox, a heifer or six goats in return for his ox. After a few years, only occasionally after some decades, he (or his heirs) will visit the beneficiary and ask for the settlement of the debt. This may be postponed or given in instalments, but I have never heard of a debt being refused outright. If the person asking for the settlement of an outstanding debt obtains a heifer, which, of course, has a higher reproductive value than the ox he gave, he is entitled to use its milk and, after some years, its meat. However, he has to return at least some of its calves. This form of stockfriendship ties households together for several decades (see Map 15).

On average the 15 men interviewed in detail on stockfriendships accounted for 10.6 TLU as outstanding debts from stockfriendship relations. If this abstract figure is transformed into its equivalent in goats or cattle this means that each man reckoned about 60 goats or 10 cattle as outstanding debts. The number of stockfriendship relationships of one actor rises with his age indicating that the investment of surplus stock into social ties is a lifelong strategy. Old men possessed up to 125 goats in debt relations. If not forced by economic necessity the herders accumulate obligations arising from stockfriendships like promissory notes. Friends see mutual indebtedness as one expression of their emotional ties. Equivalent mutual debts are not cancelled but are seen as proof of trust. Stockfriendship is inherited from father to son. It is not rare for debts resulting from a stockfriendship to be handed down for generations before final repayment.

Rarely does a man take up such a relationship with several men in one community. It seems to be an intentional act to spread stockfriendships beyond the neighbourhood. Informants invariably claimed to maintain stockfriendships in at least 6 to 10 different places. Yet, rarely did anybody have two stockfriends in one place. This is advantageous when droughts and epidemics only hit certain parts of Pokot land. In cases of hazard or labour shortage, part of the household herd may be entrusted to a stockfriend. While a clear majority of stockfriendships are between Pokot herders, occasionally Pokot herders had stockfriends among Turkana, Njemps and Samburu too.

Generalized Reciprocal Exchange between and within Descent Groups: Despite a clear tendency to spread affinal relationships, most lineages have strong ties with at least one other affinally related lineage. A relation founded on frequent mutual marriages and generalised reciprocity is called *kapkoyogh* (from *kaa pö koyogh*, the house where the final marriage ceremony takes place). It is claimed that the livestock herds of both lineages in a *kapkoyogh* relationship are "harmonising". Frequently bulls from one herd are used to breed with the other herd; the offspring of these, so it is said, are always healthy and productive. Ties between lineages created by adoption resemble

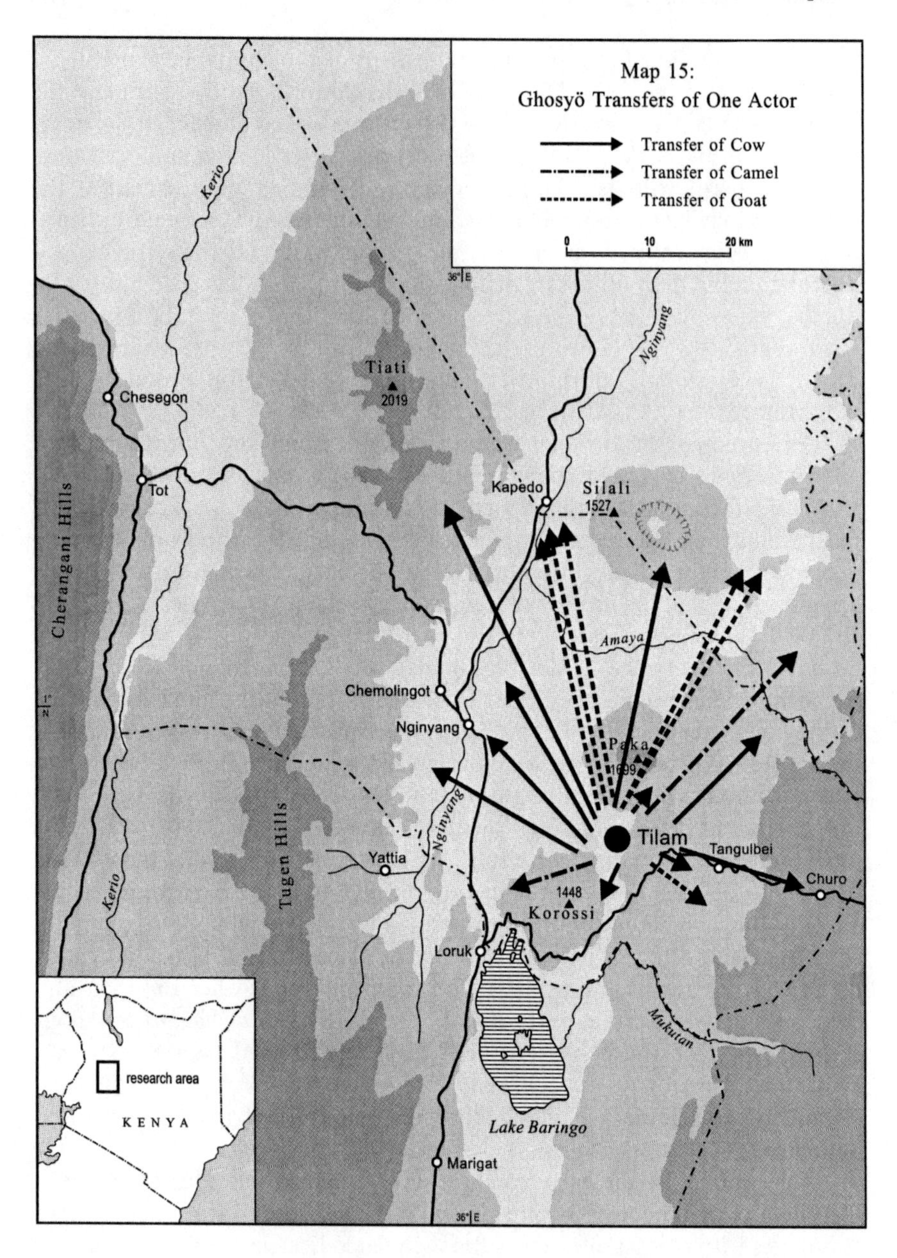

the *kapkoyogh* relationship in its intensity. Right down to the present, the relation between adopting and adopted lineages is deemed so "heavy" (*nïkïs*) that intermarriage is prohibited. Generalised reciprocity between both lineages ideally moulds transactions.

Other Reciprocal Exchange Relations: There are numerous other institutions which oblige herders to enter into reciprocal exchange. Engaging in reciprocal exchange is a major cultural topos.[12] When young men are initiated (*sapana*), they have to slaughter an ox. Rarely is this animal taken from one's father's own herd. Usually (in about 70 per cent of all cases observed, ca. 180 initiations) the father borrows the ox from a friend. Once again this transfer will result in an obligation which has to be paid back by the initiated later on. When young men are circumcised, (Bollig, 1990c) they have to wear women's clothing during the period of seclusion (see Photographs 14, 15). They borrow these skin clothes from a befriended elderly female. In return for lending the clothes, the woman receives a female sheep later on. However, this sheep is regarded as yet another token. At a later stage the woman's husband (or eldest son) will hand over one cow or a heifer to the former initiant.

6.2.1.1. Structural, Emotive, and Normative Correlates to Reciprocal Exchange

The Pokot claim that it is beneficial for every herder to have as many exchange partners as possible in as many different areas of Pokot land as is feasible. Informants were proud and self-confident when explaining the intriguing complexity of livestock exchange. They frequently finished their

Photograph 14. Pokot Men before Circumcision (*tum* ritual)

Photograph 15. Pokot Men shortly after Circumcision in Women's Clothes

accounts with pointing out that "livestock exchange networks act as a bank for the Pokot". Social institutions are geared to establish opportunities for making exchange partners. However, livestock exchange is not only about institutions; it is, at the same time highly emotional and circumscribed by a complex set of norms and values.

Social Institutions: The marriage system is a social institution (next to many others such as neighbourhood-based rituals or age-set celebrations) organised in such a way as to permit the actor to maximise the number of his affinal relatives. Next to lineage and clan exogamy, there are a number of other rules governing the spread of affinal relationships (see Figure 21). A man may neither marry a woman from his real or classificatory mother's lineages nor women from the same lineages as the wives of his brothers (of the same mother). A man should neither marry women from the same lineages as the wives of his half-brothers (different mothers, but the same father); nor women from the lineage of his paternal and maternal grandmothers; nor women from the lineage of his mother's mother's mother; nor women from the same lineages as his sons' wives (of course an elder may still marry further wives after his sons have married); nor women from lineages his sisters have been married into. Hence, every marriage ideally initiates a relationship with a set of people that one's own descent group has not yet had any relations with.[13]

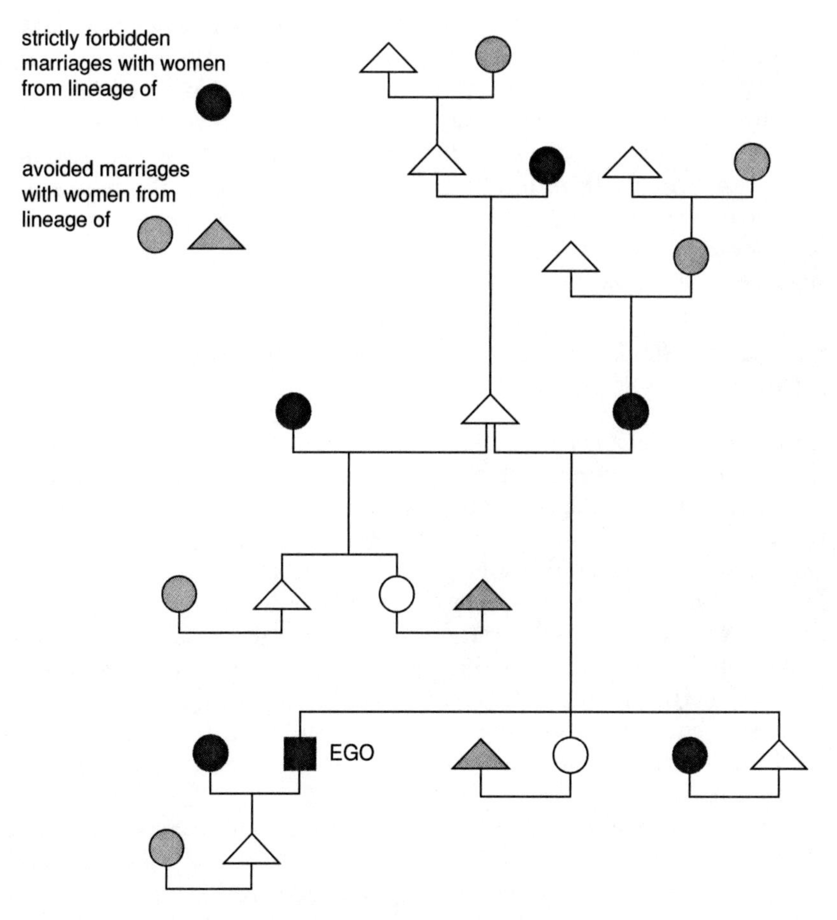

Figure 21. Marriage Prohibitions and Constraints

The diversification of exchange relations is enhanced by numerous other institutions: e.g. one should not engage in stockfriendship with in-laws. This prevents the doubling of affinal and friendship ties. The rules of spreading stockfriendship partners guarantee the spatial distribution of exchange relations.

Normative and Emotive Frames: The individual network is structured by the concept *tilyai* (cf. Schneider, 1953), whereby the *tilyai* of one actor includes those relatives and friends he has exchange relations with. The derived abstract, *tilyontön*, is the set of norms and values conferring mutual solidarity between exchange partners. If a man speaks about his *tilyai* he refers to a group of men he is related to, befriended with and has exchanged with. The exchange of livestock implies close emotional ties. Exchange partners are invariably addressed as friends (*kongot*). The absence of exchange partners gives rise to *choykonöt* (loneliness) - a feeling every person fears.

And, if after a long dry season, one goes to visit one's stockfriends in distant places, one feels *emö* (a strong longing for close friends). Herders who do not easily give in to the demands of exchange partners might be accused of meaness and even witchcraft (Bollig, 1992a:162ff). Men who are often sick, claim that the deeper reason for their disease is the bad will (*ghöityö*) and envy (*ngatkong*) of others, to whom they have denied livestock presents. The co-ordination of exchanges is deeply rooted in the concepts of the emotional and moral self and the distribution of livestock and the collecting of livestock debts are cornerstones of male identity.

6.2.1.2. The Exchange Network Put to the Test: Transactions during a Drought

In the previous chapter I described how the exchange of livestock and the sharing of food contributed significantly to food security during a crisis. The existence of a gift-exchange relation paved the way for asking a partner or a relative for a goat or directly for food. Bridewealth donations, bridewealth distributions, and stockfriendships create a framework for co-ordination and facilitate exchange in a crisis. *Ghosyö* (indebtedness among friends) features as a conceptual frame for minor transactions. The data shows that it is important to differentiate if household survival or livestock herds are to be insured. It is the social capital which helps people to make it through actual food shortages; but it is the economic capital as stored in relations of mutual indebtedness that may be used for herd reconstitution at a later stage (see Figure 22).

A look at figures for 36 herders suggests that herd reconstitution took place mainly through bridewealth donations and distributions and the collection of livestock debts. The case study of Chadar's efforts at reconstituting his depleted herd in 1992/93 may serve as an example. Turkana raiders had stolen most of Chadar's livestock in August 1992. All the cattle and goats

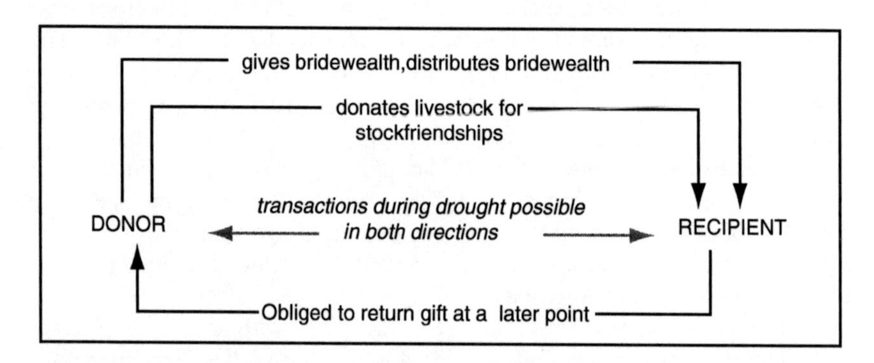

Figure 22. Nesting of Minor Transactions within the Framework of Major Gift Exchanges

were taken by raiders and only a few sheep remained. Chadar is a poor herder who had not donated many cattle to other herders.

> Soon after the devastating raid, Chadar went to his son-in-law in Seruni where he had left two heifers. These he took back to his homestead. A week later the father of another son-in-law donated 1 she-goat and 2 male goats. Some time later (March) he obtained goats from yet another son-in-law and from a long standing friend. His son-in-law gave him two lactating goats with kids and two male goats. The friend gave one lactating goat with a kid plus two males. In the same month, one brother gave two lactating goats with kids and the son of the elder he had stayed with for years gave one male and one female goat, the old man himself donated a ram. In April and May he obtained 4 more goats from outstanding debts. In 1988 it had been Chadar who had circumcised the young men. Each initiant owed him a goat for his services and many had not yet paid their debt. In September 1993, a little more than a year after the raid, his goats herd had grown to sixty animals once again. 10 were directly from donations from relatives and friends and 17 had been bought with money obtained from the sale of the male goats he had received from friends. 16 had been born by the ewes he received in 1993. Chadar was adamant that the next thing he would do would be to devote his energy to the reconstitution of his cattle herd.

Most households who had lost livestock in 1991/92 only gradually restocked their herds. Those households who had built up extensive social ties fared somewhat better. The households of Wiapale and Lomirmoi for example, both senior men of the youngest age-set, received some animals from exchange partners in the post-drought period. The effect of social relations, however, is partially overcome by simple life-event chances. Several herders married off daughters[14] in the post-drought period and reconstituted their herds via livestock loans.

Herd reconstitution, especially the reconstitution of cattle herds, takes place over a period of years. A herder has to "follow the cattle" as the Pokot say to ensure that he gets a share in bridewealth distributions and that he obtains some animals from stockfriends. Typically a herder is given one or two animals by a number of different donors. It is nearly impossible to obtain larger numbers of livestock from one single donor. The overall decline of cattle numbers makes the restocking of herds ever more difficult. However, the acquisition of cattle via social exchange is still the only opportunity to recoup losses. As yet there are only very few Pokot with access to sufficient money to buy cattle. Chadar's example shows that an internal market for small stock does exist. In the foreseeable future Pokot will depend mainly on this tightly woven network of livestock exchanges which basically institutes the continuous redistribution of livestock. However, they are acting in a downward spiral. Ever more Pokot have to distribute from a stagnant or even declining herd.

6.2.2. Networks of Dominance among the Himba of Northwest Namibia

Exchange networks among the Himba are also based on the exchange of livestock. However, livestock transfers take a different form than among Pokot herders: a herder preferably engages in exchanges with kin. Exchange relations with non related, or distantly related people are rare. Next to single animals, entire herds are transferred from wealthy herders to poor herders. Strongly unequal livestock exchanges establish patron-client relations and are the basis of political dominance.

6.2.2.1. Inheritance and Livestock Loans among the Himba of Northwest Namibia

Livestock loans of single animals and of entire herds and livestock transfers at inheritance are the major institutions for distributing and transferring use rights and property rights in livestock. Bridewealth is low (only two to three head of cattle) and does not play a major role in exchange relations. Livestock exchanges between friends, which are so prominent among East African herders, are virtually absent in Himba pastoral economy. Almost a fourth of the off-take goes into internal exchanges. When transferring livestock the Himba differentiate clearly between use rights and ownership rights. Only in the case of inheritance and livestock presents do ownership rights change alongside use rights. In most cases, however, when livestock are transferred, only use rights are exchanged while ownership rights stay with the original possessor. Livestock loaning as practised among the Himba leads to a situation whereby numerous herders rely almost completely on loaned stock. For young herders this is the normal situation: they start off with a herd which consists mainly of borrowed livestock and gradually expand their own property. For impoverished herders the dependence on borrowed stock may circumscribe a lifelong problem.

The recent distribution of livestock wealth reflects a post-disaster scenario. After the major drought of 1981, it was first of all the established households which had the capacity to allocate cattle, goats and sheep to other households. The three richest households of the sample had put substantial numbers of cattle in three to five other households (usually the household of a younger member of their matriline) who were more or less dependent on their stock loans. Furthermore they had given single heads of livestock to numerous other households. About ten other households had loaned small numbers of livestock to other households. All households herded considerable numbers of loaned stock: of a total of 4,630 animals surveyed in 35 households, 42.6 per cent were loaned stock, that is they were not herded by their owner and their milk was not used by their owner's families.

Himba differentiate two major forms of livestock loans. Many herders, even poorer ones, share out single animals and loan them to relatives. These animals are addressed as *ozondisa* - herded animals. Next to single animals, rich people allocate entire livestock camps to young and poor herders. Informants were generally of the opinion that if more than ten cattle were given by a rich herder to one household, one should rather talk about establishing a livestock camp (*ohambo*) than about giving out livestock loans (*ozondisa*). However, the differentiation between livestock loans (*ozondisa*) and an allocated livestock camp (*ohambo*) is vague. If livestock loans are repeatedly taken from one wealthy relative, the receiving household will gradually lose its independence and become a livestock camp of the rich household.

Matrilineal Kinship, Inheritance and Exchange

Exchange and Matriclan: Literature on the Himba (Crandall 1991, 1992, Malan 1972, Steyn 1977) emphasises that livestock loans are habitually donated to members of one's own matriline (*omuhoko*). Looking at quantitative data obtained from 23 informants on exchanges a clear dominance of intra-matriline exchanges over extra-matriline exchanges is not corroborated. Only 41 per cent (55 out of 135) of all recorded exchanges fell within one's own matriline. For the two matrilines dominating the area[15], the two subclans of *Omukwendata* (*ondjuwo onene* and *ondjuwo ondiṯi*), the percentage of intra-matriline exchanges is higher (44 per cent and 63 per cent respectively) than for the rest. If both subclans of *Omukwendata* are lumped together the percentage of intra-matriclan exchanges is at 67 per cent. Members of both clans do not only dominate numerically in the region. Some of their members belong to the wealthiest group of households and they also hold the main positions of power. The data presented in Table 26 suggests that clan membership is one important institution for organising livestock exchange. If a

Table 26. Intra- and inter-matriclan livestock exchanges among the Himba

	1	2	3	4	5	6	7	8	9	10	11	Total	%
1	16	6	2	2		3	4	1			2	36	26.7
2	3	15	1	1	1		1				2	24	17.8
3												0	0
4	1	1	3	2		2	2	1		1	1	14	10.4
5	2		1		4	1	1				1	10	7.4
6	1	6		1		5	2				5	20	14.8
7	3	3	1	1		1	12		1			22	16.3
8	4	1		3				1				9	6.7
9												0	0
10												0	0
Total	30	32	8	10	5	12	22	3	1	1	11	135	100.1
%	22.2	23.7	5.9	7.4	3.7	8.9	16.3	2.2	0.7	0.7	8.1	135	99.8

Note: (1) Omukwendata wondjuwo onditi (5 households), (2) Omukwendata wondjuwo onene (6), (3) Omukwendjandje (0), (4) Omukwenatja (3), (5) Omukwandongo (1), (6) Omukwatjivi (2), (7) Omukweyuva (4), (8) Omukwenambura (2), (9) Omukwambo (0), (10) Omukwahere (0). Read cell 1/1: members of Omukwendata wondjuwo onene exchanged 16 cattle amongst themselves, gave 6 cattle to members of Omukwendata wondjuwo onditi, 2 to members of Omukwenatja etc.

matriclan is numerically dominant in an area there is a tendency that members of this clan concentrate on intra-clan exchanges. If a clan is only sparsely represented in an area its members will engage in livestock exchanges with members of dominant clans. Apparently spatial proximity and political power come in here as a further ordering factor.

Exchange and Inheritance: A further look at the exact kinship relation between donor and recipient of livestock exchanges shows that livestock loans follow a more complicated path than just an intra-matriclan pattern. Exchanges are not conditioned so much by social structures as they rely on personal considerations on the access to livestock. There are several exchange partners which are clearly preferred to others: these are ZS (14 out of 91 exchanges for which kinship relation could be established, i.e. 15.4 per cent), MBS and MBSS (19, 20.9 per cent), of further importance are MZS, MZD, MZDS, MZDDS (11, 12.1 per cent), and distantly matrilineally related people (9, 9.9 per cent). On the other hand FZS and FBSS (9, 9.9 per cent), FBDS (4) and FBDD (6, 6.6 per cent) and FFBS and FFBSS (9, 9.9 per cent) as well as FFDS and FBDSS (5, 5.5 per cent) are of some importance. Only a few animals are donated to one's daughters (2, 2.2 per cent) and sons (3, 3.3 per cent). Three animals were given to brothers and brothers' sons, one to a sister and a further one to a wife's brother. Friends (defined as non-related but emotionally very close companions) were not given anything.

When allocating livestock loans, herders apparently take into account the relation between past and future inheritance transfers and livestock exchanges; ZS will inherit one's entire herd one day. The 14 (15.4 per cent) animals donated to ZS's could be regarded as an advance on the future inheritance (see Figure 23). If one has inherited from one's MB, frequently some animals are left or are given back at some stage to MBS and MBSS (19, 20.9 per cent), that is to the children of the person who leaves the inheritance. FZS is the legal heir of one's father, the person who has inherited one's father's herd after his death. Relations to FZS and FZSS are obviously close through the inheritance of one's father's property (11). FMBS and FMBSS (9 animals) are related to F via inheritance transfers too: the father of FMBS (i.e. FMB) has inherited his herd to F. FFZS and FFZSS (5 animals) are connected to FF via an inheritance transfer (FF inherited to FFZS). MZS (and by extension MZD, MZDS and MZDDS) is structurally in thc same position as oneself, as he will also inherit from MB (11 animals). Relations are strengthened through the common claim to one herd. In total 63 (69 per cent) transfers can be explained by reference to inheritance exchanges. Himba try to strengthen ties to potential heirs, to offspring of one's MB (whose herd one will inherit) and to people one has a common inheritance claim with towards one herd. Livestock loans are governed by a fusing of matriclan solidarity and personal interest: they support individual inheritance claims more than simply enforcing clan solidarity. An individual's wish to gain and maintain access to an inheritance may be defined as the main motive of livestock exchanges.

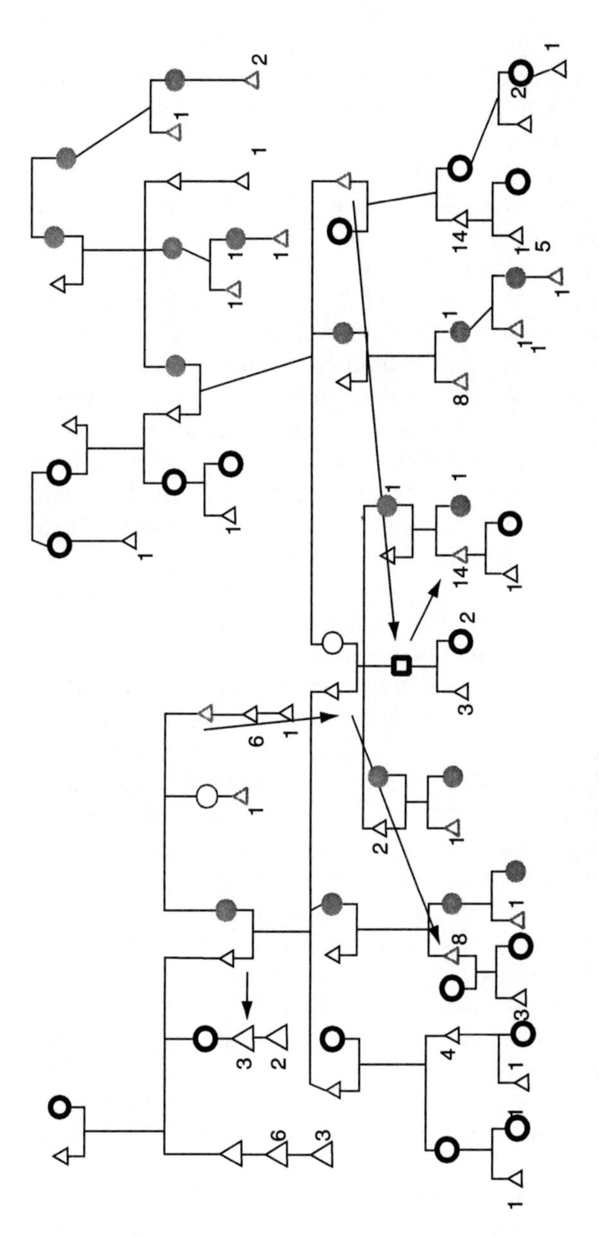

Note: ◻ Ego, ● △ members of ego's patriline, ● △ members of ego's matriline, dashes indicate inheritance transfers; the numbers indicate the number of cattle received or given by ego to a specific relative; the figure shows that factual and potential inheritance transfers as well as matriclan membership are important for directing livestock loans. The data base were 98 livestock loans for which kinship relations could be specified exactly.

Figure 23. Kinship Relation between Donor and Recipient of Livestock Loans

Conditions of Exchange: Usually only heifers are exchanged on a loan basis, male stock are given as presents. The recipient keeps the borrowed livestock; he drinks their milk and may, after some time, even take out a male animal for sale or slaughter. If he wants to barter an animal for maize or to slaughter it at a celebration, he must wait until the animal that he initially borrowed has given birth to several calves. He does not have to ask the owner of these animals beforehand; however, he should tell him about the fate of the animal afterwards. All animals descending from the borrowed heifer - no matter how long in the past - still belong to the original owner of that heifer or his legal heir. Occasionally the owner may present an animal to the borrower thereby enabling him to gradually build up his own herd. Although property rights in loaned animals stay with the original owner, livestock loans are relatively safe capital. Moral obligations prevent the owner of the cattle asking for a lot of his cattle back at the same time. Conditions are different for livestock camps. The owner of a camp may take several animals from his camp at once. Only if the person who borrowed the animals neglects them, then the donor is entitled to ask for his cattle back. Although this possibility was considered theoretically, I have never heard of a case in which an owner took back all his animals at once. While borrowed cattle are safe capital, it is deemed appropriate to obtain borrowed animals from many different recipients. To spread one's benefactors is seen as the only way to prevent dependence on one wealthy patron.

Livestock Presents (*ozohakera*): Of course it is preferable to obtain livestock presents instead of livestock loans. About a third of all livestock exchanges (excluding the transmission of livestock camps) are presents and two thirds are loans. Informants unanimously stated that it is difficult to obtain livestock presents. There are two principal ways of obtaining gifts. One may ask a close relative directly for a gift or one may try to transform a loaned animal into a livestock gift. The first case seems to be more frequent. The donation of livestock gifts underlines the close emotional contact between two people. Several herders claimed that rich people in particular are hesitant in giving livestock as presents and thus giving up ownership rights to animals. I did not find any quantitative proof of this assumption. However, progeny histories showed that rich people did not give more presents than poorer households. While they gave out significantly more stock as loans than others, the number of livestock presents was similar.

Herders clearly have a better chance of obtaining livestock presents after they have reached a specific age (see Table 27). While of the animals obtained by herders younger than 40 only 5.8 per cent were presents, the percentage of animals given as a present grows to more than 25 per cent for people between 40 and 50.

Livestock Camps: To borrow a cattle camp from a wealthy household is a strategy of young men, as well as of impoverished herders in order to continue a pastoral livelihood. Basically the transaction comprises the exchange of labour for the use of milk and, to a very limited extent, meat. The donation of a livestock camp establishes a patron-client relationship. Quantitative

Table 27. The acquisition of livestock presents in relation to age

Age of Herder	<35	%	36-40	%	41-45	%	46-50	%	51-60	%	61-70	%	>70	%	Total	percent
All other form of exchange	36	94,7	68	73.1	80	73.4	84	81.6	91	84.3	21	70.0	8	88.9	229	69.6
Animals presented	2	5,3	25	26.9	29	26.6	19	18,4	17	15,7	9	30.0	1	11.1	100	30,4
Total	38	100	93	100	109	100	103	100	108	100	30	100	9	100	329	100

evidence shows that livestock presents from patron to client are few. A herder, however, may well decide that he wants to slaughter or sell one tollie from the borrowed herd in order to upgrade the nutritional standard of his household. He can do so without consulting the owner of the herd, as long as he tells him about the sale or slaughter at a later stage. Many young men start their career as dependent herders of a livestock camp of a wealthy senior family member. Structurally, there is little difference between herding one's father's camp and herding a close relative's camp. In neither herd does one have ownership rights. One's father's herd will one day be taken by his brother or his nephew and the relative's herd will be frequently taken by somebody else. Both herds offer the opportunity of feeding an incipient household and of increasing one's own herd gradually. Day to day decisions on grazing and migrating are taken independently from the household of the camp owner. He should - and in reality usually does - show little interest in the daily routines of his camp. The following two case studies show two forms of livestock camp transactions. In the first case the camp is taken on by a young successful herder, eager to increase his own share in the herd, in the second case the camp is taken on by an impoverished herder.

Case Study: Zoowenda, a young successful herder

Zoowenda, about 30 years old, was given a cattle camp by his maternal uncle Tjandero (his MMB). Zoowenda is not Tjandero's direct heir and his chances of inheriting his uncle's herd one day are slim. Zoowenda is married to a 12 year old girl who usually does not stay with him. He stays with another relative of his age, who has an adult wife.

The distribution of ownership rights in Zoowenda's herd is typical for a cattle camp: out of 171 animals, 103 (60 per cent) were owned by Tjandero and another 15 (9 per cent) by Tjandero's daughter. Only 8 (5 per cent) animals were owned by Zoowenda himself. Remarkably, none of them was a present from Tjandero: there was one cow plus two calves presented to him by a paternal uncle, a cow plus one calf presented to him by a cousin with whom he herded together for quite some time and a cow he had exchanged for an ox (which he owned himself) plus two calves. Some 44 animals were borrowed by Zoowenda himself (26 per cent) from other destinations: from his own father, from another paternal uncle, another cousin (MBS) and a maternal uncle. A closer survey of all the animals which were loaned by Zoowenda showed that all the recipients were not only related to Zoowenda but closely related to Tjandero as well. All of them would have been eligible for direct livestock loans from Tjandero.

Case Study 2: Kahima, an impoverished herder

Kahima, an impoverished herder, is about 45 years old. He maintains a very small household which consists of his wife, his wife's daughter and her toddler, and himself. All 14 cattle he herds belong to his maternal uncle Kapira.

Although he will not inherit Kapira's herd, he relies fully on livestock loans from his uncle. Kahima apparently has always been poor and even his late father belonged to the poorest strata of the society (so his neighbours claimed). Kapira's donation (five female cows, plus seven calves and two heifers) enables him to live as a herder and to supply his small household with some milk. The absence of further livestock loans in Kahima's herd shows that he will find it difficult to become an independent herder in his own right. At his age, herders have usually built up an incipient herd on their own while still relying on borrowed stock, ideally from different owners.

The disadvantages of extensive livestock loans in the form of cattle camps come from the realm of values. Somebody who is mainly subsisting on loaned cattle (and is well above 40 years of age) is not regarded as a successful herder. Hence, individual strategies aim at spreading the number of destinations they borrow from. Every herder asked about this topic, preferred five animals from five different donors to five animals from one donor. Even if one does not get along with one donor, one can still rely on a number of others.

Structural and Normative Correlates to Reciprocal Exchange

Himba prefer to exchange livestock with their (or their father's) close matrilineal kin. Marriage rules and exchanges at marriage tend to reinforce corporate ownership of livestock by the matriline.

Marriage Patterns and Exchange: The Himba preferably marry cross-cousins, *ovaramwe*. They argue that cross-cousin marriage is the best way to ensure marital stability. Cross-cousins are formally speaking MBD and FZD, but the Himba have a very broad understanding of cross-cousinage[16] and include many kin relations they regard as structurally equivalent into this category. While factually there is a clear tendency to define this concept very broadly, everybody is keen on explaining how closely related both spouses are. The majority of Himba marriages conform to the prescribed cross-cousin pattern (see Table 28).[17]

Table 28. Cross-cousin marriage among the Himba

Matrilateral cross-cousins	n	Patrilateral cross-cousins	N
MBD	19 (38%)	FZD	33 (47%)
MBDD, MBDDD, MBDSD, MBSD	11 (22%)	FZDD, FZDDD	24 (34%)
MMBD, MMZD, MMZDD, MMMBDDD	10 (20%)	FZSD, FZSDD	2 (3%)
MZD, MZDD	3 (6%)	FFZD, FFFZDDD, FFDD, FFZDD	5 (7%)
MFZSD, MFBD, MFZDD	7 (14%)	FBD, FBSD	4 (6%)
		FMBD	1 (1%)
		ZD	1 (1%)
	50		71

The Himba generally claim that marriages should fit broadly into an inheritance strategy. The following reasoning is behind MBD marriages. Ideally the daughter of the person leaving the inheritance marries her father's heir: that is, ego transfers his livestock property to his ZS, who at the same time stands in an MBD relation to his daughter and is an ideal marriage partner. Modelled over the span of two generations, one could claim that the grandson of the person leaving the inheritance (his SS) will inherit the herd from the ZS of his grandfather (see Figure 24). While this second consideration is quite theoretical, the first one is claimed to be the major idea behind marriage strategies. However, factual data on inheritance shows that due to frequent divorces, primogeniture and the complexity of the Himba inheritance system, marriages frequently do not link up with inheritance chains. It rather seems that arranged marriages try to manifest an option for a linkage. Whether this potential link will be realised one day is open for discussion and to a great deal of chance.

If preferential marriage patterns do not pathe the way for distribution, do exchanges at marriage facilitate the redistribution of livestock property? Again the answer is no. The bridewealth payments are low among the Himba. At marriage, the wife-takers just have to transfer one ox (*otjitunia*) and one cow (*orutombe*). Additionally one male sheep, usually a castrated one, is transferred.[18] At rituals preparing the marriage, a sheep is slaughtered in the homestead of the bride-giver and another one in the homestead of the wife-taker. During the marriage ceremony an ox may be slaughtered additionally at the household of the wife-takers.

6.2.2.2. *The Exchange Network Put to the Test: Exchange and Recovery after the Drought of 1981*

Rules and practice of livestock exchange have been described in the preceding paragraphs in detail. But what do livestock exchange and inheritance have to do with risk minimisation? How do they help actors to reconstitute their herds after serious losses? The data on herd composition obtained from 35 households and on about 4,600 animals offers an interesting case study in this respect. In 1981 the herds of the Himba decreased by 90 per cent during a centennial drought. While all the herders had been massively affected by the drought, there were differences: rich herders still maintained more numerous herds than poor herders did. Households which had shifted to southern Angola early on, were left with slightly more cattle than those which had remained in Namibia throughout. Figure 25 shows that restocking via internal transfers of livestock gained momentum only in the late 1980s when the overall number of cattle in the system had grown. While in 1982 only 16 animals were exchanged, in 1988 some 61 animals were exchanged. Livestock numbers had reached a historical low in 1982 when only 15,000 cattle were counted in the entire Kaokoland. Five years later in 1987 the number had grown to roughly 60,000, and about another ten years later it had reached the pre-drought level.

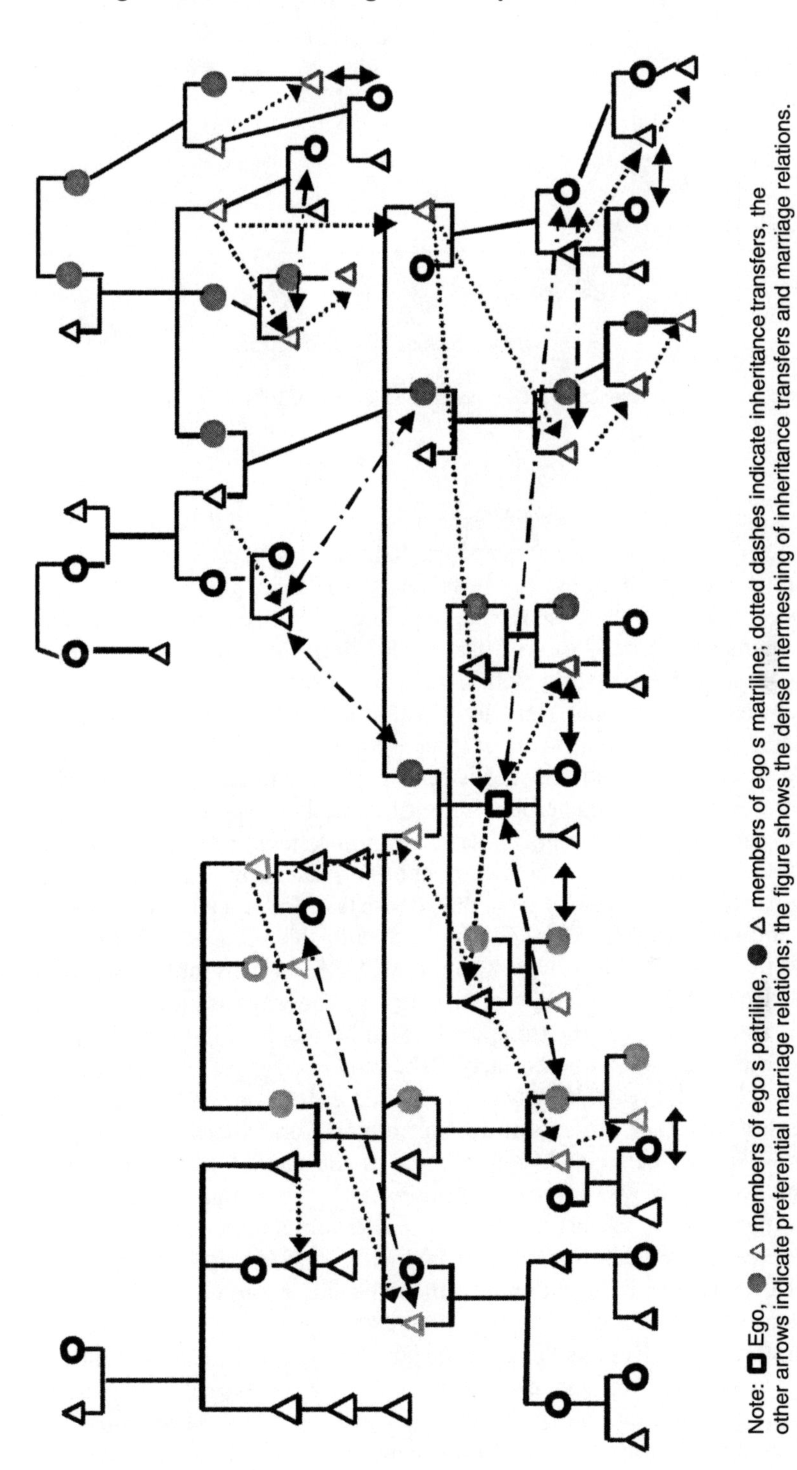

Note: □ Ego, ● △ members of ego s patriline, ● △ members of ego s matriline; dotted dashes indicate inheritance transfers, the other arrows indicate preferential marriage relations; the figure shows the dense intermeshing of inheritance transfers and marriage relations.

Figure 24. Himba Marriage and Inheritance

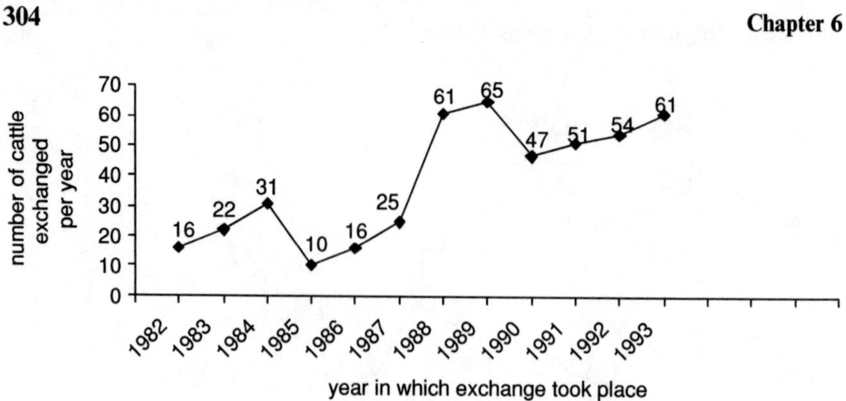

Figure 25. The Restocking of Herds after 1981

The major form of livestock exchanges in this crucial period of herd reconstitution (1982-1993) was livestock loaning, about 51 per cent of all cows and heifers exchanged were transferred as loans. 18.4 per cent were given as presents and 15 per cent were inherited. 12 per cent of all the heifers were exchanged internally or with traders for oxen. Most households herds were reconstituted successfully within a period of ten years.

Individual Herders and Herd Reconstitution via Exchange: Households take very different approaches to restocking and gain from exchanges to a different extent. A look at the reconstitution of individual herds shows different strategies: the young herder Koriautuku, perhaps 40 years of age, borrowed 44 cattle mainly from his maternal grandmother and his father. As yet (in 1995) he has not gained a single animal as a present or as an inheritance. Consequently his considerable herd only consists of livestock loans. His aged father's animals will fall to his FZS on his death; the animals of MMZ will go to his mother on her death. However, both MMZ and his mother were living with Koriautuku at the time of the survey. The way his stock loans are distributed does not pose the threat of sudden herd depletion through inheritance of stock loans to another party. Vahenuna and Mutaambanda are two further herders who relied heavily on stock loans. However, both were able to acquire some livestock presents and some animals from inheritance transfers. Both are exposed persons: Mutaambanda is the son of the former chief Mwinimuhoro who was famous for his wealth, Vahenuna is the son of Mwinimuhoro's heir. Both acquired a considerable number of stock loans from the respective heirs of their father's herds. The formal heir frequently leaves many cattle with the children (usually the eldest son) of the testator on a loan basis.

Another fact strikes as being important: the two richest households of the sample, Katjira and Mirireko, both headed by old experienced herders, only acquired a few animals from exchange: both only took 11 animals each from exchange. On the other hand poor herders like Tjandira and Kaeretire also only got little from exchange: Tjandira only got one animal as a present, he did not inherit livestock and he did not take any livestock loans. Kaeretire took 5 animals; he borrowed 4 and got one more as a present. The bulk of exchanged animals goes to the large number of household heads which are in between these two extremes - neither very poor nor very rich: Hikuminwe for example acquired 6 cows on a loan basis and 5 as livestock presents, but received 14 animals in various inheritance cases. Two herders, Tako and Mbangauiye acquired substantial numbers of animals via direct exchanges of oxen for heifers from other herders and traders. Traders from the white ranching areas, from Kwanyama and Okakarara came to exchange their heifers for Himba oxen. Additionally Tako had bred horses with the aim of exchanging these for heifers. The customary exchange rate was one heifer and one tollie per horse. In three cases he succeeded to exchange the tollie he received for a heifer as well after some years. The data shows that approaches to herd constitution differ a lot. The following paragraphs explore whether certain patterns are observable.

Age and Herd Reconstitution via Exchange: While there is a clear preponderance among younger herders to rely on livestock loans, among older herders the inheritance of livestock becomes more important. The figures basically show that herders under 45 are almost excluded from inheritance while people over 60 gain a major portion of all the inheritance available for themselves (see Table 29). Only 1.5 per cent of all the animals obtained via exchange by herders under the age of 45 originated from inheritance transfers. The major part (74.4. per cent) originated from livestock loans, and some 20.3 per cent from livestock presents (and only some 2.6 per cent from inheritance). Herders between 45 and 59 obtained 16.5 per cent of all animals through inheritance, 42 per cent through livestock loans, 22.6 per cent through livestock presents and some 18.9 per cent through other forms of exchange. Herders of 60 and above obtained nearly equal shares through the three modes of exchange: 27.9 per cent in inheritance transfers, 27.9 in livestock loans, 32.7 per cent in livestock presents and 25.9 per cent through other forms of exchange. Of all the animals transferred in inheritance, herders of 60 and above obtained 52 per cent, people between 45 and 59 obtained 44.9 per cent while herders under the age of 45 obtained just 2.6 per cent of all the inheritance transfers.

The majority of livestock presents go to middle-aged herders. 27 per cent of all livestock presents transferred went to herders of 60 and above, 38 per cent to herders between 45 and 59 and 27 per cent to herders younger than 45. Younger herders garner the majority of all livestock loans: 43.2 per cent go to herders below 45, 38.9 per cent to herders between 45 and 59 and only 17.9 per cent to herders of 60 and above. These figures indicate that different age groups rely on different forms of stock acquisition. Only people of 45 and above are able to obtain ownership rights in a substantial number of animals. Younger herders may obtain use rights through borrowing cattle from others but fail to acquire ownership rights. In sheer numbers, however,

Table 29. Livestock exchange and the age of herders

Age of herder	<35	%	36-40		41-45	%	46-50	%	51-60	%	61-70	%	>71	%	Total	%
Animals loaned	34	89.4	65	69.8	48	44.0	41	39.8	28	25.9	11	36.7	2	22.2	229	46.9
Given as presents	2	5.3	25	26.9	29	26.6	19	18.4	17	15.7	9	30.0	1	11.1	100	20.5
Animals inherited	0	0	2	2.2	24	22.0	11	10.7	37	34.3	4	13.3	0	0	78	16.0
All other	2	5.3	1	1.1	8	7.3	32	31.1	26	24.1	6	20.0	6	66.6	81	16.6
Total	38	100	93	100	109	99.9	103	100	108	100	30	100	9	99.9	488	100

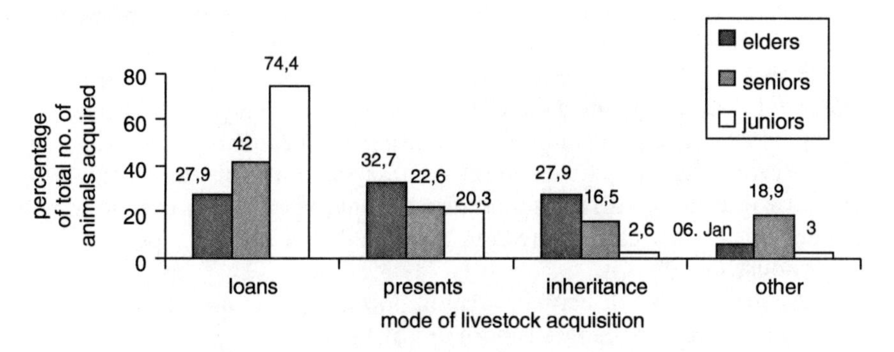

Figure 26. Mode of Livestock Acquisition amongst Elders, Seniors and Juniors

younger households (below 45) obtained 26.8 per cent of all the cattle exchanged, herders between 45 and 59 obtain 43.4 per cent and herders of 60 and above 29.8 per cent. This indicates that to some extent middle-aged herders are able to manipulate the internal exchange system to their benefit (see Figure 26).

Among seniors almost one fifth of all heifers and cows were acquired via the exchange of oxen for female stock. These exchanges took place among Himba herders as well as among Himba and traders. Occasionally these direct and balanced exchanges take the form of barter: a Himba buys a load of alcohol and tries to barter it for a heifer with a fellow herder. Other forms of exchange were negligible: only 9 out of 488 animals were returned loans, only two animals were actually bought with money and only one cow stemmed from a bridewealth payment.

Wealth and Herd Constitution via Exchange: It is mainly the poorer households that rely on richer households (see Figure 27). Poor households

Figure 27. Modes of Exchange in Relation to Wealth

obtained 35.4 per cent of all 229 animals borrowed, 48 per cent went to medium households while rich households obtained only 16.6 per cent. This indicates a net flow of livestock loans from rich households to poorer households.[19] Livestock presents were distributed more evenly. Rich households obtained 24 per cent, medium households 37 per cent and poor households 39 per cent. Rich households on the contrary obtain more animals than the other two wealth classes from inheritance exchanges: 46.2 per cent go to rich households, 41 per cent to medium households but only 12.9 per cent to a poor household.[20]

Again the relation between wealth, political power and the capacity to control individual exchanges is born out by the data.

The donation of livestock loans brings prestige to a herder. Crandall (1992:105) aptly reports: "The superior man is one perceived to give generously; he does not expect reciprocation of the gift in kind, but in the form of respect, loyalty and quiet affection."

Kinship and Herd Constitution via Exchange: Himba clearly express the idea that exchange should take place between closely related people. For 387 animals exchanged, the kinship relation between donor and recipient could be established. Mother's brothers were the most important donors. They alone accounted for 30.2 per cent (117) of all exchanges. The father was the second most important person to donate livestock, accounting for 15.2 per cent (59) of all exchanges.

The fact that mothers (together with mother's sisters, grandmothers, grandmothers' sisters) are of considerable importance as donors of cattle is highly interesting. Himba women do not act as livestock owners in public. However, compared to other pastoral groups, Himba women own substantial numbers of livestock. They receive the stock as livestock presents or as minor shares of inheritance. In many instances I found that women do not put their animals in their household's herd but allocate the animals to a young matrilineally-related herder - the closest matrilineal relative being their own son. Animals owned by one's grandmother are inherited by one's mother (if she has no brother). Hence, livestock loans coming from women are very safe capital. Sisters' sons are fairly unimportant as donors; they only gave about 2.3 per cent (9) of all cattle exchanged. A nephew is a major recipient of donations, but gifts from nephew to uncle are highly exceptional. They are regarded as shameful, disclosing the poverty of thc mother's brother. Distantly related or non-related people only account for 4.9 per cent of the total exchange.

Spatial Aspects of Herd Reconstitution via Exchange: A major proportion of animals acquired for herd reconstitution came from nearby places (see Figure 28). Himba networks of livestock exchange are not per se far flung and spatially extended. 59.9 per cent (224 animals) were exchanged between actors that were not living further than 30 kilometres away; 27 per cent (101) of all animals were more or less exchanged in the immediate neighbourhood. Only 35 (9.3 per cent) animals were exchanged over a distance of 100 kilometres or more (see Map 16).

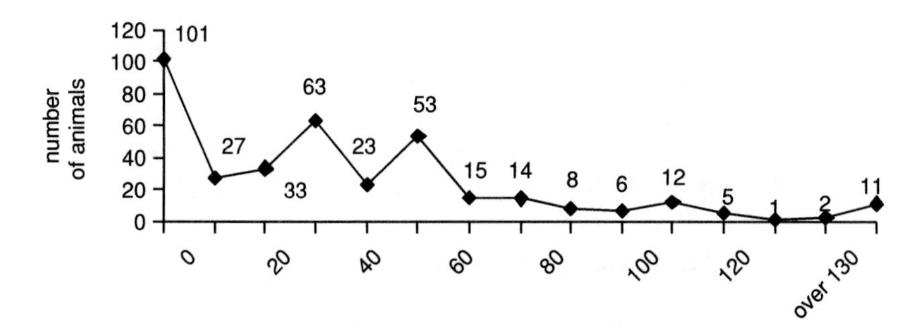

Figure 28. Distance and Livestock Exchange

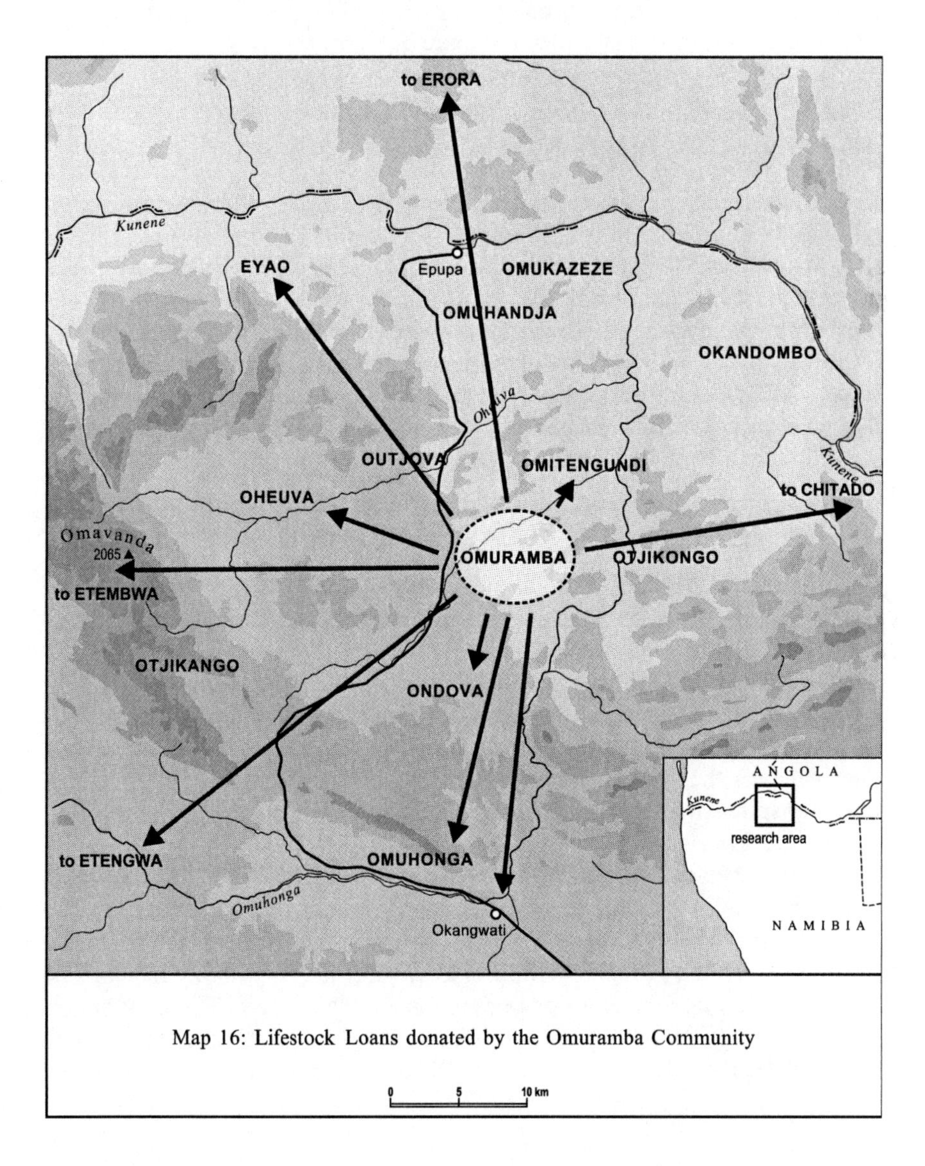

Map 16: Lifestock Loans donated by the Omuramba Community

The figures indicate that a major portion of animals are exchanged within the extended neighbourhood. However, the scope of exchange networks is wider than that: herders that live a hundred or more kilometres apart interact and exchange livestock with each other. The previous passage on kin and exchange has shown that exchange partners are usually closely related. This also holds true for exchange partners who come from a distance.

6.2.3. Comparing Exchange Networks

Exchange networks result in very specific forms of social organisation. The structures of networks condition and facilitate exchange. The subsequent chapter will, on the one hand, compare structural properties of exchange networks in both societies and then compare qualitative aspects of exchange.

6.2.3.1. A Formal Comparative Account of Livestock Exchange Networks

For both societies partial networks consisting of about 40 household heads (Himba 42 household heads, Pokot 37) were defined and all livestock based transactions and kinship relations between them were noted down. The households were taken from a wider neighbourhood in both samples.[21] The most important forms of exchange between households in both societies include livestock. Among the Pokot bridewealth transfers, bridewealth distributions and stockfriendships are the major forms of exchange, while among Himba livestock loans and presents between relatives as well as inheritance transfers are the major forms of exchange.

6.2.3.1a. Structural Qualities of Networks (density, clusters)

The analysis of network data was run with the programme UCINET (and UCINETX, both Borgatti, Everett & Freeman 1992), which was designed to detect the structural characteristics of networks. At first glance both networks are close knit. However, some differences are to be seen if one looks at the graphs representing the respective networks with some patience. The Himba network seems to have some marginal actors and some very central ones whereas the Pokot network seems to be more integrated. Differences only transpire when a detailed analysis of network properties is undertaken (see Figure 29a and 29b).[22]

Simple Measures of Structural Characteristics: Density, Centralisation Density[23]: The indices for density in both networks differ profoundly. The Pokot network shows a much higher density than the Himba network. While the density for the Himba network is at $\Delta = 0.153$ (only about 15% or one seventh of all possible relations were realised), the density of the Pokot network is at $\Delta = 0.359$ (about 36% or more than a third of all possible relations were realised) (see Table 30). There is little data on networks which are really

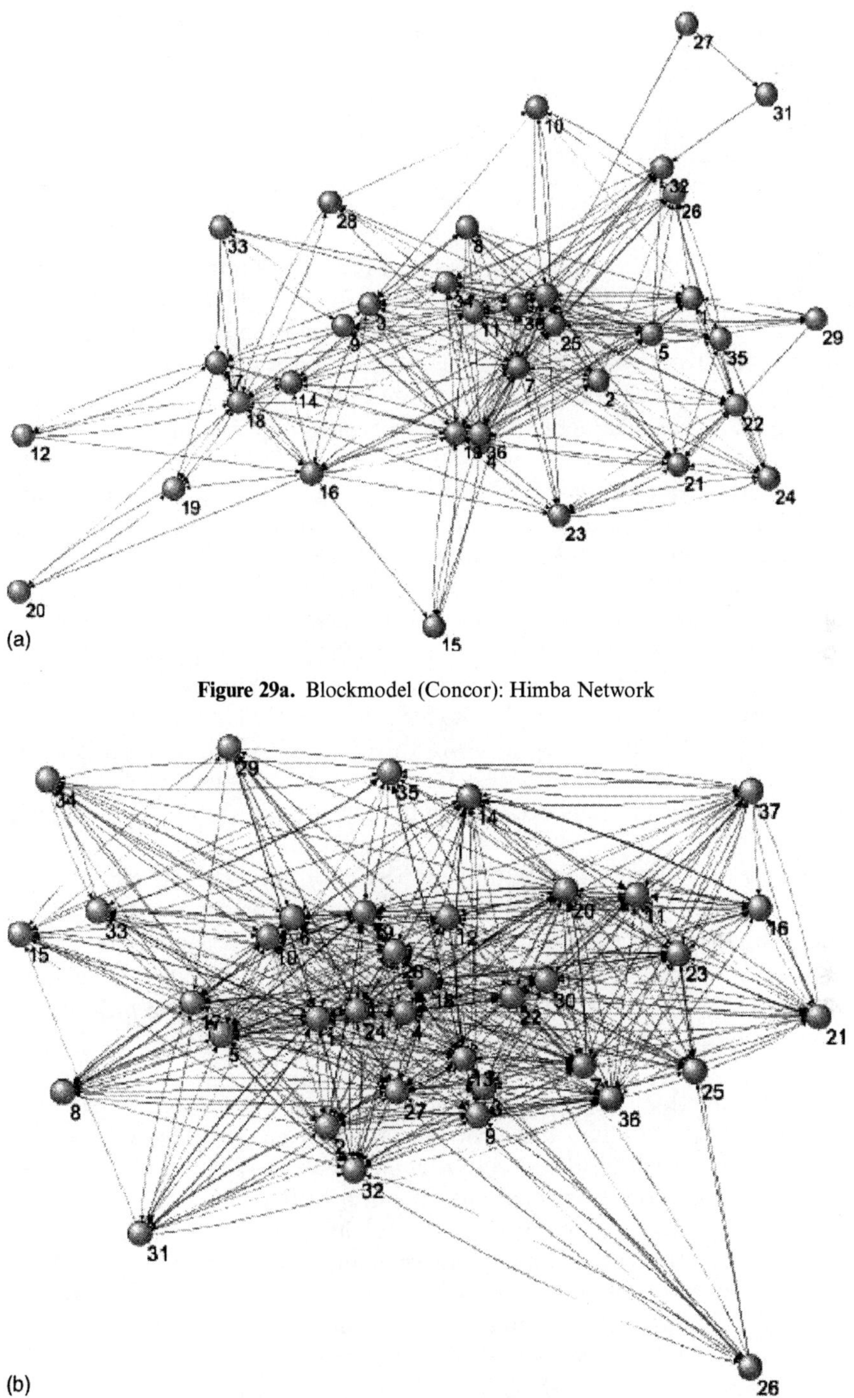

Figure 29a. Blockmodel (Concor): Himba Network

Figure 29b. Blockmodel (Concor): Pokot Network

Table 30. Structural characteristics of both Pokot and Himba networks compared

	Pokot	Himba
Density	0.359	0.153
Degree centralisation %	21.19	44.63
Closeness centralisation %	20.46	42.87

comparable. Schweizer (1997) gives figures for the !Kung network of *hxaro* exchanges and the *kula* exchange among the Trobriand islanders. While density in the *kula* network is at $\Delta = 0.20$ (in this case the islands were rated as actors), the !Kung network had a much lower density at $\Delta = 0{,}07$. These figures suggest that the cohesiveness of the Pokot network is very high. The density of the Himba network is not necessarily low, but much lower than the Pokot network.

Degree-Centralisation[24]**:** While the Pokot network is more cohesive than the Himba network, i.e. more actors interact with each other, transferring livestock, the Himba network is more centralised. This indicates that in the Himba network there are more actors who are able to concentrate exchanges upon themselves while other actors are peripheral to the exchange network.

Closeness-Centralisation[25]: The values for both networks tend in the same direction as the measure for degree-centralisation: in the Himba network there are dominant people who have more capacity to control exchange than among the Pokot. The measure for Closeness-Centralisation is about twice as high for the Himba than for the Pokot: this suggests that we find more egalitarian exchange relations in the Pokot network than in the Himba network.

This first glance at structural properties of the network suggests that while the Pokot network is dense and shows a low degree of centralisation, the Himba network is more centralised and shows less density.

6.2.3.1b. Structural Properties of Single Actors: Degree, Closeness and Betweeness Centrality

Formal network analysis does not only concern the structural properties of networks but takes a lot of its analytical power from analysing the positions of single actors within the structure of the network. Degree, closeness and betweenness-centrality are measures that differentiate the structural properties of single actors within the net. At first sight it becomes clear that the variation for these measures is much higher among the Himba than among the Pokot. Figures for standard deviations on these three measures vary profoundly (Pokot/Himba: degree: 19.68, Std Dev 4.94 and 38.52/23.20, betweenness 23.78/5.46 and 39.07/44.61; and closeness 66.78/5.00 and 56.69/6.59). The Pokot sample is much more homogenous than the Himba sample which shows a considerable degree of heterogeneity.

Table 31. Degree, betweenness, closeness in Pokot and Himba networks

	Pokot degree	Himba degree	Pokot between	Himba between	Pokot closeness	Himba closeness
Mean	19.68	38.52	23.78	39.07	66.78	56.69
Std Dev	4.94	23.20	12.96	44.61	5.00	6.59
Sum	728.00	1618.00	880.00	1641.00	2470.92	2381.01
Variance	24.38	538.44	168.00	1989.94	24.99	43.48
EucNorm	123.39	291.45	164.76	384.31	407.35	369.88
Minimum	8.00	5.00	3.32	0.00	56.25	37.27
Maximum	29.00	126.00	59.00	233.68	76.60	77.36

The actors in the Himba network show higher mean values on degree (38.52 compared to 19.68) and betweenness (39.07 compared to 23.78), while the actors in the Pokot network attain a higher value for closeness (66.78 compared to 56.69). Table 31 summarises the findings.

6.2.3.1c. Relational Analysis: Cliques, Clusters and Factions

The structural analysis of cliques looks for actors of the network who are more closely tied among themselves than with their surrounding. The comparison of the representations for Pokot and Himba show differences at a first glance (see Figure 30): While more than a third (40.5%) of all Pokot actors are involved in 22 to 40 small cliques, there is not a single Himba actor who is involved in as many groups. Only two actors (5.4%) in the Pokot network are involved in five or less cliques, in stark contrast 22 Himba actors (52.4%) are involved in five or less cliques. These figures underline the impression that the Himba network is fragmented while the Pokot network is constituted by actors who are closely connected to each other.

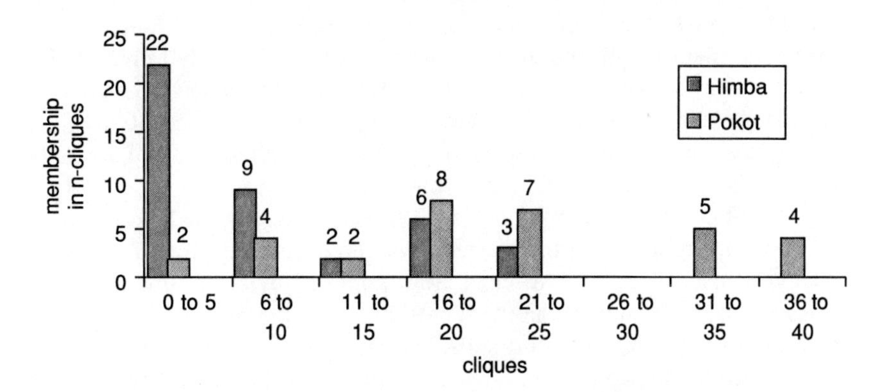

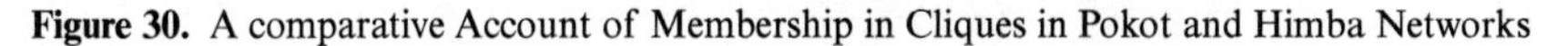

Figure 30. A comparative Account of Membership in Cliques in Pokot and Himba Networks

Livestock exchanges in both societies are embedded in network structures and at the same time create very different network structures. Himba networks are less dense than Pokot networks. Degree-centrality, closeness and betweenness measures indicate wide differences between actors. Actors central to the network are at the same time politically dominant figures among the Himba - this is not the case among the Pokot. Relational and positional analyses indicate the fragmentation of the Himba network. Fragments are tied together by influential actors. In comparison to the Himba network the Pokot network is much denser and there are fewer differences between actors. The structure corresponds to the egalitarian ideas the Pokot have about their society.

6.2.3.2. A Qualitative Comparative Account of Exchange Networks in Two Pastoral Societies

While the transfer of livestock along social relations is of importance in both societies, exchanges among the Himba take a very different form from similar exchanges in Pokot society.

Scope of Exchange: The quantities of livestock involved in internal exchanges differ profoundly. The Himba allotted significantly more cattle to internal exchanges than the Pokot did. While Pokot herders slaughtered 5.4 per cent (21), sold 13.9 per cent (54) and exchanged another 13.9 per cent (54) (386 cattle surveyed), the Himba slaughtered 11.6 per cent (177), sold 13.3 per cent (202) and transferred 23.3 per cent (354) internally (1,523 cattle surveyed). While Pokot invest as many cattle into social relations as they sell, with the Himba transfers of cattle within their own society are clearly the dominant form of investment. While the 13.9 per cent Pokot invest into exchange consist mainly of bridewealth payments and transfers to stockfriends, among the Himba livestock transactions are mainly constituted by various forms of livestock loans to relatives and inheritance transfers.

Conditions of Exchange: Among the Himba the ownership rights in animals transferred stay with the donor: if A gives an animal to B, then A does not cede his ownership rights to the animal; he still owns the original animal plus all its offspring. B may use the produce of the borrowed animals but does not achieve ownership rights. Direct transfers of ownership rights are rare. However, in contrast to the strongly hierarchised modes of livestock exchange such as the *mafisa* system of the Tswana or the *buhake* of Rwanda, Himba patrons are more confined by kinship ideology.

In stark contrast to Himba livestock transfers, among the Pokot A cedes his ownership rights to the transferred animal and its offspring. While Pokot exchange constantly reinforces egalitarian relations by atomising and redistributing wealth, Himba exchange reproduces a hierarchical society by concentrating ownership rights and consolidating access to the most important means of production. Pokot exchange of livestock creates close supportive ties between two actors. The ties founded by livestock exchange are used during droughts. It is the solidarity created through livestock exchanges which matters. In this sense Pokot livestock exchange means a conversion of economic capital into social capital. The major means of pastoral production which is a highly unstable resource is preserved as social capital. This social capital may then be tapped for support at a later stage. The Himba exchange system is organised in a very different way and to very different ends. For restocking Himba rely on support from their kin and on livestock accumulated at different places previously. The spread of herds and single animals is the patron's security - the wish of patrons to establish power via livestock loans the client's chance.

Institutions of Exchange: Among the Pokot bridewealth payments, distributions of incoming bridewealth and stockfriendship are major institutions of livestock exchange. Among the Himba livestock exchanges are mainly facilitated through inheritance transfers, the allotment of livestock camps, livestock loans and outright livestock presents. These institutions differ in many respects. First of all, the quantities of livestock involved and the volume of exchange varies. Pokot transfers rarely include more than five or six cattle at one time and usually transfers consist of just one animal. Bridewealth payments are typically paid in instalments. Even inheritance is a matter of instalments. In contrast to this livestock transfers among the Himba frequently include many heads of cattle. Livestock camps of thirty and more animals go from one benefactor to one recipient. Next to these bulk transfers of livestock, single animals may be loaned or given as presents.

Inheritance transfers by and large consist of the transfer of the entire herd from testator to heir. Bridewealth among the Himba is low (only two to three heads of cattle) and is not of major importance. Marriage is not a mechanism for redistributing property. Close kin are preferential marriage partners. Rates of polygamy are low and bridewealth is minimal: while Pokot men married 2.6 women on average, Himba men married 1.5, while Pokot transferred some 12 cattle, 3 camels and about 30 goats per wife, Himba gave just two heads of cattle - despite the fact that their herds are much bigger than those of the Pokot.

The Ideology of Egalitarian Exchange and Patronage: Pokot pastoralists exchange livestock within the framework of a moral economy as Himba herders do, although the morality in both systems takes different forms. Among the Pokot every actor is involved in a network of egalitarian exchange relations guaranteeing mutual support: a herder has to pay bridewealth and he also receives bridewealth; he distributes incoming bridewealth to relatives and friends; he engages in formal stockfriendships and presents stock to closely related patrilineal relatives and affines. There is a strong moral obligation to distribute livestock and there are numerous incentives which encourage wealthy herders to act along these lines.

Among the Himba the ideological foundations of the internal livestock exchange system are different. First of all Himba are keen to point out that preferably they transfer livestock within their matrilineal kin group. While

Pokot cherish the idea of reaching a wide range of related and non-related people with their exchanges, Himba rather opt for a concentration of wealth within a limited group of people tied together by kinship obligations. Himba morality clearly stipulates the obligations of the rich and influential to care for poor relatives and establishes the ideological foundation for patron-client networks.

6.3. RESOURCE PROTECTION IN TWO PASTORAL SOCIETIES

African pastoralists have often been blamed for causing extensive damage to the environment and of contributing significantly to desertification and an increase in vulnerability. Communal resource management has been pointed out as one of the main causes leading to the inevitable decline of the savannah environment. Hardin (1968)[26] argued that every herder profited individually from each single additional animal he herded, while all environmental damage was shared by the entire community. Hence, the accumulation of livestock and overstocking were the rational behaviour of maximising producers in systems of communal land tenure. Hardin's model has been criticised extensively (for an overview see Acheson, 1989:352f). The main thrust of criticism blamed him for confounding communal resource management with open access resource management. Open access resources are resources for which there are no defined groups of users and, if at all, only vague rules of resource exploitation. According to Acheson (1989) communal land tenure generally includes a clear definition of the people establishing the community, a definition of the resources to be protected and procedures to detect and sanction free-riding. While the major thrust of literature on communal land tenure in pastoral societies discussed whether communal land tenure is sustainable[27], there is yet another crucial problem worth considering: communal land tenure must ensure equal access to resources for all members of the community. If rules of communal management do not institutionalise rules of equal or at least predictable access they force community members into free-riding.

6.3.1. From Communal Resource Management to Open-Access Resource Management among the Pokot

The system of land tenure among the pastoral Pokot has changed repeatedly over the last two hundred years and developed from a system of clan-based resource control, to open-access grazing during periods of intense raiding, to communal resource management by local, non-kinship based communities and again to an open-access system with few rules. Communities have had to reshape their institutions of resource control due to political factors (e.g. warfare, colonial encapsulation), environmental constraints (e.g. degradation) and demographic developments.

6.3.1.1. The Development of Pokot Land Tenure in Pre-colonial and Colonial Times

It is difficult to reconstruct "traditional" land tenure management for the Pokot. Many groups which later formed the pastoral Pokot had an agricultural or agro-pastoral background. Oral traditions describe clan territories and kinship based resource management. These traditions are corroborated by more recent ethnographic descriptions of resource tenure among the Hill Pokot (Peristiany, 1954) and the neighbouring Marakwet (Kipkorir & Welbourn, 1973). The entire 19th century was marked and marred by interethnic violence. Apparently much of the territory claimed by Pokot herders was contested territory for which apparently no clear cut rules were developed. Pastures were open access resources. Until the first decade of this century Pokot herders shared the better parts of the Baringo plains with Turkana herders (Dundas, 1910; Hobley, 1906). An early observer claimed that the Pokot complained bitterly that they could not achieve arrangements on the use of wells in the Nginyang area with the Turkana living in the vicinity.

What is perceived as the traditional system of pastoral Pokot resource management - by outside observers as much as by local actors - only came into existence early on in the 20th century when colonial forces suppressed interethnic violence for some decades and the Pokot were allotted a tribal reserve with uncontested land rights and fixed boundaries. This stage was transient and probably did not last longer than a few decades. A system of resource protection developed which concentrated on the sustainable use of communal resources. The management of pasture rested on the general differentiation between rainy season pasture and dry season pasture. The mountainous areas of Paka, Korossi and Silali as well as the plains between Paka and Silali were defined as dry season grazing resort. No grazing was allowed there in the rainy season and in the early dry season (from May to September/October). The protection of Mount Paka was of special significance. The Pokot believe that the extinct volcano is the "home of the rain". While other pastures were continuously burned at the end of the dry season to encourage grass growth, it was thought of as a sacrilege to burn pastures on Mt. Paka. Besides these dry season pastures smaller areas were defined as dry season grazing areas for calves. Obviously it was of benefit for a community to have a pasture for calves nearby. This allowed herders to stay in the major settlement areas for long into the dry season. Calves could be grazed nearby on grounds which were well protected. Usually a community chose a nearby hill which was renowned for good grass-cover with highly nutritious grasses for this purpose. Two elderly informants described the traditional set up in some detail. Both also reflect on the demise of the traditional system of resource protection.

> In Parpelo the area up hill called Sekatkat was a prohibited grazing area throughout the rainy season. This area was reserved for calves as pasture during the dry season. Moruase was another hill that was spared for the dry season as a calves' pasture by the people of Kadingding and Mundi.

> As a major grazing reserve, Paka and the lands north of Paka were saved as dry season pastures. In this manner every community had a place which it saved as a dry season grazing area for calves and a wider area which was protected as dry season grazing. These areas were referred to as *kïghetöy* - prohibited. However, people have become *chepoyin* (sinners), people who do not follow any rules; because they have become *chepoyin* there is less rain and the grass does not grow well anymore. (Todokin, September 1993)

Todokin, as a member of the Chumwö generation, asserted that the emergence of unruly behaviour was the fault of the Koronkoro generation - the generation following the Chumwö as political seniors. This lack of gerontocratic rule led to a state in which even pastures for calves were already exploited during the rainy season, not to speak of the former dry season pastures. Todokin left little doubt that the environmental decline and the general non-adherence to rules were strongly connected. Wasareng an elder of roughly the same age as Todokin also commented on the decline of pasture protection.

> The protection of calf pastures was taken very serious in the past. The Kadingding and Mundi community protected Mundi and Moruase hill as dry season calf pastures. Even in the dry season it was not allowed to graze adult cattle there. These pastures were reserved for calves grazing only. Furthermore the entire Paka area was reserved as a dry season grazing resort. Places like Mötamöt, Körömwö, Kaisakat and Imönpöghet were excluded from grazing during much of the year. Each community had its dry season calves' pasture. The Korossi people had reserved Chepelion, Ngetutöy, Kamngatip and Chepkokoech. From both places Lake Baringo could be used for watering livestock. Nowadays these areas are not protected anymore. Kamngatip, one of the former reserved areas, has almost no grass at all anymore, it has been overgrazed beyond any repair. (Wasareng, September 1993)

The local system of resource protection has come under severe stress since the 1940s at least. The Pokot area, as defined by the British administration, was severely overstocked. At times there were more than 20 cattle per square kilometre (or 1 cattle per 0.5 hectare) on a range where range scientists recommend some 15 hectares for one head of cattle at least. Throughout the 20th century the number of herd units increased with the rising number of households while the total number of animals per herd decreased. Hence, in order to manage communal resources, many more peoplc had to be included into the co-ordination of resource exploitation. The system of resource protection did not change accordingly. The problems of resource management and the pressure on resources were exacerbated through a renewed increase of interethnic clashes which made any form of continuity in resource management a problem.

6.3.1.2. The Failure of the 'Traditional' System of Resource Protection: Pokot Rangelands as Open-Access Resource

Nowadays, we find different approaches of resource protection working alongside each other in northern Baringo. Basically these consist of locally formed neighbourhood councils, government sponsored grazing guards and grazing committees founded by a development agency. People working in these different institutions are occasionally identical, but not necessarily so. Informal neighbourhood councils still manage a major part of the day to day decisions of resource exploitation (grazing, water use etc.), they judge cases and organise ritual activities. The Government, and especially the Ministry of Agriculture, has some say in the organisation of communal grazing as well. Formally the government acts in accordance with local elders. In reality this proves to be difficult as the meetings of elders are usually informal and rarely planned beforehand. If resources allow (i.e. a car is available, personnel are on hand) the government undertakes to send administrative police to check on trespassing into protected areas. However, this is rarely so and towards the end of the 1980s these activities came to an end as usually no transport was available – and they did not resume until the end of the 1990s. A development organisation active in the area had its own ideas on resource protection. Again a close partnership with the so called traditional authorities was looked for, but in reality it proved to be difficult. Livestock exchange, range protection, the erection of tree nurseries, the establishment of women groups were initiated in accordance with newly formed committees, which were more encompassing than local neighbourhood councils, had no clear-cut leadership or any other form of internal organisation and formal recruitment. The problems these committees faced are exemplified by the following case study which documents a minor conflict around the protection of a communally managed dam side in February 1988. In the case study presented below the conflicting parties divide themselves into two groups. On the one hand there are the men of the neighbourhood Chepkalacha who are interested in a sustainable use of their major water reserve, the dam at Chepkalacha. On the other hand there is Rionotim and his herders. Although Rionotim had lived in the area for some time about a decade ago, he was not thought of as a man of the Chepkalacha neighbourhood. The material is based on observational data and extensive interviews with members of all three parties.

> Rionotim had lived in Tilam for only two months when at the beginning of February (1988) the nearest dam, Kadingding, became dry. This had happened at a surprisingly rapid rate. Throughout the months of December and January herds had entered the Kadingding area. To prevent a rapid depletion of the Kadingding dam the neighbourhood council had stipulated that the water of Kadingding should be used for human consumption and for calves and goat kids only, but that the adult stock should be taken to the waterholes at Nginyang river about 25 kilometres away. Many herders, however, did not adhere to the ruling of the Kadingding neighbourhood council. Frequently entire herds were watered at night at the Kadingding dam. Soon the water at Kadingding was finished and the Kadingding households were forced to disperse because of a lack of water.
>
> Rionotim decided to move his herds to Chepkalacha. As his goats herd was without water for two days in a row, he sent a boy in advance to

ask the elders of Chepkalacha to allow him to water his entire herd at the dam just once. The Chepkalacha community had reached upon a similar agreement as the Kadingding community: the water of the dam was to be used only for human consumption and young stock (calves, goat kids). All other animals were to be taken to the waterhole at Mukutani which was about 6 kilometres away. The men of Chepkalacha did not reply to Rionotim's question, neither in the positive nor in the negative. While Rionotim thought it sufficient to have informed the men of Chepkalacha, the non-answer was rather thought of as an indication of a negative stand by the people of Chepkalacha. In the morning of February 8th Rionotim ordered three herd boys to water his entire small stock herd of 350 animals at Chepkalacha. As a rich and prominent senior man who additionally had good relations with some of the seniors at Chepkalacha, Rionotim thought that he would not run into any difficulties. However, when the herd arrived at the dam, the men of the neighbourhood council denied them access. They ordered the herd to move on to Mukutani with the reasoning that the water of Chepkalacha was to be saved for human consumption and young stock. Angrily the three herders had to move on to Mukutani. According to them some 15 goats died on the way due to exhaustion.

A short while later Rionotim arrived upon the scene to look after his goats. The elders of Chepkalacha said that his herd had moved on to Mukutani but did not mention (according to Rionotim) that they were the ones who had prevented the herd from being watered in Chepkalacha. Upon questioning they said that perhaps there might have been a misunderstanding as the herd boys did not clearly express what they wanted. Only in the evening did Rionotim learn about what had happened and the fact that he had lost 15 goats within a day made him angry. The next morning he went to Chepkalacha to find out who had actually prevented his herd from being watered. He only got evasive answers and the men of Chepkalacha remained with their version of the story that it had all been a big misunderstanding.

A second incident escalated the conflict. Only a few days later Rionotim found one of his donkeys (according to him) emaciated near the Chepakalacha dam. He decided, against the agreements taken at Chepkalacha, that he would water the donkey immediately at the dam. He led the animal to a far off side of the fence surrounding the dam and opened the thornbush fence. Rionotim met strong resistance. He was questioned about why he presumed to act openly against the neighbourhood council of Chepkalacha. There were a lot of angry words exchanged and both parties went home without solving the conflict.

Rionotim had not obeyed the rules of the local neighbourhood. In the beginning the men of Chepkalacha tried to prevent any direct confrontation but after the second incident they took a more clear-cut position. Rionotim felt treated badly too. The following days were spent mediating between both parties. At the major meeting in which the contentious issues were discussed a very emotional tone set the speech. Rionotim's people (he himself did not appear) said that they felt embittered by the rude rejection of their thirsty herd and the subsequent losses. The men of Chepkalacha claimed to be bitter, strangely not so much

> because of Rionotim's transgressions of local rules, but because Rionotim accused them of something they thought they were not to blame for. Accusations and counter-accusations were exchanged but for some reason the men of the council seemed to be in the defensive. At no stage did anybody stand up and say "we acted correctly as Rionotim was transgressing the rules we put up". Obviously everybody was reticent to name culprits. Neither was Rionotim mentioned as the person who violated local standards of resource protection nor were any of Chepkalacha's men mentioned as responsible for the refusal of Rionotim's herd. Nobody talked about the fact that the situation at Chepkalacha's dam was precarious. Water was scarce and a disaster loomed if the resource was not handled properly. Factually Rionotim's behaviour threatened the entire community. However, the discussion mainly centred upon violated feelings. Anger was not the emotion most frequently mentioned but pain (*ngwönin*), the pain of having lost goats, the pain of being mentioned in name in a case and the pain of being accused of a misdeed.

The case study sheds light on the precarious state of resource protection among the Pokot. The dam committee or neighbourhood council of Chepkalacha had little means to enforce rules without violating standards of comradeship and solidarity. Hurt feelings and lack of respect were standard themes, not the vulnerability of a community and its resources.

Problems of range management are at least as intricate as problems of water protection. The sustainable management of pastures is marred by unclear definitions of who is protecting, what is being protected and how protection is to be sanctioned. Furthermore different organisations at different levels influence the process of resource management. For a long time it was the government alone which appointed grazing guards. Later on a development project concerned with range development successfully argued that grazing guards should be named by the local elders. In accordance with western views on gender equality the development organisation also argued that women should be eligible as grazing guards. In order to emphasise the importance of grazing guards the number of guards was stepped up from two to five. The protocol of interviews with two grazing guards, T. and L., on recent attempts of reforming the management system pinpoints some of the dilemmas.

Only T. had previously been a grazing guard appointed by the government and during his time as a government appointed guard he had traced several trespassers. However, many had succeeded in pacifying the elders by paying them with beers and only a very few had been punished. Some trespassers had even simply left the area and had thereby evaded any form of punishment. The guards had not been paid regularly but probably had received a bit more food aid than others. With the reshaping of the institution of grazing guards in the early 1990s the areas to be protected were redefined: however, no fixed boundaries were stipulated. Due to the high degree of climatic variability it was thought that flexible boundaries between protected and non-protected areas would be more effective. It was intended that the elders fix the boundaries from year to year anew. Only when the rainy season pastures had been exploited thoroughly, should livestock camps be allowed into the dry season grazing areas. L., in his function as a grazing guard, emphasised that some elders were to accompany the young men and boys on their move to the outlying areas in order to keep them under control.

The duties of the grazing guards were defined vaguely. Their legitimisation suffered from a badly co-ordinated effort by government institutions, a development agency and local elders. Due to unclear regulations both guards diverged somewhat on the definition of their job. While L. saw their task particularly in the supervision of far away grazing grounds (*serim*), T. mentioned the supervision of nearby grazing resources as his major task. Both also disagreed somewhat on the exact definition of the boundaries of the areas to be protected.

Ideas on sanctions against trespassing remained as unclear. L. maintained that when somebody was found to contravene grazing regulations he should be brought to the elders first. If guilty he should be either fined money or livestock. It remained unclear how much a culprit could be fined. If the accused disagreed to pay, he should be brought to the chief where the standard fine was set at 5000/-KSh after the reorganisation of the grazing guard system. Until then only the minor sum of 100 KSh had to be paid (the value of a goat was 700 KSh at the same time). However L. added that under the new system there had been no cases up to now. Before there had been a few cases: once a young man, for example, was fined as he had entered into the reserved grazing area much too early, but he succeeded in proving that he could not afford to pay the fine. The elders left it with a stern warning, probably because he was the son of an influential elder. Some seniors, however, were punished: but for some reason they only had to pay 50 KSh (although the regular fine should have been 100 KSh), i.e. the equivalent of one tenth of a goat, or about 8 kilos of maize at that time. There is little doubt that this form of "punishment" is little more than symbolical. Taking into account that the cattle of these herders were grazed on prime grazing without the competition from other herds for several weeks it was only rational to contravene against grazing regulations! T. described the treatment of wrong-doers in a different way. In his point of view wrongdoers should be reported to the respective chiefs immediately. The chief would then decide upon an adequate punishment. The chief could consult elders but he need not to do so. If needed, he would use Administrative Police to bring the culprit to justice. The fines, T. also mentioned 5000 KSh, would go to the chief who would administer the money (L. did not mention what would happen to the fines). T. thought that the money should be used to pay for shirts and shoes for the grazing guards.

The analysis of temporary resource management among the Pokot highlights various problems:

Definition of the group protecting a resource: The community using a resource consists of all households using an area over a longer period of time. The adult men of the community form a neighbourhood council (*kokwö*) in which basic decisions on resource management are taken and contraventions against such decisions are discussed. However, membership in this council is not formalised. On the one hand basically all adult men residing in an area can participate in council meetings and take part in decision making. On the other hand nobody can be forced to take part in such meetings. Two problems have contributed to the malfunctioning of neighbourhood councils in recent times. Over the last decades the number of households has increased tremendously, while at the same time the size of herds has declined. Hence, councils had to find institutional arrangements for ever more units of an ever smaller size. It is inevitable that transaction costs rise due to the increase of participants in the game. For the last two decades interethnic conflict has forced herders to leave their place of living and to use the resources of various other communities. The community taking decisions on resource management has become more vague. Basically Pokot herders are free to make use of resources within Pokot territory whenever they decide it is best. It is very hard to exclude fellow Pokot-herders from the use of specific resources. Furthermore various organisations influence the process of resource management at different levels. The neighbourhood council deals with the management of local resources. Then there is the local chief and the sub-chief who try to influence the process. Both officers are government installed and receive government salaries. They are employed and chosen by the government and they are not members of the neighbourhood council. Their authority over grazing control is not clearly defined. Officers connected to the ministries in charge are usually non-Pokot and have very little contact with and knowledge of the local community. For most of them the time in Pokot land is just an interlude in their career and jobs in central Kenya are much more attractive to them. Since the early 1980s a development project has been working in the area and has tried to influence local communities towards a more sustainable use of resources. While all parties participating in the discourse on grazing management agree upon the fact that degradation is taking place, they have very different ideas on who is to take measures against it and what measures have to be taken.

Delineation of the resource to be protected: The borders of areas to be protected are disputed as well. Even the two grazing guards questioned on the issue gave different borders for the areas to be protected. While some herders defined dry season pastures in a restrictive way and just wanted to exclude the heights of Mount Paka from early grazing, others preferred a wider definition and even included the foothills of Paka and the entire area north of Paka. Then there were different opinions on grazing reserves for calves. While some herders argued that this institution was outdated others thought that it was an institution worth maintaining. According to different ideas on the delimitation of protected areas disputes frequently entered into a chaotic phase in which both parties referred to different systems of delimitation.

Establishing contravention against grazing rules: If somebody was accused of contravening against the rules of communal resource management he was called to attend a meeting of the neighbourhood council. Frequently I observed that the council had to meet twice or thrice before the actual case was heard. It was not unusual that the accused did not come to the first meeting. If a case was then heard, discussions were extremely weary. There was no person or body guiding the proceedings. Discussions wavered between referrals to past cases, excuses and accusations. Frequently such discussions did not end in a final consent-based decision. The case simply ended when the accused party withdrew from the area and migrated to another neighbourhood.

The sanctioning of trespassing: What happens if somebody is actually found to have contravened against grazing regulations? Fines were usually paid in monetary terms. These had been marginal until the early 1990s. Officially it was said that fines were unequivocally set at 100 KSh. However, even the minimal sum of 100 KSh was rarely paid. Occasionally the offering of some beers pacified the elders and the case was dropped. Others accused alluded to the fact that they were poor and could not afford to pay the fine. The maximum that was paid was 50 KSh. These were no major sums - even for the Pokot. A reform of grazing regulations then clearly identified the lack of severity of punishment as a factor contributing to the failure of communal resource management. The new sum was fixed at 5000 KSh, which was tantamount to about seven big goats. Both grazing guards agreed that nobody had been fined yet according to the new rules in 1993. A short field stay in 2004 showed that this situation had not changed; on the contrary the power of neighbourhood councils had been further diminished and several informants claimed that the neighbourhood council had given way to meetings at beer-brewing places.

The co-management of resources: Since colonial times the government has tried to opt for a co-management of resources - with little success. Officials thought that forced de-stocking was the best help to cure the ecological problems of the pastoral areas of Baringo District. With independence all measures directed at grazing control were cancelled and for more than a decade apparently no action was taken by the government. At least since the early 1980s grazing control has again become a contended issue. Government-appointed chiefs were now held to control the protection of dry season grazing areas more strictly. Occasionally chiefs used (or threatened to use) administrative police to clear protected grazing areas from free-riders.

A local NGO pinpointed degradation and desertification as a major problem. When they first identified the problem they established their own system of (intended) grazing control. They put emphasis on educational efforts more than on direct control. Protected grazing plots were fenced off at several places to show the public how well an area develops if excluded from grazing only for a short time. However, in the early 1990s these efforts were abandoned. At some places these plots had been repeatedly opened illicitly, and the grass and bushes that had grown over a period of two or three years, were then eradicated by famished goats within one night. Such illegal action was habitually followed by numerous meetings of the local council to identify the wrong doers - usually with little success. The project defined its strategy anew. It now opted for a strengthening of the local grazing guard system. The number of guards was increased from two to five. The ideas of the international agency were to be felt clearly when they pressed for women as grazing guards. At least one of the five guards in each area had to be female. The withdrawal of the development agency from the area in late 1997 and the continued interethnic clashes in the area which sparked off major rounds of internal migration caused the system to collapse by the end of the 1990s.

6.3.2. Communal resource management among the Himba

The efficiency of communal resource management is also essential for Himba herders. In the following section I will delineate how institutions of resource protection have developed historically and ask how the group managing a resource is defined among the Himba and finally how modern agencies of the state and local herders co-manage vital resources.

6.3.2.1. Resource Protection in a Historical Context

Oral traditions do not give any indication of the land tenure system before the exodus of Kaokoland's herders to Angola in the second half of the 19th century. Cattle herds probably migrated from the settlement areas at the banks of the major rivers to pastures on the adjoining hills. Some older informants stressed the point that they hardly had to organise the protection of pastures to prevent overgrazing and competition as pasture was abundant.

The colonial government has exercised some influence on the land tenure system since the 1920s. Three chiefs were given reserves with more or less fixed boundaries in the northern parts of Kaokoland. Migrations across these boundaries were prohibited and controlled. For about twenty years (from 1925 to about 1945) settling at the Kunene river was prohibited, too. However, the colonial government did not directly influence resource protection until the 1970s. Overgrazing was diagnosed as early as the 1930s in central parts of Kaokoland but northern Kaokoland was still thought of as comparatively unspoiled grazing. Since the 1940s colonial officers acknowledged the local system of resource protection and sustained the power of the chiefs. They were allowed to punish trespassers and could fine them and even punish them physically. However, when in the middle of the 1970s northern Kaokoland became involved in the civil war, any official approach to resource protection gave way to security considerations. With independence a fresh approach to range management was adopted, but in essence not much changed on the ground.

6.3.2.2. Pasture Management in the 1990s

The group managing pastures is a neighbourhood and on a higher level the people under one chief. Day to day affairs are managed by ad hoc meetings of men from one neighbourhood, major conflicts are discussed in more formal meetings at the chief's and the councillor's home. Chiefs appoint several men as grazing guards who are responsible for scrutinising the area. Besides the division of rainy season and dry season pastures there are quite a number of rules which are meant to lead to a sustainable and equitable management of pastures. If people are found to act against these rules they are fined by a neighbourhood council or more formally, in one case at, the chief's home. Fines are paid in livestock. Minor misdeeds result in the payment of one head of small stock, repeated misdeeds or grave cases of trespassing result in the payment of a cow. A case study will elucidate the recent working of resource protection.

Case Study: K.'s case (for spatial orientation see Map 17):

On April 20th (1994) K. moved his homestead from Outjova Outiti to Ovizorombuku. At the same time he split his cattle camp and sent it to Eyao. While settlement in Ovizorombuku was controversial at that time of the year, settlement in Eyao was a grave assault on the system of grazing management. Very soon K. was severely criticised. He had not waited for a communal decision which would declare these areas as free for grazing. In the first week of May M., who had been appointed as a grazing guard, went to Eyao and called K. to a meeting in Oheuva. Apparently the meeting succeeded in convincing K. and on May 4th the cattle camp left Eyao.

However, K. played a trick on the council. He only moved his cattle camp a bit and still used the pastures of Eyao from a more hidden place. Only the household returned from Ovizorombuku to Omuhandja and thereby complied with the rules which were set up by the community. Everybody was quite angry at K.'s indolent behaviour which so obviously ignored communal decisions. A major meeting (*ombongarero*) with the chief and his councillor was quickly arranged. The meeting was not aimed at discussing the issue with K. once again but to teach him a lesson and to fine him.

At the same time the meeting had a wider scope: while K. was the main culprit, the moves of other herders were criticised too. K. was massively criticised and a lot of pressure was put on Katjira, the councillor of the chief and at the same time K.'s maternal uncle (MBDS). When K. had resettled from central Kaokoland to Omuramba in the early 1980s it had been Katjira who invited him to stay. Many men thought that it was Katjira's task to convince K. that his behaviour was seriously upsetting the community. And in fact, Katjira threatened K., that either he returned his herds to the prescribed places or he would have to leave the area altogether.

However, K. stayed on. His argument was that other herders such as Mu. and N. had moved into reserved areas and that he did not see the point in moving back if they didn't do the same. K. was correct to some degree. However, while N. and Mu. had settled at the border area between reserved and non-reserved grazing, he had chosen to settle right within the dry season grazing reserve. However, Mu. and N. felt the pressure and moved around May 20th while K. was obviously still playing for time.

On May 24th many men met with the chief and his councillors for a second time in order to discuss grazing management in general and recent

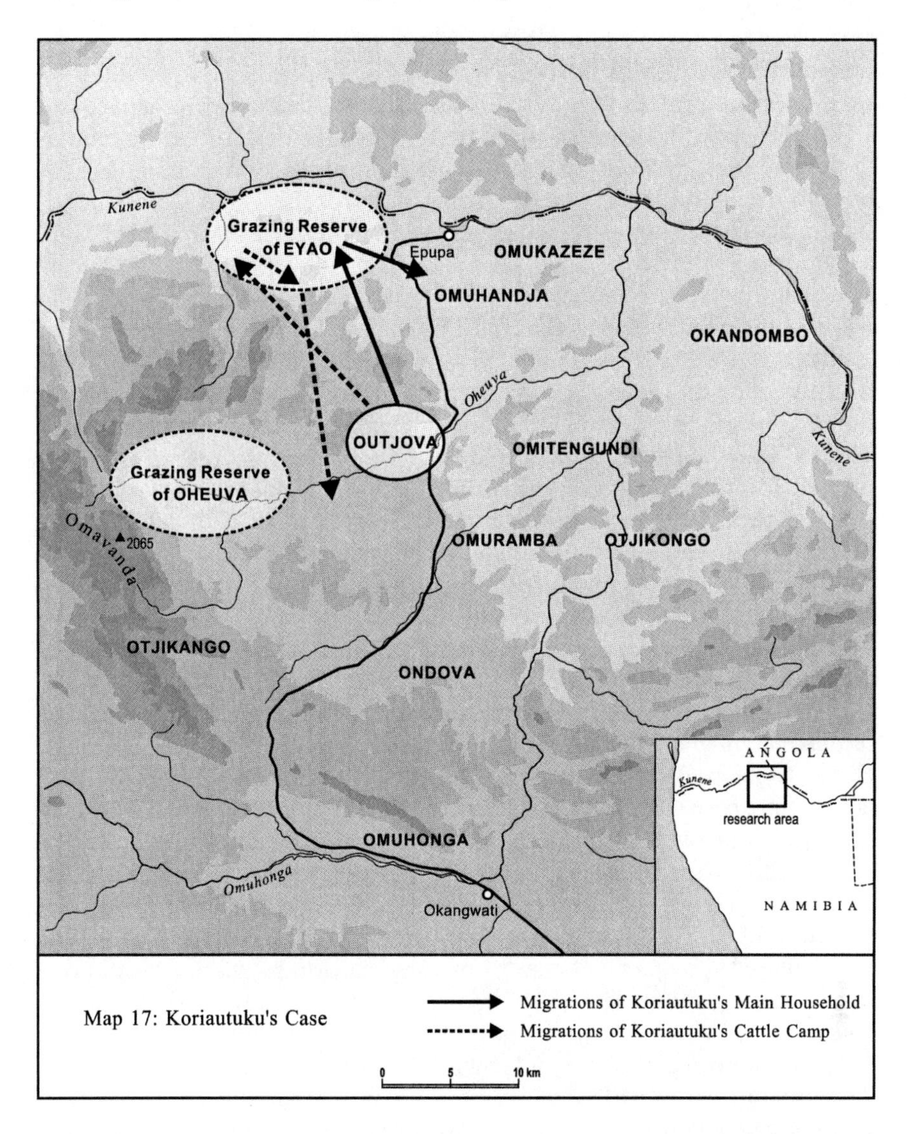

Map 17: Koriautuku's Case

transgressions of grazing rules in particular. Obviously several men had not obeyed the rules, but there was little doubt that K.'s transgression had been the gravest. While previous discussions had remained somewhat inconclusive the meeting on the 24th reached decisions swiftly. All wrongdoers were punished. Those who were found guilty of minor transgressions had to pay one sheep and K. had to pay one ox. The meeting put pressure on K. to provide the animal immediately to feed the meeting. In fact, the animal, a small ox, was brought instantly and was slaughtered early next morning. K. now resettled without further hesitation to places which were seen as open for grazing.

Cases like these crop up almost every year. The grazing behaviour of homesteads and camps is always carefully scrutinised by neighbours. An early shift into protected areas offers several advantages: the free choice of a settlement place and access to fresh pasture. Characteristically cases crop up in the late rainy and early dry season. It is especially then that herders can hope to reap advantages from free-riding. Towards the end of the dry season most places have been exploited anyhow and with the first rains movements are no longer restricted. Only when the early rainy season (*oruteni*) turns into the rainy season proper (*okurooro*) are herds to be taken back to the major settlement areas. In the case cited above K. massively contravened against the grazing rules. Two facts were criticised: first that K. went right into the centre of a reserved grazing area and not just to its margins and second that he did so on his own. It took one month to carry the affair from a state in which the men of the neighbourhood pleaded with K. to the opening of a formal case against him. Contraventions against grazing rules typically go through these two stages: first the men look for a solution at a neighbourhood level and then the matter is taken to the chief or to one of his councillors. At both levels fines can be decided upon. However, decisions taken at the chief's home are usually discussed by more men and are more binding. The following paragraphs will discuss the major building blocks of Himba pasture management.

Definition of a group protecting a resource: The community managing a resource communally is fairly clear cut. All households that are under one chief are allowed to use the pastures that fall within this chiefdom and accordingly have a say in the management of the resource.[28] Behnke (1998:4) observes that most Himba chieftaincies are organised along drainage systems and combine river frontage and upland grazing areas. A chieftaincy in this sense forms a large and loosely integrated herding and resource management system (ibid. p. 12). Although the boundaries of these chiefdoms are neither to be found on maps nor are most of these chiefdoms ancient institutions, boundaries are fairly clear. While there are several places along boundaries which are disputed, the majority of locations are clearly assigned to one chief. Factually, however, resource protection takes place at a neighbourhood level. The people of one settlement area form a neighbourhood. Several neighbourhoods come together in the use of dry season pastures (see Map 18). Several places along the Kunene are also inhabited by two to four households. Each neighbourhood has specific settlement areas, where gardens are located and permanent homestead structures are erected.

How is membership in such a resource managing community established? On the one hand there are households that have a long settlement history in one place. This historical relation between a household and a place is established by numerous graves at which ancestors are worshipped. If rules of resource management are flawed this is also interpreted as an affront against ancestor spirits. A senior male member of one of the families with a long settlement history in the area is regarded as the *omuni wehi*, the owner (or guardian) of the land.

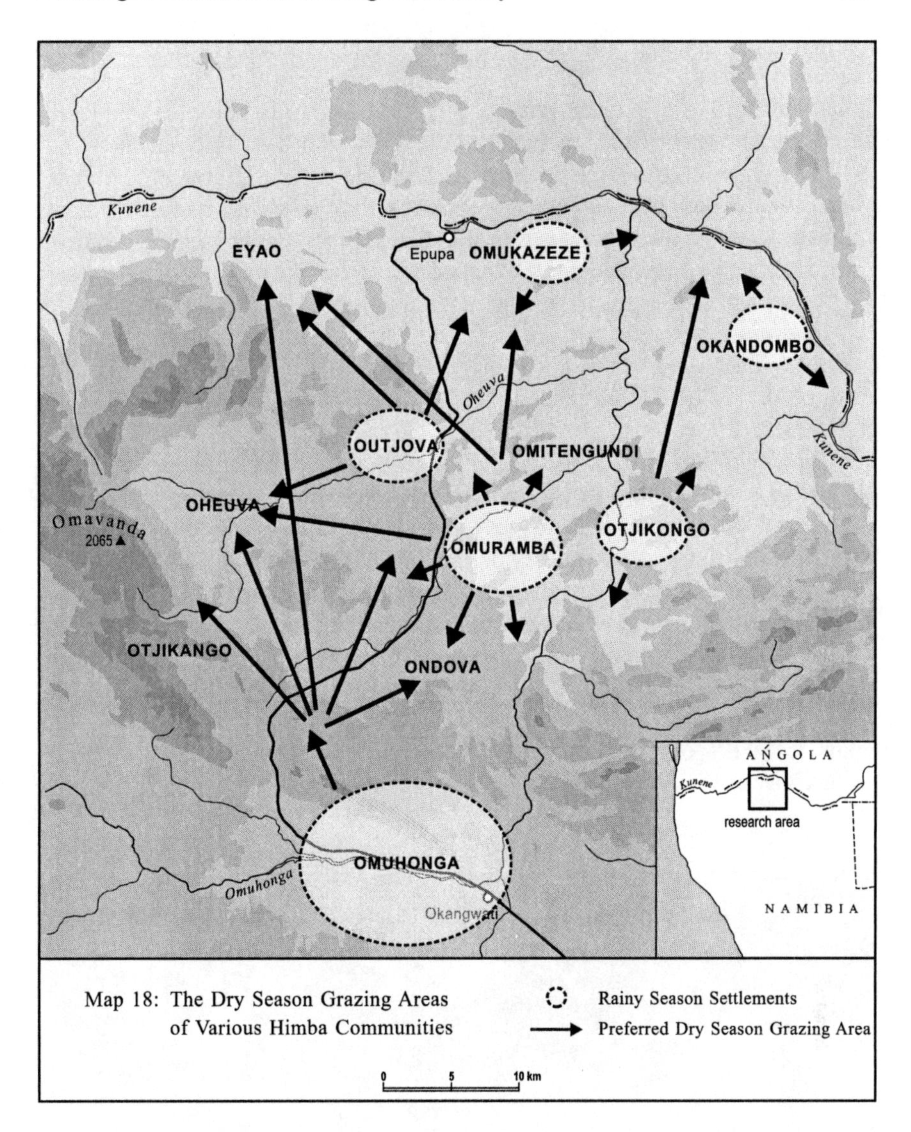

Map 18: The Dry Season Grazing Areas of Various Himba Communities

Rights to use a specific dry season grazing area are derived from the membership in one neighbourhood. Herders should use dry season grazing near to their settlements. However, for large neighbourhoods such as Omuhonga this is nearly impossible. Cattle camps have to move on to grazing areas further away.[29]

The community in charge of managing a specific resource is usually well defined and stable (see also Behnke, 1998: 8). Most herders of one locality find it easy to enumerate those households which have settlement rights in the area. Settlement areas are controlled by neighbourhoods and several neighbourhoods combine to manage dry-season pastures. Neighbourhood councils deal with matters pertaining to the locality and will also attempt to deal with contraventions against rules of good grazing. If a conflict cannot be solved at that level a case is opened at the chief's home or at the place of one of his councillors.

The Rules of Resource Protection: The rules of good grazing (*ondunino yomaryo*) are fairly easy to depict. Informants gave six major rules: (1) grazing in the dry season grazing areas is prohibited during the rainy season, (2) cattle camps must move a considerable distance away from the main settlement areas, (3) livestock camps must move together in a group, (4) herders should look for dry season grazing near to their major settlement, (5) too much movement to and fro (*okukandakanda*) is not appreciated and (6) special areas (Baynes Mts, area to the west of Oronditi) should only be used during droughts and not during normal dry seasons.

The intention of each of these rules is obvious. Rule number one ensures that there is sufficient grazing towards the end of the dry season. If this grazing is used early on during the year, herders will find it difficult to find pasture towards the end of the dry season. Rule number two guarantees that milking stock can be kept at the main settlement for a long time. If livestock camps would come closer to the main settlement the pastures adjoining the village would be soon exhausted. Rule number three ensures that camps have an equal chance to exploit a resource and guarantees co-operation between livestock camps. Obviously it would be of advantage for every camp to settle apart from the others. Cattle would find unused pastures and the herder would have a free choice where to erect his camp. However, grazing separately

Photograph 16. a, b Himba Ancestral Graves

Photograph 16. cont'd

from other camps would also entail that other herders have only a diminished chance of optimally grazing their livestock. The prohibition to settle ahead of others ensures that all herders have the same chance to exploit a resource (see Behnke, 1998:15). The fourth rule defines access to specific dry season areas and makes mobility patterns of single households more predictable, while the fifth rule tries to prevent trampling due to frequent shifts. The sixth rule guarantees that even during droughts there are some pastures to be found which have not been used during the previous year. There cattle can live for some time on grass from the last rainy season.

Several informants were asked to judge the spatial mobility of their neighbours. Did they comply with grazing rules? Comparing the rules with observed behaviour tells more about their handling and their impact. Tako judged the grazing behaviour of his neighbours critically although he consented that none of them could be fined.

- M. failed because he lived the entire year in Ozosemo, in a place where people should only live during the late dry season and the early rainy season. However, Ozosemo is not protected and some other households did the same. Apparently, the place is changing from a dry season grazing area to a settlement area.

- The places Tj. settled in were perfect. However, he changed his place too frequently and thereby destroyed a lot of grazing. First he settled in Eyayona, then in Ongorozu, returned to Eyayona and then went back back to a place near Ongorozu once again.
- P. and K., were both running livestock camps and were also behaving improperly. Both came from Omuhonga some 40 kilometres away. They should have looked for dry season grazing nearer to their main homestead, i.e. nearer to Omuhonga. The dry season pastures of Ominyandi and Ondova would have been more adequate for them than Ozosemo. Although Tako agreed that K. had some rights to live there – K.'s father was buried nearby and a close relative was living in the area - he emphatically denied P.'s rights to live there.
- K. settled at Ohengana too early. Basically there was nothing wrong with him choosing the place and as a person from the area he was obviously permitted to use it. However, he had already moved around a lot this year. First he had moved up and down the Kunene and then, for some time, had settled in the area of another chief. Tako judged his behaviour as rather unpredictable and selfish. Furthermore Tako thought that it would be reasonable that nobody should settle in Ohengana. Cattle from several settlements strolled there during the dry season. Sometimes they stayed for days in the area without being herded as the place was near water and usually had good grazing. If somebody settled there this area could not be used as dry season grazing by nearby settlements. However, Tako consented that at this stage cattle camps were allowed to go there. His complaints were directed against K.'s frequent shifts: He maintained "*K. wazepa ehozu*!" ("K. is killing the grass.") He thought that a case against K. should be opened and that he should be fined because of *okukandakanda* - too much movement.
- W. also committed some mistakes when moving his cattle. These were however not liable to open a case against him. W. has his gardens in Orieheke where he also settled for most of the year. In the dry season he placed his cattle camp in Ozosemo, about 6 kilometres away from Orieheke. Tako thought that W. should have used the grazing nearer to Orieheke first. While the grazing at Ozosemo was very good the pasture nearer to thc homestead was definitely sufficient: Tako commented "*nambano makatura kombanda yehozu*" – "now he is living in front of the grass". Tako explained that it is best to use the grazing near to the homestead first and from there exploit the pasture systematically. But jumping right into completely unused grazing means that islands of pasture are left in between which are destroyed by trampling and not by grazing.

Tako's explanations show that there are a core of rules and a wider set of guide lines. Together they form a body of standards guarding range management. While cases are only opened against those herders who do not adhere to the rules, the behaviour of others who do not consent to the wider guide lines is critically discussed as well.

Delineation of the resource to be protected: The delineation of places which are protected is fairly well defined (see Map 19). Settlement areas are well established: there are major settlement areas Omuhonga, Omuramba and Ombuku and minor areas such as Oheuva, Outjova and Ovizorombuku. A recent shift in opinion is discernible as regards to the dry season grazing area Eyao. For several years in a row people have settled in Eyao during the rainy season. Although the locality does not have any place for gardening it has permanent water from a well which makes it very attractive for more permanent

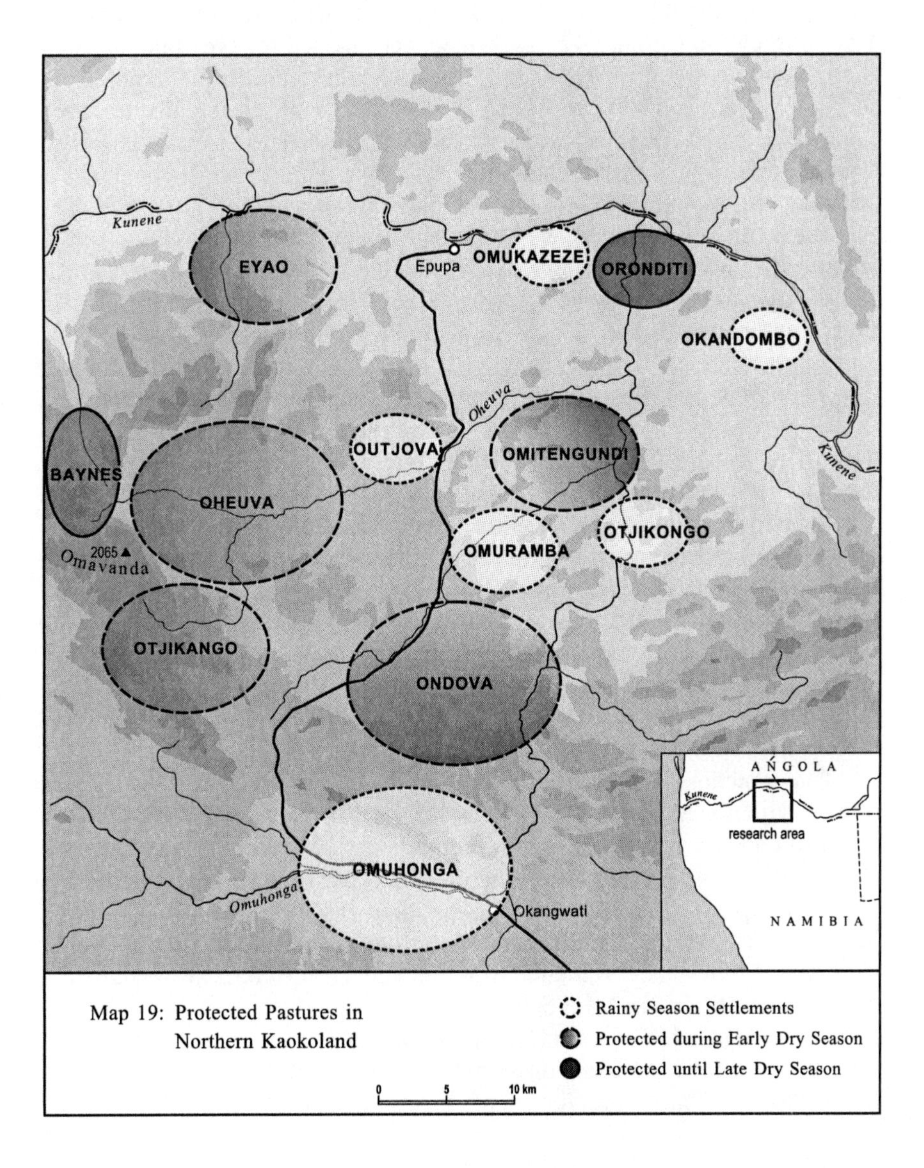

Map 19: Protected Pastures in Northern Kaokoland

settlement. Several herders argued that Eyao should be declared a settlement area and that only the pastures west, north and south of Eyao should retain the status of protected dry season grazing. Each settlement is surrounded by a safety cordon of pastures. The controlled grazing of pastures ensures good grazing for those herds that stay at the settlement long into the dry season. Map 19 shows several protected grazing areas. The non-shaded areas indicate settlement zones, shaded areas show pastures that are protected during the early rainy season and for which numerous additional rules (see above) are applied. The grey shading indicates the direction in which camps move. Dark grey areas show places that should only be used during droughts.

Establishing contraventions against grazing rules: Since 1990 grazing guards (*ovatjevere vomaryo*) have been named. These are instituted into their office by a meeting of the community. Seven men were named for the community of Omuhonga/ Omuramba While all of them were responsible for the affair as such, each guard had his special area. Formally these men are entitled to screen the area for homesteads and camps which do not adhere to the rules. However, everybody can bring up complaints against a neighbour or herder he thinks is behaving unruly. At first a meeting takes place at the neighbourhood level. The accused is summoned and he is given time to explain his case. Usually there is no punishment at these meetings and, if possible, men try to reach a consensus of how to change the situation for the better. These meetings consist of men from one neighbourhood. There is no formal recruitment. While everybody present may participate and contribute his views it is usually men from the area dominating the discussions. If these minor meetings at a neighbourhood level do not lead to a decision the case is then referred to a meeting at the chief's home. The people attending this meeting are more carefully selected. Obviously the chief or a representative (e.g. his councillor) has to be present. Then the accused and the complainants will turn up, both, ideally, with a group of supporters. A complainant or several complainants bring up the case. They are questioned by the elders and the authorities (chief, councillor).

Then the accused is given a chance to explain his behaviour. Arguments will be exchanged for some time until consent on the case emerges. It is not necessary that everybody present is convinced that the right solution has been found. If the seniors agree and a majority holds onto a specific view then there is little in the way which can obstruct a final verdict.

The sanctioning of trespassing: Non-adherence to grazing rules was already sanctioned in the past. The South African government encouraged chiefs to punish wrongdoers “according to tradition”. What tradition constituted, however, was highly questionable. Chiefs occasionally resorted to flogging in the past. Furthermore trespassers had to pay a fine, usually an ox. When several oxen had come together these were driven to the Commissioner in Opuwo. Oxen which originated from fines constituted part of the nutrition of the white officials in Kaokoland.

Nowadays, physical punishment is no longer practised. All herders interviewed on the topic of punishment for non-adherence to grazing regulations agreed that other forms of sanctioning were needed to control deviant behaviour. There were some disagreements on whether the local police should be involved or not. Most, however, thought that non-adherence to grazing regulations was a matter for the community and up to now (according to my knowledge) regular police have never been involved in conflicts on grazing. Fines are invariably paid in livestock. The following list of grazing cases for the year 1995 shows how fines correspond to specific misdeeds.

- M. had lived in Ongomberondu at the border between Oheuva and the protected grazing further west; "*watura kombanda kovakwao*", he had lived at the head of others. He was sentenced to pay a sheep.
- K. contravened profoundly against the grazing rules as he had not only moved ahead of others but had chosen a place all for himself. He had to pay an ox.
- Tj. had to pay a small ox for the same misbehaviour (although his case was not discussed as much as K.'s).
- Mb. was fined a sheep for *okukandakanda*, because he moved around too much with his stock.
- Mu. was summoned to pay a sheep as he had moved around too much and had neglected his duties as a grazing guard.
- Ng. had to pay a sheep for moving to Okaninga, a reserved area, too early.
- Tjo. had to pay a young ox as he had moved to Eyao without the consent of the elders.

Fines are either paid to the neighbourhood council or to the meeting at the chief's home. Usually the animals paid as fines are immediately slaughtered to provide those attending the meeting with meat.

The co-management of resources: Several informants mentioned that they thought the system of grazing management to be a rather recent invention. They maintained that they learnt about resource protection from the previous and the recent government and that before there had been no protection of resources. The chief's councillor remembered a major meeting in 1990 which had convened all the chiefs of Kaokoland and had discussed new grazing regulations. All the major political figures of Kaokoland were present and vowed to work for a sustainable management of pastures. Further interviews, however, corroborated that over the last fifty years rules of resource protection had remained pretty much the same. What had changed was that an additional force had stepped up its efforts to maintain resource protection. The double-edged legitimisation of resource protection transpires nicely when fines for transgressions are paid to the community and to the government, i.e. one and the same person is punished to pay a fine of two goats, one for the government and another one to the community. Usually both animals are slaughtered at the same time to feed the community. Reference to the government as the major force behind resource protection has added further legitimacy to the quest for a sustainable management of ranges.

6.3.2.3. The Protection of Trees

Numerous trees are protected as well. Trees which are used regularly are referred to as *omihupise*, plants which help people to survive. This category includes all trees that have edible parts, i.e. berries, nuts and roots or trees that are essential as livestock fodder. The major tree that falls into this category is the *omurunga* palm (Hyphenae petersiana). The palm is high yielding and its nuts are highly nutritious. Besides the palms many other trees which provide humans with food are protected.[30] If one of these trees is cut without a sound purpose the culprit has to pay a fine. Informants claimed emphatically that these trees were as valuable as cattle to them. Trees and bushes producing animal fodder, notably Faidherbia albida and Salvadora persica, are protected as well. These trees are of great importance during the drier parts of the year as they become green at the on-set of the rainy season long before there is enough grazing.

The following case study shows that contraventions against the protection of trees are followed up but that this sometimes leads to major difficulties as trees are not necessarily easier to protect than pastures.

> In January 1994 the two women of Tako, Mukaaruihi and Kaitu, started to cultivate their garden in Otjikongo on the banks of the river in the midst of palm trees. In order to make their work easier they decided to burn down the undergrowth. The fire got out of control and in the end about 30 palms were burned. (Most of them did not die but they did not yield nuts for some two to three years.) The community complained seriously about their negligent behaviour and both were fined by the neighbourhood council of Omuramba to pay six bottles of cheap liquor to the government (*ohoromende*) as a fine. While Kaitu divorced Tako a short while later and left to live in Angola (and thereby evaded paying the fine), Mukaaruihi finally exchanged a goat for six bottles of cheap liquor. One bottle Tako took for himself, as he reasoned that he was a sort of chief's policeman (*omupolise*) for the area - this was not just his personal opinion but everybody would have agreed to that. Tako sent five bottles to the chief and advised him to store these bottles to present them to the next meeting at his place. However, the chief decided that he would only store three bottles for the next meeting and sent two bottles to an elder nearby. When Tako sent the bottles to the chief he added that Kaitu would still have to pay her fine but that he could not reach her at the time.

While this case study only depicts a minor transgression and the damage both women caused was negligible, the partial destruction of the palm forest along the Kunene river is much more severe and is more difficult to trace. Along the river Himba households and households from other ethnic backgrounds (Nkumbi, Tjimbundu, Ngambwe, Thwa) live alongside each other. A common denominator of these households is that they are poor. Especially those non-Himba living at the river own hardly any livestock. They garden, live off fish and frequently tap palms in order to sap palm wine. To do so they have to climb up to the top of the palm and cut off the entire head of the tree. For about one month a juicy liquid pours from the beheaded tree into small metal or plastic containers which are fixed to the trunk of the tree. This juice ferments and is drunk as a slightly alcoholic beverage. However, the tree inevitably dies. As the practice of palm tapping has become prominent over the last twenty years several spots along the river show a severe degradation of the palm forest. Palm trees are threatened through other non-sustainable ways of resource use. Young palms are frequently cut down and the palm heart (*otjikora*) is eaten roasted.

By the end of 1995 many people had become alarmed about the state of the palms. More and more palms had been chopped down over the past year, and what irritated people most, many of them by non-Himba. In early 1996 when a major case against traders was heard at the Kunene, many men gathered as well as the chief and his councillors and the issue of the palm trees was treated in detail. The meeting made an astonishing decision: all people who were found guilty of having tapped palms in the past were fined one or two sheep. These animals had to be provided immediately and were used to supply the meeting. Then the meeting decided that from now on the control of the palm groves along the Kunene should be handled by the police. The governor and several police officers who were present at the meeting were informed that the people thought a police control would be more effective than a control by the Himba themselves. Obviously transaction costs to keep up the institution of palm-grove-protection would have been so high that everybody lost interest in the matter. Too many meetings had taken place before which had not led to an improvement of the situation.

6.3.3. Resource Protection in Two Pastoral Societies: the Comparative Perspective

Resource protection among the Himba is more effective than among the Pokot. While Himba have fairly well defined communities communally managing resources, among the Pokot these communities are fluid and membership in them changes constantly. While Himba punish non-adherence to grazing rules, Pokot frequently fail to do so.

Definition of a group protecting a resource: Pokot herders have access to the entire Pokot territory, while Himba herders only have access to the territory of the chiefdom they belong to and in a stricter sense only to the lands of the neighbourhood they belong to. Pokot define rights of access to pasture via ethnic sentiments. Himba emphasise kinship and neighbourhood as concepts regulating the access to pastoral key resources and thereby are much more specific than the Pokot. While the community protecting a resource among the Himba is a face to face community with strong kinship bonds, among the Pokot the group managing a resource is often very large and vague.

The Rules of Resource Protection and Delineation of the Resource to be Protected: Although in both communities grazing management addresses similar problems, i.e. the differential treatment of dry season and rainy season pastures, the Himba do so with a more complex and detailed catalogue of rules:

grazing of specific areas only during the dry season, safety cordons around more permanent settlements, cattle camps only moving in camp associations, prevention of excessive mobility, orientation towards dry season grounds near semi-permanent dwellings, saving of some pastures for droughts. Pokot just specify specific areas for dry season grazing and some patches as calves' pastures. There are no regulations in respect to the movement of livestock camps and the choice of dry season grazing by individual camps. Obviously a more elaborate code of grazing makes it easier to deal with contraventions among the Himba and at the same time ensures that Himba herders of one community have equal access to resources. The less differentiated rules of the Pokot neither ensure sustainable management nor equal access.

Drawn out cases on contravention show that a basic difficulty in Pokot resource protection lies in the hazy description of the spatial borders of the objects to be protected. There is much more agreement among the Himba on this point.

Establishing contravention against grazing rules and the sanctioning of trespassing: Contravention against grazing rules are diagnosed fairly similarly in both systems. At first complaints are discussed at a neighbourhood level amongst the adult males of the area. While among the Pokot the mere fact of being Pokot and being adult gives somebody the right to speak up in all matters pertaining to local resource management, the right to speak on matters of resource management in a specific place among Himba is defined via kinship and neighbourhood. This delimits the number of people who are formally entitled to comment on cases and reduces transaction costs.

In both societies council meetings can drag on for a long time. If a case is not treated adequately at a neighbourhood level, it is referred to the chief or one of his councillors among the Himba. Both, chief and councillor, belong to a face to face community of herders. Usually they are rich and influential elders and some of their prestige comes from their capacity to treat cases adequately. Pokot neighbourhood councils only rarely referred cases to the chief or sub-chief. Among the Himba punishments are invariably paid in livestock while among the Pokot punishments for wrong grazing were not paid in livestock but in money.

The co-management of resources: In both societies regional administrations influence resource protection. Governmental organisations have realised that effective resource protection can only be achieved when herders are included, but are insecure about how they are to be involved. Among the Pokot the inclusion of herders is haphazard: chiefs constitute parallel structures to supervise grazing management. Communication with local councils is sporadic. As the chief is habitually not a member of the community and is rarely present in local meetings many men doubt his rights to influence community decision making. Development agencies were also prone to develop parallel structures in the Pokot case. Dam committees and female grazing guards rather conform to their ideas of anti-desertification campaigns than to the local realities. To some extent the State made it easier for Himba chiefs and elites to have an impact on pasture management. Himba chiefs and councillors were invariably members of the local communities. Their legitimisation was derived from their place in a genealogy. Additionally the South African colonial government sustained the power of Himba chiefs.[31]

Up to now the description of the Himba case may suggest that there are relatively few resource management problems due to the authority of a few principals and the paternalistic control of their agents. This, however, is not the case. First of all, there is a great deal of competition between the leading principals of a community. In the community I worked with, the two central seniors were frequently at logger heads. At some stage their competitive quest for power resulted in severe friction within the community. During this phase open questions of resource management could not be solved and were sidelined until a truce was reached. Competition between wealthy men is apparently a key element of the system. In many Himba communities I had contact with, competition between some wealthy herders was the main topic of local politics. However, these problems are exacerbated with the inclusion of the Himba economy within the national market. After independence rich herders gained increasingly more contact with national developments. They were addressed by tour operators for example, who wanted to operate camps at the scenic Kunene river. The substantial rents for these plots were handed to the chief and his councillor – the two major leaders in our example. This happened with the tacit understanding that they would distribute the money among the community. However, both, chief and councillor, regarded the land as theirs and pocketed the money. These examples indicate that there is a real danger that the elite is being co-opted by outside political forces and that wealth and external ties may become decisive factors determining resource management in the future.[32]

6.4. FOUNDATIONS OF MORAL ECONOMIES: SOLIDARITY AND PATRONAGE

The concept *moral economy* has been coined to describe "... a network of group support, communication and interaction among structurally defined groups connected by blood, kin, community or other affinities." (Goran Hyden in Lemarchand, 1989:37). However, what "group support", "communication and interaction" means in this context remains vague. Scott (in Ensminger, 1992:2) defined *moral economy* as an economy "... in which a subsistence ethos guarantees at least minimal provisioning to all households". Here, too "subsistence ethos" remains a variable hard to disentangle. Nevertheless, both authors and with them many other social scientists are adamant, that a specific ideological frame typical for subsistence economies acts as a major means for risk buffering. In a recent contribution to the debate Lindenberg (1998) distinguished strong and weak solidarity groups. He focussed on the analysis of "solidarity frames" - cognitive schemes which act as signposts for individual decision-making. Lindenberg sees norms and values as much as emotions as key elements of cognitive frames. Group ideology and a sharp delineation of in- and out-group boundaries are important elements determining solidarity. Rituals act as catalysts for solidarity. Both economies under discussion here constitute moral economies. However, they do so in very different ways: while the Pokot combine solidarity with an egalitarian ethos, the Himba connect similar ideas to patronage. How do actors reason about exchange and social support and why do they think it is right to invest a lot of their property into social relations and why do they think it to be obligatory to render substantial help when kin or neighbours demand support? What do norms and values in such a system look like? How does ritual contribute to the maintenance of solidarity?

6.4.1. Pokot - The Ethos of Egalitarian Exchange

Pokot exchange networks are guided by an egalitarian ethos. Livestock gifts are described as acts showing solidarity between equals. Morality is frequently a topic in gatherings and ceremonies. Rituals are geared to emphasise Pokot morality. Group boundaries are clearly delineated by a number of material markers such as beadwork, coiffures and dresses.

6.4.1.1. Solidarity, Respect and Internal Peace: Norms and Values

Pokot morality rests on key concepts such as solidarity, respect, internal peace and bravery. For the discussion of exchange networks the first three concepts are of crucial importance (for the latter see Bollig, 1990a, 1992a, 1993, 1995b).

Tilyontön - Solidarity

Conceptualising their exchange networks Pokot frequently speak about their *tilyai* - i.e. the people engaging in reciprocal livestock exchanges (*tilya*) with each other (cf. Schneider, 1953).[33] The concept *tilyai* does not denote abstract kin categories but describes only those persons an actor actually has exchanged livestock with. Personal relationships are established and maintained by repeated transfers of property such as livestock, food, tobacco and occasionally pieces of clothing, beads and weaponry.[34] The abstract derived from *tilyai*, *tilyontön*, denotes a set of norms and values that imply mutual solidarity between exchange partners. Two herders who are tied by *tilyontön* practice a certain mode of exchange which is necessarily infused with feelings of personal intimacy such as sympathy and generosity. One actor who receives an animal knows that under the norm of *tilyontön* he has to return an equivalent value later on. Solidarity involves intense emotional ties between two men. The exchange of livestock implies *kongityö* (friendship). The concept *kongot* (or the abstract *kongityö*) implies very close emotional ties. If there is to be *tilyontön* between two people there must also be *chomnyogh* (affection, love), *kalya* (trust, peacefulness, ease) and - in times of need - *kisyonöt* (mercy, compassion).

Tilyontön is a gendered value. Solidarity among friends is regarded as a male domain. While men are depicted as collaborative, women are seen as individualistic. Male solidarity is founded in large initiation ceremonies in which men stay together for months. In contrast, women are initiated in small groups of two to five girls. Age-mates frequently cherish brotherly relations. As normative concomitant to these emotional identities, men of one generation set are not allowed to marry each others daughters - they are treated as if they were brothers or close relatives. The ideal warrior - the *nyakan* - not only stands out for his military prowess but also for his behaviour showing solidarity towards his compatriots. He assists them in times of need and supports them in cases. The ideal warrior finds his counterpart in the *kapolokyon,* an elder who has proven throughout his life that he lives in harmony with his social surroundings. He has never cursed nor bewitched anybody else. He has helped the needy and is honoured by his friends for his selfless behaviour and the solidarity he extended to his compatriots. Through the power of his accumulated symbolic capital a *kapolokyon* is qualified to speak blessings at major meetings and frequently acts as a mediator between quarrelling parties.

Tekotön - Respect

Respect with age is a recurrent cultural theme in East Africa (Spencer, 1998:93). In his study of individual psychological traits in four East African societies Edgerton (1971) found that pastoralists - the Pokot were one of his sample societies - tended to express respect for authority while farmers rather expressed contempt, ridicule or open disrespect for authority (Edgerton, 1971:176). Respect is the normative cornerstone of a gerontocracy. Respect - in some instances the translation "discipline" would also be correct - is demanded in interactions between junior and senior age-sets and generations-sets, between in-laws and between men and women: respectful behaviour accompanies a potentially asymmetric exchange relation. Older age-sets bestow the highly valued symbols of seniority to juniors and promote them from one generation into another. Juniors in turn offer their obedience and respect. A father-in-law gives his son-in-law a wife: as bridewealth payments are protracted the relation is deemed as asymmetric for a period of time. Only when the bridewealth is fully paid does the relation between father-in-law and son-in-law become no longer determined by respect but by mutual solidarity. In return for the respect offered by juniors, senior men are obliged to care for them materially and spiritually: they have to present and loan livestock to young men and they have to ensure their well-being by blessings and other ritual acts. Seniors nurture juniors; symbolically this relation is visualised when elders spoon-feed juniors for some time after their circumcision. At the same time seniors are entitled to withdraw cattle and spiritual assistance if respect is not forwarded to them.

Respect between specified sets of people solidifies exchange networks and mutual support. Respect means that a moral code is acknowledged in spite of the fact that there is no institution policing transactions. Obviously somebody could evade paying his full bridewealth when migrating to another edge of Pokot land; he could deny having livestock debts with somebody: however he has learnt to respect these relations. If he wants to be respected as an elder one day, he has to defer respect towards elders and in-laws himself.

Kalya - Internal Peace

The Pokot have an elaborate cultural theory on the relationship between internal conflict and its negative effects (Nyamwaya, 1987:1278; Bollig, 1992a:176ff). Curses, witchcraft and evil thoughts may cause tremendous harm with an adversary and very often symbolic aggression is thought to be caused by ill feelings instigated by the denial of food or other exchanges. There are numerous experts who specialise in the identification of witchcraft, while others treat bad will and curses, and still others identify all three types of symbolic aggression (which is subsumed under *kutĭch*i - the mouth of a person). While there is otherwise little specialisation in this pastoral society, it is remarkable that in the field of "maintenance of internal co-operation" specialisation occurs to an astonishing degree (see Table 32).

Table 32. Ritual specialists involved in the maintenance of internal peace

Ritual specialist	Medium	Diagnosis	Treatment	Oracles
Töptöpin	Tracing blockages of blood flow by feeling the body	*kutĭchi, rurwö*[a]	*Tapa*[b]-treatment	No
Kipchö	Looking for signs in a calabash with milk	*kutĭchi*	none	Yes
Kipkwegh	Throwing sandals	*kutĭchi*	none	Yes
Kipkwan	Haruspexing	*kutĭchi*	none	Yes
Liokin	Mixing and applying colours[c]	*kutĭchi*	Treatment with colours and the root *moikut*[d]	No
Kapolokyon	-	-	Blessings and applying colours	No
Pörpörin	-	*moriyon*[e]	*Parpara*[f]	No
Kolin	-	-	*Kolsyö*[g] -applying red colours	No
Werkoyon	Dream	-	-	Yes

[a] *Kutĭchi*: lit. mouth of a person includes bad will (*ghöityö*), curse (*chipö*) and witchcraft (*pan*); *rurwö* (lit shadow, ritual pollution through contact with a dead person or somebody who has touched a dead person).

[b] A ritual to clean the body from pollution and possibly blood blockages.

[c] This is done with sacred earth colours, e.g. yellow and green are said to be used for witching and counter witching.

[d] The root of Cyperus rotundus. The ritual specialists chews it and spits part of it on the breast and into the armpits of the person he is treating.

[e] *Moriyon* refers to a non-atoned blood guilt between clans

[f] This is a special blessing that is said by the *pörpörin* to alleviate blood guilt between clans.

[g] The ritual in which a person who has killed and has been transformed into a healer chases a bad shadow from the sick person by applying red earth colour.

Specialists like the *töptöpin*, the *kipchö*, *kipkwegh* and the *kipkwan* (sometimes: *kipkwen*) are frequently consulted. A *töptöpin* is called upon when somebody is seriously sick. The assumption that a disease is caused by some conflict is always nearby. By feeling the body the *töptöpin* finds blood blockages. The *kipchö* tries to trace the origins of the evils befalling his clients by looking into a calabash with milk. The patterns he sees on the milk - sometimes he even sees faces - tell him about the culprit behind his client's problems. He identifies witchcraft and cursing. Throwing sandals and analysing intestines are further ways to trace malice.

Once identified, internal quarrels can be kept at bay in communal rituals. Coloured earth is an important feature in these rituals: potential adversaries are painted in white colours to symbolise and set in action regained harmony. In order to appease conflicts within the family a father may call together his quarrelling sons, place them in the midst of the livestock enclosure and paint them all white, while he says blessings. The *parpara*, the ritual conducted by a specialist treating unatoned blood guilt between clans, is a good example of Pokot emphasis on internal peace. Pokot believe that if a murder – no matter how long ago - between two clans has not been properly atoned this results in a very specific sin called *moriyon,* which is not attached to an individual but to all members of the respective clan, perhaps a concept resembling Catholic ideas on the original sin. Whenever people of these two clans engage in marriages or other relations that involve close physical contact (like circumcision or sexual intercourse) they run the risk that their children will die or be crippled, or they themselves may become sick and die because of *moriyon*. At each marriage ceremony a *pörpörin* (a sooth-sayer) is invited to say the magic blessing to cast away *moriyon* and ground personal relations on peaceful links between descent groups. The *parpara* text, of which parts are quoted below, was spoken in a circumcision hut in December 1988 to prevent unatoned blood guilt between circumciser and circumcised affecting the outcome of the operation.[35]

we will bring unity now, ye
we are washing the sins of both sides, ye
the sins of whom? ye
they are cleansed, ye
we have detected the misdeeds of the past, ye
whose misdeeds, ye?
those of the Kopil clan, ye
they were committed, ye
they were committed, ye
with a knife, ye
with whose knife, ye
with the knife of a Kopil, ye
and who was killed? ye
somebody from the Oro-Clan, ye

somebody from Oro, ye
blood was shed, ye
and what happened then, ye
a disgrace
no, no, no ...
and now, ye
what happened now, ye
we have tracked it down, ye
and we have cleansed it, ye
and we chased it away, ye

The text refers to a case of murder between the Kopil and the Oro clan: a person from the Kopil clan killed an individual of the Oro clan with a knife. It is the community's ritual action ("we have detected the misdeeds of the past", "we have cleansed it") which purifies members of both clans from their unatoned guilt.

Pokot envision *kalya* as an emotional state that everybody is longing for. *Kalya,* is a societal as much as an individual state. *kalya* may even denote the relation between a society and its environment. The landscapes Pokot sing about in romantic songs convey *kalya*. To sing them conveys a feeling which is *anyin* - sweet. *kalya* is the precondition of *karamnyö*, well-being and of *onyinyö*, happiness. It is a state of balance and ritual cleanness and contrasts sharply with the uncleanness and disharmony of those who transgress taboos and withhold gifts and food from others. Schneider (1955:405) comments on the concept of purity among the Pokot: "The moral life is one of ritual cleanness and conformity to the laws of god and society." The ideal man (and woman) possesses numerous cattle - another sign of his ritual cleanness, of which he gives continuously, he shares food and donates oxen for public festivities. Anti-social acts, activities which originate with non-harmonious feelings, eventually end up in ritual uncleanness. Hence while a positive, peaceful and law-abiding character is a precondition for *kalya*, purification rituals help someone enter this state.

6.4.1.2. Strong Brotherly Bonds: The Reification of Identity in Rituals

Durkheim attributed communal rituals the important function of maintaining morality and communal sentiments (Durkheim, 1984:520). In Pokot society communal rituals are large gatherings. They are characteristic for rituals of initiation and rituals of the age-sets and generation-sets, as well as ceremonies which are conducted to purify a community and to pray for rain. In ritual cooperation and solidarity are enacted dramatically when people act as members of corporate groups and not as individuals. Men talking in the sacred half circle *kirket* (see Figure 31) do not talk on their own account but do so as members of generation-sets and age-sets. They will talk proudly of "we, the Koronkoro" and "we, the Kaplelach" and in the same tone will criticise "the

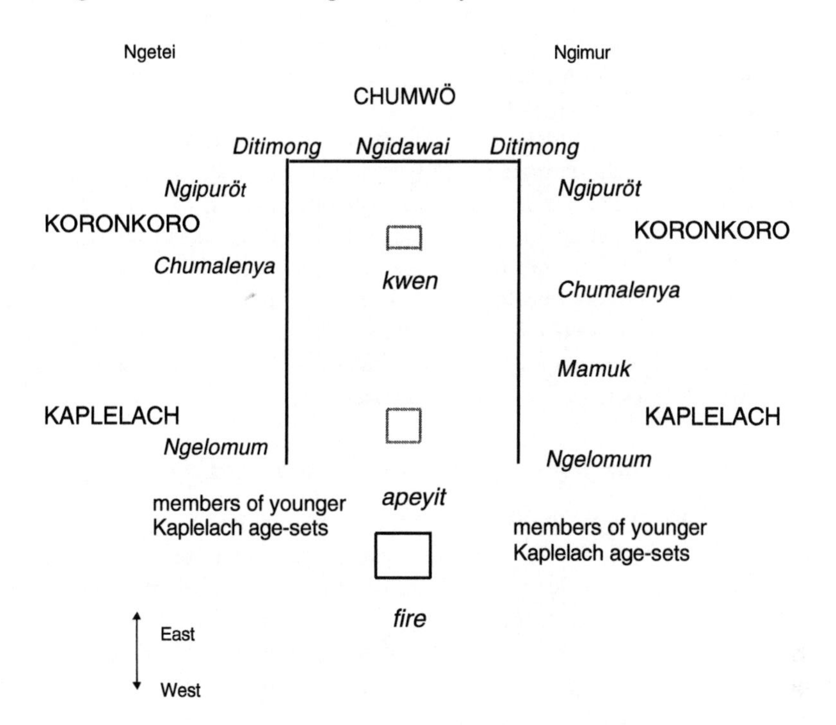

Figure 31. The Pokot Sacred Halfcircle of Men (kirket)

Koronkoro, who have spoilt the land" and "the Kaplelach, who are disobedient". While age-set relations are not of primary importance in everyday life, in ritual the sets present themselves as corporate groups (Bollig 2000). Their corporateness is underlined by the strict seating orders in the sacred half circle (*kirket*), where members of one set sit close to each other and are juxtaposed to members of other sets. At the climax of these celebrations all men sit in a half circle which opens to the west, i.e. facing the sacred Mt. Mtelo, the cradle of pastoral Pokot in oral traditions. The oldest men sit at the apex of the half circle, seniors surround them and juniors sit at the two flanks of the half circle.

Songs and dances underline unity and co-operation. In the *sapana* ritual numerous women come in long lines in the afternoon hours and bring milk to the ritual place to supply food to juniors and seniors alike. Their songs cheer up the people. Alongside the specific male age-set and generation-set specific songs, women sing their *kililyet*, songs which praise the fame of specific descent groups. Clans and lineages have common symbols (for example an animal or natural phenomena), there are common songs and specific markers (*enwait*, ear-cuttings for goats, sheep and cattle, ways of decorative cauterisation). These symbols of the descent-system give structure to the ritual activity. People of the Firestick Clan (*Koimö*) light fires in the ritual context; people of the Buffalo Clan (*Siwotoy*) are the only ones who are allowed to play the lyre at such occasions, people of the *Kiptinkö* clan conduct rituals on the tribal borders to ensure protection from raiders. Clan specific ritual specialisation suggests a society in which corporate entities are organically working together.

Internal peace is emphasised in all community rituals. The ritual purification (*mwaghat*) which precedes the prayers for rain mainly contains the enactment of harmonious relations between different sets. This ceremony is also conducted before men go on a raid: men who have disturbed internal peace profoundly and who have not atoned their sins by paying livestock to the parties damaged will neither be allowed to take part in raids nor in prayers for rain.

Group identity is visualised in these rituals and communal rituals symbolically reconfirm solidarity, comradeship, respect, discipline and internal peace - key values of the Pokot belief system.

6.4.1.3. Being Surrounded by Enemies: Visualising Ethnic Boundaries

Lindenberg (1998) hypothesised that in strong solidarity groups boundaries are well demarcated. This happens by necessity: an exchange system based on strong solidarity needs clear-cut boundaries in order to indicate what people have to do to be considered generously and to whom solidarity has not to be extended. Strong internal solidarity goes hand in hand with clearly defined group boundaries.

Marking Ethnic Boundaries

Ethnic boundaries are clearly demarcated visually in many parts of Kenya. Klumpp and Kratz (1993:195) in a contribution on Maasai and Okiek ornaments point out that beadwork is a "key visual index of ethnicity". They observed that the hardening of ethnic boundaries in the late nineteenth century was paralleled by the introduction of cloth and glass beads on a massive scale: "Colour preferences and designs became highly visible resources for advertising hardening ethnic distinctions within a broadly shared cultural understanding of the body." (Klumpp & Kratz, 1993:197). Hodder (1982:85) in an ethno-archaeological approach to the materialised culture of the Baringo basin found that objects were manipulated in social strategies to document group identity. In a survey of various artefacts (for example pottery, beadwork) he found that some styles were identical across ethnic boundaries while others were clearly used to demarcate such boundaries[36]: "Especially in the border areas where there is greatest tension and competition, material culture of many forms is used to justify between-group negative reciprocity and to support the social and economic dependencies within groups." (Hodder, 1982:56).

Pokot men and women use beads in order to express their ethnic identity as Pokot and to make their individual social status visible. The introduction of coloured plastic beads into the region in the 1960s opened up new ways of differentiation. Table 33 gives an overview of the elaborate code of adornment for Pokot males.

Table 33. Types of male jewellery and their colour idenitifications

Type of jewellery	Indigenous Term	Colour specification
Bracelet of several strings	*Lokitar*	red, dark blue, white, black
Bracelet of one long string	*Noyow*	yellow, red, dark blue
Plastic rings of upper arm	*Apokoytin*	green, blue, yellow, red
Headband of beads just above eyebrows	*Ngatöröntin*	red, dark blue, white, black
Headband of beads on forehead	*Aritaytin*	red, black, white, green
Necklace, close-fitting	*Karasighin*	green, black, red, dark-blue
Necklace, wide	*Karikat*	black, red, orange
Necklace, rather close-fitting	*Sim*	White
Belt of beads	*Karemu*	dark blue, red, black, white
Beads in ear	*Altepeyon*	red, white, dark blue
Sash (I)	*Ngeririyon*	orange, white, black
Sash (II)	*Ngeririyon*	dark blue, red
Sash (III)	*Ngeririyon*	white, red
Broad collar of beads	*Lokipir*	red, black, white, green

The colours of beads (and colour combinations) for all these items of adornment are fixed (see Table 34). The belt (*karemu*) should always have the green, red, blue combination, even the sequence should be held constant. The same holds true for the plastic rings on the upper arms (*apokoytin*): the six inner rings may be either green or blue but should always be surrounded by two red rings. There are several prohibited colour combinations: yellow/green, white/green, blue/yellow, dark-blue/yellow, orange/yellow, dark-blue/blue. Red is the only colour that is combinable with all the other colours, it goes frequently together with white, black and dark-blue.

The physical adornment of Pokot men is complimented by an elaborate head-dress. Pokot men grow their hair long in order to cover it with a mud-cap

Table 34. Colour combinations in Pokot male beadwork

	White	Black	Red	Yellow	Orange	Blue	Dark-blue	Green
White	XX							
Black	6/2	XX						
Red	6/5	7/5	XX					
Yellow	0/2	0/2	2/0	XX				
Orange	1/1	2/0	1/1	0/2	XX			
Blue	0/1	0/1	1/0	1/0	0/1	XX		
Dark-blue	4/3	4/3	7/0	1/6	0/7	0/7	XX	
Green	1/2	2/1	3/0	1/2	0/3	1/2	1/2	XX

Note: The first number indicates how frequently a colour co-occurred with another colour in several types of ornament, the second number indicates how frequently a colour occurred without a second colour, i.e. black occurred six times with white and two times without white, dark-blue did not occur with orange and blue but occurred seven times with red and not at all without red.

which is meticulously woven into the hair. This mud-cap (*siolip*) is finely painted. The patterns and colours show the status of the person in the gerontocratic system. Feathers are added to the head-dress. While younger warriors are usually not allowed to wear any ostrich feathers, older warriors may wear black and white feathers (*songol*, *palalan*). The elaborate *atöröntin* (a rhombus-like five by five centimetre rectangular mini-mudpack on the forehead, see Photograph 3) with its feathers and the big black feather pompons (*alim*) may only be worn by seniors. Turkana have similar feather-codes, although there are differences in detail. Samburu warriors anoint their long hair with red-ochre and occasionally weave beads into them, Marakwet and Tugen youths do not wear headdresses.

Women's colour codes are at least as elaborate as male codes and they are clearly different from the colour codes of surrounding groups (Biancho, 2000). Pokot women wear combinations of blue, red and light colours, while Turkana women wear predominantly blue and Samburu women tend to wear red beadwork.

Pung - Being Surrounded by Enemies and the Negligence of Cross-Cutting-Ties

Ethnic boundary markers are not only found in the material world. Pokot subsume all ethnic neighbours under the label *pung*, enemies. While life in Pokot society – according to local ideology - is shaped by solidarity, respect and internal peacefulness other ethnic groups do not reach these ideals to the same degree. Especially the agricultural neighbours, the Tugen and Marakwet, are seen as not showing solidarity and as conflict-ridden communities. Other groups lack pride, people do not adorn themselves adequately, men eat together with women or people eat unclean food. Interestingly, the Pokot see their arch-enemies, the Turkana as being the most similar to themselves.

Ideologically the Pokot conceptualise themselves as an isolated unit surrounded by enemies (*pung*). Factually however they have married from other ethnic groups and exchanged livestock with herders originating in other societies over generations. In all genealogies marriages with non-Pokot were present (see Table 35). The frequency of interethnic marriages, however, has declined over the last decades: 18 members of the most senior generation, the Chumwö, had married 34 per cent of their wives from other groups, of these

Table 35. Interethnic marriages between 1910 and 1988

Generation	Bride-Takers	Marriages	Pokot Women	Turkana Women	Tugen Women	Samburu Women	Other
Chumwö	18	54	35 (65%)	15 (28%)	2 (4%)	1 (2%)	1 (2%)
Koronkoro	34	82	72 (88%)	8 (10%)	0	1 (1%)	1 (1%)
Kaplelach	39	60	52 (87%)	2 (3%)	5 (8%)	1 (2%)	0

28 per cent were Turkana. The percentage of interethnic marriages declined rapidly: only 12 per cent of the 34 Koronkoro interviewed married from other ethnic groups, 10 per cent of all marriages were marriages with Turkana women. 13 per cent of the women of the youngest set, Kaplelach, were from other ethnic groups. Among them the percentage of Turkana had dwindled to three per cent. Young Pokot men, especially if they were educated, preferred to marry schooled women of Tugen descent, even if they had no experience in livestock herding (see Bollig, 1992a: 257).

Over the period of eighty years the pattern of interethnic marriages has changed profoundly. First of all, interethnic marriages have become less over the decades. In the past Turkana women were the preferred marriage partners. They had a lot of experience with livestock and were deemed to be extremely tough and trustworthy. However, since violence has shaped interethnic relations between Pokot and Turkana herders over the last three decades these marriages have become less frequent. Until the early 1960s Pokot and Turkana lived interspersed in the Silali and Lochorongasuwia area, young people went to dances together and seniors invited each other for rituals. In the 1990s both groups lived about one hundred kilometres away from each other in justified fear of mutual raiding.

Exchange relations across ethnic boundaries are usually very different from internal exchanges. There had been interethnic marriages but these had been one-sided: Pokot had always been bride-takers and never bride-givers. Pokot are reluctant to give cattle in bridewealth payments to members of other ethnic groups. Many of the Turkana women married by Pokot men stemmed from impoverished pastoral households and only a diminished bridewealth was paid. In most exchange relations crossing ethnic boundaries the transfer of livestock and goods takes place on a tit-for-tat basis, exchange relations entailing long term reciprocity being rare.

The apparent self-imposed isolation of the Pokot seems the more astonishing when structural links between them and their neighbours are regarded. Several Pokot lineages claim ancestry in Turkana and in fact there are parts of these lineages living in Turkana. Others claim ancestry among Marakwet, Sebei and even Karimojong. Members of these lineages had fairly close relatives in other ethnic groups. In reality, however, there was very little contact over the borders in the late 1980s and 1990s. At the same time the Pokot generation-set system is closely related to the generation-set system of neighbouring Kalenjin people such as the Tugen, Elkeyo and Marakwet. These groups also have seven or eight sets that circle in time (see also Behrend, 1987). Even the names are identical and according to Pokot ideology of intra-generation set solidarity this fact should implicate brotherly relations across the boundaries. While the fact is acknowledged that the generation-set systems are similar there are no relations based on mutual solidarity with members of the same set across ethnic boundaries.

Spencer's (1998:19) assertion that "tribal boundaries are boundaries of mistrust, setting limits to customary expectation" is born out by the data on the Pokot and their neighbours. While internal solidarity is emphasised, exchanges beyond tribal boundaries do occur, and are occasionally significant for political alliances and far ranging migrations in case of drought, but are generally shaped by mistrust. As if to underline these boundaries of trust, ethnic identities are clearly visually demarcated. Trust and mistrust become visible social facts.

6.4.2. Himba Morality: Patronage, Kinship and Ethnicity

Himba concepts of morality emphasise the corporateness and responsibilities of descent-groups. The people who are obliged to help each other are addressed as *omuhoko*, matrilineal relatives. Among these matrilineal relatives it is especially wealthy elders (*ovahona*) who donate livestock to poorer relatives.

6.4.2.1. Authority and Generosity: Norms and Values

The Himba political system is marked by the dominance of and the competition between bigmen - *ovahona*. These bigmen are remembered vividly in oral lore and they are icons of local identity. Wärnlöf (2000) writing on the history of the western Kaokoland emphasises the importance of *ovahona* for the history of this area. The key ancestor for western Kaokoland, Kakurukouye, amassed large herds of cattle when participating in Nama raids on Herero and Himba communities in southern Angola. In two other publications (Bollig, 1997a, 1998e) I described how local ideas on bigmanship were used by the South African administration in order to institute chieftaincies in Namibia's arid northwest. Local histories (Bollig, 1997c, see contributions by Katjira Muniombara, Ngeendepi Muharukwa and Kozongombe Tjingee and Tjikumbamba Tjambiru) convey the idea that political life is determined by the establishment and the competition of *ovahona* who aspire to wealth and political influence.

Himba/Herero big men do not transform easily into an elite as they are bound by the morality of kinship and patronage. Big men first of all owe allegiance to their clan-mates. Crandall (1992:128) summarises "cattle wealth, generosity, power, and good fortune from the ancestors are so interconnected that one element cannot be possessed without the other". Big men donate cattle predominantly to matrilineal relatives.[37] Kinship solidarity restricts their ambitions to transform economic obligations into political power and certainly their freedom to handle livestock as commodities. It has been pointed out before that livestock loans are safe capital for the person who loans. While there is rarely any doubt on who possesses livestock, the choices on how to deal with livestock property are restricted. A rich man cannot simply withdraw a livestock loan because he wants to buy a car. Only under very specific circumstances (e.g. neglect of borrowed animals, a case between donor and recipient) is a big man allowed to withdraw larger numbers of livestock. Usually he does not order more than one or two animals to be given back at any one time. An *omuhona* combines responsibility for poor relatives and neighbours with political aspirations. His wealth in cattle is seen as much as a sign of ancestral blessings (see also Crandall, 1992:125) as it is interpreted as proof of his personal capability. Another venerated character type of the Himba moral system is the *omuyandje*, lit. "the giver". *Ovayandje* are generous and like to host people in their homestead. In contrast to the *omuhona* the *omuyandje* does not have political aspirations. While big men give out livestock loans, "givers" frequently also donate livestock presents. A "giver" is usually also an *omunandira* - somebody who fears to confront others. *Ovayandje* do not like to attend political meetings. They are mainly concerned with the welfare of their herds. In the local community I was working in, the chief and his councillors were rated as good examples of *ovahona*, they were rich, eloquent and power-hungry. Their lust for power was frequently rated as the major motive behind their cattle loans. They were contrasted with two elders who were exceedingly rich but did not take any interest in political debates. While the *ovahona* spent their time at political meetings, they were usually to be found following their cattle or looking for lost livestock. *Ovayandje* were described as wealthy, withdrawn and even shy elderly men. According to informants they have an inclination to give because their heart is good (*vena omutima omua*). Sometimes they are also described as *ovaporise*, literally people who cool things down, and are contrasted with the "hot" *ovahona*, who stir up debates and sometimes seek conflict and competition to establish their position. Both character types, *omuhona* and *omuyandje*, are *ovanaerao*, people who have good luck because they are blessed by the ancestors. Both may also be *ovanahange*, peace-loving characters and elders who use their wealth and influence to make peace. However, several informants conceded that eventually the *omuhona* may be more valuable for internal conflict resolution. His talent to talk and to influence others is regarded as indispensable for successful conflict management.[38]

6.4.2.2. From Kinship-Based Rituals to Communal Rituals

Among the Himba rituals are kinship based. While neighbours may take part in these ceremonies the major ritual activities are conducted by a few closely related men. Usually ritual activities take place at the ancestral fire (representing ancestors in the homestead) or at the graves. Traditionally there have been no rituals in which the entire community combines its efforts. Lately, however, after independence rituals of a more comprehensive scale have been developed.

Kinship-Based Rituals: Commemoration and Graveyards

Once the activities directly connected with the burial have ceased, a grave is "closed" until the first commemoration ritual (*okuyambera*) takes place about one year later. The year between the burial and the first commemoration ritual is a period of mourning. All close relatives, male and female, wear ornaments and hairstyle in a way that distinguishes them as mourners (*ovipiriko*) (for a detailed description of the *ovipiriko* code see Bollig, 1997b). The widows are segregated from other people and spend most of their time sitting quietly dressed in festive gear in the main hut of the household. The first commemoration ritual requires careful preparation. This festivity marks an important turning point in the history of the household. The mourning dress (hair, ornaments and clothing) is changed again into the normal dress and inheritance claims are finally settled. A large number of people gather and numerous cattle are slaughtered. The entire herd of the deceased is gathered and even distant cattle camps have to return to the main homestead for the duration of the ritual. One morning, after the central actors of the ritual have been introduced to the ancestors at the holy fire, a large party sets off to the ancestral graves. Many cattle are driven to the graveyard too. Wailing in high pitched voices, the mourners march to the grave, men and women separately. At the grave they congregate and the grave is "opened" ceremoniously. Slowly, the party passes around the grave. Everybody must touch the gravestone or some stones on the grave. Then some senior male relatives squat around the grave. They put mopane leaves on it, smear the stones with butterfat and finally smear the sons of the deceased and the stones on the grave with curdled milk. Both butterfat and milk are brought from the homestead. These are produced from the milk of ancestral cattle. The whole process is accompanied by loud wailing. Occasionally the Himba trumpet is blown in order to encourage the living and to honour the dead. Finally a ritual fire is kindled using firesticks at the foot of the grave. Ashes from this small ritual fire are taken back to the holy fire of the homestead together with some mopane leaves from the grave, thereby emphasising the close link between the ancestral graves and the ancestral fire within the homestead. Once the ritual at the graveside is over, another few days pass before the question of inheritance is settled.

In subsequent years commemoration rituals take place at irregular intervals. Then again the grave is visited by many relatives and by the household herd, the grave is smeared with butterfat and milk and a ceremonial fire is lit once more. Even very old graves are still potential sites for commemoration rituals. (See Photograph 17).

Such rituals establish a specific relationship between descent groups and land. The "owners of the land" refer to numerous ancestral graves, patrilineal and/or matrilineal, that are situated in an area and they found their claim for dominance upon these graves. In that sense the chiefly Kapika patriline is connected with the Ombuku valley where many of the present chief's patrilineal relatives are buried, and the Muniombara patriline is connected with Ovizorombuku, a place with a permanent well near Epupa. The graves (or references to the graves in meetings) are the tokens of patrilineal settlement

Photograph 17. Himba Kingroup attending to the Ancestral Fire

continuity. Their presence guarantees the continued influence of the "owner of the land". References to graves and graveyards become even more important in inter-group conflicts over land. In Okangwati a long-standing dispute between local Himba and the descendants of a group of Herero, called the Toms, is frequently debated at public meetings. Here the subject of graveyards is frequently discussed. Himba speakers like to refer to the fact that their ancestors can be traced back in the region for some generations via their graves whilst the duration of the Tom's settlement in the area only spans some two or three generations. From the fact that their graves are older and more numerous than those of the competing group, the Himba derive the legitimacy of their claim for political dominance in the area. What is at stake is not clearly expressed: obviously it is not intended to contest the grazing rights of other groups or to question their rights of access to arable land. More central to the debate are issues about, for example, the right to admit traders to Okangwati, the right to name a chief for the area and the right to influence decisions on development projects (such as the Epupa hydro-electric dam) in the area.

Communal Rituals in the Making

In general, Himba rituals are strongly kin oriented. In recent years however there has been the need to define the entire community in a more concise way beyond the confines of kinship. The Herero of central Namibia have several major festivities which are catalysts of Herero identity vis à vis other ethnic groups in Namibia. At the graveyards of great chiefs of the past in Okahandja, Omaruru and Otjipawe it is not clan leaders who are commemorated but rather leaders of the Herero nation. Such expressions of ethnicity have not been in high demand among Kaokoland's Himba until recently. The encapsulation of Kaokoland during South African times (Bollig, 1998e) made it unnecessary to offer a more concise image of the community. The recent debate on a hydro-electric dam apparently urged the Himba towards clearer forms of self expression within a national context. Like the Herero of central Namibia they seek to document a community profile in ritual nowadays. Since 1995 gatherings at Mureti's place of birth have become rather frequent.[39] While there had been commemoration festivities at the grave of Mureti at Otjipawe for many years (see Ohta, 2000), these had been hampered somewhat from the point of view of Kaokolanders by the fact that the grave of Mureti was about 250 kilometres away from Kaokoland. While the festivities at Otjipawe were good to bring together descendants of Mureti from all over Namibia - it was of little use to document the common identity of people originating from Kaokoland. From the point of Kaokoland's elite the ritual was further hampered by the fact that Chief Tjakuva of Okamatapati from eastern Namibia had appropriated "the chair" of Mureti. To make things worse Tjakuva was allied with the ruling SWAPO party while most Kaokolanders were in favour of the oppositional DTA. There was a high demand for a major communal ritual within Kaokoland and after Tjakuva died in 1995, the replacement of Tjakuva by a chief from Kaokoland was envisioned. The debate on the legitimacy of royal lines among the Herero and the standing of Kaokoland's Herero vis à vis other Namibian Herero became entangled with the debate on the planned hydro-electric dam at Epupa. Several symbolic measures were taken to re-appropriate the Mureti myth. Tjakuva's main opponent and would-be successor, Muharukwa, died only a month after Tjakuva in 1995. With a lot of pomp he was buried on a hill overlooking Kaokoland's capital Opuwo. Ostensibly he was not buried in an ancestral graveyard near his homestead but in a political space. On the tombstone one could soon read "Here rests the king of Kaoko". After some time Muharukwa's successor, Tjavara, was established as the legitimate incumbent within "The Royal House of Okaoko". On a tour through Europe organised by environmentalists Tjavara was presented as the "traditional king" of Kaokoland's population. While the historical claim to a royal status was somewhat vague - basically there had never been anything like a king in Kaokoland - legitimacy was created through an emphasis on communal ritual.

Festivities at the grave of Tjavara's two successors, Muharukwa (died in 1995) and Mwinimuhoro (died in 1983) gained in complexity. More people were invited, the celebrations took more complex forms and mythical histories of Mureti and his chiefly descendants became more important. Celebrations at Mureti's birth place at Oriondjima - the rock of the baboon - were rather different from the usual ancestor commemoration rituals. People gathered irrespective of their clan membership. They congregated from all over Kaokoland. Many came by car from the regional capital Opuwo which is about 110 kilometres away, some came from as far as southern Kaokoland some 250 kilometres away. The ritual centre of the festivity was not a grave, but some stones which had allegedly been the holy fire of Mureti or his father. A major part of the festivity was made up by speeches. These had either a mythical-historical content or a direct political impetus. However, even historicising accounts had a clear relation to political topics: as one senior councillor announced in a meeting in November 1997 when opening yet another commemoration ritual at Mureti's place "we have come here to work with history and not to politicise". The key-term during these festivities is *ombazu* (custom). It is reasoned that a hydroelectric dam should not be built because it would be against *ombazu*. With *ombazu* a re-invented tradition is being put to use for political purposes. In contrast to other festivities of the Himba and Herero, in which the intricate Herero colour-code is not of importance, here colours and flags are relevant. Many Herero from Opuwo attend the meeting with the attire of their *oturuppe* (Krüger & Henrichsen, 1998). While Himba and Herero leaders express identity and the quest for political power and territorial control through ritual, they are at the same time active on a national political level. Kaokoland's leaders are closely allied to Namibia's main opposition party, the DTA. Herero and Himba leaders visit the capital Windhoek frequently and actively engage in political discussions there. On the whole Kaokoland's political elite has rapidly developed a "traditional" legitimisation of power through ritual and strong links to national political power brokers. In the Epupa debate they succeeded in engaging European environmentalists as much as senior political leaders on a national level in the debate.

The development of these community rituals defines a departure from the pattern of kinship-based rituals. Since the artificial boundaries of the South African founded Bantustans were abolished in 1990 with independence and in 1992 with a reformulation of administrative boundaries, people from all over northern Namibia were free to settle in Kaokoland. Within the period of seven years the town Opuwo grew by at least two thousand people from about 3,000 to 5,000 inhabitants. A non-Herero elite is economically dominant in Opuwo. Herero and Himba leaders point out that Ovambo farmers are squatting on their lands at the eastern border of former Kaokoland. However, the quest of the state for national resources apparently poses the greatest threat to local land tenure systems. The giant hydro-power scheme will inundate large parts of Himba land. Numerous ancestral graves will be washed by the rising tides of an artificial lake. The Namibian constitution gives the state all rights to exploit resources for the benefits of the nation. The state is the formal owner of communal lands. Mining companies comb the northern stretches of Kaokoland in search of valuable minerals. These recent developments have put local tenure systems in a limbo. Many Herero and Himba leaders are convinced that they need to combine and to emphasise their cultural integrity to withstand the quest of outsiders for their lands. They see outside interference as a major threat to their livelihood. The emphasis on ethnic identity as found in newly created rituals as much as strategic alliances with political parties operating at the national level are strategies to cope with these threatening changes.

6.4.2.3. Boundary Maintenance and Cross-Cutting Ties: Ethnic Identity and Economic Exchange

When looking at interethnic exchanges in north-western Namibia and south-western Angola two things transpire at once. The various peoples living in the area try to visualise ethnic differences. They use different forms of body anointment, beads of different colours and various forms of jewellery: Himba women anoint their bodies with red ochre, wear necklaces from ostrich egg-shells and wear broad puttees made from a long string of iron beads and on their heads they put, according to the occasion, leather caps of different degree of elaboration, the *erembe* or the *ekori*. Women of neighbouring groups prefer different beadwork, anoint their bodies with other substances and wear other head-dresses. Men also express ethnic identity via a complex code of adornment. They wear highly complex coiffures and wear distinct jewellery. While there is plenty of visual differentiation there are also close relations. The matriclans in all south-western Bantu groups are nearly identical. Matrilineal clan ties cut across ethnic boundaries. Interethnic marriages are frequent as are various livestock exchanges. Ethnic groups are specialised to some degree: while the Himba and Kuvale are expert pastoralists, Hakaona are deemed to have great ritual powers and Thwa are seen as splendid iron-workers and potters.

Maintaining Boundaries: Ornament

Himba emphasise their ethnic identity by using a particular form of dressing and jewellery. Himba men and especially Himba women are easily discernible from their neighbours. The colours they use for body painting and jewellery differ from the respective use of similar items in neighbouring ethnic groups. Women anoint their bodies with an ochre-coloured paste (*otjize*) produced from hematite, fat and different pleasant-smelling herbs. Only the puttee of iron beads on the lower parts of the legs and on the lower part of the arms are not anointed with *otjize* but with a blackish paste, *otjizumba*, that is prepared from the ashes of green twigs of the *omuzumba*-tree (Commiphora multijuga), other herbs and butter fat. Likewise the conus shell (*ohumba*), the iron beads on the necklace tying the conus shell and the lower part of the broad flap of iron beads dangling from the head onto the shoulders (*eha*) are anointed with black *otjizumba*. All other beadwork is anointed with *otjize* on a daily basis. In contrast to this Hakaona women wear colourful beadwork - yellow, red and blue are dominant. They wear characteristic head-dresses that are produced from the remains of tins. Ngambwe women prefer green and yellow beads and organise these beads in a different way. Their necklaces are thick and artificially elongate the neck. Zemba women prefer red beads and again have found different patterns to organise them. Kuvale women wear fewer beads than the other groups but wear elaborate turbans of cloth and metal earrings (see Photographs 18, 19, 20). Estermann (1960) in a coffee-table-like book on hairstyles in south-west Angola gives a photographic account on the use of hairstyles as markers of ethnic identity.

It is not only women who embody ethnicity through ornament. Men, too, are keen to point out the specific ethnic sentiment of their hairstyles and their adornments. Younger men wear the base of a conus shell (*ombandi*). These *ozombandi* are always presents of close female relatives who saw off the lower end of their conus shell when it is worn out. Older men wear four tips of ears (*omuvero*) or the two horns of a small duiker (*ozohiva*). Men from neighbouring groups developed different hairstyles and different forms of ornament. Ngambwe men wear their loin cloth in a very distinct way and arrange their hair very differently from neighbouring groups.

Photograph 18. "Nyaneka Girl" (from Estermann, 1960)

Photograph 19. "Himba Girl after the Rite" (from Estermann, 1960)

Ties Cross-Cutting Ethnic Boundaries

Estermann (1981:79f) remarked that matri-clans are largely identical in the Otjiherero speaking groups of south-western Angola and north-western Namibia. Only very few matrilineal descent groups are peculiar to just one group. All major matriclans such as the Ovakwendata, the Ovakweyuva and Ovakwendjandje are present in all groups under consideration (see Table 36). Even at the level of maximal lineages there is some correspondence, i.e. the houses of Ovakwendata and Omukweyuva are more or less the same in the Otjiherero speaking groups of the sample. Interaction between members of different ethnic groups is without many problems and - if both sides belong to the same clan - it is as intimate as with any other local person of the same ethnic group. Daily interaction is habitual as the two case-studies below point out.

> After a funeral in Angola two Himba and one Zemba came to Kambo in Omuramba. As one of the oldest members of the Omukwenatja matri-clan Kambo had been identified to transform the various necklaces into mourning gear (*okupirikisa*). The Zemba man belonged to the Omukwenatja matri-line as did Kambo, his mother's mother and Kambo's maternal grandmother had been sisters. The fact that both men

Photograph 20. "Kuvale Woman whose Father is dead" (from Estermann, 1960)

identified themselves as members of different ethnic units, Zemba and Himba, was of no importance.

Motjinduika's brother, a Ngambwe, takes part in the commemoration ritual for Vahenuna's father. Motjinduika and his brother are both children of a Ngambwe man and a Himba mother. While Motjinduika

Table 36. Common matriclans in southwestern Bantu societies

Clan-Name	Ethnic group						
	Himba	Hakaona	Zemba	Thwa	Kuvale	Ngambwe	Hinga
Ovakwendata	X	X	X	X	X	X	X
Omukwadongo	X	X		X			
Ovakwenambura	X	X	X	X	X	X	
Ovakwendjandje	X	X	X	X	X	X	
Omukwauti	X			X			
Omukwatjivi	X	X		X			
Omukwatjiti		X					
Omukwatango		X					
Omukweyuva	X	X		X			
Ovatjavikwa	X	X	X			X	X

> mainly grew up in a Himba surrounding his brother grew up in southern Angola among Ngambwe. As a close maternal relative of the deceased Motjinduika's brother qualifies as an *omusyia* (maternal nephew) and is entitled to some inheritance. During the ceremony he is easily discernible as Ngambwe: he wears a red head cloth which is characteristically tied in a way Himba would never tie their head cloth. Nobody has any doubt that he is Ngambwe, while his brother, Motjinduika, is Himba. Through his close relation to the dead (who was Himba) he has an indisputable claim to some animals of the inheritance. His claims are sustained as he has close relatives amongst local people. In the end he went home with two cows.

Ethnic boundaries among south-western Bantu groups in general and Otjiherero speaking groups in special are transient. While local identity is documented in ornament and style, cross-cutting ties are important for social and economic exchange. The matrilineal descent-group system lays the ground for co-operation across ethnic boundaries. Common membership in a matri-clan conveys trust and solidarity more than affiliation to one ethnic group.

6.4.3. A Comparative Approach to Moral Economies

While both economies are profoundly shaped by moral standards, the morality of both pastoral systems differs profoundly. The Pokot emphasise egalitarian relations. Comprehensive solidarity is extended to all fellow Pokot. Ideology constitutes "the Pokot" as a strong solidarity group. Key-values are propagated in rituals in which large numbers of people are activated. Ethnic boundaries are clearly demarcated and cross-cutting ties are few. They are "boundaries of trust". Among the Himba the matri-lineage as a segment of a matri-clan is constituted as a strong solidarity group. Descent group corporateness is emphasised in ritual. Social exchange as much as economic transactions are orientated to a limited number of closely related people. Big men dominate Himba descent groups and act as patrons loaning livestock to needy clan members. While ethnic boundaries are emphasised in ornament and style, cross-cutting ties are frequent and a common matri-clan system prepares the way for social and economic exchange.

Norms, Values and Emotions: A set of norms and values - condensed in the term *tilyontön* - defines mutual solidarity between exchange partners among the Pokot. The *tilyai* of a person are constituted by relatives (affinal and patrilineal) and friends. What matters is not so much a structural relation but the actual transaction of livestock. *Tilyontön* involves more than just a contractual relation and signifies intense emotional ties between two men. The exchange of livestock implies friendship, affection, trust, and - in times of need - mercy. A person who does not share willingly might easily be accused of envy and selfishness. Solidarity amongst the Pokot is deeply rooted in the concepts of the emotional self and of ethnic identity.

Among the Himba intense exchange relations are restricted to close kin. Considerations of matriline membership have a broad impact on economic transactions. The concepts of *omuyandje* and *omuhona* circumscribe the ideal of moral responsibility of the rich for the poorer segments of the descent-group. The *omuyandje*, literally the giver, is a perfect host to guests and does not hesitate to give out stock loans and livestock presents to others. The *omuhona*, the big man, has similar qualities as the *omuyandje* but combines them with political aspirations. As amongst the Pokot envy is emphasised as the emotional antidote to solidarity.

Rituals: In Pokot society communal rituals are large gatherings frequently involving many hundreds of people. In these rituals men of the wider neighbourhood, irrespective of lineage membership and age-set affiliation, gather and evoke their Pokotness in dances and songs. Pokot rituals are symbolic re-confirmations of solidarity, respect and ethnic identity. Many non-related people contribute to the feast: the meat of the animals is shared carefully along well defined rules; women from the wider neighbourhood bring milk to the celebration in order to feed the initiates and young men.

Amongst the Himba all major rituals like funerals and ancestor commemoration ceremonies are tied to specific kin groups. Himba rituals traditionally circled around ancestral graves thereby emphasising the corporateness of descent groups and a land tenure system based on descent and control by lineage elders. In recent years however, the Himba and Herero of Kaokoland have developed more comprehensive rituals on a regional scope. In order to defend land rights against the state and other groups who are perceived as invaders, rituals were developed that emphasise regional identity irrespective of clan membership. The graves of ancient leaders were chosen to establish a neo-tradition. The development of communal rituals has gone hand in hand with growing articulation vis à vis national authorities.

Ethnicity: The Pokot sharply delineate ethnic boundaries. Ethnic markers are well defined. Pokot dress very differently from their neighbours. Men prepare colourful head-dresses and women wear colourful bead ornaments. The arrangements of colours of beaded ornament are very specific to the Pokot. In contrast to the neighbouring Samburu and Njemps who favour reddish colours or the Turkana who favour light blue and green beads, the Pokot choose red, yellow and dark blue combinations of beads. This visualisation of ethnic identity and ethnic boundaries is carried further in their world view. Pokot see themselves as surrounded by enemies, *pung*, and see major differences in character between themselves and their neighbours.

Amongst the Himba a visualisation of ethnic boundaries is as prominent as amongst the Pokot. Himba women anoint their bodies and jewellery with red ochre. Women of neighbouring groups use black colours for the same purpose. While Himba women colour their iron-beads with read ochre, neighbouring groups prefer red, yellow and blue (Hakaona), red (Zemba) and green (Ngambwe) plastic beads. This sharp delineation of ethnic boundaries contrasts with close kinship ties that cross ethnic boundaries. Interethnic marriages and livestock transfers across ethnic boundaries are frequent. Ethnic boundaries are not boundaries of trust but rather denote a different economic specialisation. Himba do not emphasise exclusive territorial rights but respect minorities of neighbouring groups living together with them.

ENDNOTES

1. In 1992 I gathered data on 138 camel herds. An earlier survey in 1988 recorded details on 12 herds. Detailed data on Pokot camel husbandry was published in Bollig 1992b.
2. Beech (1911) who is an astute observer of other aspects of the economy does not mention camel holdings among the pastoral Pokot.
3. This reminds one of an argument forwarded by Gulliver (1955:30,39) that the Turkana kept camels very much like cattle in the past.
4. The size of the Kunene gardens was measured on 1:10,000 aerial photographs from 1995 while the Omuhonga gardens were measured on 1:50,000 aerial photographs from 1975. The Omuhonga gardens were measured by standard geographical methods, whereas the Kunene gardens were measured on the basis of rough accounts of their diameters. Gardens in places with a dense palm or tree cover were not visible on the aerial photographs and could not be accounted for.
5. While there are few differences in Himba gardens, their gardens differ significantly from Ngambwe, Zemba, Nkumbi and Hakaona gardens along the river and in the Omuhonga basin. These gardens are much bigger and people devote more work to them.
6. In a recently launched research project, Welle (2004) found that between 1975 and 1995 gardens in the Omuhonga valley had doubled in number and acreage. Presumably Himba invested more time into agriculture but also the immigration of Hakaona and Ngambwe households into the area may have contributed to this increase.
7. When estimating the wealth of their neighbours, Himba generally look at the entire herd and often do not differentiate between cattle owned and cattle loaned; i.e. somebody counts as wealthy even if half of his 300 cattle are loans.
8. The fact that all livestock are inherited along the matrilineal line is astonishing. In most Herero groups (Viehe, 1902; Irle, 1906; Crandall, 1992) and, in fact, among the western Himba too, ancestral cattle are inherited patrilineally while non-ancestral cattle are transferred in the matriline.
9. In the late 1980s there were very few marriages in which monetary payments were included in the bridewealth. Money had to be paid for brides with a secondary school education or a qualification from a college. Substantial amounts of money (obtained inevitably from the sale of livestock) were spent for the education of these young women. These expenses had to be repaid by the bridegroom, who would then profit from his wife's good education.
10. See Hakansson (1990) for a comparative analysis of systems of protracted bridewealth payment in eastern Africa.
11. In a diachronic perspective changes in the size of bridewealth payments have taken place. While around the middle of the 19th century only very few cattle had to be given, by the end of the 19th century high bridewealth payments were required. After the Rinderpest epidemic of 1891 bridewealth payments were again reduced considerably. Data gathered by Schneider in the 1940s indicates that bridewealth payments may have been considerably higher in the pastoral areas of West Pokot (see also Barton, 1921) in the 1920s than they are today. Schneider (1953:240) says that about 30 to 50 animals were paid for bridewealth.
12. Polly Wiessner (1977, 1982), in a similar vein, identified *hxaro* networks of reciprocal gifting, as a leading topic in !Kung society.
13. Paradoxically, there is one other rule which almost seems to negate the emphasis on spreading affinal relationships. Almost every lineage is able to claim a preferential marriage relationship (*kapkoyogh*) with one other lineage.
14. They did not marry them off prematurely. They simply had the luck to have several nearly adult daughters in their homesteads.
15. Crandall (1991, 1992) maintains that patrilines concentrate in one locality whereas members of matrilines are dispersed. In contrast to this, my data shows that spatial concentrations of matriclan members also occur. Regarding patriclans I only found a

concentration of households belonging to the chiefly patriline in one area whereas other patrilines were dispersed.

16. Crandall lists 24 cross-cousin marriages. In his sample the cross-cousinage includes MMBD, MMBDD, MMMBDD, FZDD, FFZD, FFZDD, FZSD, FMZD (Crandall, 1992:349). My own data indicates the same direction. The term *ovaramwe* includes a vast number of relatives who are perceived to be structurally equivalent to FZD and MBD.
17. It is of interest to note that there was a clear regional pattern in FZD and MBD (and their extensions) marriages. While the western Himba of Okauua and Otjitanda predominantly married FZD (and their extensions), the northern Himba of Omuramba and Omuhonga preferred to marry MDB and their extensions. Crandall asserts that there is a tendency to marry cross-cousins who come from further away (Crandall 1992:232). This observation is corroborated by my data.
18. Crandall (1992:242) notes that five sheep were being transferred also.
19. Wealth classes 1,2 and 3 borrowed 38 animals; 4 and 5 110 animals and, 6 and 7 81 animals.
20. Of course statistics are somewhat misleading here: some of the households became rich due to the inheritance they made.
21. It is significant that the wider neighbourhood in Pokot had a diameter of about 30 kilometres while among the Himba the neighbourhood had a diameter of about 80 kilometres.
22. I would like to thank Michael Schnegg for the visualisation of these networks.
23. Density measures the cohesiveness of the network. The measure is defined as the number of existing relations divided by the number of potential relations (Schweizer 1997:177).
24. The measure for degree-centralisation relates to the entire graph and not to single actors only. The degree measures the number of direct exchange relations one actor has. Hence degree is a measure for the activity of an actor within the network. The degree-centralisation measures the variation (scattering) of degrees in the net. UCINET transfers the measures for centralisation into percentages. A percentage of 0% indicates that all actors are active in the network to the same degree, a percentage of 100% shows that one actor had drawn all exchanges to himself (Schweizer, 1997:183ff).
25. Closeness-centralisation measures the speed of interaction within a net. Those actors are central which are connected on short paths to each other. The closer an actor stands to all others the more effective and independent of the others he can act. The measure for Closeness-Centralisation again looks at the entire graph. A value of 100% occurs if one point in the net has direct contact to all others, while all other points can reach any other point only via two steps (the typical star). If all points are the same distance from each other the measure will be 0% (Schweizer, 1997:187ff).
26. Hardin's argument was not so much aimed at pastoralists as frequent citations of his article in literature on nomadic societies suggest. Hardin was more generally discussing the problem of the world's communally owned key resources and thought that pastoral land tenure exemplified the problem in an apt way.
27. Galaty & Johnson (1990), Baxter & Hogg (1990), Hitchcock (1990) among others give overviews and case studies on the working of communal land tenure among African pastoralists.
28. Behnke (1998:5) gives several examples from the western Kaokoland for this drainage system orientation of chieftaincies.
29. The issue of who has rights to use certain resources becomes more difficult if herders come from another ethnic community. There are some Hakaona, Ngambwe and Thwa households living in the area. Most Himba herders thought that they are just tolerated but that they did not have any rights in the area. The situation in northern Namibia contrasts sharply to that in southern Angola, where Himba and other southwestern Bantu groups (Hakaona, Zemba, Ngambwe, Thwa, Ndimba, Tjavikwa, Tjilenge, Kuvale, Kuroka) intermingle. The rights of the Toms community, Herero who originated in southern Angola and moved into the area only in 1917, to pasture in the area are questioned too.

30. Informants listed the following species as *omihupise*: *omuzu* (Adansonia digitata), *omuere* (not identified), *omukongo* (Sclerocarya birrea), *omusepa* (Corida gharaf), *omuvapu* (Grewia bicolor), *omuhamati* (Grewia villosa), *omuhe* (Berchemia discolor), *omuhore* (Grewia schinzii), *omundjendjere* (Grewia tenax), *omunyandi* (Diospyros mespiliformis), *omukuyu* (Ficus sycamorus).
31. Needless to say that this strengthening of co-management did not so much originate in a genuine concern about the state of pastures but rather corresponded to the aims of the hearts-and-minds campaign formulated during the civil war.
32. Since 1998 so-called conservancies have been introduced at several places in the Kunene region (see Bollig, 2004a).
33. In Schneider's monograph on the Hill Pokot (1953) the key term *tilya* has a slightly different connotation. He describes *tilya* as "probably the most important method of exchange of cattle ... outside bridewealth." *Tilya* is based on the exchange of a cow for an ox on the provision that the one getting the cow gives back some calves at a later stage. Among the Pokot I was working with, the terms *tilya* (the exchange) and *tilyai* (the people exchanging) had a wider connotation. It contained relatives and comprised different forms of livestock exchanges. Schneider (1953:271) claims that *tilya* has a clearly pecuniary motive. While such exchanges among the pastoral Pokot are not beyond economic calculus they are much more than transactions with a solely pecuniary aim.
34. Storas (1997:60) observes the same attitudes on the maintenance of exchange relations and goods transferred among the pastoral Turkana.
35. The full text of the blessing and its original in the Pokot language is given in Bollig (1992a:190ff).
36. Material goods which pass boundaries were mainly goods attached to young men and young women. Hodder (1982:73) found that for example spears (attached to young men) and calabash designs (attached to young women) are not used as ethnic markers. They disrupt otherwise well defined boundaries.
37. Crandall (1992:124) reports that bigmen in the Opuwo area also donated a great number of cattle to non-relatives. Especially non-relatives become obliged to their benefactor.
38. When trying to differentiate a peace-loving character from a character who has the talent to make peace, Himba use *omuhanganise* instead of *omuhange* for the latter.
39. Mureti is a precolonial hero cherished in many oral traditions (see Kajira Muniombara in Bollig, 1997c).

Chapter 7

Hazards, Risk and Risk Minimisation in African Pastoral Societies

In this final chapter hazards, risks and risk-minimising strategies will be compared with reference to the two cases studied and to data recorded on other African pastoralists. What is to be gained from such a comparison? The emphasis of this final chapter will be on the causal relations between various hazards, risks and risk minimising strategies. The two cases studied represent a specific type of mobile livestock husbandry that is typical for arid lands in East and Southern Africa. In contrast to pastoral communities in Western and Northern Africa, in the Near and Middle East and in Central Asia, pastoralists in East and Southern Africa were historically peripheral to states or not integrated into states at all. Only during the course of the 20th century did a fuller integration into states occur. Typically, and again in contrast to pastoral communities in other regions of the world, overarching religious systems (such as Islam) were of no major importance. Of course, pastoral communities in Northern Africa and Asia have to cope with similar hazards: drought, degradation and market failures also occur, but the enduring presence of the state and the full-fledged integration into monetarised market systems provide conditions for improved coping mechanisms.

7.1. CHANGING HAZARDS: THE INTERPLAY BETWEEN ECOLOGY AND POLITICAL ECONOMY

The effects of hazards such as demographic growth, environmental degradation, and entitlement decline evolve over an extended period of time. These factors intensify the vulnerability of communities. Droughts, violent conflicts and livestock epidemics are shock events, which may plunge communities from a state of vulnerability into a state of famine and disaster. The comparison reveals the nature the hazards rendering pastoralists vulnerable as fairly similar. However, the degree to which pastoralists are affected depends upon the severity of such hazards and/or the varying capacity of communities to absorb such shocks.

7.1.1. Demographic Growth and Environmental Change

Demographic growth is one of the major driving factors of cultural evolution (Johnson & Earle, 1987:3,16; see also Lang, 1997:17). Unless demographic growth is accompanied by a growing, sustainable production or by territorial expansion, a system becomes increasingly vulnerable.

Demographic Growth Out-pacing Local Resources: Population growth rates of the Himba and the Pokot differ: while the Himba reached a very modest growth rate of less than 1 per cent per annum, the Pokot experienced a growth rate of 2.4 per cent per annum. The completed fertility rate for Pokot women was 4.7 that of Himba women 3.2. While the Himba population has been afflicted with STDs (especially gonorrhea) throughout most of the 20th century, this problem has been less prominent among the Pokot. A comparison of completed fertility rates of other pastoral societies reveals that figures in the Himba population are extremely low, while Pokot growth rates are not exceptionally high (see Table 37).

Population density and the ratio of people to livestock are crude indicators of the relation between human population and resources. The population density of northern Baringo District, the area inhabited by the pastoral Pokot, is unusually high for African pastoral nomadic societies (see Table 38): there were 10 to 14 persons per square kilometre during the 1990s. Pastoral neighbours of the Pokot such as the Turkana reached a density ratio of 1.7, and even the agro-pastoral Borana of the southern Ethiopian highlands only reached a density of 7.0 (both figures Galvin, Coppock & Leslie 1994:118). Figures for the densely settled Ngorongoro area (Maasai) were 5.3. For the Samburu density figures of 1.09 (Spencer 1965), and the Rendille 0.38 (Sato 1980) were noted. These figures, however, relate to the 1960s and 1970s, a period when Pokot density still stood at about 5 persons per square kilometre. Population density in the area the pastoral Pokot inhabited in the 1990s was closest to figures from the semi-arid Kitui district of Central Kenya, an area settled by sedentary agro-pastoral Kamba farmers (see O'Leary 1984:15). This means that the Pokot maintained a fairly pure pastoral livelihood despite a population density equivalent to that reached in sedentary agro-pastoral populations. Population density in Kaokoland remained at the lower end of the continuum. The population density of the Kaokoland population stood at 0.61 persons per square kilometre in the early 1990s.

While the Nginyang Division had a stocking density of 15.4 TLU/km^2 in the late 1990s, Kaokoland's stocking density rose from 2.5 TLU/km^2 during

Table 37. Comparative data on population growth in African herder communities

	Himba	Pokot	Toposa	Turkana	Rendille	Delta Fulani	Seno Fulani	Tama-sheq
Compl. fert.	3,2	4,7	6.7	7.2/5.5	4.8	7.1	6.6	5.2

Notes: Data on the Samburu, Spencer 1998:40, on all other groups besides Himba and Pokot, Roth 1994:134

the mid 1990s to 4.8 TLU/km² at the end of the 1990s. A look at stocking densities for other populations reveals that stocking numbers were even higher among the Ngorongoro Maasai of Tansania (16,8) and the agro-pastoral Borana of Ethiopia (17,4) but considerably lower for the Turkana (5.0) (see Table 38).

These figures partially reflect the higher productivity of the savannah inhabited by the Maasai and Borana. In the 1990s the Baringo savannah was in many respects ecologically more similar to the arid Turkana savannah than to the moister Maasai savannah, while stocking rates still corresponded to the ecosystem managed by Maasai and Borana pastoralists. The ratio of TLU/person shows that the Pokot were poor herders in the 1990s, while nevertheless upholding a pastoral ideal with great fervour. While the Turkana, Ngorongoro Maasai and Borana had ratios of 3.0, 3.3 and 2.5 TLU per person, the ratio was around 1 TLU per person for the Pokot. The 7.7 ratio for the pastoralists of Kaokoland was at its upper end.

Table 38. Comparative data on population density in African herder societies

	Himba	Pokot	Nama	Turkana	Rendille	Maasai	Mukog	Borana
Area km²	48,982	4,200	21,450	8,600	50,000	42,508	1,100	15,475
Tot. pop.	30,000	40,000	16,234	14,500	19,000	8,300	11,000	108,000
Cattle	101,988	45,000	11,707	11,571		115,468	12,800	247,507
Shoats	177,712	120,000	234,995	151,600		193,284	65,400	101,825
Camels	–	4,000	–	10,464		–	–	7,558
p/km²	0.61	9.5	0.76	1.7	0.38	5.12	10.0	6.98
Cat/km²	3.6	10.7	0.55	1.3		13.9	1.2	16.0
Shoat/km²	9.7	28.6	11.0	17.6		22.3	5.9	6.6
Cam/ km²	–	1.0	–	1.2		–	–	0.49
TLU/ km²	4.8	15.4	1.9	5.0		16.8	1.9	17.4
TLU/per	7.7	1.6	2.5	3.0		3.3	1.9	2.5
Cat/shoa	0.37	0.38	0.05	0.08		0.60	0.20	2.4

Note: One tropical livestock unit (TLU) = 1 cow, 1 camel = 1.25 TLUs, 1 goat or sheep = 0.125 TLU (from FAO 1967), Sato 1980:4 gives as densities for Samburu 1,09 (Spencer 1965), Turkana 1,29 (Gulliver 1955), for Kaoko I assumed some 250,000 smallstock; Data obtained from: Directorate of Rural Development, 1992. Socio-Economic Survey Southern Communal Areas. Windhoek. Ministry of Agriculture. Mukogodo: Herren 1991. Borana: Coppock, L. 1994., Turkana: McCabe, T. 1994. Data for Himba accounted for 2000.

While demographic growth in combination with a stagnant or in fact a dwindling resource base intensified vulnerability among the Pokot, the Himba at a much lower growth rate did not (yet) feel the pinch. Throughout the 1990s the rapidly expanding regional herd of Kaokoland increased faster than the human population.

African pastoral communities have developed several strategies to rectify (or prevent) a detrimental development of the relation between population growth and the economy.

(1) Communities may limit fertility, increase mortality or enforce out-migration to stabilise a population. There is little evidence for the first two practices among pastoral populations: despite Spencer's account on the growth-limiting effects of the Rendille *sepaade* (which excludes a substantial proportion of females from early reproduction) there is little evidence that fertility is inhibited (see Roth, 1994); other than Asmarom Legesse's account of Gabbra infanticide, there is little mention of any deliberate increase in mortality rates. Long-term emigration of large numbers of impoverished herders has been observed among the Turkana and has contributed to the sustainability of their herding economy (Dyson-Hudson & Meeker, 1999). However, the Pokot are an example of the opposite approach: while discouraging emigration they encouraged immigration provided that immigrants were prepared to assimilate Pokot cultural standards.

(2) A second option to limit the detrimental effects of population growth would be expansion. However, most African pastoral societies nowadays live within administratively defined boundaries, which are enforced by the state. Expansion is difficult and easily leads to conflict with neighbouring communities and governmental institutions. Expansion also necessitates military superiority.

(3) A third option is the intensification of pastoral production. Intensification is usually achieved by increasing stock numbers (and only rarely by improving the quality and the value of livestock). Consequently overstocking is a real danger and may result in a long-term reduction of bio-mass production.

(4) A fourth option is the intensification of exchange with neighbouring communities and/or the national economy. If exchange ratios are favourable, livestock can be exchanged for staple foods and other commodities. Recent increases in small stock numbers indicate that the production of livestock for sale is gaining importance.

The data presented here strongly suggest that population growth is a key problem of pastoral societies. Once a certain threshold has been reached, population growth cannot be counter-balanced by purely pastoral strategies. Currently permanent emigration of segments of the society, the diversification of economic strategies and the intensification of trade are the major options of solving this predicament.

Environmental Degradation

The sustainable use of pastures is essential for the viability of pastoral communities. Degradation means decreasing productivity and increasing vulnerability! The natural potential for livestock husbandry in the Northern Baringo plains declined throughout the 20th century. Degradation correlated with overgrazing and the demise of communal resource management. Population growth did not only pose immediate problems for supplying a population with food but did also endanger existing institutions of common property resource management.[1] In order to co-ordinate the interests of a growing number of individual economic units and a concurrent increase in transaction costs, institutions have to work more efficiently or new institutions have to be developed to manage common property resources efficiently.

Environmental changes are not as dramatic in Kaokoland. Nevertheless, a gradual replacement of perennial grasses by annuals and a decline in biodiversity is noticeable in most areas. Sustainable management of pastures is based on the mobility of herds, the protection of dry season grazing and a restrictive regime of resource access.[2] Degradation in Kaokoland is most noticeable in localities where there has been heavy pressure on woody vegetation due to intense settlement. Recent detrimental changes in plant communities apparently are as much due to decreasing rainfall since the middle of the 1970s as to high stocking numbers. Since the 1980s a decline in the capacity for sustainable communal management has led to severe degradation in places with relatively sedentary populations.

While research on pastoral ecosystems has focussed on degradation, contradictory conclusions have been drawn. A recent volume by Behnke, Scoones & Kerven (1993) differentiates between equilibrium and non-equilibrium systems.[3] Environmental changes in equilibrium systems are likely to be due to stocking rates exceeding the carrying capacity. In disequilibrium systems, environmental change seems to be linked to highly variable rainfall rather than to stocking rates. The observations on environmental degradation in both case studies conform to this model. The higher rainfall area, Pokot land, is apparently more responsive to grazing pressure than the lower rainfall area Kaokoland. Nevertheless, the non-equilibrium pastures of Kaokoland also showed signs of degradation. Herders succeeded in maintaining a relatively stable state of biomass production by protecting specific ranges, by spatial mobility and by confining grazing to limited areas during the main growing period of grasses. Nevertheless there was a definite loss of biodiversity and a comprehensive shift from perennial to annual grasses. Only when pastures become open-access resources - which is exactly what happened among the Pokot - does degradation become inevitable. It is rather the way livestock is managed than sheer numbers which matter – and this holds true for equilibrium as well as disequilibrium systems.

Entitlement Decline: Peasantisation and Encapsulation

Entitlement decline is another major factor contributing to the increased vulnerability of pastoral societies. Decline may be due to a loss of endowments (access to resources and property rights) and/or a decrease in exchange entitlements. During the course of the 20th century politically and economically independent pastoral communities were either encapsulated or they became peasantised.[4] While pastoralists used to be long-distance traders and used to rely heavily on barter exchange (Galaty & Bonte, 1991; Sobania & Waller, 1994), colonial governments (and later national governments) drastically changed the perspective for many pastoral communities (see Azarya, 1996; Klute, 1995 for an overview). In settler colonies such as Kenya and Namibia in particular pastoralists were denied access to markets to protect the interests of white farmer settlers (Kerven, 1992).

Today all African pastoral communities depend on livestock markets to obtain cereals. A major characteristic of these markets is their high degree of instability. Pastoralists are inevitably vulnerable to price fluctuations on regional markets: the prices for livestock decrease due to oversupply or rise due to the increased need for meat in rapidly growing urban areas. Studies on the Wodaabe Fulani of Niger (White, 1997) and the Hadendowa Beja of the Sudan (Manger, 1996) reveal that the dependency on market exchanges increases with every drought. The Pokot case conforms to this model. During droughts market prices for livestock plummet and herders are forced to sell more animals than usual. Soon their preferred item of exchange, namely adult male livestock, is sold out and they have to embark on sub-optimal selling strategies. Once prices for livestock recover, they are forced to sell young and female animals during the post drought phase. This inhibits the successful regeneration of herds.

However, livestock markets are an option and a hazard at the same time. Growing population numbers and stagnant or decreasing productivity force pastoralists to exchange high value proteins and fats (stored in livestock) for cheaper carbohydrates (such as maize). Livestock prices improved throughout the century. For the money he obtains for an adult goat, a herder today can buy four times more maize than in the 1920s among the Pokot. On the whole, the inclusion into a wider market was beneficial for the Pokot – and probably even essential: the decline of the overall ratio of people to livestock was compensated for by the increased value of livestock in relation to maize. For the Himba such gains were not feasible, as they were isolated from the wider economy. Traditional exchange rates of e.g. one goat for a sack of maize were maintained for decades, preventing pastoralists from reaping the benefits of increasing meat prices. In the 1990s goats were exchanged for goods at a third of the current market value!

Most researchers have emphasised the negative effects of market integration. However, there is sufficient evidence that it is exactly the ambiguous character of markets, which makes them problematic. Under favourable conditions (good rains, peace and a producer-friendly government price policy), pastoralists profit from marketing livestock. During droughts, however, the very same markets turn against them, inhibiting de-stocking as a result of price-slumps.

There were few losses from internal entitlement declines (endowment decline) in both societies under study. Ranges were still defined as communal property and a concentration of livestock had not taken place (Pokot) or was in an incipient state (Himba). There are many African pastoral societies where entitlement declines are due to a concentration of the means of production with a local elite: In pre-war Somalia (Bestemann, 1996; Nauheimer, 1991; Stern, 1991) large numbers of small-scale herders were dispossessed by herder-entrepreneurs. Ranges were fenced and market relations were dominated by a few rich farmers. Hitchcock (1990) describes comparable processes for the Tswana of Botswana. Galaty (1994) has aptly shown how Maasai livestock herders are being out-manoeuvred on their own ground by very rich Maasai and Kikuyu farmers or by urban investors.

7.1.2. Independent Factors of Stress: Droughts, Epidemics and Violent Conflicts

While demographic growth, environmental degradation and entitlement decline form a vicious triad of mutually re-enforcing hazards, droughts, epidemics and violent conflicts strike more independently. Within short periods of time – a year or in case of epidemics even days – they can destroy pastoral livelihoods.

Climatic Variability: Roughly one out of four years is a drought year in north-western Namibia and north-western Kenya. Regular cycles could not be established for either region. Bad droughts marked by a rainfall of 40 per cent below average occur in about ten per cent of all years. There is no discernible trend for declining rainfall figures in northern Kenya (Ellis, Coughenour & Swift, 1993) while the trend is negative in north-western Namibia. Drought results in massively reduced rates of fodder production and increased rates of livestock mortality. The regional herd of Kaokoland shrank by as much as 90 per cent in the early 1980s and by about 40 per cent in the late 1950s. Among the Pokot losses amounted to 60 per cent of the total regional herd.

The extreme vulnerability of herds due to climatic events is typical for all African pastoral communities. Van Dijk (1997) reports losses of 62 per cent for cattle and 55 per cent for small ruminants during Sahelian droughts, Hjort & Dahl (1991:132) describe livestock losses of 95 per cent among the Sudanese Beja during the 1984 drought. Legesse (1993:269) reports 60-70% losses of livestock among the Kenyan Borana in the 1973 drought. High stock losses have lasting effects. While a farmer will be able to reap a full harvest from his fields once it rains again, it takes years for herds to fully regenerate. White (1997) points out that Fulani herders in Niger never fully recovered from the major droughts of the 1970s (see also Curry 1993:242 for another Fulani community in Niger).

Livestock Diseases: Livestock epidemics of a disastrous scope were prominent all over Africa in the second half of the 19th century. The Pokot as well as the Himba suffered tremendous losses during the Rinderpest epidemic, which swept Africa in the 1890s. However, while such epidemics (although at lower mortality levels) have repeatedly occurred in northern Baringo until today, the Rinderpest epidemic of 1897/98 remained a singular event in north-western Namibia. Other contagious diseases are also rare in Kaokoland: the spread of CBPP, Anthrax and Foot and Mouth was successfully checked by efficient veterinary intervention. In contrast, livestock diseases in northern Baringo occur frequently and take a constant toll. The demise of the veterinary system since the late 1980s had a negative impact on the health status of herds.

Violent Conflicts: Violent conflicts have had a detrimental effect on pastoralists in many African regions. Immediate effects of violence are death of people and livestock. Indirect effects concern the loss of access to grazing lands and water points. A thorough militarisation of pastoral societies took place in the Sudan, in Ethiopia, in Kenya, in Uganda and in Somalia (for a summary of these cases see Österle & Bollig 2002). The availability of light weaponry turned pastoral areas in East and Northeast Africa into regions beset with low intensity warfare. Consequently, the Pokot live in a violent environment. They raid and they are raided. In addition, they enter into conflicts with army units, which try to restore peace but frequently escalate local conflicts.

In many African countries pastoral economies have been shaped by low intensity warfare for decades (see also Hutchinson, 2000). A look at the rather peaceful inter-community relations prevailing in south-western Africa clearly demonstrates that violence alters the entire framework for intra- and inter-community relations: only under peaceful conditions can sustainable modes of resource exploitation develop.

7.1.3. The Nature and Distribution of Damages

Demographic growth, degradation and entitlement decline result in an increased vulnerability of the entire community. It is difficult to quantify the losses resulting from these three hazards. Terms of trade deteriorate gradually. Annual population growth rates of 2.5 per cent are hardly perceptible over a period of two or even five years and a loss of some hectares more to bush encroachment is not that obvious. Losses due to droughts and epidemics are an entirely different matter. Many case studies on African (and other) pastoralists state figures for losses due to drought as a percentage of the regional herd. However, damages are typically distributed unevenly. The availability of herders, personal experience, individual quality of herdsmanship and, last but not least, luck are significant. It is tempting to look for a general patterning of losses. White (1991, 1997) showed that impoverished households of Wodaabe Fulbe who lacked adequate herders were hardest hit. Sandford (1983:75) points out that rich Maasai families usually suffer less damage from droughts than poor ones: thus rich families lost 42 per cent of their cattle during a drought, while poor families lost 52 per cent. The same holds true in Ensminger's Orma case study (1984:65): rich Orma herders lost 59 per cent of all cattle, while poor Orma families lost 74 per cent. In contrast, losses among the Kenyan Pokot were randomly distributed. Losses experienced during droughts significantly affected wealth rankings. Among the Himba losses were also rather randomly spread although the distribution seemed to be slightly skewed: rich herders lost slightly more animals than medium and poor herders did. This, however, did not markedly change wealth hierarchies as rich households had spread significant numbers of livestock over various cattle camps.

The distribution of damages due to livestock diseases is even more random than the distribution of losses due to drought. Whether a herd is affected or not mainly depends upon chance. Straying sick stock and the randomness of infections through spores which may be carried by wind or by infected water (and not necessarily by infected livestock) contribute to a wide but unpredictable spread of damages. Losses due to theft and raiding are randomly distributed as well. Neither rich nor poor herders have any advantages in the face of violent raiding commandos. Literature on the Pokot and the Turkana underlines the overall impact of raiding and its random effects (Bollig, 1990a; Johnson, 1990:182f; McCabe, 1990).

Differences in wealth which among the Pokot were never marked in any case, changed repeatedly during the 1990s. This conforms to Pokot ideas on the transitoriness of wealth. An aphorism says: "never laugh at a pauper, tomorrow it may be you who is poor". Among the Himba major wealth differences are not subject to massive changes during droughts. Himba hold that "men of long arms" will remain wealthy and powerful, and poor herders, "men of short arms" will stay clients of wealthy patrons. Nevertheless wealthy farmers suffered heavily during the major droughts of the 20th century.

7.2. THE PERCEPTION OF HAZARDS

While there is a general feeling among anthropologists that local concepts of the nature and causation of natural hazards need more scientific attention (see for example Gudeman & Rivera, 1990; Göbel, 1998; Jungerman & Slovic, 1993), there is still a surprising lack of research concerning local perceptions of risks in pastoral communities.

Comprehensive Interpretations of Risks and Misfortune

Pokot ideas on the causation of hazards are detailed and consensus based. Major disasters as well as minor misfortunes befalling individual households were habitually traced to internal conflicts. Typically these were either generational conflicts or blatant breaches of solidarity. Even major adverse environmental changes were explained in these terms: the massive degradation of the Pokot savannah was traced to the curse of an ancestor. By attributing misfortune to intra-societal conflicts, Pokot have found a powerful way of socially appropriating the hazards of an extremely unpredictable environment. The social appropriation of problems of very different scales empowers and consolidates the local community. Seniors as representatives of the society gather and conduct rituals in which they conjure up the power of their gathering. The following quote from a blessing to counter an epidemic goats' disease clarifies this point:

> *We say now "Stop now!", these words, the words of the council of men are powerful, this council which has been purified which is sitting on cleaned ground. We chased the bad things which were in our midst. Which are where? Which are in the council of men. Whose council? The council of the Pokot. We are powerful and chase this thing. And what was said? It was said "Vanish to the far end side."*

The egalitarian Pokot regard their society as being shaped by respect and solidarity. Breaches of these norms and values lead to a demise of internal harmony. The loss of harmony in turn results in disasters, which take the form of metaphysical punishments. Major communal ritual efforts are needed to rebalance a disorderly world (see Figure 32).[5]

Himba ways of interpreting disasters are less detailed and less consensus based. In contrast to the Pokot, the Himba seldom speculate about the metaphysical causes of major disasters. No human sin and no breaking of taboos is believed to cause disasters of a wider scope. Individual misfortune, however, is considered as being caused by ancestral wrath and frequently also attributed to witchcraft. Perceptions of hazards and damages directly relate to descent group ideology and help must be sought in clan-based ancestral rituals or from ritual experts (see Figure 33).

The two interpretations presented here may stand out as different approaches, but undoubtedly there are yet other ways to conceptualise hazards. The southern African !Kung, for example, have few theories on disasters: they do not ascribe them to any supernatural being and accept them as given. In a personal communication Polly Wiessner, who worked among the !Kung of Botswana and the Enga of Papua New Guinea, compared the attitude of the !Kung to Enga ideas. The Enga feel that natural disasters are caused by social problems and inappropriate communication with the ancestors. Similar to the Pokot, the Enga handle disasters in ritual - !Kung do not. These preliminary thoughts suggest that the way hazards are integrated into

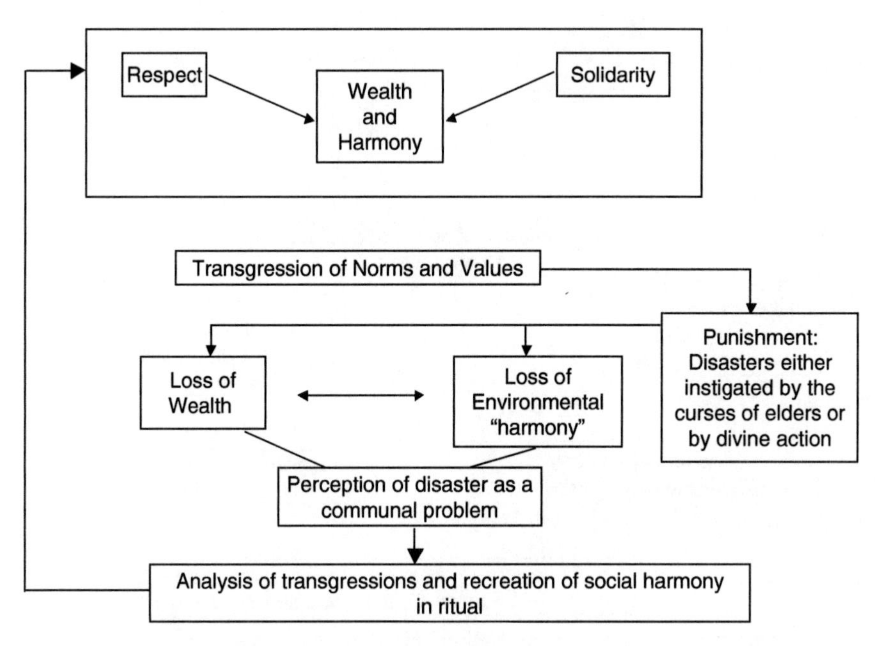

Note: The idea of representing the relation between key values, moral transgressions and punishment in a flow-diagramm is taken from Casimir 1987.

Figure 32. The Origin and treatment of disasters in Pokot world view

the *Weltanschauung* of a culture loosely corresponds to social complexity. For the Pokot the entire society is the realm for co-operation and solidarity, among the Himba co-operation and solidarity is restricted to kin. Thus these very groupings are the source of as well as the solution to many problems.

Specific Classifications of Hazards

Is there any relation between this rather general interpretation of hazards and more specific views on individual hazards? There is a very clear answer in the negative: the Pokot who tend to relate drought and degradation to transgressions of a moral code are astute observers of natural processes and have very detailed and complex procedures to describe and analyse for example environmental degradation and livestock diseases. In order to trace specific aspects of hazard perception I refer to those hazards discussed in previous chapters.

Population Growth: Demographic growth outstripping local production has been identified as a major factor rendering Pokot pastoralism ever more vulnerable. Surprisingly, the Pokot do not make this a topic at all. Quite the

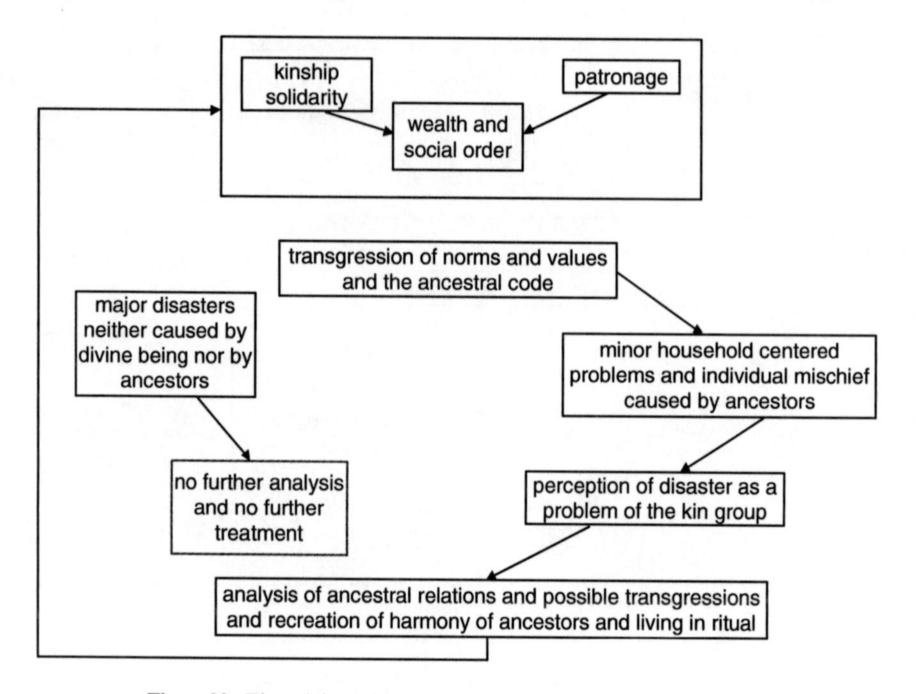

Figure 33. The origin and treatment of disasters in Himba worldview

contrary, the Pokot cherish fertility, are keen on extending their families and in general hold that the Pokot population should grow rapidly. Ideology prescribes a positive stand towards population growth and forcefully *decouples* population growth from limited resources. For the Himba population growth – which is comparatively low in any case – is not a major concern. However, the Himba and not the Pokot mentioned that grazing management had become difficult due to an increasing number of households.

Degradation: In striking contrast to demographic growth, the Pokot frequently discuss environmental decline. They tell stories to explain degradation and substantiate their observations by detailed descriptions of environmental change (see Bollig & Schulte, 1999). However, the Pokot do not attribute this phenomenon to overgrazing but rather to the decline of communal management and to spiritual processes. The Himba, just as the Pokot, have developed a detailed knowledge of the ecological dynamics of pastures. They hold that the quality of grazing is determined by the variability of rainfall and not by stocking rates. Only in 2004, when cattle and small stock numbers reached a climax of roughly 190,000 cattle and half a million small stock, did I hear Himba herders talking about the ecological consequences of overstocking. The randomness of disequilibrium systems may be the reason why the Himba - in contrast to the Pokot - were reluctant to point out any clear direction in vegetation change.

Entitlement decline: Both communities experienced entitlement restriction due to action taken by colonial governments. Surprisingly, oral traditions in both communities do not treat these losses in great detail. Quite the contrary, both the Himba and the Pokot take a very sceptical stand towards the states they are currently living in and to a certain extent glorify the colonial past. The Himba report that they feel as marginalised by the present independent government as by the former colonial government. Neither the Himba nor the Pokot fully acknowledge the benefits they gain from the inclusion into a state. Drought relief programmes have been essential to survive droughts and food aid is the order of the day in northern Kenya as much as in north-western Namibia. The rejection of the state emphatically negates this factual dependency. Both groups cherish the self-image of an independent, and in the Pokot case, of a militarily superior group.

Livestock diseases: the Pokot have developed a complex veterinary system (see Bollig, 1995a). Livestock diseases are expertly differentiated and termed. They also have an intimate knowledge on the aetiology of most of these diseases. They describe vector-borne diseases, contagious diseases, and deficiency diseases. In contrast, Himba ideas on livestock diseases are less differentiated. There is an obvious reason for this: livestock diseases are simply less prevalent. The three most important cattle diseases in north-western Namibia, contagious CBPP, Botulism and Anthrax, are diagnosed and there is some local treatment.

Droughts: In both pastoral groups, accounts of droughts feature prominently in oral lore. However, when ideas on the predictability of droughts are discussed, the Pokot become more detailed, while the Himba simply state that one cannot account for droughts. The Pokot hold that droughts can be predicted and that there are material signs predicting rains: if termites leave their hills, it will rain soon, and if there are many termites, the rains will be heavy.[6]

Intergroup Violence: Pokot men seek fame in raiding and pre-colonial warfare has been constitutive for a pastoral Pokot identity. All Pokot - even ferocious young warriors - are cautious enough to identify *tapan*, the danger of being raided, as the major threat to their livelihood. Large areas of land have had to be evacuated because of *tapan*. While the Pokot see themselves as actors in a violent world, the Himba rather regard themselves as victims of militarily superior enemies.

No Shortcut to Generalisation

It is hardly possible to generalise from these findings. In a paper on hazard perception in Western culture Jungermann & Slovic (1993:181) point out that perceptions are associated with the physical properties of hazards as well as with more general worldviews and the specific social embeddedness of individuals. They find that the degree of control humans have over a certain hazard and their experience with this hazard tends to shape concepts of risk. Hazards of a chronic nature are rated as being more dangerous than hazards cropping out of the blue. On the one hand, there is little doubt that the conceptualisation of hazards among the Himba and the Pokot relates to physical reality. The diagnosis of hazards is based on personal observation and on collective memory. On the other hand, the conceptualisation of hazards is clearly embedded in and influenced by ideology. The Pokot deny that population growth and high stocking rates might be relevant to their rating of hazards. Both processes are slow to show an impact. Moreover, to seek the root-cause of problems in population growth or overstocking would question local values. Future research should study indigenous knowledge of hazards and vulnerability but should also look into the ideologically grounded decoupling of hazards and social processes.

7.3. THE DEVELOPMENT OF RISK MINIMISING STRATEGIES AND BUFFERING INSTITUTIONS

In previous chapters strategies to minimise losses during a disaster and buffering institutions to reduce vulnerability were qualified. Where possible, they were also quantified and placed into historical context. The subsequent paragraph will summarise the results and relate them to data on other African pastoral populations.

7.3.1. Crisis Management: From Local Resources to Food Aid

Strategies to cope with disaster range from the intensified sharing of food within a defined group of people to extended spatial mobility and ritual. Due to the development of closer links to markets, the increased sale of assets is an additional option. Reliance on international food-aid is becoming increasingly important in many African arid regions.

Increased Slaughter: Many authors have pointed out that African pastoralists "store livestock on the hoof". McCabe (1990) describes how the Turkana of Kenya prevented starvation by slaughtering famished livestock. Grandin (1987), Herren (1991) and Spencer (1998) report on the same strategy among Maa-speaking pastoralists. The Fulbe and the Tuareg pastoralists of Niger define herd accumulation as a major "strategy against uncertainty" (Scott & Gormley, 1980:98). During drought related famines pastoralists were better able to evade starvation than neighbouring agriculturalists because of the "walking larder" (Clutton-Brooke, 1989) they command.

Both, the Himba and the Pokot stepped up the frequency of slaughtering during periods of stress and thereby prevented starvation. Among the Himba the slaughtering of livestock is still a major strategy of coping with food shortages. Accounts of droughts are filled with references to slaughtered small stock. Oral traditions on drought leave little doubt that the Pokot also increased slaughter during periods of stress in the past. However, during the 1990s, the opportunities for the Pokot to obtain extra food by means of slaughtering drastically decreased: their herds were depleted by droughts, epidemics and raiding. Furthermore, a steadily growing population combined with a decline in livestock numbers contributed to a dwindling ratio of livestock to people. Even during the serious drought of 1992, only little meat was made available. During a drought the Himba live off meat, store-bought maize and substitute food. Among the Pokot store-bought maize clearly dominates, leading to widespread malnutrition, especially among children as documented by a recent report on the medical status of the population (McGovern, 1999, 2000). The Pokot still share meat intensively; but the sharing seems to contribute more to the reification of group solidarity than to an improvement of the nutritional status!

The declining importance of slaughter during droughts has been observed in various other African pastoral communities (Coppock, 1993:181; Sperling, 1989; Grandin, 1989). Declining herds and growing populations as well as relatively favourable market exchange rates are given as major reasons. Large herds are increasingly regarded as repositories of marketable wealth rather than as "walking larders".

Substitute Foods: During severe food shortages many rural populations of Africa's dry belts rely on bush-food. Bernus (1980, 1988) and Gast (1987) on the Nigerien and Algerian Turareg, Klute (1992:202) on the Malian Tuareg, Legge (1989) and Scott & Gormley (1989:98) on the Fulbe of Niger, Holy (1988:145) on the Sudanese Berti, and Campbell (1986) on San and Tswana in Botswana give ethnographic examples on the widespread use of bush-food during droughts.

Oral traditions report that the Pokot relied heavily on tuberous food substitutes in the past. However, nowadays such substitute foods are only very rarely used. During the 1992 drought I did not find any of these plants being eaten by the Pokot. In stark contrast, substitute foods are still of importance for the pastoral Himba. They rely on the abundant palms and their highly nutritious nuts along the Kunene and its major tributaries. In general, de Garine's assumption (1994) that in recent decades African farmers have used secondary foods less frequently is corroborated by the Pokot case. These foods, though sometimes of high quality, are usually low yielding, widely dispersed and labour intensive to procure. Moreover, nowadays they are deemed to be old-fashioned in many regions.

Opportunities to avert starvation simply by slaughtering more livestock than usual or by intensifying the use of bushfood are rapidly diminishing. Four strategies gain more prominence: sharing, increased sales, diversification and reliance on food aid. Principally, there is a shift away from organised communal strategies to exchanges with the national market and the global economy.

Sharing Food: While hospitality is prescribed among the Pokot and the Himba and a moral code urges those who have to share with those who have not, there are some notable differences. Among the Pokot the sharing of meat is much more standardised and ritualised than among the Himba. When an animal is slaughtered, its parts are carefully divided. While Himba rules of meat-sharing are rudimentary - they mainly apply to the fact that some specific pieces of meat must not be eaten by non-kinsmen - the Pokot have designed an elaborate code of meat sharing. The intensity with which they treat the topic obviously makes it more difficult for individuals to defect from exchange obligations during stress.

Observations on food sharing under stress conform broadly to the hypotheses put forward by Hames (1990). The scarcity of a specific food item tends to determine the scope of its distribution. The scarcer a specific food item the less dominant is a kin-bias when sharing it. However, while the scarce resource meat is shared more intensively than maize, milk, which is very scarce in any drought, is not often shared in either community. Meat is scarcer among the Pokot than it is among the Himba; and indeed the Pokot share meat with a wider range of people than the Himba do. While the Himba restrict meat sharing to relatives and perhaps some neighbours, the Pokot send meat to (spatially) distant relatives and friends.

When is sharing given up in favour of more exclusive forms of consumption? There is ample evidence that during periods of intense food stress food is withheld from exchange circuits (Watts, 1988; Colson, 1979; Amborn, 1994). Dirks (1990) reports on a comparative basis that hoarding and withdrawal of food from exchange circuits (i.e. the breakdown of sharing mechanisms) is a last effort to save the lives of close kin. The Himba recount at least two droughts during the 20th century when food was no longer shared. Among the Pokot there is no historical evidence of food hoarding during periods of intense stress. According to several older informants food was shared among the Pokot even during periods of the most serious famine.

Increased Sales of Assets: While food-sharing only taps locally available resources, the increased sale of assets is a means of acquiring food and goods from other food-producing systems. Similar to many other African pastoral communities, both the Himba and the Pokot step up the sale of livestock during droughts in order to cope with food shortages. The rates of exchange between livestock and maize are similar in both systems. While Pokot obtain the money for buying cereals by selling small stock and cattle, the Himba barter livestock for food and commodities with itinerant traders. In the Namibian case in particular, pastoral producers are grossly disadvantaged.

The livestock-maize exchange circuit in both systems comes under stress during droughts when an abundance of poor quality livestock is offered on the market. The Pokot have probably felt this effect to a greater degree than the Himba because of a more comprehensive monetarisation of their livestock trade. Compared to national livestock pricing and customary prices, Himba exchange rates are low in any case and customary prices offer some security. If prices decline below a certain level due to oversupply, itinerant traders simply discontinue their arduous journeys into Kaokoland. The Pokot felt the vicissitudes of the market as in many other droughts during the calamities of 1992. Cattle prices dropped by 50 per cent as prices for small stock did. At the height of the drought, Pokot pastoralists found it hard to sell any livestock at all.

Despite all the problems livestock marketing is beset with, stepping up the sale of livestock is a coping strategy of increasing relevance. The Pokot case reveals that by exchanging meat for maize, herders may gain up to four times as much energy in the form of cereals. However, the potential for markets to limit the effects of disasters are clearly restricted. Increased sales are an apt strategy during the onset of disasters. However, such sales may become ineffective and even detrimental to long-term economic survival, as large numbers of animals have to be offered at low prices and the potential for the recovery of the household herd is jeopardised.

Diversification of Production: The more diverse and independent economic strategies are, the better can shortfalls in one field be balanced by surpluses gained in another field. Highly specialised livestock husbandry involves the risk of all major resources being endangered by the same set of hazards. Diversification in pastoral household economies can range from multi-species herding to so-called "ten-cent-jobs" such as brewing, charcoal burning, selling of traditional medicine (see for example Browman, 1987; Hjort & Dahl, 1991; Legge, 1989; Odegi-Awuondo, 1990: 123ff). Two types of diversification can be differentiated. Firstly there is an ad hoc diversification in the face of drought. Poor herders resort to "ten-cent-jobs" like charcoaling and beer-brewing frequently because their herds are partially or totally destroyed. These income-generating activities usually require only very little financial input and raise a modest amount of extra-capital, which can be devoted to the purchase of food. Secondly there are more permanent forms of diversification, which reduce the vulnerability of households. These are labour intensive, depend upon access to specific resources and tend to be well integrated within the overall economy.

Intermediate trading and brewing were ad-hoc activities adopted by the Pokot during the early 1990s when the pastoral economy suffered from a series of disasters. Many Pokot realised that unabated sales would jeopardise the viability of their pastoral households. Consequently, they tried to procure extra-money from small-scale trading. Women went into brewing and running small-scale restaurants during the drought. The money generated was almost totally devoted to food purchases. In northern Kaokoland there are fewer opportunities for spontaneous diversification beyond the traditional coping mechanisms of gathering bush-food. The Himba gather bush-food but there is little they can sell or offer as services to any other population. Only when livestock was lost in large numbers, did many people opt for full-scale agriculture in Angola and did young men join the South African army to become professional soldiers.

The integration into wider commodity and labour markets is becoming increasingly important for pastoral communities in Eastern and Southern Africa and offers room for diversification. Many Kenyan pastoralists partially live on remittances family members send home from migrant labour and on food produced by other family members on agricultural schemes. The Mukogodo (Herren, 1991), the Beja of the Sudan (Hjort & Dahl, 1991) and the Fulbe of Niger (White, 1994, 1997) have become dependent on the market for economic and social reproduction. Livestock husbandry is reduced to a secondary activity while the major household income is generated outside the pastoral sector.

Food aid: During recent decades the amount of food aid donated to the inhabitants of African arid lands has increased tremendously. So far, anthropologists have paid little attention to the impact of famine relief on local economies. Reports usually originate with political scientists or people professionally concerned with the delivery of food aid (Shipton, 1990:376). Both populations under study here have received enormous amounts of food aid over the last thirty years. Food relief has become an important part of the drought diet of both Himba and Pokot. Relief food was first distributed among the Pokot in the early 1940s. Delivery of large quantities of food aid prevented loss of human lives during the catastrophic drought of 1984 . Since the middle of the 1980s famine relief to Kenya's pastoral north has been institutionalised: all through the late 1980s and 1990s food was donated, no matter whether it rained well or not. In recent years, it has become a major duty of Pokot politicians to channel food aid to their constituencies. Famine relief has been donated to Kaokoland pastoralists since the 1920s. While the South African Administration was very reluctant to give food for free, it did not hesitate to deliver subsidised food for sale. The disastrous drought of 1958/62 necessitated major food aid deliveries. During the crash of the livestock economy in 1981, it was only food aid that prevented massive starvation. Since independence in 1990, food aid has been distributed fairly regularly. When I started my fieldwork in 1994, even livestock was receiving molasses as "food-aid" during a drought.

At the turn of the millennium it has become increasingly clear that many pastoral societies are increasingly dependent on international food aid. In all likelihood this growing dependence on food-aid and other subsidies will continue. The question is, does food aid delay economic reorientation and further economic differentiation in specialised pastoral economies? While food-aid is justified on humanitarian grounds, the economic consequences of this form of aid need to be studied in detail.

Spatial Mobility: Colonial as well as modern national governments tend to be critical of mobility (Sobania, 1988; Salzman, 1996). This has not necessarily made pastoral nomads more sedentary but has caused a reduction in the scope and flexibility of spatial mobility. In recent decades communal land tenure crumbled in many pastoral areas and gave way to more individualised forms of land-ownership further limiting the scope of mobility.

While the basic patterns of mobility, such as the differentiation between less mobile households and highly mobile livestock camps are similar among the Pokot and the Himba, both groups differ in the scope of their migrations.

Single Himba households make use of a much larger area than Pokot households do. The Pokot are highly mobile because they have to react to shortages of fodder and water as much as to considerations of security. While historical accounts on Pokot mobility indicate that during the first half of the 20th century nomadic movement covered larger areas, mobility became limited by a growing population and by borders established by colonial authorities. Population density increased from about 1.5 persons per square kilometre in the early 20th century to about 14 persons per square kilometre towards the end of that century. Due to confinement and "crowding", long distance moves were replaced by series of short distance moves within a restricted area. In contrast, Himba spatial order is less restricted. During a drought the Himba limit their movements to dry season grazing and drought-grazing reserves. Moves covering some 20 to 30 kilometres are normal. Borehole-drilling since the 1960s has made additional pastures available. Preconditions for this kind of mobility are a low population density and non-violent forms of conflict management. Many African pastoralists find themselves in situations broadly similar to those depicted above for the Pokot. National and regional boundaries as well as the privatisation of rangelands reduced the scope of mobility of a growing number of households.

Rituals: Rituals to counteract disasters have not been widely treated in ethnography. Shipton's (1990) comprehensive bibliography of crisis-management strategies has only very few entries focussing on ritual. Pokot crisis management rituals are communal acts. Communal purification is a necessary first step to enter into prayers for rain. Blessings and proclamations which continuously refer to the magical strength of a community bound by solidarity are the final step of such rituals. In contrast, the Himba do not try to cope with droughts or other disasters in communal rituals. Problems befalling single households are countered by rituals directed towards the ancestors. Ancestral rituals are kinship-based and the main ritual activities are controlled and conducted by only a few, frequently by just one elder. Power stems from the discourse between the ritual specialist and his ancestors and the public does not participate in the sacred parts of the ritual.

It seems significant that the Pokot perceive themselves as one densely knit social unit in which exchanges are shaped by solidarity and consequently interpret the group, i.e. the ethnic community as being responsible for and capable of performing curing rituals. The Himba locate these powers with a much smaller group of closely related people. This dichotomy obviously relates directly to the basic perception of the origin of disasters. If the metaphysical origins of a disaster are interpreted as a mirror of disharmonious social relations, then re-balancing these social relations will be essential. If problems arise due to disturbed relations between the living and their ancestors, ritual activity will engage only those persons with immediate kinship links.

In both communities Christian churches towards the end of the 1990s have gained some ground. Although Christians are still minorities in both communities and an astonishing variety of Christian denominations exist, community leaders and young educated people frequently have close relations to churches. Churches present new interpretations of disasters and offer new coping rituals which in the future may influence local conceptualisations of crises.

7.3.2. Economic Change and the Development of Buffering Institutions

While the immediate management of disasters is necessary to prevent the loss of human lives and excessive loss of livestock, institutionalised buffering mechanisms reduce the vulnerability of populations and enhance their potential for quick recovery. Individual benefits are forsaken to create security. Resource protection is a good case in point: herders forsake their opportunity to exploit virgin grazing for the common good of a regulated and predictable range management.

Permanent Forms of Economic Diversification: Permanent diversification of an economy reduces the overall vulnerability of households. The diversification of herds is a distinct risk minimising strategy employed by pastoral communities (Dahl & Hjort, 1976; Dyson-Hudson & Dyson-Hudson, 1980; Browman, 1987, 1997; Legge, 1989; Curry, 1989). Since droughts affect the grass layer and the bush layer to different degrees, a diversified herd acts as a buffer against shortfalls. In many East African societies, camels have become an important asset to buffer the worst impacts of a drought as they continue to lactate even during very lean months (Spencer, 1998: 265). Both the Himba and the Pokot make full use of the risk reducing capacities of diversified herds.

Among the Pokot diversification beyond livestock husbandry is only beginning to develop. They continue to be highly specialised mobile livestock herders. The Himba on the other hand, have combined pastoralism with agriculture and gathering for a long time. Throughout the 1990s many Pokot men tried with varying success to earn an income as self-employed livestock traders or merchants of veterinary drugs. Young educated men frequently established farms in the south-eastern parts of the district where rain-fed agriculture is less risky. In recent years numerous households have planted small gardens. Women engaged in off-herd activities such as brewing were very successful. There is a general and very noticeable trend towards a more diversified livelihood, a phenomenon also observed among other East African pastoralists.

In general, transaction costs of newly instituted strategies of income generation are higher among the Pokot than among the Himba. Pokot livestock traders, for example, were accused of envy and witchcraft. They had appropriated gains as intermittent traders and were not prepared to redistribute these monetary gains in the same manner as e.g. livestock from bridewealth payments. In an economy so utterly directed towards the redistribution of property they became a major obstacle in the system. The "reluctance to change" which Harold Schneider observed among the Pokot in the late 1940s (Schneider 1959), apparently describes institutional brakes to system-change. Nevertheless, these brakes were overcome in the 1990s and monetary gains were generally not redistributed. Some traders whom I followed up throughout the 1990s had built up sizeable and diversified businesses by 2004.

In consequence of the diversification of economic activities labour allocation within households is newly defined. Pokot women found new ways of procuring an income from selling beer and food at wells or at small restaurants. Although their income was low it contributed significantly to the food security of the household. Perhaps more importantly, the women stayed in control of the money they earned. They could decide if they wanted to allocate it to purchasing food or to paying school fees for one of their children.

The comparative wealth of some Himba herders makes it easier for them to opt for self-determined forms of economic diversification. Among the Pokot young traders, frequently poor and with little power in the gerontocratic system, often shoulder the burden of changing institutions. Among the Himba the big men are the agents of change in the system. But how can poor people among the Himba improve their fate? Stockless Himba do not have to leave the system, many become dependent herdsmen in the service of wealthy relatives. Others opt for employment in the army or in other sectors. This honourable exit option does not exist among the Pokot. The few Pokot men who have migrated to farms and urban centres in search of labour, find it very difficult to maintain ties to their community.

For many pastoral societies labour outside the pastoral sector has become crucial. Fulbe have moved to coastal west African countries in order to earn an income from wage labour (White 1997) and feed money back into the rural economies of Mali and Niger. The same holds true for the Beja of the Sudan as reported by Manger (1996) and Dahl & Hjort (1991). They gain progressively larger percentages of income from labour in nearby urban centres, while the importance of livestock husbandry as the sole source of income decreases. Berzborn (2004) details that Nama herders in the South African Richtersveld only gain a fraction of their income from their herds, while wages for labour in nearby mines are their mainstay. Nevertheless, they perceive themselves as herders and pastoralism is intensely linked to Nama identity.

Transfer and Flexible Distribution of Property Rights: Flexibility of the distribution of property rights is another important factor in buffering shortfalls and losses. Is it easy to obtain access to the means of production within a given economic system? The transfer of property rights is a problem very peculiar to pastoralists as they inevitably start off their professional careers with livestock, which belongs to somebody else. They inherit animals, loan stock and receive livestock presents from friends and relatives.

Himba and Pokot systems of livestock property rights are organised in very different ways. The Pokot opt for a constant redistribution of ownership rights in livestock, thus preventing the emergence of large accumulations of livestock property in individual herds. Herders constantly share out livestock to marrying sons, to sons taking out their inheritance, to their own in-laws and to stock-friends. The moral obligation to redistribute property is a cornerstone of Pokot ethics and underscores their egalitarian political system. Transaction costs of the institutions facilitating exchange are low. The transfer of ownership rights in livestock usually concern two individuals only. It is fairly easy for young Pokot men and impoverished herders to gain property rights in livestock and to become livestock owners. In contrast, young Himba men find it hard to obtain property rights in livestock. While for a dependent herdsmen it is easy to obtain the right to use substantial numbers of animals (milk and occasionally slaughter), it is much harder to obtain ownership to livestock: most men between the ages of 25 to 40 are still largely depend on animals they herd for their father or on animals they have obtained on loan from maternal relatives. Inheritance among the Pokot is managed almost in passing, involving only very few people as young men take out several animals from the paternal herd over a period of time. In contrast, inheritance transfers among the Himba may require week-long negotiations.

A general analysis of both case studies suggests that both economies provide young and/or poor herders with ample opportunities to receive livestock through institutionalised channels. However, among the Pokot, the transfer of livestock involves a transfer of ownership rights thereby preventing the emergence of patron-client relations. Among the Himba the inheritance falls upon one person only and the transfer of livestock from one generation to the next is delayed. Consequently, men are obliged to accept patron-client relations. Literature on pastoral societies suggests that these two basic patterns recur in many other African pastoral societies. Agro-pastoral communities like the Tutsi of Rwanda and Burundi or the Hima of Uganda restrict access to ownership rights in cattle and establish livestock-loaning systems which keep clients dependent on wealthy cattle owning patrons. In contrast, pastoralists such as the Fulbe and the Turkana split paternal herds equally among sons.

Networks of Livestock Exchange: While inheritance transfer is one means of attaining livestock property, loans, presents and institutionalised gift exchanges (like bridewealth and compensation) are other means. Livestock-based economies constantly produce a surplus, which may be invested into social exchange.[7] While exchange is strongly egalitarian among the Pokot, it has distinctly asymmetrical tendencies among the Himba. Pokot donors give up ownership rights in transferred animals, while Himba donors give away animals mainly on a loan basis. The way Pokot handle livestock transfers underscores their egalitarian political system. The exchange system generates a dense, cohesive network. A structural analysis of a Himba exchange network revealed a different structure: there were central as well as marginal positions and the network was less cohesive than the Pokot network. Himba and Pokot livestock exchanges also differ in the number of animals involved. The

Himba invest almost double the number of cattle into intra-societal exchange networks as the Pokot do. While transfers of livestock among the Pokot usually involve only a few animals at a time, among the Himba both minor transfers of one or two animals or major transfers of entire herds may occur

The transfer of livestock to create obligations and/or to disperse livestock is a strategy applied by many African pastoral communities. Hereby herders make full use of the specific characteristics of their major means of production. Bonte (1991:56) argues that the rationality of stock-raisers must not be reduced to the production of use values (the use of livestock for nutrition) but that it "... involves as well the fact that domestic animals constitute in these societies the principal means of production". In all societies in which the transfer of livestock involves large-scale exchanges of ownership rights (Pokot, Turkana, Nuer, Somali) egalitarian conditions prevail. In systems in which ownership rights remain with the donor, the transfer of livestock promotes the formation of hierarchies (Tutsi, Tswana, Herero).

Institutionalised Resource Protection: Pastoralists' intimate awareness of the vulnerability of savannah grasslands has been noted in several ethnographic accounts. A comparison of both cases, however, suggests that the "indigeneous knowledge approach" is limited in its potential to explain sustainable modes of production. Instead, in this analysis, the efficiency of common-property-resource-management institutions has been highlighted as the major foundation of sustainable pasture management.

The transaction costs of such institutions of resource protection can be reduced if (a) the group responsible for protection is limited and easily identifiable, (b) the resource to be protected is clearly defined and (c) if rules and procedures for quantifying and sanctioning free-riding are specified (see Orstom 1990, design principles for common property resource management). The Pokot have problems on all three counts. While ideas about range protection exist, concepts of responsibility are vague. Basically all Pokot are entitled to graze everywhere in Pokot land. The situation was aggravated in recent decades when intertribal raiding forced the Pokot to give up large tracts of land, resulting in an increase of population density in some areas and the depopulation of others. The Pokot still try to solve resource management issues by means of egalitarian neighbourhood councils with fluctuating membership. Decisions must be consensus based and even otherwise influential men are rarely able to steer discussions in a specific direction. Sanctions for free-riding are lenient and enforcement is difficult. In contrast to the Pokot the Himba link access to resources with kinship and neighbourhood and find it much easier to define the group of people entitled to use a certain resource. Himba pastoralists mainly exploit areas within their chieftaincy but access to resources within a chieftaincy is not open either. Herders belong to a grazing unit which uses common rainy season settlements and common dry season pastures. Usually "an owner" or "guardian of the land" can be named for every place with permanent settlements. The longevity of his and his family's claims to the land is corroborated by the number of family graves in the area.

While decision-making among the Himba is also consensus based, influential elders and office holders (chiefs, councillors, big men, owners of the land) have much more opportunity to channel discussions and are able to guarantee the implementation of sanctions. The costs of enforcement are mainly borne by wealthy patrons.

Himba rules regarding resource protection are more complex and detailed than Pokot rules. The boundaries of Himba grazing reserves are fairly clear-cut and can be pointed out even by young herders. In contrast, Pokot boundaries of dry season grazing areas are only vaguely defined. Institutions of common property resource management can neither guarantee a sustainable use nor can they ensure equal access to resources. In such a situation an opportunistic, self-serving strategy is probably the best way to gain access to resources. In contrast, Himba resources are communal in the proper sense of the word (see also Behnke, 1998). Decisions are taken by a well defined community on the basis of an explicit set of rules, ensuring a degree of sustainability and, perhaps even more important, predictable and equal access.

Communal resource control is deteriorating in many African pastoral communities. Among the Kenyan Orma (Ensminger, 1990, 1992) internal forces led to the demise of communal tenure, among the Maasai (Galaty, 1994), state policy was instrumental in demolishing the commons. A trend towards more privatised forms of resource control is also reported for the pastoral areas of Botswana (Hitchcock, 1990), Somalia (Nauheimer, 1991; Stern, 1991) and central Namibia (Stahl, 2000; Werner, 2000). Case studies from pastoral communities reveal that pastoral commons are not simply changing into private ranches. There are various intermediate solutions and combinations: the Herero of central Namibia combine private ranches, individualised calf pastures adjacent to households and village grazing territories (see also Bollig 2001b, 2002). The Kenyan Orma manage pastures as a combination of private rangelands and village pastures. The Himba system shows that communal tenure may still work under specific conditions. These examples refer to low population density, a clear definition of the resource protecting community and explicit standards regarding communal management rules. Both case studies underline Orstom's common property design principles (1990:90) regarding the efficient management of self-governing common-property resources.

Solidarity and Moral Economy: During recent decades the moral economy of peasants (Scott, 1976; Hyden, 1985; for a critical review of relevant literature see Lemarchand, 1989) and pastoralists (Spencer, 1998:233; Bollig, 1998a) has been a controversial issue. Scott defines moral economy as an economy in which access to resources and the provisioning of all members of society is guaranteed by a moral system (Scott, 1976; in Ensminger, 1992: 2). Lindenberg (1998) discusses the emergence of strong solidarity groups as a reaction to uncertainty. Strict group boundaries are enforced and common identity is emphasised in rituals. Ideology is a central rallying point in strong solidarity groups (Linbenberg, 1998:28).[8] In this sense the maintenance of a moral economy and the support of strong solidarity groups contribute directly to risk management and the reduction of vulnerability.

The pastoral Pokot may be regarded as a strong solidarity group in Lindenberg's sense. The Pokot cherish an ideology, which emphasises intra-ethnic cohesion and support. Solidarity, respect and internal peace are regarded as key values. The Pokot are beset by the fear that internal friction might ruin their personal livelihoods and the integrity of their society. Pokot key values - co-operation, solidarity and respect - are dramatically enacted in large rituals. In these rituals, men of the wider neighbourhood, sometimes of the entire Pokot area (irrespective of lineage membership and age set affiliation) gather to evoke their Pokotness in dance and song.

Among the Himba solidarity is a key value as well, but it is restricted to matrilineal relatives (*omuhoko*). Economic support such as livestock loans is mainly confined to a set of closely related people. Himba rituals focus on descent-group corporateness and solidarity rather than emphasising ethnic identity. Only in recent years have the Himba developed political rituals of a more comprehensive scope. Lately commemorative rituals at the graves of deceased chiefs have developed into true community rituals. The Himba suddenly feel the need to emphasise common identity in the face of potential competitors for land and governance. Local activists developed an understanding of the relevance of global networking, when the Namibian government set forth to plan a hydroelectric power scheme on the Kunene River, which would have inundated important natural resources. NGOs and capital-based human rights activists connected the Himba to foreign solidarity groups and a number of Himba men attended international meetings of indigenous communities in Arusha, Geneva and Washington (Bollig, 2001a; Bollig & Heinemann, 2002; Bollig & Berzborn, 2004). The case studies suggest that moral economies should be regarded in their historical context, as moralities and reference groups change. Lindenberg's assumption that the emphasis on group boundaries and internal solidarity are direct reactions to increased stress, has been corroborated by both case studies.

7.4. RISK MINIMISATION AND ECONOMIC CHANGE

In the final paragraph I would like to place risk management into a larger perspective and discuss its relation to socio-economic change. The historian Waller (1988:111) tells us that highly specialised forms of livestock husbandry are a relatively recent phenomenon in East Africa. He reports that various forms of mixed pastoralism, agro-pastoralism and pastro-foraging were characteristic for the Rift Valley before 1800 and that "the extreme specialisation of the Maasai is both relatively recent and, probably, short lived". Galaty (1991), describing Maasai expansion about two hundred years ago mentions the "new East African pastoralism", and hints at the emergence of highly specialised, mobile livestock breeders. New principles of social and political organisation accompanied economic specialisation.[9] Today at the turn of the millennium, however, specialised pastoralism seems to be on the retreat. Demographic growth, degradation, increased stratification and peasantisation within a broader economy, and increasingly violent conflicts associated with and partially caused by the massive proliferation of small arms contribute to the demise of a mode of production which has dominated arid and semi-arid Africa for several centuries. Comparative data leaves little doubt that livestock husbandry still is the major form of production in Africa's drylands, however, its organisational features are changing profoundly (Spencer, 1998: 242). How did the two economies investigated in the present study fare during the course of past centuries and what are their future prospects?

The Pokot and Pastoralism in East Africa

The pastoralisation of certain sections of the Pokot took place during the first half of the 19th century. During this period several groups, which are nowadays regarded as being prototypical representatives of East African pastoral nomadism, re-orientated their economies towards specialised livestock husbandry (for the Turkana see Lamphear 1976a, 1988 for the Maasai Galaty 1993). Economic change and social evolution went hand in hand. Pokot tradition emphasises that the earliest specialised herders among them introduced to their society the age-set system, the symbolism of warriorhood and certain modes of sharing. In order to compete successfully for large herds of cattle, military strategies had to be fundamentally changed. Large-scale raids involving hundreds of warriors became the most successful strategy to acquire other peoples' livestock. Clan-based territories became non-functional and were replaced by tribally-defined territories. With the demise of clan control over the means of production, the clan's significance as the central organising principle of society decreased. Age-sets, unifying men from many non-related households, became the focal organisational feature of society. Gerontocratic leadership and the control of marriageable spouses were based on the complex age-grade system. At the same time internal conflict management changed. Violent conflicts between descent groups were ruled out and substituted by complex mechanisms of compensation. Conflicts between age-groups were staged as highly ritualised, dramatic events. Non-violent internal conflict management contrasted sharply with violent external strategies.

The pastoralisation of the Pokot was accompanied by an intense increase in productivity.[10] The number of cattle kept by pastoral Pokot was much greater than that of the agro-pastoral Pokot. Comparing both sections, Schneider in the 1940s noted a ratio of livestock to people of 1.2:1 among the agro-pastoral Pokot and of 5:1 among the pastoral Pokot. During the 1920s and 1940s the ratio for the pastoral Pokot of Baringo District of about 15:1 was even higher. The emergence of a fully-fledged livestock exchange system was closely related to the immense increase in productivity. While these new resources brought about a whole set of new options for surplus production and social exchange, they also involved new risks: devastating losses of livestock during droughts, epidemics and raids, degradation due to overstocking and dependency on grain producers and trade are only some of the hazards a highly specialised pastoral economy had to cope with.

Authority and power were vested in senior age-sets and were de-coupled from wealth. Institutions aimed at diffusing wealth developed: bridewealth payments increased from just one or two head of cattle to ten or twenty and more.[11] Barton (1921:96) reports that bridewealth among the Pokot varied "... from a few goats with the Hill Suk [chiefly agricultural] to up to 80 head of cattle with the pastoral Suk" in the early 20th century. Stock-friendship channelled exchanges between non-related persons. While the institution of stock-friendship also exists among the agro-pastoral Pokot, it is of less economic and social significance.[12] Stepping up bridewealth requirements and moral obligations to marry ever more women guaranteed redistribution and prevented the trans-generational accumulation of wealth. Polygynous households had greater numbers of children. Hence the division of wealth according to inheritance claims contributed to the redistribution of livestock property. This principle apparently is a common denominator among East African pastoralists: the higher the degree of pastoral specialisation, the higher the rates of polygyny. High rates of polygyny create a relative shortage of marriageable women, which in turn has an impact on bridewealth payments (Goldtschmidt 1974:316). Spencer (1998:88) gives a polygyny profile of 34 pastoral communities in East Africa and points out the relation between pastoral specialisation, polygyny and size of bridewealth. Goldtschmidt (1974:326) presents a sample of 27 pastoral societies in which he found a positive correlation between the size of bridewealth and pastoral wealth. Hence, increasing bridewealth demands encourage a personal interest in herd growth while at the same time opening up paths for redistribution.

From the 1930s onwards the Pokot economy ceased to grow. British colonialism put an end to violent expansion and predatory warfare. Herds decreased due to devastating environmental degradation while population numbers increased. The ratio of cattle to humans fell from 15:1 to 1:1 and below. In the late 1960s, interethnic raiding resumed and during the 1990s deteriorated into a state of low intensity warfare. Many Pokot herders were forced to leave pastures they had used for decades and had to settle in densely populated areas. The Pokot tried to cope with this decline by enforcing mutual exchange. Big men lost much of their power and influence: the British colonial administration had employed big men as chiefs but largely failed to establish a group of hereditary traditional leaders. Highly egalitarian conditions developed as a result of internal processes and external regulation. The Pokot became an increasingly "strong solidarity group" which acted on the basis of an egalitarian ideology, affirmative rituals and strong group boundaries. During the 1980s and 1990s the Pokot were successful when they relied on internal solidarity. Where they depended on an institutional framework to ensure sustainable resource management or to facilitate innovation they ran into serious trouble. The cohesive networks created among the Pokot through livestock exchange clearly had a "double edge".[13] Dense networks created by kinship, friendship and livestock exchange helped to co-ordinate dyadic exchanges and act as instruments for exerting social pressure. However, the very same networks prevented group members from effectively reaching decisions on group level, from carrying out group action for the communal good and from identifying and sanctioning transgressors. Dyadic ties frequently ruled in favour of consanguinal, affinal and friendship ties and against common action towards a common goal. While this may sound very theoretical, from the Pokot perspective it is a hard reality. The case reported in chap. 6.3 is a case in point: A free-rider was not punished because everybody had a more immediate interest in the relation to the accused than in the common good. Strong social ties undermined social control. Polly Wiessner (1998:358) analysed this dilemma in a theoretical treatise on emergency decisions, egalitarianism and group selection arguing that "... the very egalitarian relations that evolved in the service of the evolution of far-flung networks of kinship and reciprocity may have crumbled again under the dilemma that the latter posed for group action".

Since the late 1990s new forms of power and authority have been gaining ground. An internal elite of traders, teachers and administrators has emerged. They all maintain a stake in livestock husbandry and manage their herds through kinsfolk and employed herdsmen. The significance of neighbourhood councils continued to decline while administrative chiefs gained more prominence. However, this did not noticeably improve common property resource management. Livestock marketing increased as small market centres developed "in the bush". During a short survey conducted in March 2004 a wealth ranking revealed that in contrast to the 1980s a group of stockless paupers had emerged and the number of poor herders – herders with less than ten head of cattle – had significantly increased. A survey conducted in the same region in 1987 had found no stockless persons or persons owning less than ten head of cattle. In 2004 a small group of very rich, mobile livestock owners still existed but did not dominate politically. The integration into national structures was facilitated by the advent of multiparty politics. Different candidates for parliamentary seats were competing for a local voter base and noticeable rifts occurred within a hitherto fairly homogenous society. The increased relevance of churches and rapidly increasing school enrolment since about 2000 contributes to an ongoing differentiation of Pokot society.[14]

The Himba and Pastoralism in South-western Africa

Pastoral specialisation in south-western Africa is also a recent phenomenon.[15] While livestock husbandry became the main subsistence base about 200 years ago, foraging strategies remained important. Oral accounts suggest that the herders of Kaokoland were wealthy pastoralists at that time. Big men (*ovahona*) are prominent in oral accounts on the history of the nineteenth and early twentieth century. They founded large homesteads with numerous dependants, youths, impoverished herders and foragers with a client status. While Himba big men dominated local politics, they did not succeed in transforming the system into a clearly stratified hierarchy in which status was conferred by birth. However, in contrast to the Pokot, their wealth was not constantly redistributed through bridewealth payments (which stayed low) or inheritance as the entire herd was transferred from the original owner to just one heir. While matrilineal inheritance (combined with patrilocality) largely prevented the build-up of localised power bases, the inheritance system as such ensured that there were always wealthy men who had the material means to strive for political dominance. While the wealth of Pokot big men crumbled when herds were divided among many heirs, the wealth of a Himba big man stayed intact and was transferred to his heir as an entity.

In the middle of the 19th century the wealth of the Himba and the Herero did not only attract traders but also raiders. The raids of Nama commandos in the second half of the 19th century brought local livestock husbandry to the brink of annihilation. Most herders retreated to southern Angola where the cattle plague epidemic of 1897 further decimated their herds. In Angola the Himba subsequently reconstituted their herds by diversifying their economy. They tilled gardens, worked on plantations, traded as commercial hunters and joined the Portuguese colonial army as mercenaries. Big men re-emerged as local warlords under the command of the Portuguese army. By 1910 the Himba were prosperous herders well integrated into the predatory Portuguese colonial system. Due to changes in the Portuguese colonial administration a large part of the Himba population re-crossed the Kunene River between 1910 and 1930 and again settled in Kaokoland. Himba economy changed substantially under South African colonial rule. The herders of Kaokoland were cut off from the major markets in southern Angola and central Namibia. Local economies became artificially separated through administrative regulations and herders were reduced to subsistence production. While local big men were prevented from transferring livestock wealth into other fields of the economy (e.g. trade), they were otherwise sustained by the South African government. Some of them were declared chiefs, others became councillors. Existing hierarchies were stabilised by decree and the big man system of pre-colonial times was restructured into a political system dominated by administrative chiefs.

Population growth rates remained low due to high rates of infection with sexually transmitted diseases and the ratio of people to cattle fluctuated around a high mean. Small-scale agriculture increased and with it the self-sufficiency of Kaokoland's herders. In contrast to the Pokot, the Himba found it easier to manage natural key-resources. Clear definitions of what was to be protected and who could take part in decision making reduced the transaction costs of common property resource management. However, an investigation of natural resource management in a historical perspective reveals a long-term trend towards pasture degradation: high stocking numbers were reached in 1958 when a two year drought reduced herds by 50 per cent. Natural resources recovered. Very high stocking rates were reached in the mid 1970s and again followed by a widespread degradation of pastures. The centennial drought of 1980/81 reduced cattle numbers by 95 per cent, while the 1980s were a period of ecological recovery. Most of the field-work for this study (1994-96) was conducted towards the end of this period of recovery when stock numbers had reached a new peak. Very high stocking rates between 1995 and 2004 led to an impoverishment of pastures. Perennial grasses were replaced by annuals and even a pasture of diverse annual grasses was replaced by a pasture mainly characterised by a low yielding Aristida species.[16] While these dynamics may indicate a long-term equilibrium, there are clear signs of irreversible degradation. Even when livestock numbers were once again drastically reduced, valleys much damaged by soil degradation such as Omuhonga (Sander, Bollig, Schulte, 1998) did not recover. It is likely that further commoditisation and stratification will lead to a dismantling of the commons and a disintegration of patron-client networks. Problems of resource control became very obvious recently when tour operators asked local chiefs and councillors for permission to set up tourist facilities in their areas. Several chiefs accepted payments which they pocketed without redistribution. The Epupa debate showed the precarious state of land tenure: when the Namibian government set out to plan and build a large hydroelectric power plant on the Kunene River, the Himba had to discover that formally the ground they were living on was owned by the state. The planned Epupa scheme sparked off forceful opposition from the Himba and brought them closer to opposition politicians, environmental activists and NGOs in the capital. Since 1998 the conservancy movement (see Bollig, 2004) has also reached north-western Namibia. Pastoral communities constitute themselves as resource management units, with a democratically elected leadership, fixed membership and a concise management plan. Conservancies are generally meant to foster local development and conservation at the same time. With governmental endorsement conservancies gain the right to enter into lucrative contracts with the private tourist sector. The widespread acceptance of this programme in Kaokoland leads to a further diversification of income generating opportunities, to political change and and to exposure to national and international agencies. The programme may ensure a tighter control of land tenure through local communities, but it also entails the integration of local economy and politics into national and even international structures.

The Future of African Pastoralists

In a recent overview on pastoral nomadic societies, Fratkin (1997:252) writes: "Pastoralists ... are moving into the twenty-first century with less ability to maintain their subsistence livestock economies than at any time in their past." Demographic growth, degradation of rangelands, commoditisation of livestock, increasing inequalities in wealth, state integration and low intensity warfare as a result of massive small arms proliferation are perhaps the main problems for African pastoralists. How do African herders react to these new challenges? Diversification of production (wage-based labour, agriculture), changing labour arrangements, individuation of common property resources, increased reliance on markets and co-management of resources with government bodies and non-governmental organisations seem to be the major solutions. McCabe (1997) observed that the adoption of cultivation among the Ngorongoro Maasai did not lead to the demise of pastoralism but rather reinvigorated livestock husbandry. McCabe concludes that the diversification of strategies has led to a situation in which Maasai pastoralism has become more sustainable. Ensminger (1992) reports on the individuation of land tenure rights among the Kenyan Orma. Elites were keen on seeing their investments rewarded by more exclusive access to territory. The individuation of property rights lowered the transaction costs in institutions of local land tenure, defining territories anew, specifying the users more clearly, restricting their numbers and establishing rules to trace transgressors. Livestock husbandry among the Orma became more stable as a result of this transition. Galaty (1984) shows how the individuation of land rights among the Kenyan Maasai is dominated by local elites who steer land allocation to their own advantage. While livestock husbandry in Kenya's Maasai land is still the dominant mode of production, it became organised on individual farms and group ranches. During recent years property rights in pastoral lands in Botswana and Namibia have also become more individualised (Hitchcock, 1990; Stahl, 2000; Werner, 2000). Despite the commoditisation of livestock production and the engagement of extremely wealthy town dwellers in livestock production on peripheral rangelands, it is still mobile livestock husbandry which to a large extent shapes relations between man and the environment. The demise of common property resource management frequently results in the definition of a smaller community which then manages a specific natural resource. In many regions of arid Africa, livestock husbandry is currently accommodated "as a semi-autonomous fringe activity within an underdeveloped pastoral economy" (Spencer, 1998:242). The process of commoditisation and peasantisation will not only alter the economic basis of pastoralists but also transform the organisation and control of labour. An increase in labour migration is noticeable in many cases. Women then shoulder the major burden of household and herding labour. In addition they contribute significantly to family income by earning money from ten-cent jobs like brewing, charcoaling and handicrafts. Coping strategies within communities are becoming more diversified and more market-orientated. While peasant herders rely increasingly on the sale of small stock, on the diversification of productive resources, and on waged herding and cash income from labour in non-pastoral activities, wealthy herders diversify their assets by investing in trade, intensifying livestock production (increased use of veterinary medicine and supplementary fodder) and by consolidating their personal claims to the means of production (fencing of pasture, exclusive use of privately drilled boreholes). In this process they gain more political bargaining power in national politics and dominate the fate of local traditions of livestock husbandry.

Throughout the 1990s another process has gained prominence. Increasingly, African pastoralists have come to understand themselves as indigenous communities within a regional but also within the global setting. Tanzanian Maasai campaigned against the transformation of rangelands into wheat farms with the help of international supporters while the Himba campaigned against the Epupa hydroelectric power plant. In Kenya pastoralists from different ethnic denominations are campaigning against governmental measures and for development within the Pastoral Network Forum – despite the fact that at the local level the lives of herders are soured by inter-community raiding. Pastoral elites from Kenya, Tansania and Namibia have come to understand their traditions as cultural capital within a global context. During the 1990s pastoral communities in southern Africa have engaged in projects of communal natural resource management. While up until the 1980s they had been regarded as the natural enemies of wildlife, now they are heralded as the natural guardians of game and savannah. While the state is still important, globalisation has undeniably reached pastoralists at the margins of national states. Even if in many pastoral areas the state has proven ineffective either in maintaining security or in implementing positive forms of development, the inclusion into national and international structures is progressing swiftly and will determine many aspects of pastoral livelihoods during the 21st century.

ENDNOTES

1. Douglas North (1987, 1990) described the institutional problems connected to population growth.
2. Botanical research with the Cologne based ACACIA programme has underlined this observation. Recently Becker & Jürgens (2002) and Schulte (2002) have outlined the sustainability of Himba herding practices from a botanical point of view.
3. Generally equilibrium systems are characterised by rainfall of more than 300mm and a rate of climatic variability of less than 30 per cent, while disequilibrium systems have a precipitation below 300 mm and rates of variability above 30 per cent.
4. For well documented case studies see Manger 1996 on the Sudanese Beja, Herren 1991 on the Kenyan Mukogodo and Silvester 1998 on the Namibian Nama.
5. The idea of representing the relation between key values, moral transgressions and punishment in a flow-diagramm is taken from Casimir 1987.
6. Termites have receptors for humidity. The swarming queens need loose wet soil to found a new colony.
7. Bourdieu (1979:348) points out that under unstable conditions social and symbolic capital is a more endurable resource than economic capital stored in livestock or grains.
8. Lindenberg (1998:18) states "emotions have an important role within rational behavior for creating credible commitments". "Good standing" in a sharing group refers to the history

of a member's social history. Solidarity frames are strengthened through mechanisms of building group identity (p.32).

9. However, archaeologists hint at earlier forms of specialised livestock husbandry. Ambrose (pers. com.) assumes that the Maasai entered East Africa as specialised livestock herders some 1200 years ago.
10. Wiessner & Tumu (1998a, b) give a fascinating account on the intricate interdependencies between production and exchange in the Papua New Guinea highlands. After the introduction of the sweet potatoe households were able to produce sizable surplusses which in turn resulted in the rise of two systems of ceremonial exchange: the Great Wars and the Te exchange. There are ethnographic examples which describe how the pastoralisation or repastoralisation of a group meant an immense increase in productivity. Bonte (1991) gives the example of the pastoralisation of populations in the Zagros Mountains in central Iran, Henrichsen (2000) describes the impact that repastoralisation had on central Namibian economies in the 19th century.
11. Goldtschmidt (1974:315) described how an increase in pastoral wealth brought about an increase in bridewealth among the Kenya/Ugandan Sebei of Mt. Elgon, a group closely related culturally to the Pokot and hypothesised: "In conformity with economic theory we should expect a gradual secular increase in brideprice and such is clearly shown, both for cattle and for total value of brideprice ... While cattle payments increased gradually throughout this period in the pastoral sector of the Sebei, they levelled off at about six in the farming community."
12. Institutionalised livestock loaning to non-relatives is more frequent among highly specialised, wealthy herders than among poor herders and agro-pastoralists (Johnson, 1990; McCabe, 1997). Herren (1991) reports that among the poor Mukogodo herders livestock loaning was virtually unknown. White (1997) reports that livestock loaning was widespread among Wodaabe Fulbe herders in Niger, but since their herds decreased, livestock loans were rare and have been replaced by commercial shepherding arrangements.
13. In an experimental study on decision making problems within highly cohesive networks, Flache (1996) coined the term "the double edge of networks" and found that, on the one hand, they are conducive to high solidarity, but on the other hand, they are detrimental to the management of communal goods.
14. These recent trends of socio-political change are currently researched by Matthias Österle a PhD student from Cologne University. The data of the 2004 field trip have not been fully analysed here.
15. There are no archaeological sites with large quantities of cattle bones in south-western Africa, despite a considerable amount of surveying for them (pers. com. A.Smith, University of Cape Town) and a Cologne based archaeology project working in Kaokoland (Vogelsang 2002) did not identify any pastoral sites older than 200 years.
16. A survey in 2004 indicates that population growth rates are on the increase. The extension of health care to remote areas has led to a reduction of sexually transmitted diseases which in turn has led to higher birth rates. At the time this book is going to the press, however, this survey has not been properly evaluated.

References

Abel, H., 1954, Beiträge zur Landeskunde des Kaokoveldes (Südwestafrika). *Deutsche Geographische Blätter* 47:7-123.

Abel, N.O.J. and Blaikie, P., 1989, Land Degradation, Stocking Rates and Conservation Policies for the Communal Rangelands of Botswana and Zimbabwe. *Land Degradation and Rehabilitation* 1:101-123.

Acheson, J. M., 1989, Management of Common-Property Resources. In *Economic Anthropology,* edited by St. Plattner, pp. 351-378. Stanford University Press, Stanford.

Ahuya, C.O. and Odongo, S.O., 1991, *Livestock Marketing in Nginyang and Tangulbei Divisions. A report for Kenya Freedom from Hunger Council.* Manuscript. Nakuru.

Almagor, U., 1979, Raiders and elders. A confrontation of generations among the Dassanetch. *SENRI Ethnological Studies* 3:119-146.

Almagor, U. and Baxter, P.T.W. (eds), 1978, *Age, Generation and Time. Some Features of East African Age Organisation.* Hurst, London.

Amborn, H., 1994, Wirtschaftliche und soziale Stabilisierungsstrategien Südäthiopischer Feldbauern. In *Überlebensstrategien in Afrika*, edited by M. Bollig and F. Klees, pp. 159-177, Heinrich Barth Institut, Cologne.

Anderson, J., 1863, Neuere Reisen nach dem Kunene und Okavango Flusse. In *Entdeckungen an der Westküste Afrikas.* Bearbeitet von Hermann Wagner, edited by P. Chaillu, L. Magyar, and J. Anderson, pp. 269-288. Otto Spamer, Leipzig.

Anderson, D., 1984, Depression, Dust Bowl, Demography and Drought. The Colonial State and Soil Conservation in East Africa during the 1930s. *African Affairs* 83:321-343.

Anderson, D., 1981, *Some Thoughts on the Nineteenth Century History of the Il Chamus of Baringo District.* Institute of African Studies, University of Nairobi. Seminar Paper No. 149.

Anonymous, 1878, Herero-Land, Land und Leute. *Petermanns Mittheilungen* 24:306-311.

Arhem, K., 1984, Two Sides of Development: Maasai Pastoralism and Wildlife Conservation in Ngorongoro, Tanzania. *Ethnos* 49:186-210.

Austin, H.H., 1903, A Journey from Omdurman via Lake Rudolph. *Geographical Journal* 19: 669-688.

Azarya, V., 1996, Pastoralism and the State in Africa: Marginality or Incorporation? *Nomadic Peoples* 38:11-36.

Baksh, M. and Johnson, Al., 1990, Insurance Policies among the Machiguenga: An Ethnographic Analysis of Risk Management in a Non-Western Society. In *Risk and Uncertainty in Tribal and Peasant Economies*, edited by E. Cashdan, pp.193-228. West View Press, Boulder.

Barber, J., 1968, *The Imperial Frontier. A Study of Relations between the British and the Pastoral Tribes of North East Uganda.* East Africa Publishing House, Nairobi.

Barrow, A. and Long M., 1981, *Grasses, Bushes and Trees in East Pokot. East Pokot Agricultural Project*. Kositei. Manuscript.

Barton, J., 1921, Notes on the Suk Tribe of Kenya Colony. *Journal of the Royal Anthropological Society* (1921) pp.82-99.

Baumann, H.. 1975, Die Südwest-Bantu-Provinz. In *Die Völker Afrikas und ihre traditionellen Kulturen,* edited by H. Baumann, pp. 473-511. Franz Steiner Verlag, Wiesbaden.

Baxter, P.T.W., 1994, The Creation and Constitution of Oromo Nationality. In *Ethnicity and Conflict in the Horn of Africa,* edited by K. Fukui and J. Markakis, pp. 167-186. James Currey, London.

Baxter, P.T.W. and Hogg. R. (eds), 1990, *Property, Poverty and People: Changing Rights in Property and Problems of Pastoral Development*. Department of Social Anthropology, Manchester.

Bechmann, G., 1993, Risiko als Schlüsselbegriff in der Gesellschaftstheorie. In *Risiko und Gesellschaft. Grundlagen und Ergebnisse interdisziplinärer Risikoforschung*, edited by G. Bechmann, pp. 237-276. Westdeutscher Verlag, Opladen.

Beck, U., 1986, *Risikogesellschaft. Auf dem Weg in eine andere Moderne*. Suhrkamp, Frankfurt.

Becker, T. and N. Jürgens, 2002, Vegetationsökologische Untersuchungen im Kaokoland, Nordwest-Namibia. In *Interdisziplinäre Perspektiven zu Kultur-und Landschaftswandel im ariden und semiariden Nordwest Namibia*, edited by Bollig, M., Brunotte, E. and T. Becker, pp. 101-118. Kölner Geographische Arbeiten 77.

Beech, M., 1911, *The Suk. Their Language and Folklore*. 2nd edition. The Negroe Press, New York. (Original: The Clarendon Press, Oxford)

Behnke, R., Scoones, I., and Kerven C. (eds), 1993, *Range Ecology at Disequilibrium. New Models of Natural Variability and Pastoral Adaptation in African Savannas*. Overseas Development Institute. London.

Behrend, H., 1987, *Die Zeit geht krumme Wege. Raum, Zeit und Ritual bei den Tugen in Kenia*. Campus, Frankfurt.

Bell, D., 1995, On the Nature of Sharing: Beyond the Range of Methodological Individualism. *Current Anthropology* 36:826-830.

Belsky, A.J., 1990, Tree/Grass Ratios in East African Savannas: A Comparison of Existing Models. *Journal of Biogeography* 17:483-489.

Berntsen, J., 1979, *Pastoralism, Raiding and Prophets. Maasailand in the Nineteenth Century*. Ph.D. University of Wisconsin. Ann Arbor. UMI.

Bernus, E., 1988, Seasonality, climatic fluctuations, and food supplies (Sahelian nomadic pastoral societies). In *Coping with Uncertainty in Food Supply*, edited by I. de Garine and G. Harrison, pp. 318-336. Oxford University Press, Oxford.

Bernus, E., 1980, Famines e Sécheresses chez les Tourages Sahéliens. Les Nourritures de Substitution. *Africa* 50:1-7.

Best, G., 1978, *Vom Rindernomadismus zum Fischfang. Der Sozio-Kulturelle Wandel bei den Turkana am Rudolfsee, Kenia*. Franz Steiner Verlag, Stuttgart.

Berzborn, S. 2004, *Haushaltsökonomie, soziale Netzwerke und Identität. Risikominimierende Strategien von Pastoralisten und Lohnarbeitern im Richtersveld, Südafrika*. PhD Dissertation. University of Cologne. http://kups.ub.uni-koeln.de/volltexte/2004/1318 .

Besteman, C., 1996, Violent Politics and the Politics of Violence: the Dissolution of the Somali Nation-State. *American Ethnologist* 23:579-596.

Bianco, B., 2000, Gender & Material Culture in West Pokot, Kenya. In *Rethinking Pastoralism in Africa*, edited by D. Hodgson, pp 29-42. James Currey, Oxford.

Boholm, A., 1996, Risk Perception and Social Anthropology: Critique of Cultural Theory. *Ethnos* 61:64-84.

Bollig, M., 2004, Landreform in Namibia: Landverteilung und Transformationen kommunalen Landbesitzes. In: *Namibia – Deutschland: Eine Geteilte Geschichte. Widerstand – Gewalt – Erinnerung*, edited by L. Förster, D. Henrichsen, and M. Bollig. Ethnologica, New Series 24:106-121.

Bollig, M., 2002, Problems of Resource Management in Namibia's Rural Communities: Transformations of Land Tenure between State and Local Community. *Die Erde* 133: 155-182.

Bollig, M., 2001a, Kaokoland - Zur Konstruktion einer Kultur-Landschaft in einer Globalen Debatte. In *Spuren des Regenbogens*, edited by St. Eisenhofer, pp. 474-483. Arnold'sche Verlagsbuchhandlung, Stuttgart.

Bollig, M., 2001b, Probleme kommunalen Besitzes: Lokales Wissen, Recht und Dynamik pastoralen Weidemanagements in Süd-und Ostafrika. In *Mensch und Umwelt. Gedanken aus Sicht der Rechtswissenschaften, Ethnologie, Geographie. Laudationes und Vorträge gehalten aus Anlass der Verabschiedung von Frau Ursula Far-Holländer,* edited by E. Ehlers, pp. 36-49, Sankt Augustin.

Bollig, M., 2000, Staging Social Structures: Ritual and Social Organisation in an Egalitarian Society - the Pastoral Pokot of Northern Kenya. *Ethnos* 65:341-365.

Bollig, M., 1998a, Moral Economy and Self-Interest: Kinship, Friendship, and Exchange among the Pokot (N.W.Kenya). In *Kinship, Networks, and Exchange*, edited by Th. Schweizer and D. White, pp. 137-157. Cambridge University Press, Cambridge.

Bollig, M., 1998b, Zur Konstruktion ethnischer Grenzen im Nordwesten Namibias (zwischen 1880-1940). Ethnohistorische Dekonstruktion im Spannungsfeld zwischen indigenen Ethnographien und kolonialen Texten. In *Afrikaner Schreiben Zurück. Texte und Bilder Afrikanischer Ethnographen,* edited by H. Behrend and Th. Geider, pp. 245-271, Köppe, Köln.

Bollig, M., 1998c, Power and Trade in Precolonial and Early Colonial Northern Kaokoland. In *Namibia under South African Rule. Mobility and Containment, 1915-1946*, edited by P. Hayes et al., pp. 175-193. James Currey, London.

Bollig, M., 1998d, Framing Kaokoland. In *The Colonising Camera. Photographs in the making of Namibian history*, edited by W. Hartmann, J. Silvester, and P. Hayes, pp. 164-170. University of Cape Town Press, Cape Town.

Bollig, M., 1998e, The Colonial Encapsulation of the North-Western Namibian Pastoral Economy. *Africa 68* (4):506-536.

Bollig, M., 1997a, Söldnerführer, Chiefs und Indigenous Rights Aktivisten - Intermediäre der Macht im Nordwesten Namibias. *Peripherie* 67:67-84.

Bollig, M., 1997b, Contested Places - Graves and Graveyards in Himba Culture. *Anthropos* 92:35-50.

Bollig, M., 1997c, *'When war came the cattle slept ...' Himba Oral Traditions*. Köppe, Cologne.

Bollig, M., 1996a, Krieger und Waffenschieber in der Ostafrikanischen Savanne. In *Krieg und Kampf. Die Gewalt in unseren Köpfen,* edited by E. Orywal, A. Rao, and M. Bollig, pp. 147-156. Reimer, Berlin.

Bollig, M., 1995a, The Veterinary System of the Pastoral Pokot. *Nomadic Peoples* 36/37:17-34.

Bollig, M., 1995b, Zur Legitimation von Gewalt bei ostafrikanischen Hirtennomaden. In *Töten im Krieg*, edited by H. v. Stietencron and J. Rüpke, pp. 363-398. Veröffentlichungen des 'Instituts für Historische Anthropologie', Tübingen/Freiburg.

Bollig, M., 1994a, Krisenmanagement und Risikominimierung bei den pastoralnomadischen Pokot Nordwestkenias. In: *Überlebensstrategien in Afrika,* edited by M. Bollig and F. Klees, pp. 125-157. Heinrich Barth Institut, Cologne.

Bollig, M., 1994b, Pokot Social Organisation. Structures, Networks and Ideology. In *Sprachen und Sprachzeugnisse in Afrika*, edited by Th. Geider and R. Kastenholz. Festschrift Wilhelm J.G. Möhlig, pp. 63-87. Köppe, Cologne.

Bollig, M., 1993, Intra- and Interethnic Conflict in Northwest Kenya - A Multicausal Analysis of Conflict Behavior. *Anthropos* 87:176-184.

Bollig, M., 1992a, *Die Krieger der Gelben Gewehre. Intra-und Interethnische Konfliktaustragung bei den Pokot Nordwestkenias*. Lit, Münster.

Bollig, M., 1992b, East Pokot Camel Pastoralism. *Nomadic Peoples* 31:34-50.

Bollig, M., 1990a, Ethnic Conflicts in North-West Kenya. Pokot - Turkana Raiding 1969-1984. *Zeitschrift für Ethnologie* 115:73-90.

Bollig, M., 1990b, An Otline of Precolonial Pokot History. *Afrikanistische Arbeitspapiere* 23:73-92.

Bollig, M., 1990c, Der Kampf um Federn und Farben - Promotion von Altersgruppen bei den Pokot Westkenias. In *Männerbande und Männerbünde. Zur Rolle des Mannes im Kulturvergleich,* edited by G. Völger, G and K. von Welck, pp. 259-266. Rautenstrauch Joest Museum, Cologne.

Bollig, M., 1987a, The Imposition of Colonial Rule in Northwest Kenya: Interethnic Conflicts and Anticolonial Resistance. *Afrikanistische Arbeitspapiere* 11:5-39.

Bollig, M., 1987b, Ethnic Relations and Spatial Mobility in Africa: A Review of the Peripatetic Niche. In *The Other Nomads. Peripatetic Minorities in Cross-Cultural Perspective,* edited by A. Rao, *pp. 179-228*. Böhlau, Leipzig.

Bollig, M. and Casimir, M., 1993, Pastorale Nomaden. In *Handbuch der Ethnologie*, edited by Th. Schweizer, M. Schweizer, and W. Kokot, pp. 521-560. Reimer, Berlin.

Bollig, M. and Lang, H., 1999, Demographic Growth and Resource Exploitation in Two Pastoral Communities. *Nomadic Peoples* N.S. 3,2:16-34.

Bollig, M. and Österle, M., (in press), 'We Turned our Enemies into Baboons': Warfare, Ritual and Pastoral Identity among the Pokot of Northern Kenya. *In The Practice of War: The Production, Reproduction and Communication of Armed Violence*, edited by A. Rao (et al.). Berghahn, New York.

Bollig, M. and Schulte, A., 1999, Environmental Change and Pastoral Perceptions: Degradation and Indigenous Knowledge in two African Pastoral Communities. *Human Ecology* 27:493-514.

Bollig, M. and Gewald, J.-B., 2000, Introduction: People, Cattle and Land – Transformations of Pastoral Society. In *People, Cattle and Land – Transformations of a Southwest African Pastoral People*, edited by M. Bollig & J.B. Gewald, pp. 1-52. Köppe, Köln.

Bollig, M., Brunotte, E. and Becker, T., 2002, Interdisziplinäre Perspektiven zu Kultur- und Landschaftswandel im ariden und semiariden Nordwest Namibia. *Kölner Geographische Arbeiten 77.*

Bollig, M. and Vogelsang, R., 2002, Naturraum und Besiedlungsgeschichte im Nordwesten Namibias. In *Interdisziplinäre Perspektiven zu Kultur-und Landschaftswandel im ariden und semiariden Nordwest-Namibia*, edited by M. Bollig, E. Brunotte, and T. Becker, pp. 145-158. Kölner Geographische Arbeiten 77.

Bollig, M. and Heinemann, H., 2002, Nomadic Savages, Ochre People and Heroic Herders: Visual presentations of the Himba of Namibia's Kaokoland. *Visual Anthropology* 15:267-312.

Bollig, M. and Berzborn, S., 2004, The Making of Local Traditions in a Global Setting. Indigenous Peoples' Organisation and their Effects on the Local Level in Southern Africa. In *Between Resistance and Expansion: Explorations of Local Vitality in Africa*, edited by P. Probst and G. Spittler, pp. 297-329. LIT, Münster.

Bonte, P., 1991, French Marxist Perspectives on Nomadic Pastoral Societies. In *Nomads in a Changing World,* edited by C. Salzman and J. Galaty, pp. 49-101. Instituto Universitario Orientale, Dipartimento di Studi Asiatici, Naples.

Booysen, J.M., 1982, *Otjiherero. n' Volledige Grammatika met oefeninge en sleutels in Afrikans.* Gamsberg, Windhoek.

Borgatti, S.P., Everett, M.P., and Freeman, L., 1992, *UCINET IV*. Analytic Technologies, Columbia.

Borgerhoff-Mulder, M., 1991, Behavioral Ecology of Humans: Studies of Foraging and Reproduction. In *Behavioral Ecology*, edited by J.R. Krebs and N.B. Davies, pp. 69-98. 3rd edition. Blackwells, Oxford.

Boserup, E., 1965. *The Conditions of Agricultural Growth. The Economics of Agrarian Change under Population Pressure*. Aldine Publishing Company, Chicago.

Bourdieu, P., 1979, *Entwurf einer Theorie der Praxis auf der ethnologischen Grundlage der kabylischen Gesellschaft*. Suhrkamp, Frankfurt.

Boyd, R. and Richerson, P.J., 1985, *Culture and the Evolutionary Process*. University of Chicago Press, Chicago.

Brainard, J., 1981, *Herders to Farmers. The Effects of Settlement and the Demography of the Turkana Population of Kenya*. Ph. D. Dissertation. State University of New York. Binghamton.

Broch-Due, V., 2000, A Proper Cultivation of Peoples: the Colonial Reconfiguration of Pastoral Tribes and Places in Kenya. In *Producing Nature and Poverty in Africa*, edited by V. Broch-Due and R.A. Schroeder, pp. 53-93. Nordiska Africainstitutet, Uppsala.

Broch-Due, V. and Anderson D., 1999, *The Poor are not Us: Poverty and Pastoralism in Eastern Africa*. James Curry, Oxford.

Brokensha, D.W., Warren D.M., and Werner O. (eds), 1989, *Indigenous Knowledge Systems and Development*. University of America Press, Lanham.

Browman, D.L., 1997, Pastoral Risk Perception and Risk Definition for Altiplano Herders. *Nomadic Peoples*. N.S. 1: 23-35.

Browman, D. L., 1987, Agro-Pastoral Risk Management in the Central Andes. *Research in Economic Anthropology* 8:171-200.

Brunotte, E., and Sander H., 2002, Die Morphosequenz des Randschwellenberglandes im zentralen Kaokoland von der Kalkkrustenstufe zum Namibrand anhand eines Transektes. In *Interdisziplinäre Perspektiven zu Kultur-und Landschaftswandel im ariden und semiariden Nordwest Namibia*, edited by Bollig, M., Brunotte, E. and T. Becker, pp. 13-22. Kölner Geographische Arbeiten 77

Campbell, A., 1986, The Use of Wild Food Plants and Drought in Botswana. *Journal of Arid Environments* 11:81-91.

Cashdan, E., 1990a, *Risk and Uncertainty in Tribal and Peasant Economies*. West View Press, Boulder.

Cashdan, E., 1990b, Introduction. In *Risk and Uncertainty in Tribal and Peasant Economies*, edited by E. Cashdan, pp. 1-16. West View Press, Boulder.

Cashdan, E., 1990c, Information Costs and Prices. In *Risk and Uncertainty in Tribal and Peasant Economies*, edited by E. Cashdan, pp.259-278. West View Press, Boulder.

Cashdan, E., 1985, Coping with Risk: Reciprocity among the Basarwa of Northern Botswana. *Africa* 55:454-476.

Cashdan, E., 1983, Territoriality among Human Foragers: Ecological Moduls and an Application to Four Bushman Groups. *Current Anthropology* 24:47-66.

Casimir, M., 1991, *Flocks and Foods. A Biocultural Approach to the Study of Pastoral Foodways*. Böhlau, Cologne.

Casimir, M., 1987, In search of guilt. Legends on the Origin of the Peripatetic Niche. In *The Other Nomads: Peripapetic Minorities in Cross-Cultural Perspective*, edited by A. Rao, pp. 373-390. Böhlau, Cologne.

Casimir, M. and M. Bollig, 2002, Ökologische Grundlagen der mobilen Weidewirtschaft der Himba in Nord-Namibia. In *Interdisziplinäre Perspektiven zu Kultur-und Landschaftswandel im ariden und semiariden Nordwest Namibia*, edited by Bollig, M., Brunotte, E. and T. Becker, pp. 207-218. Kölner Geographische Arbeiten 77

Catley, A. and Mohammed, A.A., 1995, Ethnoveterinary Knowledge in Sanaag Region, Somaliland Part I: Notes of the Local Descriptions of Livestock Diseases and Parasites. *Nomadic Peoples* 36/37:3-16.

Chapman, W., n.d., *The Angola Boers*. Unpublished Manuscript. Nambian National Archives.

Clarence-Smith, G., 1978, The Myth of Uneconomic Imperialism. The Portuguese in Angola, 1836-1926. *Journal of Southern African Studies* 4:165-223.

Clarence-Smith, G., 1976, Slavers in Coastal Southern Angola. *Journal of Southern African Studies 2:*214-223.

Clarence-Smith, G. and Moorsom, R., 1975, Underdevelopment and Class Formation in Ovamboland, 1844-1917. In *The Roots of Rural Poverty in Central and Southern Africa*, edited by R. Palmer and Parsons, pp.365-381. University of California Press, Berkeley.

Clutton-Brock, J. (ed.), 1989, *The Walking Larder. Patterns of Domestication, Pastoralism, and Predation*. Unwin Hyman, London.

Coale, A.J. and Demeny, P., 1966, *Regional Model Life Tables and Stable Populations*. Princeton.

Colson, E., 1979, In Good Years and in Bad: Food Strategies of Self-Reliant Societies. *Journal of Anthropological Research* 35:18-29.

Conant, F., 1982, Thorns Paired, Sharply Recurved: Cultural Controls and Rangeland Quality in East Africa. In *Anthropology of Desertification*, edited by B. Spooner, pp. 111-122. London, Academic Press.

Conant, F., 1965, Korok - A Variable Unit of Physical and Social Space among the Pokot of East Africa. *American Anthropologist* 67:429-434.

Coppock, L., 1994, *The Borana Plateau of Southern Ethiopia: Synthesis of Pastoral Research, Development and Change, 1980-91*. ILCA, Addis Ababa.

Crandall, D., 1992, *The Ovahimba of Namibia. A Study of Dual Descent and Values*. Ph. Diss. University of Oxford.

Crandall, D., 1991, The Strength of the Ovahimba Patrilineage. *Cimbebasia* 13:45-51.

Craven, P. 2002, Plant Species Diversity in the Kaokoveld, Namibia. In *Interdisziplinäre Perspektiven zu Kultur-und Landschaftswandel im ariden und semiariden Nordwest Namibia*, edited by Bollig, M., Brunotte, E. and T. Becker, pp. 75-80. Kölner Geographische Arbeiten 77

Crazzolara, P., n.d., *A Study of the Pokot (Suk). Language Grammar and Vocabulary*. Bologna.

Curry, J., 1993, Occupation and Drought Vulnerability: Case studies from a Village in Niger. In *African Food Systems in Crisis, Part One: Microperspectives,* edited by R. Huss-Ashmore and S.H. Katz, pp. 239-260. Gordon and Breach, Yverdon. (2nd edition)

Dahl, G., 1979, *Suffering Grass. Subsistence and Society of Waso Borana*. Studies in Social Anthropology, Stockholm.

Dahl, G. and Hjort, A., 1976, *Having Herds. Pastoral Herd Growth and Household Economy*. Studies in Social Anthropology 2, Stockholm.

De Almeida, J., 1935, *Sul de Angola. Relatório de um Govêrno de Distrito (1908-1910)*. 2nd edition. Divisão de Publicaçoes e Biblioteca Agência Geral das Colónias, Lisbon.

De Garine, I.,1994, Contribution of Wild Food Resources to the Solution of Food Crises. In *Coping with Vulnerability and Criticality,* edited by H.G. Bohle et al, pp. 339-359. Breitenbach, Saarbrücken.

De Garine, I. and Harrison, G. (eds), 1988, *Coping with Uncertainty in Food Supply*. Oxford University Press, Oxford.

De Leeuw, P. et al., 1993, An Analysis of Feed Demand and Supply for Pastoral Livestock: The Gourma Region of Mali. In *Range Ecology at Disequilibrium. New Models of Natural Variability and Pastoral Adaptation in African Savannas*, edited by R. Behnke, I. Scoones, and C. Kerven, pp. 136-152. Overseas Development Institute, London.

De Ridder, N., and Breman, H., 1993, A New Apporach to Evaluating Rangeland Productivity in Sahelian Countries. In *Range Ecology at Disequilibrium. New Models of Natural Variability and Pastoral Adaptation in African Savannas*, edited by R. Behnke, I. Scoones, and C. Kerven, pp. 104-117. Overseas Development Institute, London.

Dietz, T., 1987, *Pastoralists in Dire Straits. Survival Strategies in a Semi-Arid Region at the Kenya/Uganda Border. Western Pokot 1900-1986*. Netherlands Geographical Studies 49. Amsterdam.

Dirks, R., 1990, Social Responses during Severe Food Shortages and Famines. *Current Anthropology* 21: 21-44.

Donaldson, T.J.,1986, *Pastoralism and Drought. A Case Study of the Borana of Southern Ethiopia*. MScThesis. Department of Agriculture and Horticulture. University of Reading.

Douglas, M., 1994, *Risk and Blame. Essays in Cultural Theory*. (1992 1st edition). Routledge, London.

Douglas, M., 1985, *Risk Acceptability According to the Social Sciences*. Russell Sage Foundation, New York.

Douglas, M., 1978, *Cultural Bias*. Royal Anthropological Institute of Great Britain and Ireland. Occasional Papers 35.

Douglas, M., and Wildavsky, A., 1982, *Risk and Culture: An Essay on the Selection of Technical and Environmental Dangers*. University of California Press, Berkeley.

Downs, R.E. et al.,1991, The Political Economy of African Famine. Food and Nutrition. In *History and Anthropology*, Vol. 9. Gordon and Breach, Philadelphia.

Dundas, K.R., 1910, Notes on the Tribes Inhabiting the Baringo District, East Africa Protectorate. *Journal of the Royal Anthropological Institute* 40: 49-72.

Durkheim, E., 1984, *Die Elementaren Formen des Religiösen Lebens*. Suhrkamp, Frankfurt, 3rd edition.

Dyson-Hudson, N. 1966. *Karimojong Politics*. Oxford. The Clarendon Press.

Dyson-Hudson, R., 1983, Understanding East African Pastoralism. An Ecosystem Approach. In *The Keeping of Animals,* edited by R. Berleant-Schiller and E. Shanklin, pp. 1-11. Osmum Publishers, New York.

Dyson-Hudson, R. & Meekers, D. 1999, Migration across Ecosystem Boundaries. In: *Turkana Herders of the Dry Savanna. Ecology and Biobehavioural Response of Nomads to an Uncertain Environment*, edited by M.A. Little and P.W. Leslie, pp. 303-313. Oxford University Press, Oxford.

Dyson-Hudson, R. and Dyson-Hudson, N., 1980, Nomadic Pastoralism. *Annual Reviews in Anthropology* 9: 1-33.

Dyson-Hudson, R. and McCabe, T., 1985, *South Turkana Nomadism. Coping with an Unpredictably Varying Environment*. HRAF Files, Ann Arbor.

Edgerton, R., 1971, *The Individual in Cultural Adaptation. A Study of Four East African Peoples*. University of California Press, Berkeley.

Ellis, J.E., Coughenour, M., and Swift, D., 1993, Climate Variability, Ecosystem Stability, and the Implications for Range and Livestock Development. In *Range Ecology at Disequilibrium. New Models of Natural Variability and Pastoral Adaptation in African Savannas,* edited by R. Behnke, I. Scoones, and C. Kerven, pp. 31-41. Overseas Development Institute, London.

Ellis, J. and Swift, D., 1988, Stability of African Pastoral Ecosystems. Alternate Paradigms and Implications for Development. *Journal of Range Management* 41: 450-459.

Ensminger, J., 1992, *Making a Market. The Institutional Transformation of An African Society*. Cambridge University Press, New York.

Ensminger, J., 1990, Co-Opting the Elders. The Political Economcy of State Incorporation in Africa. *American Anthropologist* 92:662-675.

Ensminger, J., 1984, *Political Economy among the Pastoral Galole Orma: The Effects of Market Integration*. Ph.D. Northwestern University.

Esser, M., 1897, Meine Reise nach dem Kunene im nördlichen Grenzgebiet von Deutsch Südwest-Afrika. *Verhandlungen der Gesellschaft für Erdkunde zu Berlin* 24: 103-113.

Estermann, C., 1981, *The Ethnography of South-Western Angola*. Vol3. The Herero People. G.D. Gibson (ed. and transl.). Africana Publishing, New York.

Estermann, C., 1969, Beiträge zur Geschichte der Erforschung von Land und Leuten des unteren Kunene. *Ethnological and Linguistic Studies in Honour of N.J. van Warmelo*, pp. 63-80. Pretoria.

Estermann, C., 1960, *Album de Penteados do Sudoeste de Angola*. Junta de Investigações do Ultramar, Lisbon.

Flache, A.,1996, *The Double Edge of Networks. An Analysis of the Effect of Informal Networks on Cooperation in Social Dilemmas*. University Centre for Social Sciences, Groningen.

Fleisher, L., 1998, Cattle Raiding and its Correlates: the Cultural-Ecological Consequences of Market-Oriented Raiding among the Kuria of Tanzania. *Human Ecology* 26:547-572.

Fleuret, A., 1986, Indigenous Responses to Drought in Sub-Saharan Africa. *Disasters* 10:224-229.

Forbes, H., 1989, Of Grandfathers and Grand Theories. The Hierachised Ordering of Responses to Hazard in a Greek Rural Community. In *Bad Year Economics. Cultural Responses to Risk and Uncertainty,* edited by P. Halstead and J. O'Shea., pp. 87-97. Cambridge University Press, Cambridge.

Fratkin, E., 1997, Pastoralism: Governance and Development Issues, *Annual Review of Anthropology* 26: 235-261.

Fratkin, E., 1991, *Surviving Drought and Development: Arial Pastoralists of Northern Kenya.* Westview Press, Boulder.

Fratkin, E. and Smith, K., 1994, Labour, Livestock and Land. The Organisation of Pastoral Production. In *African Pastoralist Systems,* edited by E. Fratkin, K. Galvin, and E.A. Roth, pp. 91-112. Lynne Rienner Publishers, Boulder.

Fukui, K., 1994, Conflict and Ethnic Interaction. The Mela and their Neighbours. In *Ethnicity and Conflict in the Horn of Africa,* edited by K. Fukui and J. Markakis , pp. 33-47. James Currey. London.

Fukui, K., 1979, Cattle Colour Symbolism and Intertribal Homicide among the Bodi. *SENRI Ethnological Studies* 3: 147-177.

Fukui, K. and Markakis, J., 1994, *Ethnicity and Conflict in the Horn of Africa.* James Currey, London.

Fukui, K. and Turton, D., 1979, Warfare among East African Herders. *SENRI Ethnological Studies* 3.

Galaty, J., 1994, Rangeland Tenure and Pastoralism in Africa. In *African Pastoralist Systems,* edited by E. Fratkin, K. Galvin, and E.A. Roth, pp. 185-204, Lynne Rienner Publishers, Boulder.

Galaty, J., 1993, Maasai Expansion and the New East African Pastoralism. In *Being Maasai. Ethnicity and Identity in East Africa,* edited by Th. Spear and R. Waller, pp. 61-85. James Currey, London.

Galaty, J. and Bonte, P., 1991, *Herders, Warriors and Traders. Pastoralism in Africa.* Westview, Boulder.

Galaty, J. and Johnson, D. 1990, Introduction. In *The World of Pastoralism*, edited by J. Galaty and D. Johnson, pp. 1-31. Guildford Press, London.

Galvin, K., 1987, *Food Procurement, Diet and Nutrition of Turkana Pastoralists in an Ecological and Social Context.* UMI. Ann Arbor.

Galvin, K. and Little, M, 1999, Dietary Intake and Nutritional Status. In: *Turkana Herders of the Dry Savanna. Ecology and Biobehavioural Response of Nomads to an Uncertain Environment*, edited by M.A. Little and P.W. Leslie, pp. 125-145. Oxford University Press, Oxford.

Galvin, K., Coppock, L., and Leslie, P., 1994. Diet, Nutrition and the Pastoral Strategy. In *African Pastoralist Systems*, edited by E. Fratkin, K. Galvin, and E.A. Roth, pp. 113-131. Lynne Rienner Publishers, Boulder.

Gartrell, B., 1988, Prelude to Disaster: the Case of Karamjoa. In *The Ecology of Survival. Case Studies from Northeast African History*, edited by D. H. Johnson and D. M. Anderson, pp. 193-193. Westview Press, Boulder.

Gast, M., 1987, *Manger à Tamanrasset. Travaux du Laboratoire d'Anthropologie et de Préhistoire des Pays de la Méditerranée Occidentale.* Université de Provence.

Gebre-Mariam, A., 1994, The Alienation of Land Rights among the Afar in Ethiopia. *Nomadic Peoples* 34/35:137-146.

Gewald, J., 2000, Colonisation, Genocide and Resurgence: The Herero of Namibia 1890-1933. In *People, Cattle and Land: Transformations of a Southwest African Pastoral People*, edited by M. Bollig and J.B. Gewald, pp. 187-226. Köln, Köppe.

Gewald, J., 1996, *Towards Redemption. A Socio-Political History of the Herero of Namibia between 1890 and 1923.* Research School at CNWS. Leiden.

Gibson, G., 1977, Himba Epochs. *History in Africa* 4:67-120.

Gibson, G., 1956, Double Descent and its Correlates among the Herero of Ngamiland. *American Anthropologist* 58:109-139.

Glantz, M.H., 1987, *Drought and Hunger in Africa. Denying Famine a Future.* Cambridge University Press, Cambridge.

Glantz, M.H., 1993, Drought, Famine, and the Seasons in Sub-Saharan Africa. In *African Food Systems in Crisis. Part One: Microperspectives*, edited by R. Huss-Ashmore and S.H. Katz, pp. 45-72.

Goldschmidt, W., 1976, *Culture and Behavior of the Sebei.* University of California Press, Berkeley.

Goldtschmidt, W. 1974, The Economics of Brideprice among the Sebei (Uganda) and in East Africa. *Ethnology* 13: 311-333.

Goldschmidt, W., 1971, Independence as an Element in Pastoral Social Systems. *Anthropological Quarterly* 44:132-142.

Göbel, B., 1998, Risk, Uncertainty, and Economic Exchange in a Pastoral Community of the Andean Highlands (Huancar, N.W. Argentia). In *Kinship, Networks, and Exchange,* edited by Th. Schweizer and D. White, pp. 158-177. Cambridge University Press, Cambridge.

Göbel, B,. 1997, 'You have to exploit luck': Pastoral Household Economy and the Cultural Handling of Risk and Uncertainty in the Andean Highlands. *Nomadic Peoples* N.S. 1: 37-53.

Government of Kenia, 1989, *Baringo District Development Plan.* Ministry of Development and Planning. Nairobi.

Government of Namibia 1991, *Census 1991*. Government Printers, Windhoek.

Government of Namibia, Directorate of Rural Development, 1992. Socio-Economic Survey Southern Communal Areas. Ministry of Agriculture, Windhoek.

Grandin, B., 1989, Labour Sufficiency, Livestock Management, and Time Allocation on Maasai Group Ranches. *Research in Economc Anthropology* 11:143-178.

Grandin, B., 1987, Pastoral Culture and Range Management: Recent Lessons from Maasailand. *ILCA Bulletin* 28:7-13.

Gray, S., Sundal, M., Wiebusch B., Little M., Leslie P., and Pike, I., 2003, Cattle Raiding, Cultural Survival, and Adaptability of East African Pastoralists. *Current Anthropology* 44:3-30

Gudeman, S. and Rivera, A., 1990, *Conversations in Colombia. The Domestic Economy in Life and Text*. Cambridge University Press, Cambridge.

Gulliver, P., 1955, *The Family Herds*. Kegan, London.

Hakansson, Th., 1990, Descent, Bridewealth, and Terms of Alliance in Eastern African Societies. *Research in Economic Anthropology* 12:149-173.

Halstead, P. and O'Shea, J., 1989, *Bad Year Economics. Cultural Responses to Risk and Uncertainty. New Directions in Archeology*. Cambridge University Press, Cambridge.

Hames, R., 1990, Sharing among the Yanomamö. Part I. The Effects of Risk. In *Risk and Uncertainty in Tribal and Peasant Economies*, edited by E. Cashdan, pp. 89-106. West View Press, Boulder.

Hardin, G., 1968, The Tragedy of the Commons. *Science* 162:1243-1248.

Hartmann, G., 1902/03, Meine Expedition 1900 ins nördliche Kaokofeld und 1901 durch das Amboland. *Beiträge zur Kolonialpolitik und Kolonialwirtschaft* 4:1-29.

Hartmann, G., 1897, Das Kaokogebiet in Deutsch Südwest Afrika aufgrund eigener Reisen und Beobachtungen. *Verhandlungen der Gesellschaft für Erdkunde zu Berlin* 24:113-141.

Haugerud, A., 1995, *The Culture of Politics in Modern Kenya.* Cambridge University Press, London.

Hayes, P., 1998, The 'Famine of the Dams'. Gender, Labour and Politics in Colonial Ovamboland, 1929-1930. In *Namibia under South African Rule. Mobility and Containment 1915-146*, edited by P. Hayes, pp. 117-146. James Currey, London.

Hayes, P., 1997, *"Healing the Land" – Kaulinge's history of Kwanyama*. Köln, Köppe.

Hayes, P., 1992, *A History of the Ovambo of Namibia, ca 1880-1930*. PhD Thesis, University of Cambridge, Cambridge.

Helten, E., 1994, *Die Erfassung und Messung des Risikos*. Gabler, Braunschweig.

Hendrickson, D., Mearns, R. and Armon, J., 1996, Livestock Raiding Among the Pastoral Turkana of Kenya: Redistribution, Predation and the Links to Famine. In *War and Rural Development in Africa,* edited by J. Swift. Institute of Development Studies. [IDS Bulletin 27 (3)]

Hereros Baroja, Th. and Sikamoy, P., 1989, *Analytical Grammar of the Pokot Language. Kitapu ngala Pokot nyo Kikir*. Bibliotheca Africana, Trieste.

Herren, U., 1991, *Socioeconomic Strategies of Pastoral Maasai Households in Mukogodo Kenya*. Ph.D. dissertation, University of Bern, Bern.

Herskovitts, M., 1926, The Cattle Complex in East Africa. *American Anthropologist* 28:230-272, 361-388, 494-528, 633-664.

Hitchcock, R., 1990, Water, Land, and Livestock. The Evolution of Tenure and Administration in the Grazing Areas of Botswana. In *The World of Pastoralism. Herding Systems in Comparative Perspective*, edited by J. Galaty and D. Johnson, pp. 195-215. Belhaven, London.

Hjort, A., 1979, *Savanna Town. Rural Ties and Urban Opportunities in Northern Kenya.* Stockholm Studies in Social Anthropology in cooperation with Nordiska Afrikainstitutet, Uppsala.

Hjort, A. and Salih, M. (eds), 1989, *Ecology and Politics. Environmental Stress and Security in Africa.* Stockholm Studies in Social Anthropology in cooperation with Nordiska Afrikainstitutet, Uppsala.

Hjort, A. and Dahl, G., 1991, *Responsible Man. The Atmaan Beja of North-Eastern Sudan.* Stockholm Studies in Social Anthropology in cooperation with Nordiska Afrikainstitutet, Uppsala.

Hobley, C.W., 1906, Notes on the Geography and People of the Baringo District of the East Africa Protectorate. *Geographical Journal* 26:471-481.

Hodder, I., 1982, *Symbols in action. Ethnoarcheological Studies of Material Culture*. Cambridge University Press, Cambridge.

Hodgson, D. (ed), 2000, *Rethinking Pastoralism in Africa. Gender, Culture and the Myth of the Patriarchal Pastoralist*. Currey, Oxford.

Hogg, R., 1989, Settlement, Pastoralism and the Commons. The Ideology and Practice of Irrigation Development in Northern Kenya. In *Conservation in Africa. People, Policies and Practice*, D. Anderson and R. Grove, pp. 293-306. Cambridge University Press, London.

Hogg, R., 1986, The New Pastoralism: Poverty and Dependency in Northern Kenya. *Africa* 56:519-555.

Holy, L., 1988, Cultivation as a long-term strategy of survival: the Berti of Darfur. In *The Ecology of Survival. Case Studies from Northeast African History,* edited by D.H. Johnson and D.M. Anderson, pp. 135-154. Westview Press, Boulder.

Homewood, K. and Lewis, J., 1987, Impact of Drought on Pastoral Livestock in Baringo, Kenya. *Journal of Applied Ecology* 24:615-631.

Hutchinson, S., 2000: Nuer Ethnicity Militarized. *Anthropology Today* 16 (3): 6-13.

Hutchinson, S., 1996, *Nuer Dilemmas. Coping with Money, War, and the State*. University of California Press, Berkeley.

Hyden, G., 1985, *Beyond Ujamaa in Tanzania. Underdevelopment and an Uncaptured Peasantry*. Heinemann, London.

Immelman, D., 1978, Kaoko-Otavi. *SWA Annual*:131-132.

Irle, J., 1906, *Die Herero*. Bertelsmann, Gütersloh.

Jacobs, A., 1979, Maasai Intertribal Relations: Belligerent Herdsmen or Peaceable Pastoralists? *SENRI Ethnological Studies* 3:33-53.

Jacobs, A., 1968, A Chronology of the Pastoral Maasai. *Hadith* 1:10-31.

Jacobsohn, M., 1990 *Himba Nomads of Namibia*. Photographs by Peter Pickford and Beverly Pickford. Text by Margaret Jacobsohn. Struik, Cape Town.

James, W., 1994, War and 'ethnic visibility': The Uduk on the Sudan-Ethiopia border. In *Ethnicity and Conflict in the Horn of Africa,* edited by F. Katsuyoshi and J. Markakis, pp. 140-165. James Currey, London.

Jätzhold R. and Schmidt, H., 1982/83, *Farm Management Handbook of Kenya*. Nairobi.

Johnson, A., 1991, Regional Comparative Field Research. *Behavior Science Research* 25:3-22.

Johnson, B.R., 1990, *Nomadic Networks and Pastoral Strategies. Surviving and Exploiting Local Instability in South Turkana*. UMI, Ann Arbor.

Johnson, D. and Anderson, D., 1988, *The Ecology of Survival. Case Studies from Northeast African History*. Westview Press, London.

Johnson, A. & Earle T. 1987, *The Evolution of Human Societies: From Foraging Group to Agrarian State*. Standford University Press, Stanford.

Jungermann, H. and Slovic, P., 1993, Die Psychologie der Kognition und Evaluation von Risiko. In *Risiko und Gesellschaft. Grundlagen und Ergebnisse interdisziplinärer Risikoforschung*, edited by G. Bechmann, pp. 177-208. Westdeutscher Verlag, Oppladen.

Kaplan, H., Hill, K., and Hurtado, A.M., 1990, Risk, Foraging and Food Sharing among the Aché. In *Risk and uncertainty in tribal and peasant economies*, edited by E. Cashdan, pp. 107-143. Westview Press, Boulder.

Kent, S., 1993, Sharing in an Egalitarian Kalahari Community. *Man* 28:479-514.

Kerner, D.O. and Cook, K., 1991, Gender, Hunger and Cirisis in Tanzania. In *The Political Econom,y of African Famine. Food and Nutrition in History and Anthropology* Vol.9., edited by R.E. Downs, D.O. Kerner, and S. Reyna, pp. 257-272. Gordon and Breach, Philadelphia.

Kerven, C., 1992, *Customary Commerce. A Historical Reassessment of Pastoral Livestock Marketing in Africa*. ODI, London.

Kipkorir, B and Welbourn, A., 1973, *The Marakwet: A Preliminary Study*. East African Literatue Bureau, Nairobi.

Klumpp, D. and Kratz, C., 1993, Aesthetics, Expertise, and Ethnicity. Okiek and Maasai Perspectives on Personal Ornament. In *Being Maasai*, edited by Th. Spear and R.Waller, pp. 195-221. James Currey, London.

Klute, G., 1992, *Die schwerste Arbeit der Welt. Alltag von Tuareg-Nomaden*. Trickster, München.

Klute, G. (ed), 1996, Nomads and the State. *Nomadic Peoples* 38 (*Nomadic Peoples Special Issue*)

Krings, Th., 1991, Nomaden und Staat in der Republik Mali. In *Nomaden. Mobile Viehhaltung*, edited by F. Scholz, pp. 55-72. Das Arabische Buch, Berlin.

Krüger, G. and Henrichsen, D., 1998, We have been Captives Long Enough. We Want to Be Free. Land, Uniforms and Politics in the History of Herero in the Interwar Period. In *Namibia under South African Rule. Mobility and Containment 1915-146*, edited by P. Hayes et al., pp. 149-174. James Currey, London.

Krzywinski, K., Vetaas, O.R., and Manger, L., 1996, Vegetation Dynamics in the Red Sea Hills - Continuities and Changes. In *Survival on Meagre Resources. Hadendowa Pastoralism in the Red Sea Hills*, edited by L. Manger, pp. 59-80. Nordiska Afrikainstitutet. Uppsala.

Kuntz, J. *1912. Ovatschimba im nördlichen Kaokofeld (Deutsch-Südwestafrika). Petermanns* Geographische Mittheilungen 25: 206.

Kurimoto, E., 1994, Civil War and Regional Conflicts: The Pari and their Neighbours in South-Eastern Sudan. In *Ethnicity and Conflict in the Horn of Africa*, edited by F. Katsuyoshi and J. Markakis, pp. 95-111. James Currey, London.

Lamphear, J., 1994, The Evolution of Ateker 'New Model' Armies: Jie and Turkana. In *Ethnicity and Conflict in the Horn of Africa*, edited by K. Fukui and J. Markakis, pp. 63-92. Currey, London.

Lamphear, J., 1993, Aspects of 'Becoming Turkana'. Interactions and Assimilation between Maa- and Ateker-Speakers. In *Being Maasai. Ethnicity and Identity in East Africa*, edited by Th. Spear and R. Waller, pp. 87-104. Currey, London.

Lamphear, J., 1988, The People of the Grey Bull. The Origin and Expansion of the Turkana. *Journal of African History* 29:27-39.

Lamphear, J., 1976a, *The Traditional History of the Jie of Uganda*. The Clarendon Press, Oxford.

Lamphear, J., 1976b, Aspects of Turkana Leadership during the Era of Primary Resistance. *Journal of African History* 29:225-243.

Lamprey, H., 1983, Pastoralism Yesterday and Today. The Overgrazing Problem. In *Tropical Savannahs, Ecosystems of the World*, edited by F. Boulière, Vol 13. pp. 643-666 Elsevier, Amsterdam.

Lane, C., 1994, Pastures Lost: Alienation of Barabaig Land in the Context of Land Policy and Legislation in Tanzania. *Nomadic Peoples* 34/35:81-94.

Lang, H., 1997, Ethnodemographie und die Bedeutung von ethnographischen Zensuserhebungen. In *Geburt und Tod. Ethnodemographische Probleme, Methoden und Ergebnisse*, edited by W. Schulze, H. Fischer and H. Lang, pp. 4-36. Reimer, Berlin.

Lau, B., 1987, *Namibia in Jonker Afrikaner's Time*. Windhoek National Archives (Archeia 8) Windhoek.

Lee, R.B., 1979, *The !Kung San. Men, Women and Work in a Foraging Society*. Cambridge University Press, Cambridge.

Legesse, A., 1993, Adaptation, Drought, and Development: Boran and Gabra Pastoralists of Northern Kenya. In *African Food Systems in Crisis. Part One: Microperspectives*, edited by R. Huss-Ashmore and S. H. Katz, pp. 261-280. Gordon and Breach, Yverdon. (2nd Edition)

Legge, K., 1989, Changing Responses to Drought among the Wodaabe of Niger. In *Bad Year Economics. Cultural Responses to Uncertainty*, edited by P. Halstead and J. O'Shea, pp.81-86. Cambridge University Press, Cambridge.

Leslie, P.W. and Dyson-Hudson, R. 1999. People and Herds. In: *Turkana Herders of the Dry Savanna. Ecology and Biobehavioural Response of Nomads to an Uncertain Environment*, edited by M.A. Little and P.W. Leslie, pp. 233-247. Oxford University Press, Oxford.

Lemarchand, R., 1989, African Peasantries, Reciprocity and the Market. The Economy of Affection Reconsidered. *Cahiers d'Etudes Africaines* 113:33-67.

Lindenberg, S., 1998, Solidarity: Its Microfoundations and Macro Dependence. A Framing Apporach. In *The Problem of Solidarity: Theories and Models*, edited by P. Doreian and T. Fararo, pp.282-328. Gordon and Breach, New York.

Little, M., 1989, Human Biology of African Pastoralists. *Yearbook of Physical Anthropology* 32:215-247.

Little, M. and Leslie, P.W., 1990, *The South Turkana Ecosystem Project*. Dept of Anthropology. State University Binghampton.

Little, M. and Mugambi, M., 1983, Cross-Sectional Growth of Nomadic Turkana Pastoralists. *Human Biology* 59:695-707.

Little, M., Galvin, K. and Leslie, P.W., 1988, Health and Energy Requirements of Nomadic Turkana Pastoralists. In *Coping with Uncertainty in Food Supply*, edited by de I. Garine and G. Harrison, pp. 290-315. Oxford University Press, Oxford.

Little, P., 1992, *The Elusive Granary: Herder, Farmer, and State in Northern Kenya*. Cambridge University Press, Cambridge.

Little, P., 1987, Land Use Conflicts in the Agricultural/ Pastoral Boderlands. The Case of Kenya. In *Lands at Risk in the Third World. Local Level Perspectives,* edited by P. Little and M. Horowitz, pp. 195-211. Westview, New York.

Luhmann, N., 1993, *Die Moral des Risikos und das Risiko der Moral*. In Risiko und Gesellschaft. Grundlagen und Ergebnisse interdisziplinärer Risikoforschung, edited by G. Bechmann, pp. 327-338. Westdeutscher Verlag, Oppladen.

Mainguet, M.,. 1994, *Desertification. Natural Background and Human Mismanagement*. Springer Verlag, Berlin. (2nd Edition)

Malan, St., 1974, The Herero-Speaking Peoples of Kaokokland. *Cimbebasia* B 2:113-129.

Malan, St., 1973, Double Descent among the Himba of South West Africa. *Cimbebasia* B 2:81-112.

Malan, St., 1972, *Dubbele afkomsberekenings van die Himba, 'n hererosprekende volk in die Suidwest Afrika*. Ph.D Dissertation, Johannesburg.

Malan, St. and Owen-Smith, G., 1974, The Ethnobotany of Kaokoland. *Cimbebasia Series* B 2:131-178.

Manger, L. (ed), 1996, *Survival on Meagre Resources. Hadendowa Pastoralism in the Red Sea Hills*. Nordiska Afrikainstitutet, Uppsala.

Manger, L., 1988, Traders, Farmers and Pastoralists: Economic Adaptations and Environmetal Problems in the Southern Nuba Mountains. In *The Ecology of Survival. Case Studies from Northeast African History,* edited by D. H. Johnson and D. M. Anderson, pp. 155-171. Westview Press, Boulder.

Matson, A.T., 1972, *Nandi Resistance to British Rule. 1890-1906*. East Africa Publishing House, Nairobi.

Mazrui, A., 1977, The Warrior Tradition and the Masculinity of War. In *The Warrior Tradition in Eastern Africa,* edited by A. Mazrui, pp. 69-81. Brill, Leiden.

McCabe, T., 1997, Risk and Uncertainty among the Maasai of the Ngorongoro Conservation Area in Tanzania: A Case Study in Economic Change. *Nomadic Peoples* N.S. 1: 54-65.

McCabe, T., 1994, Mobility and Land Use Among African Pastoralists. Old Conceptual Problems and New Interpretations. In *African Pastoralist Systems,* edited by E. Fratkin, K. Galvin, and E.A. Roth, pp. 69-89. Lynne Rienner Publishers, Boulder.

McCabe, T., 1990, Success and Failure: the Breakdown of Traditional Drought Coping Institutions among the Pastoral Turkana of Kenya. *Journal of African and Asian Studies* 25 (3-4):146-160.

McCabe, T., R. Dyson-Hudson and J. Wienpahl, 1999, Nomadic Movements. In *Turkana Herders of the Dry Savanna: Ecology and Biobehavioural Response of Nomads to an Uncertain Environment*, edited by M. Little & P. Leslie, pp. 109-122. Oxford University Press, Oxford.

McGovern, S., 2000, *East Pokot Medical Project. Annual Progress Report January-December 2000*. Manuskript. Kositei.

McGovern, S., 1999, *East Pokot Medical Project. Annual Progress Report January-December 1999*. Manuskript. Kositei.

McKittrick, M., 1995, *Conflict and Social Change in Northern Namibia*, 1850-1954. PhD Dissertation, Stanford University.

Miescher, G. and Henrichsen D., 2000a, *New Notes on Kaoko. The Northern Kunene Region (Namibia) in Texts and Photographs*. Basler Afrika Bilbliographien, Basel.

Miescher, G. and Henrichsen D. 2000b, Epilogue. In *New Notes on Kaoko. The Northern Kunene Region (Namibia) in Texts and Photographs*. Pp. 239-247. Basler Afrika Bilbliographien, Basel.

Miller, J. C., 1988, *Way of Death. Merchant Capitalism and the Angolan Slave Trade, 1730-1830*. James Currey, London.

Mink, L. and Smith, K., 1989, The Spirit of Survival. Cultural Responses to Resource Variability in North Alaska. In *Cultural Responses to Risk and uncertainty. New Directions in Archeology. Cambridge,* edited by P. Halstaead, and J. O'Shea, pp. 8-39. Cambridge University Press, Cambridge.

Möhlig, W., 1981, Die Bantusprachen im engeren Sinn. In *Die Sprachen Afrikas,* edited by B. Heine, Th. Schadeberg, and E. Wolff, pp. 77-116. Buske, Hamburg.

Mortimore, M., 1998, *Roots in the African Dust. Sustaining the Drylands*. Cambridge University Press, Cambridge.

Muriuki, G., 1974, *A History of the Kikuyu, 1500-1900*. Oxford University Press, Nairobi.

Nauheimer, H., 1991, Der Umgang mit dem Mangel. Produktionsstrategien in Trockengebieten und Trockenzeiten. In: *Nomaden. Mobile Viehhaltung*, edited by F. Scholz, pp. 213-1232. Das Arabische Buch, Berlin.

Ndagala, D., 1994, Pastoral Territory and Policy Debates in Tanzania. *Nomadic Peoples* 34/35:23-36.

Ndagala, D., 1991, The Unmaking of the Datoga: Decreasing Resources and Increasing Conflict in Rural Tanzania. *Nomadic Peoples* 28:71-82.

Nogueira, A.F., 1880, *A Raça Negra*. Lisbon.

North, D., 1990, *Institutions, Institutional Change and Economic Performance*. Cambridge University Press, Cambridge.

North, D., 1987, Transaction Costs, Institutions, and Economic History. In *The Institutional Economics*, edited by E. Furubotn and R. Richter, pp. 203-213. Mohr, Tübingen.

Notkola, V. and Siiskonen, H., 2000, *Fertility, Mortality, and Immigration in Subsaharan Africa. The Case of Ovamboland in North Namibia, 1925-90*. St. Martin's Press, New York.

Nyamwaya, D., 1987, A Case Study of the Interaction between Indigenous and Western Medicine among the Pokot of Kenya. *Social Science Medicine* 25:1277-1287.

Odegi-Awuondo, C., 1990, *Life in the Balance. Ecological Sociology of Turkana Nomads*. ACTS Press, Nairobi.

Österle, M. and M. Bollig, 2003, Continuities and Discontinuities of Warfare in Pastoral Societies: Militarization and The Escalation of Violence in East and North East Africa. *Zeitschrift für Entwicklungsethnologie* 12: 109-143.

Ohly, R., 1990, *Herero I. Phonetics, Morphology, Syntax, Literature*. Manuscript. Windhoek.

Ohta, I., 2000, Drought and Mureti's Grave: The 'We/Us' Boundaries between Kaokolanders and People of Okakarara area in Early 1980s. In *People, Cattle and Land – Transformations of a Southwest African Pastoral People*, edited by M. Bollig & J.B. Gewald, pp. 299-317. Köppe, Köln.

O'Leary, M., 1984, *The Kitui Akamba. Economic and Social Change in Semi-Arid Kenya*. Heinemann Educational Books, Nairobi.

Orstom, E., 1990, *Governing the Commons. The Evolution of Institutions for Collective Action*. Cambridge University Press, Cambridge.

Osogo, J., 1970, The Significance of Clans in the History of East Africa. *Hadith* 2:30-41.

Page, D., 1975, *Evaluasie van die hulpbronne van Kaokoland en ontwikkelingsvoorstelle*. Universiteit van Stellenbosch.

Palmer, W.C., 1965, *Metereological Drought*. Research Paper 45. US Weather Bureau, Washington DC.

Pankhurst, R. and Johnson, D. H., 1988, The Great Drought and Famine of 1888-92 in Northeast Africa. In *The Ecology of Survival. Case Studies from Northeast African History*, edited by D. H. Johnson and D. M. Anderson, pp. 47-72. Westview Press, Boulder.

Pennington, R. and Harpending, H., 1993, *The Structure of an African Pastoralist Community. Demography, History and Ecology of the Ngamiland Herero*. Clarendon Press, Oxford.

Peristiany, J.C., 1975, The Ideal and the Actual. The Role of Prophets in the Pokot Political System. In *Studies in Social Anthropology*, edited by W. Beattie and R.G. Lienhardt, pp. 167-212. Clarendon Press, London.

Persitiany, J.C., 1954, Pokot Sanctions and Structure. *Africa* 24:17-25.

Persitiany, J.C., 1951, The Age-Set System of the Pastoral Pokot. *Africa* 21:188-206.

Pflaumbaum, H., 1994, Futterressourcen in der Butana (Rep. Sudan). Zur Problematik der Dynamik ökologischer Tragfähigkeit. In *Überlebensstrategien in Afrika,* edited by M. Bollig and F. Klees, pp. 67-79. Heinrich Barth Institut, Cologne.

Pratt, D.J. and Gwynne, M.D., 1977, *Rangeland Management and Ecology in East Africa*. Hodder and Stoughton, London.

Reckers,U., 1992, *Nomadische Viehhalter in Kenya. Die Ost-Pokot aus Human-Ökologischer Sicht. Nomadische Viehhalter in Kenya*. Institut für Afrika-Kunde, Hamburg.

Roth, E. A., 1994, Demographic Systems: Two East African Examples. In *African Pastoralist Systems,* edited by E. Fratkin, K. Galvin, and E. A. Roth, pp.133-145. Lynne Rienner, Boulder.

Roth, E. A., 1986, The Demographic Study of Nomadic Peoples. *Nomadic Peoples* 20:73-76.

Rottland, F., 1982, *Die Südnilotischen Sprachen. Beschreibung, Vergleichung und Rekonstruktion*. Reimer, Berlin.

Salih, H. M., 1994, Struggle for the Delta: Hadendowa Conflict over Land Rights in the Sudan. *Nomadic Peoples* 34/35:147-158.

Salih, M., 1994, The Ideology of the Dinka and the Sudan People's Liberation Movement. In *Ethnicity and Conflict in the Horn of Africa,* edited by K. Fukui and J. Markakis, pp. 187-201. James Currey, London.

SALTLICK, 1991, A Baseline Data Survey in the Nginyang and Tangulbci Divisions of Baringo District. Manuskrict. Isolo.

Salzman P. C., 1996, Peasant Pastoralism. In *The Anthropology of Tribal and Peasant Pastoral Societies*, edited by U. Fabietti and P. C. Salzman, pp. 149-166. Collegio Ghislieri., Pavia.

Salzman, P. C., 1971, Comparative Studies of Nomadism and Pastoralism. Special Issue. Introduction. *Anthropological Quarterly* 44:104-108.

Sander, H. and Becker T., 2002, Klimatologie des Kaokolandes. In *Interdisziplinäre Perspektiven zu Kultur-und Landschaftswandel im ariden und semiariden Nordwest-Namibia*, edited by Bollig, M, Brunotte, E. and T. Becker, p. 57-68. Kölner Geographische Arbeiten 77.

Sander, H., Bollig, M., and Schulte, A., 1998, Himba Paradise Lost. Stability, Degradation and Pastoralist Management of the Omuhonga Basin (Namibia). *Die Erde* 129:301-315.

Sandford, S. 1983, *Management of Pastoral Development in the Third World*. Wiley, Chichester.

Sato, S., 1980, Pastoral Movements and the Subsistence Unit of the Rendille of Northern Kenya. With Speical Reference to the Camel Ecology. *SENRI Ethnological Studies* 6:1-78.

Schlee, G., 1985, Interethnic Clan Identities among Cushitic Speaking Pastoralists. *Africa* 55:17-37.

Schmuck-Widman, H., 1997, Leben mit der Flut. Lokale Wahrnehmungen und Strategien zur Bewältigung der Flut in Bangladesh. *Sociologus* 46:130-158.

Schneider, H., 1994, *Animal Health and Veterinary Medicine in Namibia*. Agrivet, Windhoek.

Schneider, H., 1979, *Livestock and Equality in East Africa. The Economic Basis for Social Structure*. Indiana University Press, Bloomington.

Schneider, H., 1959, Päkot Resistance to Change. In *Continuity and Change in African Cultures,* edited by W. Bascom and M. Herskovitts, pp. 144-167. University of Chicago Press, Chicago.

Schneider, H., 1955, The Moral System of the Pakot. In *Encyclopedia of Morals*, edited by V. Ferm, pp. 403-409. Peter Owen, London.

Schneider, H., 1953, *The Päkot (Suk) of Kenya with Special Reference to the Role of Livestock in their Subsistence Economy*. Michigan, Ann Arbor.

Schulte, A., 2002, Stabilität oder Zerstörung? Veränderungen der Vegetation des Kaokolandes unter pastoralnomadischer Nutzung. In *Interdisziplinäre Perspektiven zu Kultur-und Landschaftswandel im ariden und semiariden Nordwest-Namibia*, edited by Bollig, M, Brunotte, E. and T. Becker, pp. 101-118. Kölner Geographische Arbeiten 77.

Schweizer, T., 1998, Epistemology: The nature and validation of anthropological knowledge. In *Handbook of Methods in Cultural Anthropology*, edited by H.R. Bernard, pp. 457-479. Altamira Press, Walnut Creek.

Schweizer, T., 1997, *Muster sozialer Ordnung. Netzwerkanalyse als Fundament der Sozialethnologie*. Reimer, Berlin.

Scoones, I., 1996, *Hazards and Opportunities. Farming Livelihoods in Dryland Africa. Lessons from Zimbabwe*. Zed Books, London.

Scott, J., 1976, *The Moral Economy of the Peasant. Rebellion and Subsistence in South-East Asia*. Yale University Press, New Haven.

Scott, M. and Gormley, B., 1980, The Animal of Friendship. An Indigenous Mode of Sahelian Pastoral Development. In *Indigenous Knowledge Systems and Development,* edited by D. W. Brohensha, D. M. Warren, and O. Werner, pp. 92-110. University of America Press, Lanham.

Sellen, D., 1996, Nutritional Status of Sub-Saharan African Pastoralists. A Review of the Literature. *Nomadic Peoples* 39:107-134.

Sen, A. 1985, Women, Technology, and Sexual Divisions. *Trade and Development* 6:195-223.

Sen, A., 1981, *Poverty and Famines. An Essay on Entitlements*. Oxford University Press, Oxford.

Shipton, P., 1990, African Famines and Food Security. *Annual Reviews in Anthropology* 19:353-394.

Siiskonen, H., 1990, *Trade and Socioeconomic Change in Ovamboland. 1850-1906*. SHS, Helsinki.

Silvester, J., 1998, Beasts, Boundaries and Buildings. The Survival and Creation of Pastoral Economies in Southern Namibia 1915-1935. In *Namibia under South African Rule. Mobility and Containment, 1915-1946*, edited by P. Hayes et al, pp. 95-116. Currey, Oxford.

Smith, E. 1988. Risk and Uncertainty in the 'Original Affluent Society': Evolutionary Ecology of Resource Sharing and Land Tenure. In *Hunters and Gatherers 1. History, Evolution and Social Change*, edited by T. Ingold, D. Riches, and J. Woodburn, pp. 222-251. Berg, Oxford.

Smith, K., 1996, *Environmental Hazards. Assessing Risk and Reducing Disaster*. Routledge, London.

Sobania, N. W., 1991, 'Feasts, Famines and Friends' Nineteenth Century and Exchange and Ethnicity in the Eastern Lake Turkana Region. In *Herders, Warriors and Traders. Pastoralism in Africa*, edited by J. Galaty and P. Bonte Boulder, pp. 132-150. Westview, Boulder.

Sobania, N. W., 1988, Pastoralist Migration and Colonial Policy: A Case Study from Northern Kenya. In *The Ecology of Survival. Case Studies from Northeast African History,* edited by D. H. Johnson and D. M. Anderson, pp. 219-240. Westview Press, Boulder.

Sollod, A., 1990, Rainfall Variability and Twareg Perceptions of Climate Impacts in Niger. *Human Ecology* 18:267-281.

Souci, S.W. et al., 1979/82, *Food Composition and Nutrition Tables 1981/82*. Wissenschaftliche Verlagsgesellschaft, Stuttgart.

Spear, T.,. 1978, *The kaya Complex. A History of the Mijikenda Peoples of the Kenya Coast to 1900*. Kenya Literature Bureau, Nairobi.

Spencer, P., 1998, *The Pastoral Continuum. The Marginalization of Tradition East Africa* Clarendon Press, London.

Spencer, P., 1973, *Nomads in Alliance. Symbiosis and Growth among the Rendille and Samburu of Kenya*. Oxford University Press, London.

Spencer, P., 1965, *The Samburu. A Study of Gerontocracy in a Nomadic Tribe*. Routledge, London.

Sperling, L., 1987, The Adoption of Camels by Smaburu Cattle Herders. *Nomadic Peoples* 23:1-17.

Spittler, G., 1989a, *Dürren, Kriege und Hungerkrisen bei den Kel Ewey (1900-1985)*. Lang, Stuttgart.

Spittler, G., 1989b, *Handeln in einer Hungerkrise. Tuaregnomaden und die große Dürre von 1984*. Westdeutscher Verlag, Opladen.

Spooner, B., 1971, Towards a Generative Model of Nomadism. *Anthropological Quarterly* 44:198-210.

Spooner, B., 1993, Desertification: The Historical Significance. In Rebecca Huss-Ashmore and Solomon H. Katz (eds) *African Food Systems in Crisis*. Part One: Microperspectives. pp. 73-111, Gordon and Breach, Yverdon. (2nd Edition).

Stals, E. and Otto-Reiner, A., 1990, *Oorlog en Vrede aan die Kunene. Die Verhaal van Kapteil Vita Tom, 1863-1937*. Windhoek. Manuscript.

Stahl, U., (2000), 'In the end of the Day we will Fight' - Communal Landrights and Illegal Fencing amongst Herero Pastoralist in Otjozondjupa Region. In *People, Cattle and Land – Transformations of a Southwest African Pastoral People*, edited by M. Bollig & J.B. Gewald, pp. 319-346. Köppe, Köln.

Stern, W., 1991, Mobile Viehhaltung als Hauptwirtschaftsfaktor - Der Fall Somalia. In *Nomaden. Mobile Viehhaltung,* edited by F. Schiolz, pp. 113-130. Das Arabische Buch, Berlin.

Steyn, H. P., 1977, *Pastoralisme by die Himba n'kulturelle-ekologiese Studie*. Ph.Dissertation University of Stellenbosch.

Storas, F., 1997, The Nexus of Economic and Political Viabilities among Nomadic Pastoralists in Turkana, Kenya. *Research in Economic Anthropology* 18:115-163.

Sutton, J. 1990. *A Thousand Years of East Africa*. Nairobi. British Institute in East Africa.

Swift, J. 1977, Sahelian Pastoralists: Underdevelopment, Desertification, Famine. *Annual Revue of Anthropology* 6: 457-479.

Talavera, P., Katjimune, J., Mbinga, A., Vermeulen, C., and Mouton, G., 2000, *Farming Systems in Kunene North. A Resource Book*. Ministry of Agriculture, Windhoek.

Tanaka, J., 1984, List of Plants Colllected from Baringo, Elgeyo/Marakwet and West Pokot Districts. In *Kerio Valley. Past, Present and Future,* edited by B. Kipkorir et al., pp. 147-164. Institute of African Studies, Nairobi.

Timberlake, J., 1987, *Ethnobotany of the Pokot of Northern Kenya*. Manuscript.

Tournay, S., 1979, Armed Conflicts in the Lower Omo Valley, 1970-1976. An Analysis from within Nyangatom Society. *SENRI Ethnological Studies* 3:97-117.

Tully, D., 1985, *Human Ecology and Political Process. The Context of of Market Incorporation in West Pokot District*. UMI, Ann Arbor.

Turton, D., 1994, Mursi Political Identity and Warfare. The Survival of an Idea. In *Ethnicity and Conflict in the Horn of Africa,* edited by K. Fukui and J. Markakis, pp. 15-32. Currey, London.

Turton, D., 1979, War, Peace and Mursi Identity. *SENRI Ethnological Studies* 3:179-210.

Überall, P.H.,1963, *Herero Grammatik*. Manuscript. Döbra.

van der Jagt, K., 1989, *Symbolic Structures in Turkana Religion*. Van Gorcum, Maastricht.

van Dijk, H., 1997, Risk, Agro-pastoral Decision Making and Natural Resource Management in Fulbe Society, Central Mali. *Nomadic Peoples* N.S. 1: 108-132.

van Dijk, H., 1995, Farming and Herding after the Drought: Fulbe Agro-Pastorlaists in Dryland Central Mali. *Nomadic Peoples* 35:65-84.

Vansina, J., 1985, *Oral Tradition as History*. Heinemann, Nairobi..

van Warmelo, N. J., 1951, *Notes on the Kaokoland (South West Africa) and its People*. Ethnological Publications 26. Government Printers, Pretoria.

Vedder, H., 1934, Das alte Südwestafrika. Südwestafrikas Geschichte bis zum Tode Mahareros 1890. Berlin.

Vedder, H., 1928, The Herero. In *Native Tribes of South West Africa*, edited by C. H. Hahn, H. L. Vedder, and L. Fourie, pp.153-211. Cape Times Ltd., Cape Town.

Vedder, H., 1914, *Reisebericht des Missionars Vedder an den Bezirksamtmann von Zastrow*. Namibia National Archives J XIIIb5. Geographische und Ethnographische Forschungen im Kaokoveld 1900-1914

Viehe, G., 1902, Die Omaanda und Otuzo der Ovaherero. *Mittheilungen des Seminars für Orientalische Sprachen zu Berlin. Dritte Abteilung* 5:109-117.

Viljoen, J. J. and Kamupingene, T. K., 1983, *Otjiherero. Woordeboek. Dictionary. Embo Romambo*. Gamsberg, Windhoek.

Vogelsang, R. 2002, Migration oder Diffusion? Frühe Viehhaltung im Kaokoland. In *Interdisziplinäre Perspektiven zu Kultur-und Landschaftswandel im ariden und semiariden Nordwest-Namibia*, edited by Bollig, M, Brunotte, E. and T. Becker, pp. 101-118. Kölner Geographische Arbeiten 77.

von Hoehnel, L., 1897, *Rudolph-See und Stephanie-See. Die Forschungsreise des Grafen Samuel Teleki in Ost-Aequatorial-Afrika, 1887-1888*. Alfred Hölder, Wien.

von Koenen, H. and von Koenen, E., 1964, Grassaat als Feldkost. *S. W.A. Annual*: 87.

Waller, R., 1990, Tsetse Fly in Western Narok, Kenya. *Journal of African History* 31:81-101.

Waller, R., 1988, Emutai: Crisis and Response in Maasailand 1883-1902. In *The Ecology of Survival. Case Studies from Northeast African History*, edited by D. H. Johnson and D. M. Anderson, pp. 73-112. Westview Press, Boulder.

Waller, R. and Sobania, N., 1994, Pastoralism in Historical Perspective. In *African Pastoralist Systems*, edited by E. Fratkin, K. Galvin, and E.A. Roth, pp. 45-68. Lynne Rienner, Boulder.

Warren, M. and Rajasekaran, B., 1995, Using Indigenous Knowledge for Sustainable Dryland Management. A Global Perspective. In *Social Aspects of Sustainable Dryland Management*, edited by D. Stiles, pp. 193-209. Wiley, New York.

Watts, M., 1991, Heart of Darkness: Reflections on Famine and Starvation in Africa. In *The Political Economy of African Famine,* edited by R. E. Downs, D. O. Kerner, and S. P. Reyna, pp. 23-68. Food and Nutrition in History and Anthropology Vol. 9. Gordon and Breach, Philadelphia.

Watts, M., 1988, Coping with the Market. Uncertainty and Food Security among Hausa Peasants. In *Coping with Uncertainty in Food Supply,* edited by I. Degarine and G. Harrison, pp. 260-289. Oxford University Press, Oxford.

Watts, M., 1983, Silent Violence. Food, Famine and Peasantry in Northern Nigeria. University of California Press, Berkeley.

Weatherby, J. M., 1967, Nineteenth Century Wars in Western Kenya. *Azania* 2:133-144.

Wehmeyer, A. S. 1986, *Edible Wild Plants of Southern Africa*. National Food Research Institute. Pretoria.

Werner, W. 2000 From Communal Pastures to Enclosures – the Development of Land Tenure in Herero Reserves. In *People, Cattle and Land – Transformations of a Southwest African Pastoral People*, edited by M. Bollig & J.B. Gewald, pp. 247-269. Köppe, Köln.

Werner, W., 1998, *"No One will Become Rich". Economy and Society in the Herero Reserves in Namibia, 1915-1946*. Basler Afrika Bibliographien, Basel.

White, C., 1997, The Effect of Poverty on Risk Reduction Strategies of Fulani Nomads in Niger. *Nomadic Peoples* N.S. 1: 90-107.

White, C., 1991, Increased Vulnerability to Food Shortage among the Fulani Nomads in Niger. In *The Political Economy of African Famine,* edited by R. E. Downs, D. O. Kerner, and S. P. Reyna, pp. 123-145. Food and Nutrition in History and Anthropology Vol. 9. Gordon and Breach, Philadelphia.

White, C., 1984, *Herd Reconstitution: the Role of Credit among the WoDaaBe herders in Central Niger*. Pastoral Development Network Paper 18d. London Overseas Development Institute.

Wiessner, P., 1998, On Emergency Decisions, Egalitarianism, and Group Selection. *Current Anthropology* 39:356-358.

Wiessner, P., 1982, Risk, Reciprocity, and Social Influences on !Kung San Economics. In *Politics and History in Band Societies*, edited by E. Leacock and R. Lee, pp. 61-84. Cambridge University Press, Cambridge.

Wiessner, P., 1977, *Hxaro. A Regional System of Reciprocity for Reducing Risk among the !Kung San*. University Microfilms. UMI, Ann Arbor.

Wiessner, P. and Tumu, A., 1998a, *Historical Vines. Enga Networks of Exchnage, Ritual, and Warfare in Papua New Guinea*. Smithsonian Institution Press, Washington.

Wiessner, P. and Tumu, A., 1998b, The Capacity and Constraints of Kinship in the Development of the Enga Tee Ceremonial Exchange (Papua New Guinea Highlands). In *Kinship, Networks, and Exchange*, edited by T. Schweizer and D. White, pp. 277-302. Cambridge University Press, Cambridge.

Williams, F.-N., 1991, *Precolonial Communities of Southwestern Africa. A History of Owambo Kingdoms 1600-1920*. National Archives of Namibaia, Windhoek.

Wilmsen, E., 1989, *Land Filled with Flies. A Political Economy of the Kalahari*. Chicago University Press, Chicago.

Winterhalder, B., 1990, Open Field, Common Pot. Harvest Variability and Risk Avoidance in Agricultural and Foraging Societies. In *Risk and Uncertainty in Tribal and Peasant Economies,* edited by E. Cashdan, pp. 67-88. Westview, Boulder.

Winterhalder, B., 1981, Optimal Foraging Strategies and Hunter-Gatherer Research in Anthropology: Theory and Modells. In *Hunter-Gatherer Foraging Strategies. Ethnographic and Archeological Analyses*, edited by B. Winterhalder and E. A. Smith, pp.13-35. University of Chicago Press, Chicago.

Wipper, A., 1977, *Rural Rebels. A Study of Two Protest Movements in Kenya*. Oxford University Press, London.

Young, J., 1989, *The Kenya Freedom From Hunger Council/East Pokot Agricultural Project/Intermediate Technology Development Group Community Animal First Aid Worker Programme*. ITDG Nairobi. Manuscript.

Subject Index

D

F

I

N

S